NANOCARRIERS FOR BRAIN TARGETING

Principles and Applications

NANOCARRIERS FOR BRAIN TARGETING

Principles and Applications

Edited by
Raj K. Keservani, MPharm
Anil K. Sharma, MPharm, PhD
Rajesh K. Kesharwani, PhD

AAP | APPLE
ACADEMIC
PRESS

Apple Academic Press Inc.
3333 Mistwell Crescent
Oakville, ON L6L 0A2
Canada

Apple Academic Press Inc.
1265 Goldenrod Circle NE
Palm Bay, Florida 32905
USA

Library and Archives Canada Cataloguing in Publication

Title: Nanocarriers for brain targeting : principles and applications / edited by Raj K. Keservani, MPharm, Anil K. Sharma, MPharm, PhD, Rajesh K. Kesharwani, PhD.

Names: Keservani, Raj K., 1981- editor. | Sharma, Anil K., 1980- editor. | Kesharwani, Rajesh Kumar, 1978- editor.

Description: Includes bibliographical references and index.

Identifiers: Canadiana (print) 20190080248 | Canadiana (ebook) 20190080264 | ISBN 9781771887304 (hardcover) | ISBN 9780429465079 (ebook)

Subjects: LCSH: Drug delivery systems. | LCSH: Blood-brain barrier. | LCSH: Brain—Physiology. | LCSH: Brain—Diseases.

Classification: LCC RS199.5 .N36 2019 | DDC 616.8/0461—dc23

Library of Congress Cataloging-in-Publication Data

Names: Keservani, Raj K., 1981- editor. | Sharma, Anil K., 1980- editor. | Kesharwani, Rajesh Kumar, 1978- editor.

Title: Nanocarriers for brain targeting : principles and applications / editors, Raj K. Keservani, Anil K. Sharma, Rajesh K. Kesharwani.

Description: Toronto ; New Jersey : Apple Academic Press, 2019. | Includes bibliographical references and index.

Identifiers: LCCN 2019010969 (print) | LCCN 2019011257 (ebook) | ISBN 9780429465079 (ebook) | ISBN 9781771887304 (hardcover : alk. paper)

Subjects: | MESH: Brain--drug effects | Brain Diseases--drug therapy | Blood-Brain Barrier--drug effects | Drug Delivery Systems | Nanocapsules--therapeutic use

Classification: LCC RC350.C54 (ebook) | LCC RC350.C54 (print) | NLM WL 300 | DDC 616.8/0461--dc23

LC record available at https://lccn.loc.gov/2019010969

ABOUT THE EDITORS

Raj K. Keservani

Raj K. Keservani, MPharm, is in the Faculty of BPharm, CSM Group of Institutions, Allahabad, India. He has more than 10 years of academic (teaching) experience at various institutes of India in pharmaceutical education. He has published 35 peer-reviewed papers in the field of pharmaceutical sciences in national and international journals, 16 book chapters, two co-authored books, and 10 edited books. He is also active as a reviewer for several international scientific journals. His research interests include nutraceutical and functional foods, novel drug delivery systems (NDDS), transdermal drug delivery/drug delivery, health science, cancer biology, and neurobiology. Mr. Keservani graduated with a pharmacy degree from the Department of Pharmacy, Kumaun University, Nainital (UA), India. He received his Master of Pharmacy (MPharm) (specialization in pharmaceutics) from the School of Pharmaceutical Sciences, Rajiv Gandhi Proudyogiki Vishwavidyalaya, Bhopal, India.

Anil K. Sharma

Anil K. Sharma, MPharm, PhD, is working as an assistant professor at the Department of Pharmacy, School of Medical and Allied Sciences, GD Goenka University, Gurugram, India. He has more than 9 years of academic experience in pharmaceutical sciences. He has published 28 peer-reviewed papers in the field of pharmaceutical sciences in national and international journals as well as 15 book chapters and 10 edited books. His research interests encompass nutraceutical and functional foods, novel drug delivery systems (NDDS), drug delivery, nanotechnology, health science/life science, and biology/cancer biology/neurobiology. He received a bachelor's degree in pharmacy from the University of Rajasthan, Jaipur, India, and a Master of Pharmacy degree from the School of Pharmaceutical Sciences, Rajiv Gandhi Proudyogiki Vishwavidyalaya, Bhopal, India, with a specialization in pharmaceutics. He earned his doctorate (PhD) from the University of Delhi, India.

Rajesh K. Kesharwani

Rajesh K. Kesharwani, PhD, is affiliated with the Department of Advanced Science & Technology, NIET, Nims University Rajasthan, Jaipur, India. He has more than 7 years of research and 2 years of teaching experience at various institutes of India, imparting bioinformatics and biotechnology education. He has received several awards, including the NASI-Swarna Jayanti Puraskar–2013 by The National Academy of Sciences of India. He has authored over 32 peer-reviewed articles and 10 book chapters. He has been a member of many scientific communities as well as a reviewer for many international journals. His research fields of interest are medical informatics, protein structure and function prediction, computer-aided drug designing, structural biology, drug delivery, cancer biology, and next-generation sequence analysis. Dr. Kesharwani received a BSc in biology from Ewing Christian College, Allahabad, India, an autonomous college of the University of Allahabad; his MSc (Biochemistry) from Awadesh Pratap Singh University, Rewa, Madhya Pradesh, India; and MTech-IT (specialization in Bioinformatics) from the Indian Institute of Information Technology, Allahabad, India. He earned his PhD from the Indian Institute of Information Technology, Allahabad, and received a Ministry of Human Resource Development (India) Fellowship and Senior Research Fellowship from the Indian Council of Medical Research, India.

CONTENTS

THE PRESENT BOOK IS DEDICATED TO

OUR BELOVED

AASHNA

ANIKA

ATHARVA

AND

VIHAN

CONTRIBUTORS

Jayanthi Abraham
Microbial Biotechnology Laboratory, School of Biosciences and Technology, VIT University, Vellore, Tamil Nadu, India

Juliana Palma Abriata
School of Pharmaceutical Sciences of Ribeirao Preto, University of Sao Paulo, Sao Paulo, Brazil

B. A. Aderibigbe
Department of Chemistry, University of Fort Hare, Alice Campus, Eastern Cape, South Africa

I. A. Aderibigbe
Department of Chemical and Metallurgical Engineering, Tshwane University of Technology, Pretoria

M. M. de Araujo
Department of Pharmaceutical Sciences, School of Pharmaceutical Sciences of Ribeirão Preto, University of São Paulo, Ribeirão Preto, SP, CEP 14040-903, Brazil

Debjani Banerjee
Department of Applied Science, Indian Institute of Information Technology, Allahabad, India

Saswata Banerjee
Birla Institute of Technology and Science (BITS), Pilani, Rajasthan, India

Jaleh Barar
Research Center for Pharmaceutical Nanotechnology, Tabriz University of Medical Science, Tabriz, Iran
Department of Pharmaceutics, Faculty of Pharmacy, Tabriz University of Medical Science, Tabriz, Iran

Mst. Marium Begum
Department of Pharmacy, East West University, Dhaka, Bangladesh

M. V. L. B. Bentley
Department of Pharmaceutical Sciences, School of Pharmaceutical Sciences of Ribeirão Preto, University of São Paulo, Ribeirão Preto, SP, CEP 14040-903, Brazil

D. B. Borin
Franciscan University, Santa Maria, Rio Grande do sul, Brazil

Patrícia Mazureki Campos
School of Pharmaceutical Sciences of Ribeirao Preto, University of Sao Paulo, Sao Paulo, Brazil

Morteza Eskandani
Research Center for Pharmaceutical Nanotechnology, Biomedicine Institute, Tabriz University of Medical Sciences, Tabriz, Iran

V. O. Fasiku
Department of Biological Sciences, North-West University, Private Bag X2046, Mmabatho 2735, South Africa

Josef Jampílek
Department of Pharmaceutical Chemistry, Faculty of Pharmacy, Comenius University, Odbojárov 10, 83232, Bratislava, Slovakia

Rebecca Jonczyk
Institute of Technical Chemistry, Leibniz University of Hannover, Callinstr. 5, 30167 Hannover, Germany

Emil Joseph
Birla Institute of Technology and Science (BITS), Pilani, Rajasthan, India

Khushwinder Kaur
Department of Chemistry and Centre of Advanced Studies in Chemistry,
Punjab University, Chandigarh, India

Katarína Kráľová
Institute of Chemistry, Faculty of Natural Sciences, Comenius University, Ilkovičova 6, 84215,
Bratislava, Slovakia

Marcela Tavares Luiz
School of Pharmaceutical Sciences of Ribeirao Preto, University of Sao Paulo, Sao Paulo, Brazil

P. D. Marcato
Department of Pharmaceutical Sciences, School of Pharmaceutical Sciences of Ribeirão Preto,
University of São Paulo, Ribeirão Preto, SP, CEP 14040-903, Brazil

Juliana Maldonado Marchetti
School of Pharmaceutical Sciences of Ribeirao Preto, University of Sao Paulo, Sao Paulo, Brazil

Nidhi Mishra
Department of applied science, Indian Institute of Information Technology, Allahabad, India

E. Mukwevho
Department of Biological Sciences, North-West University, Private Bag X2046, Mmabatho 2735,
South Africa

Soumya Nair
Microbial Biotechnology Laboratory, School of Biosciences and Technology, VIT University,
Vellore, Tamil Nadu, India

Yadollah Omidi
Research Center for Pharmaceutical Nanotechnology, Tabriz University of Medical Science, Tabriz, Iran
Department of Pharmaceutics, Faculty of Pharmacy, Tabriz University of Medical Science, Tabriz, Iran

S. J. Owonubi
Department of Biological Sciences, North-West University, Private Bag X2046, Mmabatho 2735,
South Africa

A. P. I. Popoola
Department of Chemical and Metallurgical Engineering, Tshwane University of Technology,
Pretoria, South Africa

Mohammad A. Rafi
Department of Neurology, Sidney Kimmel College of Medicine, Thomas Jefferson University,
Philadelphia, PA, USA

Nagarjun Rangaraj
Department of Pharmaceutics, National Institute of Pharmaceutical Education & Research (NIPER-H),
Balanagar, Hyderabad, India

Kirti Rani
Amity Institute of Biotechnology, Amity University Uttar Pradesh, Noida, Sec-125, Noida, India

Giovanni Loureiro Raspantini
School of Pharmaceutical Sciences of Ribeirao Preto, University of Sao Paulo, Sao Paulo, Brazil

Ali Sadeghinia
Research Center for Pharmaceutical Nanotechnology, Biomedicine Institute,
Tabriz University of Medical Sciences, Tabriz, Iran

E. R. Sadiku
Department of Chemical, Metallurgical and Materials Engineering, Tshwane University of Technology,
CSIR Campus, Building 14D, Private Bag X025, Lynwood Ridge 0040, Pretoria, South Africa

Sunitha Sampathi
Department of Pharmaceutics, National Institute of Pharmaceutical Education & Research (NIPER-H),
Balanagar, Hyderabad, India

R. C. V. Santos
Federal University of Santa Maria, Santa Maria, Rio Grande do sul, Brazil

Thomas Scheper
Institute of Technical Chemistry, Leibniz University of Hannover, Callinstr. 5, 30167 Hannover, Germany

Didem Ag Seleci
Institute of Technical Chemistry, Leibniz University of Hannover, Callinstr. 5, 30167 Hannover, Germany

Muharrem Seleci
Institute of Technical Chemistry, Leibniz University of Hannover, Callinstr. 5, 30167 Hannover, Germany

Huma Shamshad
Department of applied science, Indian Institute of Information Technology, Allahabad, India

Gautam Singhvi
Birla Institute of Technology and Science (BITS), Pilani, Rajasthan, India

I. L. Suzuki
Department of Pharmaceutical Sciences, School of Pharmaceutical Sciences of Ribeirão Preto,
University of São Paulo, Ribeirão Preto, SP, CEP 14040-903, Brazil

L. B. Tofani
Department of Pharmaceutical Sciences, School of Pharmaceutical Sciences of Ribeirão Preto,
University of São Paulo, Ribeirão Preto, SP, CEP 14040-903, Brazil

Md. Sahab Uddin
Department of Pharmacy, Southeast University, Dhaka, Bangladesh

Shivani Uppal
Department of Chemistry and Centre of Advanced Studies in Chemistry,
Punjab University, Chandigarh, India

Frank Stahl
Institute of Technical Chemistry, Leibniz University of Hannover, Callinstr. 5, 30167 Hannover, Germany

V. Wiwanitkit
Visiting professor, Hainan Medical University, China

S. Yasri
KMT Primary Care Center, Bangkok, Thailand

ABBREVIATIONS

AAS	acoustic attenuation spectroscopy
Ab	antibody
ABC	ATP-binding cassette
ACH	aminocyclohexane carboxylic acid
AD	Alzheimer's disease
ADS	analytical disk centrifugation
AFM	atomic force microscopy
AMT	adsorptive-mediated transcytosis
Ap	aptamer
APP	amyloid protein precursor
AUC	area under the curve
AuNPs	gold nanoparticles
BBB	blood–brain barrier
BCEC	brain capillary endothelial cells
BCSF	blood–cerebrospinal fluid
BCSFB	blood–cerebrospinal fluid barrier
BDDT	brain drug delivery and targeting
BSE	back-scattered electrons
BTB	blood–tumor barrier
CMT	carrier-mediated transport
CNS	central nervous system
CNTs	carbon nanotubes
CPP	cell-penetrating peptide
CS	chitosan
CsA	cyclosporine-A
CSF	cerebrospinal fluid
CTAB	cetyltrimethyl ammonium bromide
CTX	chlorotoxin
DDS	drug delivery systems
DOX	doxorubicin
DSC	differential scanning calorimetry
DTE%	drug transport efficiency
DTP%	direct transport percentage
EB	Evans Blue

EC	endothelial cell
EGFR	epidermal growth factor receptor
EIP	emulsion inversion point
EIV	ethanol injection vesicles
EPR	enhanced permeability and retention
ESM	eggshell membrane
FA	folic acid
FFF	field flow fractionation
FITC	fluorescein isothiocyanate
FPV	French press vesicles
FUdr	5-fluoro-2'deoxyuridine
FUS	focused ultrasound
GBM	glioblastoma multiforme
GM1	monosialoganglioside galactose
GO	graphene oxide
GRAS	generally recognized as safe
HAuCl4	hydrogen tetrachlorocuprate
HIV	human immunodeficiency virus
HPH	high-pressure homogenization
HSA	human serum albumin
i.n.	intranasal
i.v.	intravenous
IR	insulin receptor
ISFs	interstitial fluids
JAM	junctional adhesion molecule
LCST	lower critical solution temperature
LDL	low-density lipoprotein
Lf	lactoferrin
LMVs	large multilamellar vesicles
LPHNPs	lipid-polymer hybrid nanoparticles
LRP	lipoprotein receptor-related protein
LUV	large unilamellar vesicle
MALDI–MSI	matrix-assisted laser desorption/ionization mass spectrometry imaging
MALS	multi-angle light scattering detector
MDDA	12-mercaptododecanoic acid
MH	Morin hydrate
MLV	multilamellar vesicle
MRP	resistance-associated proteins

MRT	mean residence time
MSA	mercaptosuccinic acid
MTAs	microtubule-targeting agents
MTPs	membrane transport proteins
MW	molecular weight
NaBH4	sodium borohydride
NEms	nanoemulsions
NHS	N-hydroxysulfosuccinimide sodium salt
NLCs	nanostructured lipid carriers
NMR	nuclear magnetic resonance
NPs	nanoparticles
o/w	oil in water
OA	oleic acid
OND	ondansetron
P80	polysorbate® 80
PAA	poly(acryl amide)
PBCA	poly(butyl cyanoacrylate)
PBECs	porcine brain endothelial cells
PCDA	10,12-pentacosadiynoic acid
PCL	polycaprolactone
PDI	polydispersity index
PEG	polyethylene glycol
PEI	polyethylenimine
PGO	porphyrin-functionalized graphene oxide
PGP	protein–graphene–protein
PIC	phase inversion composition
PIT	phase inversion temperature
PL	photoluminescence
PLA	poly(lactic acid)
PLA	poly(d,l-lactide)
PLGA	poly(d,l-lactide-co-glycolide)
PLs	PEGylated liposomes
PNG	porphyrin-immobilized nanographene oxide
PNIPAm	poly(N-isopropylacrylamide)
pro-THP	pro-theophylline
PS	presenilin
PSO	pomegranate seed oil
PTX	paclitaxel
PVP	poly(vinyl pyrrolidone)

QDs	quantum dots
RES	reticuloendothelial system
REV	reverse-phase evaporation vesicles
RMN	rimonabant
RMT	receptor-mediated transcytosis
RMTs	receptor-mediated transporters
SEM	scanning electron microscopy
siRNA	small interfering RNA
SLC	solute carrier
SLN	solid lipid nanoparticles
STM	surface tunneling microscopy
SUV	small unilamellar vesicle
TAM	tamoxifen
TDD	targeted drug delivery
TEM	transmission electron microscopy
Tf	transferrin
TfR	transferrin receptor
TJ	tight junction
TME	tumor microenvironment
TOPO	trioctyl phosphine oxide
TSP	tamarind seed polysaccharide
UCST	upper critical solution temperature
VEGF	vascular endothelial growth factor
VIN	vinpocetine
VIP	vasoactive intestinal peptide
VLF	venlafaxine
w/o	water in oil
WGA	wheat germ agglutinin
XPS	X-ray photoelectron spectroscopy
XRD	X-ray diffraction spectroscopy

PREFACE

The inherent anatomy and physiology renders the brain unique from other organs. The past few decades have witnessed significant research on brain ailments; nevertheless, the majority of hospitalization occurs due to age-related central nervous system (CNS) disorders. The prevalence of diverse diseases such as Alzheimer's disease, Parkinson's disease, amyotrophic lateral sclerosis, multiple sclerosis, HIV-dementia, etc., is about 1.5 billion globally, which is further anticipated to touch 1.9 billion by the year 2020.

The blood–brain barrier (BBB) is observed to serve as a rate-limiting factor in drug delivery to the brain in combating CNS disorders. The challenges at present are to design and develop drug delivery careers, which ought to be able to deliver drugs across the BBB in a safe and effective fashion. Nanotechnology in the field of pharmaceuticals assists in improving the drug-delivery approach together with the kinetics and therapeutic index to unravel the delivery problems of some drugs.

This book, *Nanocarriers for Brain Targeting*, comprises 16 chapters divided into four sections that deal with describing brain physiology that pose obstacles to the development of DDS. In addition, the variety of strategies employed or strived for TO achieve targeting to the brain are discussed in detail. The nanocarriers covered are nanoparticles, vesicular carriers, carriers having carbon as core constituent, dispersed systems, etc.

SECTION I: INTRODUCTION OF CARRIERS IN BRAIN TARGETING

Chapter 1 gives an introduction to the BBB and intricacies involved while attempting delivery to the brain. Also, an overview of different carriers employed for delivery of drugs is provided. The information is suitably tabulated in order to give a precise view.

Chapter 2 discusses the importance of the BBB on brain targeting and advanced drug delivery systems (DDS) used for targeted therapy of brain diseases. The features of BBB are first described, followed by mechanisms by which drugs travel inside the brain. A discussion of different polymeric, lipidic nanoparticles is provided in ensuing sections. The microfluidics utilized as a technology platform also find a mention in later parts of the chapter.

The general structure and functions of the BBB, on various drug carriers have been given in Chapter 3. The chapter offers a preamble of drug carriers being polymeric, metallic, or vesicular in nature.

Chapter 4 gives an exhaustive discussion of nano-based carriers reported for the delivery of drugs to the brain resulting in enhanced therapeutic effects, their mechanisms, and biological efficacy (in vitro and in vivo).

Diverse facets of drug delivery to the CNS via nanocarriers are described in Chapter 5. The chapter focuses on nanoformulations of various classes of drugs, the site of action of which is in the CNS, that were prepared using polymers of natural origin and their semisynthetic modifications representing a class of biocompatible and biodegradable carriers, ensuring better therapeutic efficiency and toxicity profile of drugs.

SECTION II: PARTICULATE CARRIERS

Chapter 6 provides an overview of various nanocarriers used in brain drug targeting. The authors have highlighted the functional advantages that each of these nanoparticles, namely, polymeric and lipid nanoparticles, provide, owing to their chemical structure and composition. In addition, possible physiological hurdles have been mentioned that formulation scientists face while designing a particular formulation for targeted delivery of drugs to the brain, so that the reader can understand and comprehend the need for such drastic and complicated technological applications.

The description of nanocapsules applications in brain therapeutics is provided in Chapter 7, which focuses on the BBB, its physiology, and the preparative methods of nanocapsules, along with the various strategies in brain targeting using its modifications. Furthermore, the review evaluates the types of nanocapsules used for drug delivery in brain targeting and the mechanism involved in the process.

Chapter 8 provides a focused view of lipid nanocarriers explored for drug delivery to the brain. The chapter begins with an introduction to the BBB and the role of several drug carriers to overcome the biological barriers, and then moves further with detailed description of solid lipid nanoparticulates embracing preparation techniques and applications. In conclusion, the authors have mentioned future prospects of lipid-based carriers.

The uses of quantum dots (QDs) as drug carriers relevant to brain targeting have been discussed in Chapter 9. The author has provided a preamble of properties and synthesis methods of QDs along with their applications in CNS targeting. The whole chapter briefs the presentation of concerned information.

Chapter 10 has exhaustively discussed numerous applications of graphene and its derivatives in targeted drug delivery (TDD) to CNS. In addition, the authors have introduced DDS and TDD, mapping out targeting of drug formulations aimed for delivery inside the body. They then focused on brain targeting and the strategies that are employed to deliver drugs to the brain, indicating properties that nanocarriers possess that allow the BBB to be transversed.

SECTION III: VESICULAR CARRIERS

The customized applications of liposomes in site-specific drug delivery to the CNS are given in Chapter 11. The authors have endeavored to provide a general introduction of liposomes, followed by history and preparation methods. The last part of chapter deals with vectorization of liposomes in brain drug targeting.

Chapter 12 discusses extensively many aspects of liposomal carriers, including a general introduction, BBB structure and functions, and drug transport mechanisms. Subsequently, the methods of liposome preparation are discussed. Lastly, their potential applications as different therapeutic modalities in brain drug delivery have been described.

An overview of niosomes and their specialized applications in brain drug delivery has been presented in Chapter 13. The chapter provides fundamental information about niosomal DDS and their recent applications in brain targeting.

SECTION IV: MISCELLANEOUS

Chapter 14 focuses on the utility and advantage of nanoemulsions as a delivery vehicle for the administration of medicaments to the brain. Nanoemulsions are thus considered as an ideal delivery system as they offer inherent advantages of site-specific drug delivery, ability to dissolve hydrophobic drugs, and protecting drug degradation, thus enhancing long-term drug stability.

The details of nanogels employed in drug delivery to the brain have been provided in Chapter 15. The chapter begins with a description of the BBB and diverse mechanisms for transport of actives in the CNS. Thereafter, nanogels are discussed, covering materials employed, preparation, and characterization, and finally, the applications directed toward brain have been mentioned.

Chapter 16 describes the potential of nanotechnology-based drug carriers in management of tropical infections. The variety of concerns on the cost effectiveness, safety, and adverse effects of the new techniques also find a place in the chapter.

SECTION I
INTRODUCTION OF CARRIERS IN BRAIN TARGETING

CHAPTER 1

CARRIERS FOR BRAIN TARGETING: RECENT ADVANCES AND CHALLENGES

MD. SAHAB UDDIN[1,*] and MST. MARIUM BEGUM[2]

[1]*Department of Pharmacy, Southeast University, Dhaka, Bangladesh*

[2]*Department of Pharmacy, East West University, Dhaka, Bangladesh*

Corresponding author. E-mail: msu-neuropharma@hotmail.com; msu_neuropharma@hotmail.com

ABSTRACT

The blood–brain barrier (BBB) and the blood–cerebrospinal fluid barrier (BCSFB) guards the brain by firmly sealing the central nervous system (CNS) from the changeable milieu of blood. Brain delivery represents a great challenge, owing to the presence of the BBB and the BCSFB. Currently, numerous types of drug-delivery methods are developed to combat brain disorders, but most of them have failed due to the existence of these biological barriers. Maintaining typical body functions as well as the transport of therapeutic agents like essential biological elements through biological membranes are extremely vital. In order to get an enhancement of drug CNS performance, copious sophisticated tactics are hastily emerging. Nanoparticulate carriers which are specifically used to deliver drugs to the CNS have obtained great importance due to nanostructure advantages. Similarly, for treating CNS disorders, microtubules-targeting agents are evolving as potential therapeutic targets. Therefore, this chapter represents an outline of barriers to brain drug-delivery and transport mechanisms as well as currently developed brain targeted drug-delivery system and associated challenges.

1.1   INTRODUCTION

In central nervous system (CNS), application of the drug-delivery process is a very critical process. Scientists face problems in formulations of drugs due to the presence of blood–brain barrier (BBB). Studies have shown that approximately 98% of CNS drugs cannot pass BBB, so accurate therapeutic drug concentration cannot be achieved (Pardridge, 2009). Various types of drugs are obstructed by BBB to enter into brain. These include some antibiotics, antineoplastic drugs, and neuropeptides which cannot pass through the capillaries of brain due to BBB. Recently, various types of drug-delivery methods are invented for CNS diseases, but most of them have failed to show therapeutic activities as cannot reach to the target area due to BBB. All drug-delivery methods which are traditional based on trials and errors. Normally, these methods are used for delivering limited numbers of drugs that have drug–receptor interactions or structure–activity relationships (Jones and Shusta, 2007). However, maintaining normal body functions and to gain transportation of essential biological elements including therapeutic agents through biological membranes are highly crucial (Senel et al., 2001). But till now, only a few methods permit drugs for effective membrane permeation. Nowadays, various drug-delivery methods are invented based on rational drug design and screening receptor–ligand interactions. Noninvasive and less toxic drugs and delivery methods have been also developed to reduce postdelivery toxicity of the drugs. Not only the binding affinity to the receptor but also structure–activity relationships of drugs, target receptor binding, and its behavior in animal system are important during selection of drugs (De Boer, 2003). In addition, there are many factors and biological obstacles which influence the drug delivery to the BBB (Schaddelee et al., 2003).

Biotechnology-based products and small molecule drugs cannot cross BBB. In addition, majority of the small molecule drugs (>98%) fail to cross BBB. But if the barrier is overcome, then there is possibility of effective treatment of several diseases of CNS. Various scientific possibilities have been tried for this purpose such as prodrugs, intracerebral injection, use of implants, disruption of BBB, etc. Prodrugs are special types of drugs which undergo a chemical or enzymatic biotransformation to convert to active drug to produce therapeutic action in the body. Lipophilic compounds are used in formulations of prodrugs so that can easily pass through BBB (Rautio et al., 2008). For this, lipophilic esters and other hydrophobic compounds which are important elements of prodrugs have been studied for formulations. Carrier-mediated prodrug transport (endogenous transporters), receptor-mediated prodrug transport (macromolecular delivery), and gene-mediated enzyme prodrug therapy are examples of sophisticated prodrug approaches (Pavan et al., 2008; Rautio et al., 2008).

Conversion of prodrug to the drug in plasma is regarded as an obstacle in prodrug. Another way to enter into BBB is reversible disruption which makes junctions of the endothelial cells leaky and thus provides access for biological elements to the brain (Gabathuler, 2010). Osmotic disruption, ultrasound disruption, and disruption by bradykinin analogue are some techniques that have been used for disruption. In the process of osmosis, it shrinks endothelial cells, thus disrupting the tight junctions (TJs). An effective example of this is administration of mannitol solution with subsequent administration of drugs that increase drug concentration in brain, which is not possible if taking of drug alone (Gabathuler, 2010). Another fruitful process is magnetic resonance imaging-guided based ultrasound technique to enter BBB. A recent study has shown that Evans Blue (EB) extravasation can be enhanced by the application of repeated sonication, where 1-MHz frequency was applied (Yang et al., 2011). Sonications were applied at an ultrasound frequency of 1 MHz and a repetition frequency of 1 Hz. EB extravasation in double sonication groups was increased approximately twofold compared with the single sonication group. In the study, it was seen that compared to groups where no sonication has been applied, extravasation was high with both the cases. In BBB, B2 bradykinin receptor agonist and cereport (i.e., RMP-7, a bradykinin B2 agonist) have increased permeability of drugs. There are confirmations of increased CNS delivery when carboplatin, loperamide, and cyclosporin-A are administered with RMP-7 (Borlongan and Emerich, 2003). The main two obstacles to enter BBB are brain uptake of plasma albumin and other protein components of blood, which are noxious to brain cells (Vykhodtseva et al., 2008).

Another useful traditional approach to bypass BBB is intracerebral injection or use of implants. The bolus injection of therapeutic agents and the placement of a biodegradable drug impregnated wafer depend on the norm of diffusion to drive the drug through BBB. In an *in vitro* study, various types of paclitaxel-loaded lipidic implants and poly(D,L-lactide-*co*-glycolide) (PLGA)-based microparticles have been prepared and characterized (Elkharraz et al., 2006). They can directly be injected into the brain tissue overcoming the restriction of the BBB (i.e., for paclitaxel) to a significant extent upon systemic administration. In this study, it was found that direct administration of controlled drug-delivery system is helpful to improve the local treatment of operable and inoperable brain tumors (Elkharraz et al., 2006). But these approaches are less successful to cross the BBB and most of them have numerous adverse effects. Also, these are harmful in nature for the body in long term. Moreover, it is a challenge to enhance drug permeability to brain for effective therapeutic efficacy with no or limited side-effects and with better patient compliance. Therefore, the purpose of this chapter is to give an overview of barriers to brain drug-delivery and transport mechanisms as well as recently developed brain targeted drug-delivery system and their challenges.

1.2   PHYSIOLOGY OF BARRIERS TO BRAIN DRUG DELIVERY

Figures 1.1 and 1.2 gives an idea of BBB and blood–cerebrospinal fluid barrier (BCSFB) and their various components. It is cleared to understand from the figures that how BBB protects the brain from most substances in the blood, supplying nutrients and restricting to enter poisonous elements from the brain. Endothelial cells which form the anatomical substrate called cerebral microvascular endothelium are the main components of BBB. BBB transfers solutes and other substances in and out of the brain, leukocyte migration and maintains the homeostasis of the brain microenvironment. Neurovascular unit is essential for the health and function of the CNS which is composed of cerebral microvascular endothelium, together with astrocytes, pericytes, neurons, and the extracellular matrix (Hawkins and Davis, 2005; Persidsky et al., 2006). The transport of solutes and other substances across BBB is firmly constrained through TJs, adherents junctions, and metabolic barriers (i.e., enzymes, diverse transport systems), thus excluding very small, electrically neutral, and lipid soluble molecules. For this reason, drugs or chemotherapeutic agents are unable to pass through the barrier.

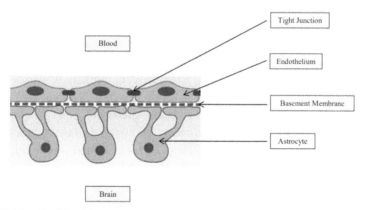

FIGURE 1.1   The blood–brain barrier that obstructs drug delivery to central nervous system.

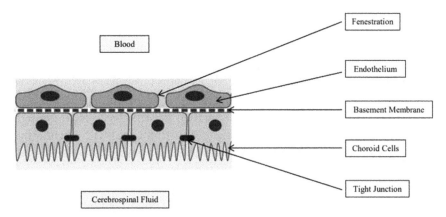

FIGURE 1.2   The blood–cerebrospinal fluid barrier that obstructs drug delivery to central nervous system.

TJs in BBB exert an intricate complex of transmembrane proteins, with cytoplasmic accessory proteins acting as physiological and pharmacological barrier, thereby preventing influx of molecules from the bloodstream into the brain. From the figures, we can see that BBB is characterized by two membranes, luminal and abluminal, facing blood capillary and brain interstitial fluids (ISFs), respectively. A difference exists between the endothelia of the brain capillaries and endothelia in other capillaries, such as TJs between adjacent endothelial cells in BBB (Brightman and Rees, 1969; Hawkins and Davis, 2005), lack of fenestrations, and pinocytotic vesicles (Reese and Karnovsky, 1967; Saunders et al., 1999a, 1999b, 2000, 2008; Stewart, 2000). Also, there exists other CNS barrier shielding the delicate brain tissue which may play a role in drug transport such as the blood–tumor barrier (BTB) and the blood–retina barrier (Burkhard and Heiko, 2005; Decleves et al., 2006) formed of pigment epithelium enclosing the retina and thus acting as a barrier interface between the systemic blood vessels of the neighboring choroid and the retina. Finally, targeting of tumor tissue is constricted by the BTB (Decleves et al., 2006).

Another important feature of CNS is BCSFB which is formed by the epithelial cells of the choroid plexus. BCSFB controls the penetration of molecules within the ISF of the brain parenchyma by controlling the exchange of molecules between the blood and CSF. Facilitated diffusion and active transport (i.e., into the CSF, from CSF to the blood) are two mechanisms of transport pertaining to the choroid plexus (Begley and Brightman, 2003; Spector, 1985; Zeuthen, 1991).

1.3   TRANSPORT MECHANISMS ACROSS BRAIN

Diffusion and active transport are the main transport mechanisms; in Figure 1.3, transports through the BBB are shown (Begley and Brightman, 2003). There are various processes through which solutes move from membrane to membrane in brain but all these mechanisms fall into two forms. Diffusion or facilitated transport across aqueous channels is the first basic form through which solutes pass. The basic mechanism of this process is a concentration gradient across the membranes, between cells or across cells. By diffusion, solutes transport through membranes, but in this case, some factors like size and lipophilicity of the substances are very important (Fischer et al., 1998). Second process is active transport which is mainly controlled by carriers like proteins. Molecular affinity, fluid streams, or magnetic fields are responsible for movement of molecules by this process (Lodish et al., 2000).

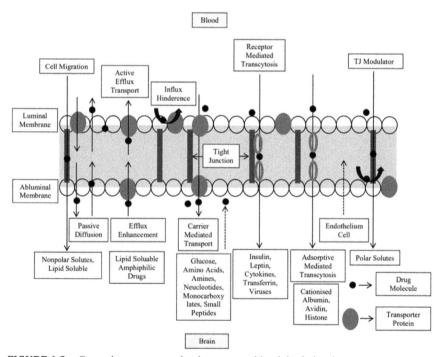

FIGURE 1.3   General transport mechanisms across blood–brain barrier.

The different mechanisms are shown through which transports of solutes, drugs, and other particles follow (Fig. 1.3). In BBB by the process of chemotaxis, cell migration from blood cells (i.e., leukocytes like mono-cytes/macrophages) can be occurred, thus modifying the functionality of TJs (Engelhardt, 2006).

Carrier-mediated transports are forms of diffusion which may be passive or active. It mainly depends on the context and unidirectional transport of drugs from the blood to the brain. It is mostly instrumental in the transport of numerous basic polar atoms, with the assistance of transporters, for example, glucose (GLUT1 glucose transporter), amino acids (the LAT1 large neutral amino acid transporter, the CAT1 cationic amino acid transporter), carbox-ylic acids (the MCT1 monocarboxylic acid transporter), and nucleosides (the CNT2 nucleoside transporter) into the brain. Active efflux transport includes extrusion of drugs from the brain by the help of efflux transporters, for example, P-glycoprotein, multidrug imperviousness protein, and breast tumor safety protein, and other transporters (Tamai and Tsuji, 2000). In contrast to the carrier-mediated transport, the active efflux transport causes the active efflux of drugs from brain back to blood. It exerts a major impedi-ment to pharmacological drug conveyance of the CNS. A previous study confirmed that endogenous peptides like Tyr-Pro-Trp-Gly-NH2, transported from the brain to the blood by peptide transport system-1, are transported via active efflux (Banks et al., 1993).

Receptor-interceded transport is mainly engaged in the transport of macromolecules like peptides and proteins over the BBB by conjugating the substance with ligands, for example, lactoferrin, transferrin, and insulin (King and Johnson, 1985; Maratos-Flier et al., 1987; Roberts et al., 1993). It is an imperative transport mechanism of transcendent enthusiasm for drug delivery. Next, adsorptive-interceded transport is a kind of endocytosis incited by conjugating the molecule to cationized ligands or peptides, for example, albumin (Kumagai et al., 1987; Sai et al., 1998). Due to the influ-ence of anionic site located in the membrane, ligand-conjugated cationized NPs utilized adsorption-mediated transport process for the entry within the brain.

The relaxation of junctions causes the modulation of TJ, ultimately enhancing the selective aqueous diffusion across the paracellular junctions in the BBB. According to Mahajan et al. (2008a), methamphetamine is used to modulate the TJ. Furthermore, the modulation of TJs was described by using morphine and human immunodeficiency virus (HIV) 1 by activating intracellular Ca^{2+} release, myosin light chain kinase, and proinflammatory cytokines (Mahajan et al., 2008b). Increased transendothelial migration and decreased transendothelial electric resistance were observed across the BBB in their studies. The observations were similar in cocaine on the permeability of BBB, which is responsible for worsening HIV dementia. Detailed studies are required to identify the development of novel anti-HIV-1 therapeutics that target specific TJ proteins, such as ZO-1, JAM-2, Occludin, Claudin-3, and Claudin-5. Using the ultrasound-mediated molecular delivery, a new approach was developed besides the normal physiological delivery methods. For example, Choi et al. (2007) demonstrated gadolinium's deposition through ultrasound-induced BBB openings in the hippocampus of the murine.

One major point which is often ignored in nanodrug delivery is the ultimate destination of the nanocarriers themselves. There are dramatic differences observed based on administration, dosage, and functionalization. Active targeting is a noninvasive way of transporting drugs by using site-specific ligands to target organ. It can highly increase the BBB permeability of drugs. Nanocarriers which are conjugated with ligands are able to recognize the brain capillary endothelial cells (BCEC) and brain tumoral cells as an important breakthrough in neuro-oncology and CNS drug delivery (Béduneau et al., 2007). Bareford and Swaan (2007) have reviewed the role of endocytosis in targeted brain delivery and the prediction was by the efficient targeting of conjugated nanocarrier systems to the endolysosomal pathway. For the treatment of lysosomal storage disease, Alzheimer's disease and cancer significant improvement of the drug delivery can be accomplished.

1.4   NANOTECHNOLOGY IN BRAIN TARGETING

1.4.1   POLYMERIC NANOPARTICLES

There are some nanoparticles which are specifically used to deliver drugs to the CNS. Those nanoparticles are composed of various biocompatible and biodegradable polymers. Some of the examples are PLGA, poly(D,L-lactide), polybutyl cyanoacrylates, etc. (Costantino and Boraschi, 2012).

Notable studies have been conducted to identify an appropriate polymeric nanoparticle system in order to ease the delivery of drugs across the BBB. Polycaprolactone (PCL) and PLGA nanoparticles, which are loaded with etoposide, are generally prepared by the pharmacokinetics and biodistribution of the radiolabeled etoposide. The loaded PCL and PLGA were carried out ultimately (Snehalatha et al., 2008). These PCL and PLGA, loaded with etoposide, were labeled as Tc-99m and they were administered via the following two routes, oral and intravenous. After the administration, the pharmacokinetic and biodistribution properties were identified. The results claimed that pure etoposide is advantageous as a carrier for PCL and PLGA nanoparticles in comparison to the high plasma residence. It also suggested that this carrier increased the bioavailability along with distributing selectively and brain permeability and decreased the chances of toxicity associated to etoposide. After 24 h of administration, it was observed that the etoposide-loaded nanoparticles showed three times more distribution than the pure form of drug. PLGA nanoparticles were loaded with imatinib mesylate in another study and were administered into a rat model to identify the pharmacokinetic and biodistribution properties (Bende, 2008). The results claimed that the extent of drug permeation in brain showed to rise up to 100% in mean residence time. It also suggested a threefold increase in AUC_∞ in comparison to the pure form of the drug (Bende, 2008).

There are a number of explanations regarding the delivery route of nose-to-brain drugs. The drugs are first transferred to the nasal cavity and then sent to the CNS directly through the trigeminal nerves or the olfactory epithelium (Mistry et al., 2009). It is considered as the best route to deliver drugs more efficiently to brain. Chitosan nanoparticles were prepared by loading venlafaxine so that the uptake of the drug to the brain from the intranasal cavity is executed appropriately (Haque et al., 2012). It has the potential to show better efficacy in treating depression if its direct transport percentage (80.34%) and drug transport efficiency (508.59%) are higher than other formulations.

Another useful approach can be the coating of the nanoparticles with surfactants. It can influence increased brain uptake. Numbers of studies have been conducted in this path, and coating with surfactant systems has been found to result in increased brain concentration of drug in comparison to uncoated systems (Ambruosi et al., 2006; Ren et al., 2009).

1.4.2   SOLID LIPID NANOPARTICLES

Colloidal particles, which are composed of biodegradable and biocompatible lipid matrix, are known as solid lipid nanoparticles (SLN). The matrix remains solid in body temperature. It shows its size and shape within a range of 100–400 nm (Joshi and Müller, 2009). Several advantages such as targeted delivery, high drug payload, increased drug stability, least biotoxicity, large-scale production, ease of sterilization, and controlled drug release are offered by SLN (Mehnert and Mäder, 2012). The regular ingredients of SLN are emulsifiers, water, and solid lipids. Lipid here is used in a broader meaning. It includes compounds, such as fatty acids (stearic acid), steroids (cholesterol), waxes (cetyl palmitate), and triglycerides (tristearin).

Due to their massive advantages and higher ability to pass BBB, SLN are widely used for the delivery of active pharmaceutical ingredients to the brain. For example, Martins et al. (2012) prepared the camptothecin-loaded SLN using cetyl palmitate, Dynasan 114, and Witepsol E85. They used hot high-pressure homogenization process. Comparison of macrophages showed a higher affinity of the SLN to the porcine BCEC. *In vivo* studies in rats showed that fluorescent-labeled SLN were detected highly in the brain after IV administration. Similar studies on SLN preparation with enhanced brain uptake were identified in the literature (Manjunath and Venkateswarlu, 2005; Zara et al., 2002).

A nanostructured lipid carrier is known as the new version of SLN. It has increased drug-loading capacity which is gaining its popularity for targeting brains. It is made of solid lipids and a certain portion of liquid lipids. The solidity of the compound is maintained both in the body temperature and at room temperature (Müller et al., 2007; Tsai et al., 2012).

1.4.3 LIPOSOMES

Liposomes are the classic examples of nanoformulations. They have phospholipid bilayer system, and in this system, the hydrophilic drugs remain inside the aqueous phase and the phase is engulfed by lipophilic drugs and phospholipid bilayers (Zahin et al., 2019). With this nature, they can easily enter and integrate into the membrane (Allen and Cullis, 2013). Studies are being conducted by the researchers for achieving better drug delivery such as long-circulating (polyethyl glycolylated) liposomes, triggered release liposomes, liposomes containing nucleic acid polymers, ligand-targeted liposomes, and liposomes containing combinations of drugs. Various clinical trials of anticancer drugs, antifungal drugs, antibiotics, gene medicines, anesthetics, and anti-inflammatory drugs have been possible for these advancements (Allen and Cullis, 2013). Liposomal systems were used in targeted brain delivery and it resulted in considerable increase of drug concentration in brain (Bellavance et al., 2010; Ramos et al., 2011).

1.4.4 DENDRIMERS

Dendrimers are a special class of synthetic polymers that possess a major role in nanotechnological advances of drug delivery (Oliveira et al., 2010; Uddin, 2019). For targeted brain delivery, novel dendrimer-based drug-delivery systems that consist G3 polyamidoamine (PAMAM) and surfactant-conjugated dendritic nanoconjugates have been successfully applied (Gajbhiye and Jain, 2011; Teow et al., 2013). According to the cytotoxicity studies, it was observed that free G3 PAMAM was relatively nontoxic while the conjugation of lauryl chains and paclitaxel molecule on the surface of G3 PAMAM dendrimer specifically increased the cytotoxicity in both human colon adenocarcinoma cell line (Caco-2) and primary cultured porcine brain endothelial cells (PBECs). The demonstration of the enhancement of the permeation of the lauryl-modified G3 PAMAM dendrimer-paclitaxel conjugates across Caco-2 cell and PBEC monolayers has also been executed. The conjugate of dendrimer had approximately 12-fold greater permeability across both cell monolayers in comparison to paclitaxel alone (Teow et al., 2013).

1.4.5 MICELLES

Polymeric micelles are derived from block copolymers. They are derived as colloidal carriers for targeting genes and drugs. They have been receiving much attention in the field of drug delivery and targeting because of the high drug-loading capacity (Kataoka et al., 2012). A variety of drugs with distinct characteristics such as genes and proteins can be incorporated into the core. Researchers have explained effective targeting of micelles systems to the brain by intranasal and intravenous route (Abdelbary and Tadros, 2013).

1.4.6 NANOEMULSIONS

Nanoemulsions have gained considerable attention in research and in therapeutics because of its advantages. Advantages such as ease of preparation, optical clarity, thermodynamic stability, and their ability to incorporate both hydrophilic and hydrophobic solutes, etc. have grabbed massive attention (Rajpoot et al., 2011). Kumar et al. (2008) studied the intranasal nanoemulsion-based brain-targeting drug-delivery system of risperidone. Higher drug transport efficiency and direct nose-to-brain drug transport for these mucoadhesive nanoemulsions were determined by the researchers. Similar results have been derived with saquinavir-loaded nanoemulsions resulting in efficient brain delivery (Vyas et al., 2008).

For drug-delivery applications, various researchers have been attracted the interest of nanoformulations of medicinal (Jeevanandam et al., 2016). These nanoformulations increase the properties of conventional drugs. They are particular to the targeted delivery site. Nanoformulations like dendrimers, polymeric nanoparticles, liposomes, nanoemulsions, and micelles are receiving massive importance in pharmaceutical industry for increased formulation of drugs. Wide varieties of synthesis methods are prevalent which are used for the preparation of nanoformulations in order to deliver drugs in biological system. The choice of methods of synthesis depends on the shape and size of biological properties of drug, particulate formulation, and the targeted site (Jeevanandam et al., 2016). In Table 1.1, list of nanoparticulate systems used for brain drug delivery and targeting is presented, and Table 1.2 represents nanoformulations investigated for better brain delivery.

TABLE 1.1   Outline of Nanoparticulate Systems Used as Brain Drug Delivery and Targeting.

Nanocarrier	Drug	Targeting ligand	Relevant results
PEGylated-liposomes (Schmidt et al., 2003)	Prednisolone	–	Brain accumulation of the injected PEGylated liposomes in rats with EAE, reaching values of up to 4.5-fold higher than in healthy control animals. No apoptosis in resident cells such as astrocytes, oligodendrocytes, or microglia in spinal cord, which rules out some important unwanted side-effects
PEGylated-liposomes (Zhang et al., 2004)	RNAi	mAbs	Weekly intravenous RNAi gene therapy directed against the human epithelial growth factor receptor by the PEGylated liposomes gene transfer technology causes an 88% increase in survival time in adult mice with intracranial human brain cancer
Immuno-liposomes (Pardridge, 2007)	DNA	TfR-mAbs, HIR-mAbs	TH gene therapy with liposomes resulted in a complete normalization of striatal TH enzyme activity ipsilateral to the lesioned nigrostriatal dopaminergic pathway of rat brain
Immuno-liposomes (Soni et al., 2008)	5-Fluorouracil	Tf	An average of 10-fold increase in the brain drug uptake was observed after the not-targeted liposomal administration, while the Tf-coupled liposomes caused a 17-fold increase

TABLE 1.1 *(Continued)*

Nanocarrier	Drug	Targeting ligand	Relevant results
Immuno-liposomes (Qin et al., 2008)	FA	RGD peptide	RGD-coated liposomes exhibiting vivobrain targeting ability with sixfold concentration FA in brain compared with FA solution and threefold in comparison of plain liposomes
SLNs clozapine (Manjunath and Venkateswarlu, 2005)	Clozapine	–	Positively charged clozapine SLNs enhanced the bioavailability of clozapine from 3.1- to 4.5-fold on intraduodenal administration. The AUC of clozapine SLNs showed higher uptake in RES organs and brain after intravenous administration than clozapine suspension
SLNs (Manjunath and Venkateswarlu, 2006)	Nitrendipine	–	Effective bioavailability of nitredipine SLNs was 2.81–5.35-folds greater after administration in comparison with drug suspension. In tested organs, the AUC of drug SLNs were higher than those of drug suspension, especially in brain, heart, and RES organs
SLNs (Bond'I et al., 2010)	Riluzole	–	Riluzole entrapped into SLNs that reach the brain was threefold greater than free riluzole, 16 h postinjection and showed a smaller accumulation in the RES organs. In addition, rats treated with riluzole-loaded SLNs showed clinical signs of experimental allergic EAE later than those treated with free riluzole
SLNs (Lockman et al., 2003)	–	Thiamine	Association of thiamine-targeted nanoparticles with the BBB thiamine transporter, and accumulation at the BBB, increasing brain uptake during perfusion time frames

TABLE 1.1   *(Continued)*

Nanocarrier	Drug	Targeting ligand	Relevant results
SLNs (Gupta et al., 2007)	Quinine dihy-drochloride	Tf	Enhancement of brain uptake of quinine dihydrochloride into Tf-targeted SLNs, as demonstrated by the recovery of a higher percentage of the dose from the brain after *in vivo* administration of Tf-coupled SLNs compared with unconjugated SLNs or drug solution
PACA nanoparticles (Das and Lin, 2005; Smith and Gumbleton, 2006)	Dalargin	Polysorbate 80 (PS80)	Measurement of *in vivo* central antinociceptive effect of dalargin along with a dose–response curve was observed after 60 min of oral administration of PS80-targeted nanoparticles to mice
PACA nanoparticles (Gulyaev et al., 1999)	Doxorubicin	PS80	High brain drug concentrations (>6 μg/g) were achieved with the PS80-coated nanoparticles, while both uncoated nanoparticles and drug saline solution were always below the detection limit (<0.1 μg/g). However, both types of nanoparticles prevented the drug accumulation in the heart
PACA nanoparticles (Friese et al., 2000)	MRZ 2/576 (*N*-methyl-D-aspartate receptor antagonist)	PS80	Intravenous administration of the drug bound to PS80-targeted PACA nanoparticles prolongs the duration of the anticonvulsive activity in mice up to 210 min and after probenecid pretreatment up to 270 min compared to 150 min with probenecid and MRZ 2/576 alone
PEG-PLA nanoparticles (Lu et al., 2005)	–	CBSA	CBSA nanoparticles did not impact the integrity of BBB endothelial tight junctions and also showed little toxicity against BCECs. The permeability of CBSA nanoparticles was about 7.76 times higher than that of BSA nanoparticles

TABLE 1.1 *(Continued)*

Nanocarrier	Drug	Targeting ligand	Relevant results
PLGA nanoparticles (Tosi et al., 2007; Vergoni et al., 2009)	Loperamide	Glyco-peptides (H_2N-Gly-L-Phe-D-ThrGly-L-Phe-L-Leu-L-Ser(O-β-D-glucose)-CONH$_2$)	13% of the injected dose of loperamide loaded into peptide-modified nanoparticles was found in the brain 4 h postinjection, while fluorescent peptide-modified nanoparticles about 9% of injected dose per gram of tissue at 0.25 h after administration
PEG CS nanoparticles (Aktas et al., 2005)	Z-DEVD-FMK peptide (Caspase-3 inhibitor)	Monoclonal antibody (OX26)	Translocation of PEG-CS/OX26 nanoparticles into the brain tissue after intravenous administration to mice
HSA nanoparticles (Ulbrich et al., 2009)	Loperamide	Tf or TfR-mAbs	Significant antinociceptive effects induced by loperamide-loaded targeted HSA nanoparticles in mice after intravenous injection, and only marginal effects with control loperamide-loaded HSA nanoparticles
			Escape from the macrophage uptake and internalization into neuroblastoma cells of PS80-targeted PHEA–PLA micelles
PHEA-PLA micelles (Craparo et al., 2008)		PS80	Brain penetration of fluorescent probe-loaded TAT-PEG-β-Chol micelles in hippocampus sections of rats 2 h after intravenous injection, while free probe did no cross the BBB
PEG-β-Chol micelles (Liu et al., 2008a, 2008b)		TAT	

BCECs, brain capillary endothelial cells; BSA, bovine serum albumin; CBSA, cationic bovine serum albumin; CS, chitosan; DNA, deoxyribonucleic acid; EAE, experimental autoimmune encephalomyelitis; FA, ferulic acid; HAS, human serum albumin; HIR, human insulin receptor; mAbs, monoclonal antibodies; PACA, poly(alkylcyanoacrylate); PEG, polyethyl glycol; PHEA, poly-2-hydroxyethyl aspartamide; PLA, poly(D,L-lactic acid); PLA, polylactic acid; PLGA, poly(lactic-*co*-glycolic acid); RES, reticuloendothelial system; RGD, arginylglycylaspartic acid; RNAi, RNA interference; TAT, transactivator of transcription.; Tf, transferring; TfR, transferrin receptor; TH, tyrosine hydrolase.

TABLE 1.2   Outline of Nanoformulations Investigated for Better Brain Delivery.

Formulation	Materials used	API/model molecule	Advantages
Polymeric nanoparticles (Ambruosi et al., 2006; Bende, 2008; Haque et al., 2012; Ren et al., 2009; Snehalatha et al., 2008)	PLGA and PCL	Etoposide	Selective distribution with higher brain permeability
	PLGA	Imatinib mesylate	Increased the extent of drug permeation to brain
	Chitosan	Venlafaxine	Better brain uptake, higher direct transport percentage
	PLA-PEG-tween 80	Amphotericin B	Drug concentration in mice brain greatly enhanced, reduced the toxicity of amphotericin B to liver, kidney, etc.
	PBCA-tween 80	Doxorubicin	Increased accumulation of nanoparticle in the tumor site and in the contralateral hemisphere
SLNs/Nanostructured lipid carriers (Martins et al., 2012; Manjunath and Venkateswarlu, 2005; Tsai et al., 2012; Zara et al., 2002)	Cetyl palmitate, dynasan, witepsol	Camptothecin	Higher affinity to the porcine brain capillary endothelial cells as compared to macrophages
	Trimyristin, tripalmitin, Tristearin	Clozapine	The AUC and MRT of clozapine SLNs were significantly higher in brain
	Stearic acid	Idarubicin	Drug and its metabolite were detected in the brain only after IDA-SLN administration
	Tripalmitin, gelucire, vitamin E	Baicalein	Brain-targeting efficiency of baicalein was greatly improved by nanostructured lipid carriers
Liposomes (Bellavance et al., 2010; Ramos et al., 2011)	DPPC, DC-Chol, DOPE, DHPE	Oregon green	Liposomes were strongly internalized in cultured cell lines within 6 h
	DSPC, cholesterol, DSPE	Citicoline	Considerable increase (10-fold) in the bioavailability of the drug in the brain parenchyma

TABLE 1.2   *(Continued)*

Formulation	Materials used	API/model molecule	Advantages
Dendrimers (Gajb-hiye and Jain, 2011; Teow et al., 2013)	Polyamidoamine	Paclitaxel	12-fold greater permeability across porcine brain endo-thelial cells
	Polypropylenei-mine	Docetaxel	Higher targeting efficiency and biodistribution to the brain
Micelles (Abdelbary and Tadros, 2013)	Block copoly-mers of ethylene oxide/propylene oxide	Olanzapine	Demonstrated higher drug-targeting index, drug-targeting efficiency, and direct transport percentage
Nanoemulsion (Kumar et al., 2008; Vyas et al., 2008)	Glyceryl monocaprylate	Risperidone	Higher drug transport effi-ciency and increased direct nose to brain drug transport
	Flax-seed, safflower oil	Saquinavir	Improved brain uptake

DC-Chol, 3-[N-(N',N'-dimethylaminoethane)-carbamoyl]cholesterol hydrochloride; DHPE, 1,2-dihexadecanoyl-sn-glycero-3-phosphoethanolamine; DOPE, 1,2-dioleoyl-sn-glycero-3-phosphoethanolamine; DPPC, dipalmitoylphosphatidylcholine; IDA, idarubicin; MRT, mean residence time; PEG, polyethyl glycol; PLGA, poly(lactic-co-glycolic acid); SLN, solid lipid nanoparticles.

1.5   OLFACTORY-TARGETED MICROPARTICLES WITH TAMARIND SEED POLYSACCHARIDE

The major challenge is the targeted delivery and retention of drug formula-tions in the olfactory mucosa, the target site for nose-to-brain drug absorp-tion because of the geometrical complexity of the nasal clearance and nose (Djupesland, 2013). Besides, the volumes of aqueous solution needed to dissolve most of the drug molecules often cannot be accommodated by the upper, neuron-containing, region of the nasal passage and are, sometimes, swallowed or lost to dripping (Kapoor et al., 2016). The percentage of drug is much less than 1% to reach the brain via the nose in animal experiments which is generally very small (Casettari and Illum, 2014). As powder formu-lations are less readily cleared from the nasal cavity, so they are preferred for intranasal drug delivery to liquid preparations.

Mucoadhesion is crucial in increasing the residence time of deposited particles in the nasal cavity. Similarly, a common approach to increase the residence time of drugs on the olfactory epithelium has been set to incorporate them into particulate carriers which have been formulated with polymers that can interact with mucin. Mucin is a major constituent of mucous. Formulations that are prepared using polymers interacting with mucin enable mucoadhesion and retention of the formulation on the olfactory mucosa (Ugwoke et al., 2005).

The size of the constituting particles influences the deposition of particles within the geometrically complex human nasal cavity. Due to high inertial impaction, particles larger than 20 μm show preferential deposition in the anterior part of the nasal cavity while particles less than 5 μm escape the nasal cavity with the air stream lines and deposit in the lungs (El-Sherbiny et al., 2015; Shi et al., 2007). It was indicated by the current modeling data that particles around 10 μm in size show maximum deposition in the olfactory region when the administration is done intranasally at normal inhalation rates of around 20 L/min (Schroeter et al., 2015; Shi et al., 2007).

Tamarind seed polysaccharide (TSP) is a highly branched polysaccharide. It has a molecular weight of 720–880 kDa and is got from the endosperm of *Tamarindus indica* seeds. TSP is generally recognized as safe approved by the Food and Drug Administration. It is generally used as an excipient in food and pharmaceutical preparations mostly as a thickener or stabilizer because of its gelling properties (Gupta et al., 2010; Kulkarni et al., 1998). For a number of drug-delivery applications, TSP has also been investigated (Avachat et al., 2013; Pal and Nayak, 2012; Sumathi and Ray, 2003), mostly because of stability and biodegrability under acidic pH conditions.

Microparticles are prepared with naturally occurring mucoadhesive polymers such as chitosan. They have shown promising nose-to-brain transfer of several drugs (Casettari and Illum, 2014). Yarragudi et al. (2017) suggest that after intranasal administration, formulation of drug into mucoadhesive microparticles of 10 μm in size can potentially target and increase the residence time of drug on the olfactory mucosa.

1.6   POLY(LACTIDE-*CO*-GLYCOLIDE) MICROSPHERES

The strategies that have been developed for drug delivery into the CNS, locally controlled drug release by the way of an implantable polymeric device have been developed in recent years. The first polymeric devices to be developed were the macroscopic implants which are needed for open surgery for implantation. In the last few years, poly(lactide-*co*-glycolide) microspheres have been shown to be safe and promising for drug delivery into the brain. It is biodegradable and biocompatible with brain tissue. These microspheres can be easily implanted by stereotaxy in discrete, precise, and functional areas of the brain without causing damage to the surrounding tissue due to their size. The treatments of brain tumor have been developed by using this approach, and clinical trials have also been conducted. Some potential applications in neurodegenerative diseases have also been explored, particularly neurotrophic cell therapy and factor delivery (Menei et al., 2005).

1.7 MICROTUBULE-TARGETING AGENTS

Microtubules are major parts of the cytoskeleton in all cells. They are dynamic macromolecular structures made up of polarized tubulin dimers with fast-growing plus ends and more stable minus ends. They are usually organized in a radial array with plus ends directed toward the cell periphery (Hur and Lee, 2014).

They have been among the most successful targets in anticancer therapy and a large number of microtubule-targeting agents (MTAs) are in various stages of clinical development for the treatment of several malignancies (Maday et al., 2014; Millecamps and Julien, 2013). As the acute or chronic disruption of the structural integrity of neurons, they accompany injury and diseases in the CNS. The microtubules provide structural support for the nervous system cellularly and intracellularly (Maday et al., 2014; Millecamps and Julien, 2013). For treating CNS disorders, microtubules are emerging as potential therapeutic targets. It has been determined that exogenous application of MTAs might prevent the breakdown or degradation of microtubules after injury or during neurodegeneration. It will thereby aid in the preservation of the structural integrity and function of the nervous system (Maday et al., 2014; Millecamps and Julien, 2013).

Neurodegenerative diseases indicate similar pathological features such as abnormal protein aggregation, mitochondrial dysfunction, and disease-specific neuronal degeneration (Maday et al., 2014; Millecamps and Julien, 2013). There are several pathogenic proteins, such as tau, α-synuclein, parkin, leucine-rich repeat kinase 2, and Huntingtin, related to neurodegenerative diseases which have been indicated to directly bind tubulin or modulate microtubule stability. Recently, increase of lines of evidence suggests that MTAs can ameliorate the pathogenic symptoms in animal models of neurodegenerative diseases (Maday et al., 2014; Millecamps and Julien, 2013). Additionally, for the administration of drugs that directly stabilize microtubules, strategies for tackling microtubule-based transport system are also under development, as impairment in the axonal transport has recently come up as an usual factor in several neurodegenerative diseases such as Alzheimer disease and Parkinson disease (Maday et al., 2014; Millecamps and Julien, 2013).

1.8 NASAL CHITOSAN MICROPARTICLES

Nasal administration constitutes a potentially efficacious way to achieve the brain uptake of neuroactive agents (Fine et al., 2014, 2015; Illum, 2000; Vyas et al., 2005). Drugs deposited on the olfactory epithelium of the nose can obtain direct access to the CNS, precisely the CSF, via transcellular transport through olfactory epithelial cells. The absorbed drugs in the CSF then diffuse into the ISF and then penetrate the brain parenchyma (Illum, 2000, 2004; Thorne and Frey, 2001). Furthermore, drugs that are deposited on the olfactory epithelium can be transported into the brain parenchyma by olfactory neurons or trigeminal nerves that reach the nasal cavity (Finger et al., 1990; Illum, 2000; Johnson et al., 2010). Ultimately, the nasally administered drugs can be absorbed into the systemic circulation from the respiratory epithelium (Cho et al., 2014), and if they are capable of crossing the BBB, they can then reach the CNS (Illum, 2000).

Generally, appropriate strategies are needed to improve the delivery of drugs in brain, such as the addition of penetration enhancers and mucoadhesive materials to formulations or the preparation of micro- and nanoparticulate delivery systems (Casettari and Illum, 2014; Dalpiaz et al., 2008; Horvát et al., 2009; Mistry et al., 2009; Rassu et al., 2015). Chitosan is a polysaccharide which is derived from the alkaline deacetylation of chitin. It used in different formulations for the nose-to-brain delivery of drugs (Casettari and Illum, 2014), because of its biocompatibility, nontoxicity, and high charge density, which confers mucoadhesive properties (Bernkop-Schnürch and Dünnhaupt, 2012; Sinha et al., 2004). Chitosan is poorly soluble in water at physiologic pH values, but it forms salts with inorganic or organic acids, such as hydrochloride and glutamic acid. They are soluble in water and possess better characteristics than chitosan itself, such as mucoadhesiveness and the ability to enhance the penetration of neuroactive agents into the CNS (Dalpiaz et al., 2008; Gavini et al., 2011; Maestrelli et al., 2004).

Zidovudine (AZT) is an antiretroviral drug. It is a substrate of active efflux transporters (AETs) that extrude the drug from the CNS and macrophages. They are considered to be sanctuaries of HIV. The conjugation of AZT to ursodeoxycholic acid is known to produce a prodrug (UDCA-AZT) which is able to elude the AET systems and indicate the potential ability of this prodrug to act as a carrier of AZT in macrophages and in the CNS. According to Dalpiaz et al. (2015), the nasal chitosan microparticles target a zidovudine prodrug to brain HIV sanctuaries.

1.9   PATENTS RELATED TO BRAIN DRUG TARGETING

A combination of physical and electrostatic barriers is presented by the BCSFB and BBB (Engelhardt and Sorokin, 2009). To circumvent this barrier, a number of drug-delivery strategies have been developed. As stated before, nanoparticles are promising brain delivery modalities that can be coated or loaded with drug or attached with various enhancers; they also show high stability in comparison to other nanocarriers. There are several pathways through which drug delivery is enabled by nanoparticles into the brain across the BBB, which does not allow the passage of most therapeutic agents. Based on assessing the recent patents, these pathways include, through attaching or containing an enhancer to pass the BBB, brain targeting that could be achieved by comprising an agent that is capable of delivering the drug into the designated brain tissue and controlling the delivery of nanoparticles by external stimuli (Martins et al., 2013). Table 1.3 presents a few patents for brain-targeted drug-delivery system.

1.10   CHALLENGES IN BRAIN DRUG TARGETING

One of the most challenging problems in drug development is not only to develop drugs to treat diseases of the CNS but also to manage to distribute them to the CNS across the BBB (Bhowmik et al., 2015). In the recent years, massive research attempts have generated to develop new drugs for brain targeting for the treatment of neurodegenerative diseases which has led to the remarkable growth in CNS drugs. The main interest has been focused on the identifying of new therapeutic molecules for the CNS than development of new approaches. The approaches for the drug-delivery systems are for targeting the drug molecule to the brain (Tiwari et al., 2012). The importance of drug-delivery system to improve therapies was understood more than 30

TABLE 1.3 Recent Patents Associated to Brain Targeted Drug-Delivery System.

Publication number	Title	Synopsis	Publication date	Inventors
US 9295728 B2	Copolymer conjugates	This application relates generally to biocompatible water-soluble polymers with pendant functional groups and methods for making them, and particularly to copolymer polyglutamate amino acid conjugates useful for a variety of anticancer drug-delivery applications	March 29, 2016	Kwok Yin Tsang, Hai Wang, Hao Bai, Yi Jin, Lei Yu
US 9572808 B2	Benzenesulfonamide derivatives of quinoxaline, pharmaceutical compositions thereof, and their use of methods for treating cancer	Chemical entities that are kinase inhibitors, pharmaceutical compositions, and methods of using these chemical entities, for example, for treatment of cancer are described	February 21, 2017	Yong-Liang Zhu, Xiangping Qian
US 9289505	Compositions and methods for delivering nucleic acid molecules and treating cancer	The present invention provides compositions and methods for the delivery of nucleic acids to a cell. The present invention additionally provides compositions and methods for the treatment of a disease or disorder, particularly cancer. In a particular embodiment, the composition comprises at least one pharmaceutically acceptable carrier and at least one liposome or dendrimer comprising at least two chemotherapeutic agents with different mechanisms of action and at least two inhibitors of cellular drug resistance	March 22, 2016	Minko Tamara, Rodriguez-rodriguez Lorna, Garbuzenko Olga B., Taratula Oleh, Shah Vatsal

TABLE 1.3 (Continued)

Publication number	Title	Synopsis	Publication date	Inventors
US 9278990 B2	Substituted nucleotide analogs	Disclosed herein are phosphorothioate nucleotide analogs, such as thiophosphoramidate prodrugs and thiohosphates (including α-thiomonophosphates, α-thiodiphosphates, and α-thiotriphosphates), methods of synthesizing phosphorothioate nucleotide analogs, such as thiophosphoramidate prodrugs and thiophosphates, and methods of treating viral infections such as hepatitis C virus, with the phosphorothioate nucleotide analogs, such as thiophosphoramidate prodrugs and thiophosphates	March 8, 2016	David Bernard Smith, Jerome Deval, Natalia Dyatkina, Leonid Beigelman, Guangyi Wang
US 20130253004 A1	Lysophosphatidic acid receptor antagonists and their use in the treatment fibrosis	Described herein are compounds that are antagonists of lysophosphatidic receptor(s). Also described are pharmaceutical compositions and medicaments that include the compounds described herein, as well as methods of using such antagonists, alone and in combination with other compounds, for treating LPA-dependent or LPA-mediated conditions or diseases	September 26, 2013	Thomas Jon Seiders, Bowei Wang, John Howard Hutchinson, Nicholas Simon Stock, Deborah Volkots
WO 2014093383 A1	Substituted 1H-pyrrolo [2, 3-b] pyridine and 1H-pyrazolo [3, 4-b] pyridine derivatives as SIK2 inhibitors	The present invention relates to compounds according to Formulas I, IA or IB: to pharmaceutically acceptable composition, salts thereof, their synthesis, and their use as SIK2 inhibitors including such compounds and methods of using said compounds in the treatment of various diseases and or disorders such as cancer, stroke, cardiovascular, obesity, and type II diabetes	June 19, 2014	Hariprasad Vankayalapati, Venkatakrishnareddy Yerramreddy, Venu Babu Ganipisetty, Sureshkumar Talluri, Rajendra P. Appalaneni

TABLE 1.3 (Continued)

Publication number	Title	Synopsis	Publication date	Inventors
WO 2011097594 A3	Therapeutic methods and compositions involving allosteric kinase inhibition	The present invention is directed to methods and compositions for suppressing lymphangiogenesis, angiogenesis, and/or tumor growth. The methods comprise contacting the tumor with a compound that stabilizes a protein kinase in the inactive state and is not an ATP competitive inhibitor of the protein kinase in the active state	December 22, 2011	Eric A. Murphy, David A. Cheresh, Lee Daniel Arnord
WO 2012166415 A1	Heterocyclic autotaxin inhibitors and uses thereof	Described herein are compounds that are inhibitors of autotaxin. Also described are pharmaceutical compositions and medicaments that include the compounds described herein, as well as methods of using such inhibitors, alone and in combination with other compounds, for treating autotaxin-dependent or autotaxin-mediated conditions or diseases	December 6, 2012	Jeffrey Roger Roppe, Timothy Andrew Parr, John Howard Hutchinson
WO 2014100505 A1	Substituted nucleosides, nucleotides, and analogs thereof	Disclosed herein are nucleotide analogs, methods of synthesizing nucleotide analogs, and methods of treating diseases and/or conditions such as a hepatitis C virus infection with one or more nucleotide analogs	June 26, 2014	Leonid Beigelman, Guangyi Wang, David Bernard Smith
US 9249111 B2	Substituted quinoxalines as B-raf kinase inhibitors	Chemical entities that are kinase inhibitors, pharmaceutical compositions, and methods of treatment of cancer are described. Specifically quinoxaline derivatives of Formula I and their use in modulating the activity of B-raf and/or mutant B-raf kinase (family of three serine/threonine-specific protein kinases) to regulate and modulate abnormal or inappropriate cell proliferation, differentiation, or metabolism are disclosed. Also, disclosed are methods of treating cancer associated with B-raf and/or mutant B-raf kinase activity in a subject, comprising administering the compounds of Formula I	February 2, 2016	Xiangping Qian, Yong-Liang Zhu

LPA, lysophosphatidic acid; SIK2, salt inducible kinase 2.

years ago. So, it is deeply disappointing that most efforts are still oriented to the sole development of drug discovery programs.

Major needs in targeting of brain drug are (Pavan et al., 2008):

- Develop therapeutics to specific brain regions or cell types.
- Improve understanding of BBB transport systems.
- Evaluate brain drug pharmacokinetics.
- Identify new brain drug targeting systems.
- Enhance development and application of molecular imaging probes and targeted contrast agents.

1.11   FUTURE OPPORTUNITIES IN BRAIN DRUG TARGETING

The overall aim of the future research regarding brain drug targeting is to expand the drug space of CNS from small lipid soluble molecules to large lipid soluble molecules which generally do not cross BBB (Misra et al., 2003). The following particular areas are identified below (Reddy et al., 2014):

- Identification of new BBB transporters that could be portals of entry for brain drug targeting systems.
- Development of brain drug targeting systems that enable the brain delivery of recombinant protein neurotherapeutics.
- Validation of new drug-targeting systems using *in vivo* models.
- Optimization of pharmacokinetics of *in vivo* brain drug-targeting systems.
- Development of genomic and proteomic discovery platforms that enable the identification of new BBB transporters.

1.12   CONCLUSION

The treatment of brain diseases is specifically challenging because the delivery of drug molecules to the brain is sometimes hindered by a number of biochemical, metabolic, and physiological barriers. The present outlook for patients who are suffering from several types of CNS diseases remains poor, but recent developments in drug-delivery techniques provide reasonable hope that the prevalent barriers shielding the CNS may ultimately be overcome. Direct drug delivery to the brain has recently been markedly increased through the rational design of polymer based drug-delivery systems. This approach has opened numerous opportunities for the formulation of scientists for the better delivery of therapeutic agents to CNS. Substantial progress will only come about, however, if continued vigorous research efforts to develop more therapeutic and less toxic drug molecules are paralleled by the aggressive pursuit of more effective mechanisms for delivering those drugs to their brain targets.

KEYWORDS

- **brain targeting**
- **blood–brain barrier**
- **blood–cerebrospinal fluid barrier**
- **brain drug delivery**
- **nanotechnology**
- **microparticles**

REFERENCES

Abdelbary, G. A.; Tadros, M. I. Brain Targeting of Olanzapine via Intranasal Delivery of Core–Shell Difunctional Block Copolymer Mixed Nanomicellar Carriers: In Vitro Characterization, Ex Vivo Estimation of Nasal Toxicity and In Vivo Biodistribution Studies. *Int. J. Pharm.* **2013**, *452*, 300–310.

Aktas, Y.; Yemisci, M.; Andrieux, K.; et al. Development and Brain Delivery of Chitosan-PEG Nanoparticles Functionalized with the Monoclonal Antibody OX26. *Bioconjug. Chem.* **2005**, *16*, 1503–1511.

Allen, T. M.; Cullis, P. R. Liposomal Drug Delivery Systems: From Concept to Clinical Applications. *Adv. Drug Deliv. Rev.* **2013**, *65*, 36–48.

Ambruosi, A.; Khalansky, A. S.; Yamamoto, H.; Gelperina, S. E.; Begley, D. J.; Kreuter, J. Biodistribution of Polysorbate 80-Coated Doxorubicinloaded [^{14}C]-Poly(Butyl Cyanoacrylate) Nanoparticles After Intravenous Administration To Glioblastoma-Bearing Rats. *J. Drug Target.* **2006**, *14*, 97–105.

Avachat, A. M.; Gujar, K. N.; Wagh, K. V. Development and Evaluation of Tamarind Seed Xyloglucan-Based Mucoadhesive Buccal Films of Rizatriptan Benzoate. *Carbohydr. Polym.* **2013**, *91* (2), 537–542.

Banks, W. A.; Kastin, A. J.; Ehrensing, C. A. Endogenous Peptide Tyr-Pro-Trp-Gly-NH2 (Tyr-W-MIF-1) Is Transported from the Brain to the Blood by Peptide Transport System-1. *J. Neurosci. Res.* **1993**, *35* (6), 690–695.

Bareford, L. M.; Swaan, P. W. Endocytic Mechanisms for Targeted Drug Delivery. *Adv. Drug Deliv. Rev.* **2007**, *59* (8), 748–758.

Béduneau, A.; Saulnier, P.; Benoit, J. P. Active Targeting of Brain Tumours Using Nanocarriers. *Biomaterials* **2007,** *28* (33), 4947–4967.

Begley, D.; Brightman, M. Structural and Functional Aspects of the Blood–Brain Barrier. *Prog. Drug Res.* **2003,** *61,* 39–78.

Bellavance, M. A.; Poirier, M. B.; Fortin, D. Uptake and Intracellular Release Kinetics of Liposome Formulations in Glioma Cells. *Int. J. Pharm.* **2010,** *395,* 251–259.

Bende, G. *Design, Development and Pharmacokinetic Studies of Nanoparticulate Drug Delivery Systems of Imatinib Mesylate*; Birla Institute of Technology and Science: Pilani, India, 2008.

Bernkop-Schnürch, A.; Dünnhaupt, S. Chitosan-Based Drug Delivery Systems. *Eur. J. Pharm. Biopharm.* **2012,** *81,* 463–469.

Bhowmik, A.; Khan, R.; Ghosh, M. K. Blood Brain Barrier: A Challenge for Effectual Therapy of Brain Tumors. *BioMed. Res. Int.* **2015,** *2015,* 1–20.

Bond'I, M. L.; Craparo, E. F.; Giammona, G.; Drago, F. Brain-Targeted Solid Lipid Nanoparticles (SLNs) Containing Riluzole: Preparation, Characterisation and Biodistribution. *Nanomedicine* **2010,** *5,* 25–32.

Borlongan, C. V.; Emerich, D. F. Facilitation of Drug Entry into the CNS via Transient Permeation of Blood Brain Barrier: Laboratory and Preliminary Clinical Evidence from Bradykinin Receptor Agonist, Cereport. *Brain Res. Bull.* **2003,** *60,* 297–306.

Brightman, M. W.; Reese, T. S. Junctions between Intimately Apposed Cell Membranes in the Vertebrate Brain. *J. Cell Biol.* **1969,** *40* (3), 648–677.

Burkhard, S.; Heiko, S. The Blood–Brain Barrier and the Outer Blood–Retina Barrier. *Med. Chem. Rev.* **2005,** *2,* 11–26.

Casettari, L.; Illum, L. Chitosan in Nasal Delivery Systems for Therapeutic Drugs. *J. Control. Release* **2014,** *190,* 189–200.

Cho, W.; Kim, M. S.; Jung, M. S.; Park, J.; Cha, K. H.; Kim, J. S.; Park, H. J.; Alhalaweh, A.; Velaga, S. P.; Hwang, S. J. Design of Salmon Calcitonin Particles for Nasal Delivery Using Spray-Drying and Novel Supercritical Fluid-Assisted Spray-Drying Processes. *Int. J. Pharm.* **2014,** *478,* 288–296.

Choi, J. J.; Pernot, M.; Brown, T. R.; et al. Spatio-Temporal Analysis of Molecular Delivery through the Blood–Brain Barrier Using Focused Ultrasound. *Phys. Med. Biol.* **2007,** *52* (18), 5509–5530.

Costantino, L.; Boraschi, D. Is There a Clinical Future for Polymeric Nanoparticles as Brain Targeting Drug Delivery Agents? *Drug Discov. Today* **2012,** *17,* 367–378.

Craparo, E. F.; Ognibene, M. C.; Casaletto, M. P.; Pitarresi, G.; Teresi, G.; Giammona, G. Biocompatible Polymeric Micelles with Polysorbate 80 for Brain Targeting. *Nanotechnology* **2008,** *19,* 485603.

Dalpiaz, A.; Fogagnolo, M.; Ferraro, L.; Capuzzo, A.; Pavan, B.; Rassu, G.; Salis, A.; Giunchedi, P.; Gavini, E. Nasal Chitosan Microparticles Target a Zidovudine Prodrug to Brain HIV Sanctuaries. *Antiviral Res.* **2015,** *123,* 146–157.

Dalpiaz, A.; Gavini, E.; Colombo, G.; Russo, P.; Bortolotti, F.; Ferraro, L.; Tanganelli, S.; Scatturin, A.; Menegatti, E.; Giunchedi, P. Brain Uptake of an Antiischemic Agent by Nasal Administration of Microparticles. *J. Pharm. Sci.* **2008,** *97,* 4889–4903.

Das, D.; Lin, S. Double-Coated Poly(Butylcynanoacrylate) Nanoparticulate Delivery Systems for Brain Targeting of Dalargin via Oral Administration. *J. Pharm. Sci.* **2005,** *94,* 1343–1353.

De Boer, A. G.; Van der Sandt, I. C. J.; Gaillard, P. J. The Role of Drug Transporters at the Blood–Brain Barrier. *Annu. Rev. Pharm. Toxicol.* **2003**, *43*, 629–656.

Decleves, X.; Amiel, A.; Delattre, J. Y.; Scherrmann, J. M. Role of ABC Transporters in the Chemoresistance of Human Gliomas. *Cur. Can. Drug Target.* **2006**, *6*, 433–445.

Djupesland, P. G. Nasal Drug Delivery Devices: Characteristics and Performance in a Clinical Perspective—A Review. *Drug Deliv. Transl. Res.* **2013**, *3* (1), 42–62.

Elkharraz, K.; Faisant, N.; Guse, C.; Siepmann, F.; Arica-Yegin, B.; Oger, J. M.; Gust, R.; Goepferich, A.; Benoit, J. P.; Siepmann, J. Paclitaxel-Loaded Microparticles and Implants for the Treatment of Brain Cancer: Preparation and Physicochemical Characterization. *Int. J. Pharm.* **2006**, *314*, 127–136.

El-Sherbiny, I. M.; El-Baz, N. M.; Yacoub, M. H. Inhaled Nano- and Microparticles for Drug Delivery. *Glob. Cardiol. Sci. Pract.* **2015**, *2015*, 2.

Engelhardt, B. Molecular Mechanisms Involved in T Cell Migration across the Blood–Brain Barrier. *J. Neural Transm. (Vienna)* **2006**, *113* (4), 477–485.

Engelhardt, B.; Sorokin, L. The Blood–Brain and the Blood–Cerebrospinal Fluid Barriers: Function and Dysfunction. *Semin. Immunopathol.* **2009**, *31* (4), 497–511.

Fine, J. M.; Forsberg, A. C.; Renner, D. B.; Faltesek, K. A.; Mohan, K. G.; Wong, J. C.; Arneson, L. C.; Crow, J. M.; Frey, W. H.; Hanson, L. R., 2nd. Intranasally-Administered Deferoxamine Mitigates Toxicity of 6-OHDA in a Rat Model of Parkinson's Disease. *Brain Res.* **2014**, *1574*, 96–104.

Fine, J. M.; Renner, D. B.; Forsberg, A. C.; Cameron, R. A.; Galick, B. T.; Le, C.; Conway, P. M.; Stroebel, B. M.; Frey, W. H.; Hanson, L. R., 2nd. Intranasal Deferoxamine Engages Multiple Pathways to Decrease Memory Loss in the APP/PS1 Model of Amyloid Accumulation. *Neurosci. Lett.* **2015**, *584*, 362–367.

Finger, T. E.; Jeor St, V. L.; Kinnamon, J. C.; Silver, W. L. Ultrastructure of Substance P- and CGRP-Immunoreactive Nerve Fibers in the Nasal Epithelium of Rodents. *J. Comp. Neurol.* **1990**, *294*, 293–305.

Fischer, H.; Gottschlich, R.; Seelig, A. Blood–Brain Barrier Permeation: Molecular Parameters Governing Passive Diffusion. *J. Membr. Biol.* **1998**, *165* (3), 201–211.

Friese, A.; Seiller, E.; Quack, G.; Lorenz, B.; Kreuter, J. Increase of the Duration of the Anticonvulsive Activity of a Novel NMDA Receptor Antagonist Using Poly(Butylcyanoacrylate) Nanoparticles as a Parenteral Controlled Release System. *Eur. J. Pharm. Biopharm.* **2000**, *49*, 103–109.

Gabathuler, R. Approaches to Transport Therapeutic Drugs across the Blood–Brain Barrier to Treat Brain Diseases. *Neurobiol. Dis.* **2010**, *37*, 48–57.

Gajbhiye, V.; Jain, N. K. The treatment of Glioblastoma Xenografts by surfactant conjugated dendritic nanoconjugates. *Biomaterials* **2011**, *32*, 6213–6225.

Gavini, E.; Rassu, G.; Ferraro, L.; Generosi, A.; Rau, J. V.; Brunetti, A.; Giunchedi, P.; Dalpiaz, A. Influence of Chitosan Glutamate on the ± Intranasal Absorption of Rokitamycin from Microspheres. *J. Pharm. Sci.* **2011**, *100*, 1488–1502.

Gulyaev, A. E.; Gelperina, S. E.; Skida, A. S.; Antropov, G. Y.; Kreuter, J. Significant Transport of Doxorubicin into the Brain with Polysorbate 80-Coated Nanoparticles. *Pharm. Res.* **1999**, *16*, 1564–1569.

Gupta, V.; Puri, R.; Gupta, S.; Jain, S.; Rao, G. Tamarind Kernel Gum: An Upcoming Natural Polysaccharide. *Syst. Rev. Pharm.* **2010**, *1* (1), 50.

Gupta, Y.; Jain, A.; Jain, S. K. Transferrin-Conjugated Solid Lipid Nanoparticles for Enhanced Delivery of Quinine Dihydrochloride to the Brain. *J. Pharm. Pharmacol.* **2007,** *59,* 935–940.

Haque, S.; Md, S.; Fazil, M.; Kumar, M.; Sahni, J. K.; Ali, J.; Baboota, S. Venlafaxine Loaded Chitosan NPs for Brain Targeting: Pharmacokinetic and Pharmacodynamic Evaluation. *Carbohydr. Polym.* **2012,** *89,* 72–79.

Hawkins, B. T.; Davis, T. P. The Blood–Brain Barrier/Neurovascular Unit in Health and Disease. *Pharmacol. Rev.* **2005,** *57* (2), 173–185.

Horvát, S.; Fehér, A.; Wolburg, H.; Sipos, P.; Veszelka, S.; Tóth, A.; Kis, L.; Kurunczi, A.; Balogh, G.; Kürti, L.; Eros, I.; Szabó-Révész, P.; Deli, M. A. Sodium Hyaluronate as a Mucoadhesive Component in Nasal Formulation Enhances Delivery of Molecules to Brain Tissue. *Eur. J. Pharm. Biopharm.* **2009,** *72,* 252–259.

Hur, E.-M.; Lee, B. D. Microtubule-Targeting Agents Enter the Central Nervous System (CNS): Double-Edged Swords for Treating CNS Injury and Disease. *Int. Neuro J.* **2014,** *18* (4), 171–178.

Illum, L. Is Nose-to-Brain Transport of Drugs in Man a Reality? *J. Pharm. Pharmacol.* **2004,** *56,* 3–17.

Illum, L. Transport of Drugs from the Nasal Cavity to the Central Nervous System. *Eur. J. Pharm. Sc*i. **2000,** *11,* 1–18.

Jeevanandam, J.; Chan, Y. C.; Danquah, M. K. Nano-Formulations of Drugs: Recent Developments, Impact and Challenges. *Biochimie* **2016,** *128,* 99–112.

Johnson, N. J.; Hanson, L. R.; Frey, W. H. Trigeminal Pathways Deliver a Low Molecular Weight Drug from the Nose to the Brain and Orofacial Structures. *Mol. Pharm.* **2010,** *7,* 884–893.

Jones, A. R.; Shusta, E. V. Blood–Brain Barrier Transport of Therapeutics via Receptor-Mediation. *Pharm. Res.* **2007,** *24* (9), 1759–1771.

Joshi, M. D.; Müller, R. H. Lipid Nanoparticles for Parenteral Delivery of Actives. *Eur. J. Pharm. Biopharm.* **2009,** *71,* 161–172.

Kapoor, M.; Cloyd, J. C.; Siegel, R. A. A Review of Intranasal Formulations for the Treatment of Seizure Emergencies. *J. Control. Release* **2016,** *237,* 147–159.

Kataoka, K.; Harada, A.; Nagasaki, Y. Block Copolymer Micelles for Drug Delivery: Design, Characterization and Biological Significance. *Adv. Drug Deliv. Rev.* **2012,** *64,* 37–48.

King, G. L.; Johnson, S. M. Receptor-Mediated Transport of Insulin across Endothelial Cells. *Science* **1985,** *227* (4694), 1583–1586.

Kulkarni, D.; Dwivedi, A.; Singh, S. Performance Evaluation of Tamarind Seed Polyose as a Binder and in Sustained Release Formulations of Low Drug Loading. *Ind. J. Pharm. Sci.* **1998,** *60,* 50–52.

Kumagai, A. K.; Eisenberg, J. B.; Pardridge, W. M. Absorptive-Mediated Endocytosis of Cationized Albumin and a Beta-Endorphin-Cationized Albumin Chimeric Peptide by Isolated Brain Capillaries. Model System of Blood–Brain Barrier Transport. *J. Biol. Chem.* **1987,** *262* (31), 15214–15219.

Kumar, M.; Misra, A.; Babbar, A. K.; Mishra, A. K.; Mishra, P.; Pathak, K. Intranasal Nanoemulsion Based Brain Targeting Drug Delivery System of Risperidone. *Int. J. Pharm.* **2008,** *358,* 285–291.

Liu, L.; Guo, K.; Lu, J.; et al. Biologically Active Core/Shell Nanoparticles Self-Assembled from Cholesterol-Terminated PEG-TAT for Drug Delivery across the Blood–Brain Barrier. *Biomaterials* **2008a,** *29,* 1509–1517.

Liu, L.; Venkatraman, S. S.; Yang, Y. Y.; et al. Polymeric Micelles Anchored with TAT for Delivery of Antibiotics across the Blood–Brain Barrier. *Biopolym. Pept. Sci. Sect.* **2008b,** *90*, 617–623.

Lockman, P. R.; Oyewumi, M. O.; Koziara, J. M.; Roder, K. E.; Mumper, R. J.; Allen, D. D. Brain Uptake of Thiamine-Coated Nanoparticles. *J. Control. Release* **2003**, *93*, 271–282.

Lodish, H.; Berk, A.; Zipursky.; S. L.; Matsudaira, P.; Baltimore, D.; Darnell, J. *Molecular Cell Biology*, 4th ed.; W. H. Freeman: New York, NY, 2000.

Lu, W.; Tan, Y. Z.; Hu, K. L.; Jiang, X. G. Cationic Albumin Conjugated Pegylated Nanoparticle with Its Transcytosis Ability and Little Toxicity against Blood–Brain Barrier. *Int. J. Pharm.* **2005**, *295*, 247–260.

Maday, S.; Twelvetrees, A. E.; Moughamian, A. J.; Holzbaur, E. L. Axonal Transport: Cargo-Specific Mechanisms of Motility and Regulation. *Neuron* **2014**, *84*, 292–309.

Maestrelli, F.; Zerrouk, N.; Chemtob, C.; Mura, P. Influence of Chitosan and Its Glutamate and Hydrochloride Salts on Naproxen Dissolution Rate and Permeation across Caco-2 Cells. *Int. J. Pharm.* **2004**, *271*, 257–267.

Mahajan, S. D.; Aalinkeel, R.; Sykes, D. E.; Reynolds, J. L.; Bindukumar, B.; Adal, A.; Qi, M.; Toh, J.; Xu, G.; Prasad, P. N.; et al. Methamphetamine Alters Blood Brain Barrier Permeability via the Modulation of Tight Junction Expression: Implication for HIV-1 Neuropathogenesis in the Context of Drug Abuse. *Bras. Res.* **2008a,** *1203*, 133–148.

Mahajan, S. D.; Aalinkeel, R.; Sykes, D. E.; Reynolds, J. L.; Bindukumar, B.; Fernandez, S.; Chawda, R.; Shanahan, T.; Schwartz, S. Tight Junction Regulation by Morphine and HIV-1 Tat Modulates Blood–Brain Barrier Permeability. *J. Clin. Immunol.* **2008b,** *28* (5), 528–541.

Manjunath, K.; Venkateswarlu, V. Pharmacokinetics, Tissue Distribution and Bioavailability of Nitrendipine Solid Lipid Nanoparticles after Intravenous and Intraduodenal Administration. *J. Drug Target.* **2006**, *14*, 632–645.

Manjunath, K.; Venkateswarlu, V. Pharmacokinetics, Tissue Distribution and Bioavailability of Clozapine Solid Lipid Nanoparticles after Intravenous and Intraduodenal Administration. *J. Control. Release* **2005**, *107*, 215–228.

Maratos-Flier, E.; Kao, C. Y.; Verdin, E. M.; King, G. L. Receptor-Mediated Vectorial Transcytosis of Epidermal Growth Factor by Madin–Darby Canine Kidney Cells. *J. Cell. Biol.* **1987**, *105* (4), 1595–1601.

Martins, P.; Rosa, D. R.; Fernandes, A.; Baptista, P. V. Nanoparticle Drug Delivery Systems: Recent Patents and Applications in Nanomedicine. *Rec. Pat. Nanomed.* **2013**, *3* (2), 105–118.

Martins, S.; Tho, I.; Reimold, I.; Fricker, G.; Souto, E.; Ferreira, D.; Brandl, M. Brain Delivery of Camptothecin by Means of Solid Lipid Nanoparticles: Formulation Design, In Vitro and In Vivo Studies. *Int. J. Pharm.* **2012**, *439*, 49–62.

Mehnert, W.; Mäder, K. Solid Lipid Nanoparticles: Production, Characterization and Applications. *Adv. Drug Deliv. Rev.* **2012**, *64*, 83–101.

Menei, P.; Montero-Menei, C.; Venier, M. C.; Benoit, J. P. Drug Delivery into the Brain Using Poly(Lactide-*co*-Glycolide) Microspheres. *Expert. Opin. Drug Deliv.* **2005**, *2* (2), 363–376.

Millecamps, S.; Julien, J. P. Axonal Transport Deficits and Neurodegenerative Diseases. *Nat. Rev. Neurosci.* **2013**, *14*, 161–176.

Misra, A.; Ganesh, S.; Shahiwala, A. Drug Delivery to the Central Nervous System: A Review. *J. Pharm. Pharm. Sci.* **2003**, *6* (2), 252–273.

Mistry, A.; Stolnik, S.; Illum, L. Nanoparticles for Direct Nose-to-Brain Delivery of Drugs. *Int. J. Pharm.* **2009,** *379,* 146–157.

Müller, R. H.; Petersen, R. D.; Hommoss, A.; Pardeike, J. Nanostructured Lipid Carriers (NLC) in Cosmetic Dermal Products. *Adv. Drug Deliv. Rev.* **2007,** *59,* 522–530.

Oliveira, J. M.; Salgado, A. J.; Sousa, N.; Mano, J. F.; Reis, R. L. Dendrimers and Derivatives as a Potential Therapeutic Tool in Regenerative Medicine Strategies—A Review. *Prog. Polym. Sci.* **2010,** *35,* 1163–1194.

Pal, D.; Nayak, A. K. Novel Tamarind Seed Polysaccharide-Alginate Mucoadhesive Microspheres for Oral Gliclazide Delivery: In Vitro–In Vivo Evaluation. *Drug Deliv.* **2012,** *19* (3), 123–131.

Pardridge, W. M. Alzheimer's Disease Drug Development and the Problem of the Blood–Brain Barrier. *Alzheimer Demen.* **2009,** *5,* 427–432.

Pardridge, W. M. shRNA and siRNA Delivery to the Brain. *Adv. Drug Deliv. Rev.* **2007,** *59,* 141–152.

Pavan, P.; Dalpiaz, A.; Ciliberti, N.; Biondi, C.; Manfredini, S.; Vertuani, S. Progress in Drug Delivery to the Central Nervous System by the Prodrug Approach. *J. Mol.* **2008,** *13* (5), 1035–1065.

Persidsky, Y.; Ramirez, S.; Haorah, J.; Kanmogne, G. Blood–Brain Barrier: Structural Components and Function under Physiologic and Pathologic Conditions. *J. Neuroimmunol. Pharmacol.* **2006,** *1* (3), 223–236.

Qin, J.; Chen, D.; Hu, H.; Cui, Q.; Qiao, M.; Chen, B. Surface Modification of RGD-Liposomes for Selective Drug Delivery to Monocytes/Neutrophils in Brain. *Chem. Pharm. Bull.* **2007,** *55,* 1192–1197.

Rajpoot, P.; Pathak, K.; Bali, V. Therapeutic Applications of Nanoemulsion Based Drug Delivery Systems: A Review of Patents in Last Two Decades. *Rec. Pat. Drug Deliv. Formul.* **2011,** *5,* 163–172.

Ramos, C. P.; Agulla, J.; Argibay, B.; Pérez, M. M.; Castillo, J. Serial MRI Study of the Enhanced Therapeutic Effects of Liposome Encapsulated Citicoline in Cerebral Ischemia. *Int. J. Pharm.* **2011,** *405,* 228–233.

Rassu, G.; Soddu, E.; Cossu, M.; Brundu, A.; Cerri,G.; Marchetti, N.; Ferraro, L.; Regan, R. F.; Giunchedi, P.; Gavini, E.; Dalpiaz, A. Solid Microparticles Based on Chitosan or Methyl-β-Cyclodextrin: A First Formulative Approach to Increase the Nose-Tobrain Transport of Deferoxamine Mesylate. *J. Control. Release* **2015,** *201,* 68–77.

Rautio, J.; Laine, K.; Gynther, M.; Savolainen, J. Prodrug Approaches for CNS Delivery. *AAPS J.* **2008,** *10,* 92–102.

Reddy, K. V. R.; Reddy, V.; Loya, P. C. Molecular Aspects of BBB. *Res. J. Pharm. Dos. Forms Technol.* **2014,** *6,* 105–109.

Reese, T. S.; Karnovsky, M. J. Fine Structural Localization of a Blood–Brain Barrier to Exogenous Peroxidase. *J. Cell Biol.* **1967,** *34* (1), 207–217.

Ren, T.; Xu, N.; Cao, C.; Yuan, W.; Y u, X.; Chen, J.; Ren, J. Preparation and Therapeutic Efficacy of Polysorbate-80-Coated Amphotericin B/PLA-b-PEG Nanoparticles. *J. Biomater. Sci.* **2009,** *20,* 1369–1380.

Roberts, R. L.; Fine, R. E.; Sandra, A. Receptor-Mediated Endocytosis of Transferrin at the Blood–Brain Barrier. *J. Cell Sci.* **1993,** *104* (2), 521–532.

Sai, Y.; Kajita, M.; Tamai, I.; Wakama, J.; Wakamiya, T.; Tsuji, A. Adsorptive Mediated Endocytosis of a Basic Peptide in Enterocyte-Like Caco-2 Cells. *Am. J. Physiol. Gastrointest. Liver Physiol.* **1998,** *275* (3), G514–G520.

Saunders, N. R.; Ek, C. J.; Habgood, M. D.; Dziegielewska, K. M. Barriers in the Brain: A Renaissance? *Trends Neurosci.* **2008**, *31* (6), 279–286.

Saunders, N.; Habgood, M. D.; Dziegielewska, K. Barrier Mechanisms in the Brain, II. Immature Brain. *Clin. Exp. Pharmacol. Physiol.* **1999a**, *26* (2), 85–91.

Saunders, N.; Habgood, M. D.; Dziegielewska, K. Barrier Mechanisms in the Brain, I. Adult Brain. *Clin. Exp. Pharmacol. Physiol.* **1999b**, *26* (1), 11–19.

Saunders, N.; Knott, G.; Dziegielewska, K. Barriers in the Immature Brain. *Cell. Mol. Neurobiol.* **2000**, *20* (1), 29–40.

Schaddelee, M. P.; Voorwinden, H. L.; Groenendaal, D.; et al. Blood–Brain Barrier Transport of Synthetic Adenosine A1 Receptor Agonists in Vitro: Structure Transport Relationships. *Eur. J. Pharm. Sci.* **2003**, *20* (3), 347–356.

Schmidt, J.; Metselaar, J. M.; Wauben, M. H.; Toyka, K. V.; Storm, G.; Gold, R. Drug Targeting by Long-Circulating Liposomal Glucocorticosteroids Increases Therapeutic Efficacy in a Model of Multiple Sclerosis. *Brain* **2003**, *126*, 1895–1904.

Schroeter, J. D.; Tewksbury, E. W.; Wong, B. A.; Kimbell, J. S. Experimental Measurements and Computational Predictions of Regional Particle Deposition in a Sectional Nasal Model. *J. Aer. Med. Pulm. Drug Deliv.* **2015**, *28* (1), 20–29.

Senel, S.; Kremer, M.; Nagy, K.; Squier, C. Delivery of Bioactive Peptides and Proteins across Oral (Buccal) Mucosa. *Cur. Pharm. Biotechnol.* **2001**, *2* (2), 175–186.

Shi, H.; Kleinstreuer, C.; Zhang, Z. Modeling of Inertial Particle Transport and Deposition in Human Nasal Cavities with Wall Roughness. *J. Aer. Sci.* **2007**, *38* (4), 398–419.

Sinha, V. R.; Singla, A. K.; Wadhawan, S.; Kaushik, R.; Kumria, R.; Bansal, K.; Dhawan, S. Chitosan Microspheres as a Potential Carrier for Drugs. *Int. J. Pharm.* **2004**, *274*, 1–33.

Smith, M. W.; Gumbleton, M. Endocytosis at the Blood–Brain Barrier: From Basic Understanding to Drug Delivery Strategies. *J. Drug Target.* **2006**, *14*, 191–214.

Snehalatha, M.; Venugopal, K.; Saha, R. N.; Babbar, A. K.; Sharma, R. K. Etoposide Loaded PLGA and PCL Nanoparticles II: Biodistribution and Pharmacokinetics after Radiolabeling with T c-99m. *Drug Deliv.* **2008**, *5*, 277–287.

Soni, V.; Kohli, D. V.; Jain, S. K. Transferrin-Conjugated Liposomal System for Improved Delivery of 5-Fluorouracil to Brain. *J. Drug Target.* **2008**, *16*, 73–78.

Spector, R. Thymidine Transport and Metabolism in Choroid Plexus: Effect of Diazepam and Thiopental. *J. Pharmacol. Exp. Ther.* **1985**, *235* (1), 16–19.

Stewart, P. A. Endothelial Vesicles in the Blood–Brain Barrier: Are They Related to Permeability? *Cell. Mol. Neurobiol.* **2000**, *20* (2), 149–163.

Sumathi, S.; Ray, A. R. Role of Modulating Factors on Release of Caffeine from Tamarind Seed Polysaccharides Tablets. *Trends Biomater. Art. Org.* **2003**, *17*, 41–46.

Tamai, I.; Tsuji, A. Transporter-Mediated Permeation of Drugs across the Blood–Brain Barrier. *J. Pharm. Sci.* **2000**, *89* (11), 1371–1388.

Teow, H. M.; Zhou, Z.; Najlah, M.; Yusof, S. R.; Abbott, N. J.; D'Emanuele, A. Delivery of Paclitaxel across Cellular Barriers Using a Dendrimer-Based Nanocarrier. *Int. J. Pharm.* **2013**, *441* (1), 701–711.

Thorne, R. G.; Frey, W. H. Delivery of Neurotropic Factors to the Central Nervous System. *Clin. Pharmacokinet.* **2001**, *40*, 907–946.

Tiwari, G.; Tiwari, R.; Sriwastawa, B.; Bhati, L.; Pandey, S.; Pandey, P.; Bannerjee, S. K. Drug Delivery Systems: An Updated Review. *Int. J. Pharm. Invest.* **2012**, *2* (1), 2–11.

Tosi, G.; Costantino, L.; Rivasi, F.; et al. Targeting the Central Nervous System: In Vivo Experiments with Peptide-Derivatized Nanoparticles Loaded with Loperamide and Rhodamine-123. *J. Control. Release* **2007,** *122,* 1–9.

Tsai, M. J.; Wu, P. C.; Huang, Y. B.; Chang, J. S.; Lin, C. L.; T sai, Y. H.; Fang, J. Y. Baicalein Loaded in Tocol Nanostructured Lipid Carriers (Tocol NLCs) for Enhanced Stability and Brain Targeting. *Int. J. Pharm.* **2012,** *423,* 461–470.

Uddin MS (2019). Nanoparticles as nanopharmaceuticals: a smart drug delivery system. In: Keservani RK, Sharma AK (eds) Nanoparticulate Drug Delivery Systems. 1st ed. Apple Academic Press, USA.

Ugwoke, M. I.; Agu, R. U.; Verbeke, N.; Kinget, R. Nasal Mucoadhesive Drug Delivery: Background, Applications, Trends and Future Perspectives. *Adv. Drug Deliv. Rev.* **2005,** *57* (11), 1640–1665.

Ulbrich, K.; Hekmatara, T.; Herbert, E.; Kreuter, J. Transferrin- and Transferrin-Receptor Monoclonal Antibody-Modified Nanoparticles Enable Drug Delivery across the Blood–Brain Barrier (BBB). *Eur. J. Pharm. Biopharm.* **2009,** *71,* 251–256.

Vergoni, A. V.; Tosi, G.; Tacchi, R.; Vandelli, M. A.; Bertolini, A.; Costantino, L. Nanoparticles as Drug Delivery Agents Specific for CNS: In Vivo Biodistribution. *Nanomed. Nanotechnol. Biol. Med.* **2009,** *5,* 369–377.

Vyas, T. K.; Shahiwala, A.; Amiji, M. M. Improved Oral Bioavailability and Brain Transport of Saquinavir upon Administration in Novel Nanoemulsion Formulations. *Int. J. Pharm.* **2008,** *347,* 93–101.

Vyas, T. K.; Shahiwala, A.; Marathe, S.; Misra, A. Intranasal Drug Delivery for Brain Targeting. *Curr. Drug Deliv.* **2005,** *2,* 165–175.

Vykhodtseva, N.; McDannold, N.; Hynynen, K. Progress and Problems in the Application of Focused Ultrasound for Blood–Brain Barrier Disruption. *Ultrasonics* **2008,** *48,* 279–296.

Yang, F. Y.; Lin, Y. S.; Kang, K. H.; Chao, T. K. Reversible Blood–Brain Barrier Disruption by Repeated Transcranial Focused Ultrasound Allows Enhanced Extravasation. *J. Control. Release* **2011,** *150,* 111–116.

Yarragudi, S. B.; Richter, R.; Lee, H.; Walker, G. F.; Clarkson, A. N.; Kumar, H.; Rizwan, S. B. Formulation of Olfactory-Targeted Microparticles with Tamarind Seed Polysaccharide to Improve Nose-to-Brain Transport of Drugs. *Carbohydr. Polym.* **2017,** *163,* 216–226.

Zahin, N; Anwar, R; Tewari, D; Kabir, MT; Sajid, A; Mathew, B; Uddin, MS; Aleya, L; Abdel-Daim, MM. Nanoparticles and its biomedical applications in health and diseases: special focus on drug delivery. *Environ. Sci. Pollut. Res.* **2019,** In Press, 1–18.

Zara, G. P.; Bargoni, A.; Cavalli, R.; Fundaro, A.; Vighetto, D.; Gasco, M. R. Pharmacokinetics and Tissue Distribution of Idarubicin-Loaded Solid Lipid Nanoparticles after Duodenal Administration to Rats. *J. Pharm. Sci.* **2002,** *91,* 1324–1333.

Zeuthen, T. Secondary Active Transport of Water across Ventricular Cell Membrane of Choroid Plexus Epithelium of *Necturus maculosus*. *J. Physiol.* **1991,** *444* (1), 153–173.

Zhang, Y.; Zhang, Y. F.; Bryant, J.; Charles, A.; Boado, R. J.; Pardridge, W. M. Intravenous RNA Interference Gene Therapy Targeting the Human Epidermal Growth Factor Receptor Prolongs Survival in Intracranial Brain Cancer. *Clin. Can. Res.* **2004,** *10,* 3667–3677.

CHAPTER 2

BLOOD–BRAIN BARRIER AND ADVANCED CARRIERS FOR BRAIN DRUG DELIVERY AND TARGETING

ALI SADEGHINIA[1,†], JALEH BARAR[1,2,†], MORTEZA ESKANDANI[1], MOHAMMAD A. RAFI[3], and YADOLLAH OMIDI[1,2,*]

¹Research Center for Pharmaceutical Nanotechnology, Biomedicine Institute, Tabriz University of Medical Sciences, Tabriz, Iran

²Department of Pharmaceutics, Faculty of Pharmacy, Tabriz University of Medical Sciences, Tabriz, Iran

³Department of Neurology, Sidney Kimmel College of Medicine, Thomas Jefferson University, Philadelphia, PA, USA

*Corresponding author. E-mail: yomidi@tbzmed.ac.ir

†These authors have equal contribution in this chapter.

ABSTRACT

The blood–brain barrier (BBB), formed by brain capillary endothelial cells (BCECs) connected through tight junctions, provides excellent barrier functionalities, preventing penetration of blood-circulating compounds into the brain parenchyma. Excellent restrictiveness of BBB controls the entrance of drugs into the brain discriminately; hence, brain drug delivery and targeting (BDDT) needs to be advanced to circumvent the tight barrier restrictiveness of BCECs. The transport machineries of BCECs can be utilized for efficient delivery of drugs into brain. Once inside, the drugs need to be accumulated in the target site through targeting diseased cells. Of the molecular machineries, carrier- and receptor-mediated transporters provide great possibility for drug transportation into the brain. Modified drug-delivery systems (DDSs) appear to be success paradigms for BDDT. Chemically modified drugs with inherent properties (e.g., drug–polymer nanoconjugates, prodrugs, drug–ligand conjugates) have been successfully used. Further, surface-modified/functionalized nanosystems (NSs) encapsulated/loaded with designated drug molecules that are categorized into organic (e.g., polymer and lipid) and inorganic (e.g., gold, iron oxide) nanoparticles have also been utilized. Of these, dual-targeting NSs and multifunctional nanomedicines and theranostics have undergone for translation into clinical applications. The main aim of this chapter is to discuss the importance of BBB on BDDT and advanced drug delivery systems (DDSs) used for targeted therapy of brain diseases.

2.1 INTRODUCTION

The traverse of blood-circulating nutrients and substances into the brain parenchyma occurs in an extremely selective manner. In fact, the entry of blood-borne substances into the brain is selectively/specifically maintained by a unique biological barrier so-called blood–brain barrier (BBB) that is created by the brain capillary endothelial cells (BCECs) in association with other neighboring cellular entities, including astrocytes and pericytes (Abbott et al., 2006; Omidi and Barar, 2012b; Willis, 2012). The selective/specific barrier restrictiveness of BCECs is largely dependent upon formation of tight junctions (TJs) among these cells in association with cooperating cells, astrocytes and pericytes. As shown in Figure 2.1, within such coop setting, the astrocytes are projected as the astrocytic foot process investing approximately 99% of the abluminal surface of the capillary, and pericytes associate with the basement membrane of BCECs. BBB functions to retain the brain hemostasis and restrict/control the trafficking of blood-borne molecules, nutrients, and drugs into the brain parenchyma selectively (Barar et al., 2016; Omidi and Barar, 2012b). Such cellular coop appears to associate with the functional expression of BBB transport machineries necessary for inward and/or outward selective transportation of nutrients and drugs (Omidi et al., 2008a). Travers of blood circulating substances across the BBB are largely dependent on the physicochemical characteristics of compounds [e.g., molecular weight (MW), lipophilicity, pK_a, hydrogen bonding, etc.] and also the biological features (e.g., functional expression of transporters and enzymes) of BCECs (Barar et al., 2016).

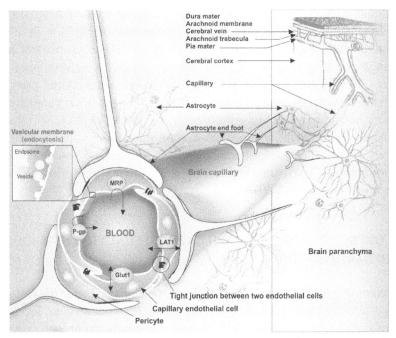

FIGURE 2.1   Schematic illustration of the BBB. The BCECs form BBB through tight junctional connection and functional expression of specialized transport machineries. There is a coop among BCECs, astrocytes, and pericytes in terms of the function of BBB. BBB, blood–brain barrier; BCECs, brain microvasculature endothelial cells. Image was adapted with permission from our previously published work (Omidi and Barar, 2012b). Note: Not drawn to scale.

2.2   BIOLOGICAL FEATURES OF BLOOD–BRAIN BARRIER

The TJ-based highly restricted connection of the BCECs is the main feature of the BBB, whereas these cells display functional expression of transport machineries (Krizbai and Deli, 2003; Liu et al., 2012). The main molecular components of TJ include occludins (MW 60–65 kDa), claudins (claudin-5, MW 20–245 kDa), and zonula (ZO-1, MW ~225 kDa) (Liu et al., 2012; Stevenson et al., 1986).

Despite being somewhat enigmatic in terms of TJ-based barrier functionality of BBB (Ge et al., 2005), the functional expression of series of membrane transport proteins (MTPs) such as carrier-mediated transporters (CMTs) together with receptor-mediated transport machineries for traverse of macromolecules seems to be absolutely crucial for the holistic functions of BBB. Of MTPs, for example, ATP-binding cassette (ABC) transporters (e.g., multidrug resistance protein 1 or *P*-glycoprotein 1/ABCB1) are predominantly localized on the luminal membrane of BCECs, in which they are dealing with trafficking a wide range of different substrates back into the blood circulation—an efflux function of BCECs (Mahringer and Fricker, 2016). Further, other MTPs are involved in functional presence of BBB and selective transportation of solutes into the brain parenchyma, including solute carrier (SLC) families (e.g., Glut1/SLC2A1) as integral transmembrane proteins, channels/pores (e.g., aquaporin, voltage-gated ion channels, and ligand-gated ion channels), electrochemical potential-driven transporters (e.g., excitatory amino acid transporters), and primary active transporters (Barar et al., 2016).

2.3 OVERCOMING RESTRICTED BARRIER FUNCTION OF BLOOD–BRAIN BARRIER

To overcome the BBB restrictive hindrance and hence enhance the entrance of drugs into the brain, various approaches have been undertaken, including (1) loosening of the TJs among BCECs using the osmotic agent such as hypertonic solution of arabinose or mannitol, or active agents such as bradykinin and histamine (Abbott, 2000; Rapoport, 2001); (2) using specific mechanisms related to the carrier-mediated transportation (Pardridge, 2015, 2012); and (3) exploiting mechanisms related to the receptor-mediated vesicular transportation such as clathrin-coated pits and membranous caveolae or lipid rafts (de Boer and Gaillard, 2007; Preston et al., 2014; Wang et al., 2009).

2.4 MOLECULAR TRAFFICKING ACROSS BLOOD–BRAIN BARRIER

Trafficking of blood-borne molecules and drugs across BBB may be divided into several transport mechanisms, including (1) passive diffusion for traverse of small molecules, depending on lipophilicity of the solutes; (2) facilitated transport/diffusion via specific transmembrane integral proteins and/or SLCs/CMTs; (3) active transport via SLCs and/or CMTs using ATP; and (4) endocytosis and transcytosis for trafficking of macromolecular by vesicular transportation machineries (Barar et al., 2016; Omidi and Barar, 2012a, 2012b). The trafficking of hydrophilic small molecules (e.g., carbohydrates and amino acids) appears to be via catalyzed transport systems of CMTs, while the transportation of macromolecules like insulin and transferrin (Tf) as well as most of nanoscaled targeted drug-delivery systems

(DDSs) occur via receptor-mediated transporters (RMTs) (Pardridge, 2007, 2012). In addition to these mechanisms, some other types of extracellular vesicular systems such as exosomes and ectosomes may be involved in the transportation of macromolecules, which can also be exploited for brain drug delivery and targeting (BDDT) (Rafi and Omidi, 2015; Wood et al., 2011; Yang et al., 2015).

2.5 BRAIN DRUG DELIVERY

A large number of studies have been performed to address the BDDT using both in vitro cell-based BBB models and animal models. In BDDT, it is crucial to make sure upon the integrity of the BBB, that is to say that the capillary endothelia of brain need to be safely and efficiently crossed. Figure 2.2 schematically shows trafficking mechanisms of various blood-borne substances, nutrients, and drugs across the BBB, including (1) paracellular transportation (e.g., small hydrophilic substances), (2) transcellular passive diffusion (e.g., small lipophilic substances), (3) CMTs (e.g., various amino acids), (4) endocytosis through RMTs [e.g., monoclonal antibodies (Abs) targeting different receptors at the BBB such as Tf receptor], and (5) fluid phase adsorption through endocytosis or pinocytosis (e.g., albumin) (Barar et al., 2016; Omidi and Barar, 2012b).

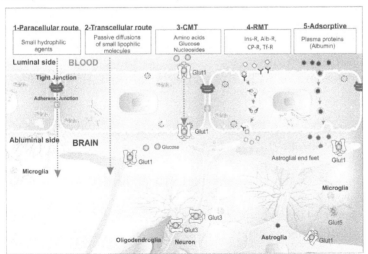

FIGURE 2.2 Schematic demonstration of different mechanisms involved in the brain drug delivery and targeting. Image was adapted with permission from our previously published work (Omidi and Barar, 2012b). Note: Not drawn to scale.

2.5.1   TRANSCELLULAR PASSIVE DIFFUSION

Of the transport mechanisms at BBB, the passive diffusion is a main route for traverse of small molecules through the traverse of drug molecules down to a concentration or electrochemical gradient without consumption of cellular energy (Omidi and Barar, 2012b). Based on the first Fick's law, in the passive diffusion, the overall flux (J) of small solutes in one dimension can be described by eq (2.1).

$$J = -D \cdot K_{\mathrm{p}} \cdot A \cdot \left(\frac{\mathrm{d}C}{\mathrm{d}x} \right)_t \qquad (2.1)$$

where, J is the flux of drug; D is the diffusion coefficient of a given drug molecules across the cellular barrier; K_{p} refers to a global partition coefficient (cell membrane/aqueous fluid); A denotes the surface area of barrier available for absorption; x is the thickness of absorption barrier, and $(\mathrm{d}C/\mathrm{d}x)_t$ is the concentration gradient of drug across the absorption barrier. It should be noted that the passive transcellular diffusion is indeed a process which is solely possible for compounds/molecules with small MW (<500 Da) that are unionized with partition coefficient value (log P) around 2 and hydrogen bonds less than 10 (Lai et al., 2013). In short, various factors appear to influence the trafficking of substances across BBB and their entrance into the central nervous system (CNS), including (1) physicochemical properties such as lipophilicity/hydrophilicity, MW, hydrogen bonding, and pK_{a}; (2) morphology of particulates (e.g., size, shape, and surface charge); and (3) biological features of BCECs (e.g., functional expression of transporters) (Barar et al., 2016; Omidi and Barar, 2012b).

2.5.2   CARRIER-MEDIATED TRANSPORT

It is very interesting to know that over 10% of the total proteins of BCECs specifically belong to solute transporters such as amino acids and glucose transporters. CMT can be performed by cells bi- and/or uni-directionally, which may function as symporters or antiporters. This process can be an active transportation against concentration gradient (ATP-dependent) or a facilitated concentration-gradient transport (Barar et al., 2016; Omidi and Barar, 2012b).

2.5.3 ENDOCYTOSIS AND TRANSCYTOSIS

Endocytosis and transcytosis are the main route for internalization of macromolecules such as proteins and nanoscaled particulates. For instance, nanoconjugates of drug–ligand [Ab, peptide, aptamer (Ap)] and nanoscaled targeted DDSs may cross the BBB through receptor-mediated endocytosis and transcytosis (Monaco et al., 2017; Ruozi et al., 2016; Yemisci et al., 2015). The vesicular trafficking at BBB is shown in Figure 2.3.

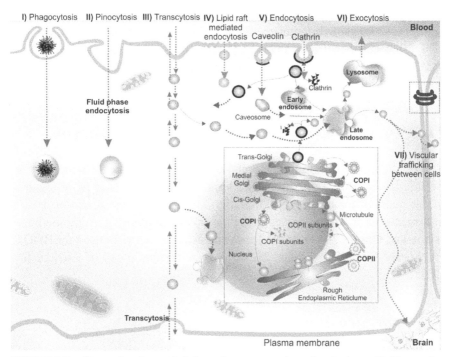

FIGURE 2.3 (See color insert.) Schematic representation of vesicular trafficking at the BBB. Macromolecules are internalized through phagocytosis (I). Engulfed particles in the phagosomes may be subjected to fusion with lysosomes. The surrounding substances are often internalized by pinocytosis/fluid phase endocytosis (II). The main route for trafficking of large molecules is transcytosis (III). Lipid rafts are involved in signal transduction and internalization of some macromolecules like cholera toxin (IV). The membranous caveolae and CCPs are engaged in endocytosis process (V). Various biomolecules such as Tf, LDL may be colocalized with CCPs. The internalized macromolecules within vesicles may interact with endosomes and lysosomes and exocytose out of the cell (VI) or even associate with the neighboring cells (VII). BBB, blood–brain barrier; CCPs, clathrin-coated pits; Tf, transferrin; LDL, low-density lipoprotein. Image was adapted with permission from our previously published work (Barar et al., 2016). Note: Not drawn to scale.

In addition to the fluid phase endocytosis, specialized trafficking systems such as receptor-mediated endocytosis/transcytosis mechanisms are recruited by the BCECs for transportation of macromolecules mainly by means of clathrin-coated pits and membranous caveolae/lipid rafts (Barar et al., 2016; Omidi and Barar, 2012b). Some different receptors are involved in RMT, including homodimer (180 kDa) type II transmembrane-associated glycoprotein transferrin receptor (TfR) whose specific Ab is OX-26, transmembrane insulin receptor (IR), low-density lipoprotein (LDL) receptor-related protein (LRP) 1, insulin-like growth factor receptor, LDL LRP 2, diphtheria toxin receptor, heparin-binding epidermal-like growth factor receptor.

Further, cell-penetrating peptides (CPPs) with short amino acid sequences (10–27 amino acids) and positive surface charge can be also used in transportation of macromolecules such as proteins.

2.6 CARRIERS FOR BRAIN DRUG DELIVERY AND TARGETING

By far, different organic and inorganic materials have been exploited for BDDT, including (1) polymers, (2) lipids, (3) proteins, and (4) inorganic [noble metal nanoparticles (NPs)] carriers. The NPs can be further decorated or armed with homing devices using various conjugating/grafting methods (Barar and Omidi, 2014). In the following contexts, carriers used for BDDT are briefly discussed. Figure 2.4 represents various types of carriers for BDDT.

2.6.1 POLYMER-BASED CARRIERS

Polymeric advanced carries have been developed for delivery of genes, drugs, or peptides/proteins to the target tissue/cells with controlled-release possibility (Orlu-Gul et al., 2014). The most important prerequisite in polymeric nanoparticulate brain delivery systems is their biodegradablity, though cationic polymers have also been used for the delivery nucleic acids despite showing some degrees of toxicities. As shown in Figure 2.4, drugs can be entrapped, encapsulated, incorporated, or grafted with polymers.

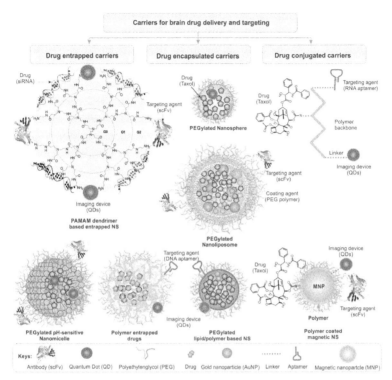

FIGURE 2.4   **(See color insert.)** Schematic representation for various types of carriers for brain drug delivery and targeting. Image was adapted and modified with permission from our previously published work (Barar and Omidi, 2014).

Short polyamines such as linear polyethylenimine (PEI), triethylene-tetramine, and spermine were shown to induce low toxicity with reducible disulfide backbones in favor of packaging of nucleic acids and their internalization and consequent release (Zhang and Vinogradov, 2010). In a study, polyethylene glycol (PEG)-PEI, PEG-PEI, was engineered as a carrier for gene delivery to brain by conjugating it with a cyclic tripeptide Arg-Gly-Asp (RGD) sequence, c(RGDyK) (cyclic arginine–glycine–aspartate–D-tyrosine–lysine). The engineered glioblastoma-targeting gene delivery system was shown to increase therapeutic efficacy of plasmid open reading frame-tumor necrosis factor-related apoptosis-inducing ligand (pORF-hTRAIL), resulting in substantially improvement in the survival of the intracranial glioblastoma-bearing nude mice (Zhan et al., 2012).

Of various polymeric advanced materials, several biodegradable polymers have used for BDDT, including (1) poly(alkyl cyanoacrylates) like poly(butyl cyanoacrylate) (PBCA) and poly(isohexyl cyanoacrylate),

(2) poly(lactic acid) (PLA) and its copolymer poly(lactide-*co*-glycolide) (PLGA), and (3) human serum albumin (Kreuter, 2014). Among them, PBCA is the fastest biodegrading polymer, PLGA is an FDA and European Medicine Agency approved polymer that is the most widely studied one and provides possibility for development of sustained release DDSs and surface functionalization for active targeting (Danhier et al., 2012; Matthaiou et al., 2014; Mooguee et al., 2010). It can be used for codelivery of macromolecular nucleic acids and chemotherapy small molecules. Further, PLGA NPs can be armed with different homing agents such as Aps, Abs, and peptide ligands such as CPPs.

Once used, most of these polymeric carriers must cross a biological barrier such as BBB and passive plasma membrane. Given that BBB serve as an excellent hindering barrier, polymeric carriers need to meet certain criteria to be able to cross BBB efficiently. For instance, polymeric nanosystems (NSs) can be an effective carrier to circumvent BBB enhancing therapeutic efficiency at brain with reduced toxicity and side-effect. Of the polymeric nanocarriers, polymersomes have been used for active targeting of BCECs (Chen et al., 2014b; Dieu et al., 2014; Yu et al., 2012). For example, diblock copolymer poly(dimethylsiloxane)-block-poly(2-methyl-2-oxazoline) and polymersomes were conjugated with antihuman insulin receptor (IR) specific monoclonal Ab (i.e., mAb 83-14). Polymersomes were found to perform a self-assembly of 200 nm vesicles after extrusion, and once conjugated with mAb 83-14, they were able to actively target the IR-positive hCMEC/D3 cells as a BBB model (Dieu et al., 2014). In another study, to efficiently treat brain tumors with reduced side-effects, ligand-conjugated polymersomal doxorubicin (DOX) was developed (Chen et al., 2014a). To this end, poly(ethylene glycol-g-glutamate)-*co*-poly(distearin-g-glutamate) polymersomes encapsulated with DOX (P-D) were conjugated with both des-octanoyl ghrelin (G) and folate (F) DOX (green fluorescent protein (GFP)-D). In this NS, ligands G and F were able to target the BCECs and brain tumor cells, respectively. The engineered NS (85 nm) was shown to be able to efficiently cross the BBB and actively target the tumor cells in vitro and in vivo. As compared to free DOX, liposomal DOX, P-D, or single-ligand conjugated P-D, the engineered GFP-D NS showed an improved antiglioma effect, and significantly improved the overall survival of the tumor-bearing mice.

Overall, the polymeric NPs have been satisfactorily used for transportation of small and large molecules into the CNS. The most improved polymeric BDDT systems are those functionalized with targeting entities such as Ab/Ap specific to TfR. Such advanced systems, so-called molecular Trojan horses (MTHs), are able to take advantage of RMT mechanisms across the BBB (Kreuter, 2013). In addition to Abs and Aps, the polymeric BDDT has been achieved by conjugation of other proteins and peptides such as apolipo-proteins A-I and/or E that are internalized through Low-density lipoprotein (LDL)/scavenger receptors expressed by the BCECs, which is the same for Tf and insulin. Ideally, the BDDT NSs should possess several important characterizations, including (1) suitable morphology and size (~100 nm); (2) excellent biocompatibility and biodegradability, (3) induction of no toxic impacts, (4) generation of no immunologic reaction, (5) no impacts of cargo molecules, (6) protection of cargo molecules during blood circulation, (7) no accelerated metabolism within blood, and (8) capability of surface function-alization for conjugation with targeting agents.

2.6.2   STIMULI-RESPONSIVE POLYMERIC CARRIERS

Stimuli-responsive polymers for drug delivery are considered as an effec-tive means for active targeting, by which designated cargo-drug molecules are transported to the target site in an on-demand manner (Li et al., 2013). Once reached the target site, these systems can be literally stimulated by certain pathological changes occurring in tissues or local stimulus upon which the release of loaded drug molecules is triggered. For example, solid tumor tissues exhibit a dropped pH in tumor microenvironment (TME) as a permissive milieu and a higher temperature as compared to most normally tissues (Barar and Omidi, 2013a; Li et al., 2014; Torchilin, 2009). Further, a greater redox potential has been reported for these tissues (Russo et al., 1986). These factors are called "internal stimuli" because they are local stimuli in the tumor cells/tissues. In contrast, some other external stimuli can also be applied in the region of specific. For instance, magnetic field, near infrared, light, ultrasound, and temperature have been applied as "external stimuli" to trigger the liberation of cargo molecules from carriers (Mura et al., 2013; Wang et al., 2012).

2.6.3   THERMOSENSITIVE POLYMERIC CARRIERS

Pathophysiological conditions like inflammation at tumor tissue may lead to somewhat inevitable temperature rise. It should be stated that higher temperatures around 40°C can cause slight greater blood flow in the tumor tissue, and hence possibly enhance the vasculature penetrability in comparison with the healthy normal tissues whose temperature are about 37°C. Polymeric carriers can be accumulated within the hyperthermic areas, in large part due to increased permeability and penetrability of these areas. As a result, polymer/lipid based with suitable phase transition temperature, so-called lower critical solution temperature (LCST), can be formulated as thermo-sensitive system and liberate its cargo-drug molecules in response to the raised temperatures (Kono, 2001; Zhu and Torchilin, 2013). Thermo-responsive polymers tend to undergo a phase transition (generally at over 40°C), which is called critical transition point of these polymers (Schwerdt et al., 2008). Below this temperature, the polymers present in a water-soluble hydrophilic form, mainly because of the formation of hydrogen bond between the polymer chains and water molecules. The hydrogen-bond network can aggregate at temperatures above the LCST, resulting in the rejection of water molecules from the polymer structure, and hence forming a hydrophobic form of polymer. The latter structure of polymer is insoluble in water and can lead to further aggregation within the hyperthermic tissues (Yang et al., 2007). One of the well-studied thermo-responsive polymers is poly(N-isopropylacrylamide) (PNIPAm) that has a LCST around 32°C (Fathi et al., 2017; Sahoo et al., 2013). Often a hydrophilic monomer is added to PNIPAm to increase its LCST and improve thermosensitivity (Cheng et al., 2009). For example, addition of copolymers such as acrylamide and vinyl-pyrrolidone to PNIPAm was shown to increase its LCST to the human body temperature (37°C) that provides great thermo-sensitivity function (Chung et al., 1999; Kohori et al., 1998). This grants possibility for sol–gel behavior of these types of polymers, which can be exploited for the BDDT. Jose et al. synthesized thermosensitive gels encompassing lorazepam microspheres for intranasal BDDT using Pluronics (e.g., PF-127 and PF-68). These polymers are thermo-sensitive that form solution at temperatures around 4–5°C and gel at body temperature (37°C). Therefore, these polymers can be used as thermo-responsive hydrogels. Use of the mucoadhesive microspheres in the gel carrier via nasal route could provide a dual purpose of improved bioavailability and prolonged drug release. These researchers showed a prolonged release profile for the dispersion of microspheres in the viscous environment, with no toxic effect nasal mucosa (Jose et al., 2013).

2.6.4   pH-SENSITIVE POLYMERIC CARRIERS

It been reported that pH plays a critical role in various diseased tissues, including inflammation, infection, and solid tumors. For instance, solid tumors exhibits lower pH values in TME as compared to the physiological pH in normal tissues (Barar and Omidi, 2013a). Such unique biological trait can be exploited for designing DDSs responsive to acidic pH, at which they can specifically release the loaded drug molecules at the target site (Gao et al., 2010).

Polymers with ionizable groups such as amines and carboxylic acids are deemed to be ideal candidates for designing pH-sensitive carriers. Upon pH variations, these polymers can undergo some solubility changes and/or conformational changes (Dai et al., 2004). In fact, at acidic conditions, these carriers can efficiently liberate release their loaded drug molecules through changing their structure via swelling or expansion. Further, a pH-sensitive linker can be grafted between drug molecules and polymeric carrier as reported for a hydrazone linker between DOX and PEG-b-(aspartate) copolymer forming micelles from which drug molecules are released in the endosomal compartments at pH about 5.0–6.0 (Bae et al., 2005).

2.6.5 LIPID-BASED CARRIERS

Lipid-based carriers have also been used to serve as nanoscaled DDSs for transportation of a variety of drugs into the brain (Lai et al., 2013). The lipid-based carriers can be categorized into liposomes, solid lipid nanoparticles (SLNs) and nanostructured lipid carriers (Muller et al., 2002), and nonionic surfactant-based vesicle so-called niosomes. We will briefly discuss some of these carriers.

2.6.6 LIPOSOMAL CARRIERS

Liposomes (LIPs), as spherical vesicles with lipid bilayer, are composed of phospholipids such as phosphatidylcholine and phosphatidylethanolamine. From morphological view point, the liposomes can be formulated in different types, including (1) large multilamellar vesicles (LMLVs) having several lamellar bilayers, (2) large unilamellar vesicles, (3) small unilamellar vesicles (SULVs), and (4) multivesicular liposomes. As a result, liposomes differ in size, from nanometer range to micrometer. Liposomes can be formulated by solubilizing phospholipids as well as lipophilic drugs in an appropriate organic solvent. Using a rotary evaporator, the solvent is then evaporated resulting in a thin layer lipid. Then, aqueous phase containing proper surfactant and hydrophilic drugs is introduced to the formed lipid layer under sonication (the so-called thin-film hydration method). This results in formation of LMLVs that can be extruded to attain SULVs with uniform size distribution. Formulation of liposomal DDSs is largely dependent upon cargo-drug molecules' physicochemical properties, nature of the solvent, concentration of the entrapped/incorporated substance, nature of the solvent, types of phospholipids, size, and polydispersity rate (Lai et al., 2013; Ramos-Cabrer and Campos, 2013; Samad et al., 2007; Speiser, 1991). As shown in Figure 2.5, lipophilic drugs can be incorporated within the lipid bilayer of liposomes (panel A), while hydrophilic drugs are encapsulated in the core of liposomes (panel B).

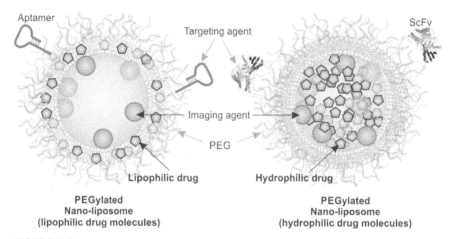

FIGURE 2.5   Liposomal formulations. (A) Incorporated lipophilic drugs within the lipid bilayer. (B) Encapsulated hydrophilic drugs within the core of liposome. Antibodies (scFv, Fab, Fv), aptamers, and peptide ligands can be used as targeting agents. Imaging devices can be also used to develop liposomal theranostics. To avoid immune clearance, liposomes are usually covered with PEG—a technique so-called PEGylation. PEG, polyethylene glycol.

Different types of liposomal DDSs have been engineered including stimuli-responsive formulations (Kneidl et al., 2014; Kono, 2001; Paliwal et al., 2015; Zhao et al., 2015). In a study, peptide H7K (R2)2-modified pH-sensitive liposomes were loaded with DOX (DOX-PSL-H7K (R2)2). Overall results showed that a specific targeting effect could be triggered under acidic pH status in vitro tests in C6 and U87-MG gliomas cells. An antitumor activity of DOX-PSL-H7K(R2)2 was also observed in C6 tumor-bearing mice as well as U87-MG orthotopic tumor-bearing nude mice (Zhao et al., 2016). Polymer-modified hybrid liposomes have been engineered with unique characteristics. They are formulated using lipids with gel-to-liquid

crystalline phase transition at temperatures above physiological temperature (Kono, 2001). The liposomal formulations can be further functionalized
using different entities such as PEG and homing devices to make them long
circulating targeted therapies (Accardo and Morelli, 2015). For example, it
has been reported that synergistic dual-ligand DOX liposomes conjugated
with both CPPs trans-activating transcriptional activator (TAT) and Tf (T7)
(i.e., DOX-T7-TAT-LIP) were able to improve the targeting and therapeutic
efficacy of DOX in brain-tumor bearing mice (Zong et al., 2014).

2.6.7 SOLID LIPID NANOPARTICLES

In the early 1990s, different groups have focused on the establishment
of safe colloidal lipidic particles for drug delivery and developed SLNs
(Müller et al., 2000). SLNs are biocompatible NPs that are promising DDSs
for improving bioavailability and bioactivity of lipophilic and hydrophilic
bioactive compounds (Eskandani and Nazemiyeh, 2014; Geszke-Moritz
and Moritz, 2016; Shanmukhi, 2013). SLNs (50–1000 nm) are composed
of lipids that are solid at both room and body temperatures and provide
sustained/controlled drug release (Attama and Umeyor, 2015). SLNs usually
fabricated using simple homogenization, solvent emulsification/evaporation,
and microemulsion and stabilized by a GRAS (generally recognized as safe)
emulsifier (Ezzati Nazhad Dolatabadi and Omidi, 2016). Despite advantages
of SLNs (biocompatibility, biodegradability, passive drug targeting, and
sustained drug delivery), like another types of NPs, they have some short-
comings such as issues related with hydrophobicity of lipids (Ezzati Nazhad
Dolatabadi et al., 2015). It should be also pointed out that the types of formulation and factors related to the formulation (e.g., surface modification and
functionalization, etc.) determine the final fate of SLNs within the body in
terms of blood circulation as well as the time of contact with the target site
(Sinha et al., 2010). The interaction of hydrophobic surface of SLNs with the
plasma protein triggers the phagocytosis of NPs by macrophages—a process
called opsonization during systemic drug delivery that can drastically reduce
the efficacy of the drugs incorporated in SLNs (Shanmukhi, 2013). This
weakness is overcome by modifying of the surface using hydrophilic molecules (e.g., PEG) and/or surfactants such as polysorbate.

In addition to the problems associated with drug delivery of bioactive
molecules to the target site, BDDT using SLNs appear to be problematic
due to the functional presence of BBB. In the case of brain drug delivery,
the surface of SLNs is usually modified by polysorbate which enables them

to escape from the immune clearance and provide possibility for grafting/ linking with the apolipoproteins (Cacciatore et al., 2016). Apolipoprotein-linked SLNs are able to interact with their specific receptors on the BCECs and hence cross BBB through active targeting mechanism (Patel et al., 2013). Moreover, the surface of SLNs could be further decorated with targeting moieties such as Ab, protein, peptide, Aps, and small molecules. This can resulted in improved targeted delivery of drug-loaded SLNs across the BBB (Patel et al., 2013). Table 2.1 represents some selected studies related to SLN-based BDDT used for treatment of several neurodegenerative diseases including Alzheimer, and Parkinson, and brain tumors.

2.6.8 NIOSOMES

As vesicular systems, niosomes are incorporation of nonionic surfactants (amphiphilic molecules with a hydrophilic end and a lipophilic end) and cholesterol/derivatives of cholesterol. They show similar structure to lipo-somes, that is, lamellar structures with lipid bilayers. In comparison with liposomes, niosomes are much more chemically stable, osmotically active, biodegradable, nonimmunogenic, and less toxic due to having nonionic nature. Similar to liposomes, niosomes can be prepared using different methods like thin-film hydration, microfluidization, reverse-phase evapo-ration, membrane extrusion, lipid injection, and ether injection methods (Moghassemi and Hadjizadeh, 2014).

By far, several niosome-based DDS have been developed for delivery of drugs to the brain. For instance, to establish a brain-targeted delivery system, glucose-bearing niosomes were developed for delivery of the vasoactive intestinal peptide (VIP). After i.v. injection of the VIP- or ^{125}I-VIP-loaded glucose-bearing niosomes to mice, the brain uptake was analyzed using gamma counting. It was found that the encapsulation of drug molecules in the glucose-bearing niosomes could significantly improve the VIP brain uptake as compared to the control niosomes (up to 86%, 5 min after treatment), at which point the engineered niosomes were proposes as a novel BDDT system with capability of crossing BBB (Dufes et al., 2004). Likewise, a niosomal formulation was developed through functionalization of niosomes (i.e., Span 60, cholesterol, and Solulan C24) with N-palmitoylglucosamine for targeted delivery of dynorphin-B as selective agonist of k-opioid receptor(s). The i.v. administration of the niosomal formulation to mice (100 mg/kg) was shown to result in significantly higher antinociceptive effect as compared to the peptide alone (Bragagni et al., 2014). Recently, a niosomal formulation

TABLE 2.1 Solid Lipid Nanoparticles for Treatment of Neurodegenerative Diseases and Brain Tumors.

Disease	Modification	Size (nm)	Drug	Type of delivery	Targeting moiety	Reference
Alzheimer's disease	Chitosan-coated	470	BACE1 siRNA	Nasal delivery	Cell-penetrating peptide	Rassu et al. (2017)
	—	200	Tarenflurbil	Nasal delivery	—	Muntimadugu et al. (2016)
	—	240	Chrysin	ICV injection	—	Vedagiri and Thangarajan (2016)
	—	142	Resveratrol and grape extract	In vitro model of the human blood–brain barrier	OX26 mAb	Loureiro et al. (2017)
	Polysorbate	275	Piperine	Intraperitoneal	Polysorbate	Yusuf et al. (2013)
	—	94-140	Ferulic acid	In vitro evaluation	—	Bondi et al. (2009)
Parkinson's disease	—	108	Levodopa	In vitro evaluation	—	Shui-ming et al. (2010)
	—	197.5	Bromocriptine	Intravenous administration	—	Esposito et al. (2008)
	—	50–200	Apomorphine	Orally administration	—	Tsai et al. (2011)
	—	66	Ropinirole hydrochloride	Nasal delivery	—	Pardeshi et al. (2013)
Brain tumor	—	170–270	Etoposide	In vitro/ex vivo evaluation	Melanotransferrin antibody and tamoxifen	Kuo and Wang (2016)
	—	120–210	Etoposide	In vitro/ex vivo evaluation	83-14 MAb and anti-epidermal growth factor receptor	Kuo and Lee (2016)
	PEGylation	113–163	Doxorubicin	In vitro/ex vivo evaluation	Phenylalanine	Kharya et al. (2013)

OX26 mAb, the antitransferrin receptor monoclonal antibody; MAb, monoclonal antibody.

using ionizable glycerol-based cationic lipids comprising a primary amine, a triglycine, and a dimethylamino ethyl pendent group were prepared. In vitro experiments in HEK-293, ARPE-19, and PECC cells showed a marked higher transfection efficiency in ARPE-19 and PECC cells mainly due to integration of dimethylamino ethyl pendent group. In vivo studies in rat brain after cerebral cortex administration revealed promising high transfection efficiency, upon which this niosomal formulation was suggested as nonviral vector for delivery of nucleic acids to the brain (Ojeda et al., 2016). This is quite an important issue because nonviral vectors may inevitably induce undesired genotoxicity and toxicogenomics (Barar and Omidi, 2013b; Hollins et al., 2007; Omidi et al., 2003, 2005a, 2005b, 2008b).

2.7 DUAL-TARGETING NANOCARRIER

Given that most of drugs are not able to cross BBB and enter to the brain and target the right cells within the CNS, dual-targeting nanocarriers have been developed displaying two homing agents, consisting of (1) a ligand to target BCECs for safe-crossing BBB and (2) a targeting ligand related to designated biological entities within the CNS. For example, effective targeted therapy of brain glioma or any neurodegenerative diseases such as Alzheimer's disease (AD) demands such dual-targeting nanocarriers.

In 2009, a dual-targeting liposomal carrier loaded with topotecan was developed through incorporation of tamoxifen (TAM) within the lipid bilayer of liposomes and conjugation of wheat germ agglutinin (WGA) onto the liposomal surface. After in vitro optimization, the engineered liposomal carrier was examined in vivo by its i.v. administration to the brain C6 glioma-bearing rats. The TAM- and WGA-modified liposomes loaded with topotecan exhibited effective dual-targeting capacity crossing the BBB first and then targeting the brain tumor cells. In the brain tumor-bearing animals, the dual-targeting liposomal DDS substantially improved the overall survival of the tumor-bearing rats in comparison with free drug and nontargeting drug-loaded liposomes (Du et al., 2009). Xin et al. prepared a dual-targeting nanocarrier by conjugating Angiopep (ANG) with poly(ethylene glycol)-*co*-poly(e-caprolactone) Angiopep nanoparticles (ANG-NP). The engineered NPs were able to cross BBB by the ANG-NP that could target the LDL LRP as well as the glioma cells. Paclitaxel (PTX) embeds into the ANG-NP to form ANG-NP-PTX. In comparison with nontargeting NPs, rhodamine isothiocyanate-labeled dual-targeting NPs showed significantly higher endocytosis by U87 MG cells. According to Xin et al. (2011), ANG-NP-PTX exhibited improved inhibitory effects on the U87 MG glioma cells. Li et al. synthesized a pH-sensitive dual-targeting nanocarrier based on the fourth generation (G4) polyamidoamine (PAMAM) dendrimers loaded with DOX. TAM and Tf are conjugated on the interior and in the exterior of the dendrimers (PAMAM-DOX-PEG-Tf-TAM). Actually, these are selected as the targeting groups for increasing the BBB-carrying ability and accumulation drugs in the glioma cells. They found that the pH-responded DOX release was 32% at pH 4.5 and 6% at pH 7.4. The formulated PAMAM-DOX-PEG-Tf-TAM nanosystem showed higher BBB carrying ability with the transporting ratio of 6.06% in 3 h (Li et al., 2012). Gao et al. modified the surface of DOX-loaded liposomal formulation with both Tf and folic acid (FA) to establish a dual-targeting DDS. The formulated liposomes showed enhanced BBB transportation and improved distribution mainly into the glioma cells resulting in antitumor effects (Gao et al., 2013). Zhang et al. developed dual-functional NPs based on PEGylated poly(lactic acid) whose surface was functionalized with QSH, a D-enantiomeric peptide (QSHYRHISPAQV); and TGN, a 12 amino acids containing peptide (TGNYKALHPHNG). They used it for delivery of NPs to AD brain lesions. TGN has chosen to specifically targets ligands at the BBB and QSH has high tendency to $A\beta_{1-42}$, which is the predominant species of amyloid plaque (Zhang et al., 2014). Recently, PEGylated liposomes (PLs) decorated with anti-TfR antibody (OX26) and chlorotoxin (CTX) have been established for transporting plasmid DNA (pC27) across the BBB and targeting the brain glioma. The dual-targeting OX26/CTX-PL/pC27 system showed therapeutic efficacy in vitro and in vivo. OX26/CTX-PL/pC27 treated cells (C6, F98, and HEK293T cells and BMVECs/C6 cells coculture model) proved the transfection efficiency and improved intracellular distribution of the gene delivery system. The dual-targeting OX26/CTX-PL/pC27 system showed markedly high accumulation in the brain diminishing tumor volumes significantly and extending the survival time (46 days) in C6 glioma-bearing rats (Yue et al., 2014). In a recent study, nanoliposomes (NLPs, 124 nm) were armed with a alphavbeta3 ligand, cyclic arginine–glycine–aspartate–tyrosine–cysteine peptide [c(RGDYC)] by means of thiol–maleimide conjugation. The formulation showed substantial uptake by the alphavbeta3-positive glioblastoma (U87) cells and human umbilical vein endothelial cells as compared to nontargeted liposomes. As a result, the researchers proposed it as a new dual-targeting system for an improved boron neutron capture therapy of glioblastoma (Kang et al., 2017).

2.8   STRATEGY TO PROLONG BRAIN MAINTENANCE OF POLYMERIC NANOCARRIERS

There exist several strategies for prolonging blood circulation of nanocarriers. These strategies are mainly based on the surface modification of DDSs to give them potential to circumvent the immune clearance functions of reticuloendothelial system. These strategies can help nanocarriers to remain in the blood circulation for a longer period of time that favors the accumulation of targeted carriers in the right biological sites depending on homing devices used. Of these methods, PEGylation and ligand conjugation are briefly discussed in the following sections.

PEGylation method is performed to prolong the circulation period of NPs in the blood. Coating carriers with PEG improves the solubilization of DDSs and provides great hydrophilicity, electrical neutrality, and possibility for further functionalization. Such features are deemed to reduce elimination of nanocarriers and hence enhance their accumulation in the brain by both active and passive-targeting mechanisms. Such modification can also improve the pharmacokinetics profile of particles (Hamidi et al., 2006; Harris et al., 2001). In a recent study, two new PEGlyated dimeric RGD peptide conjugates were developed by grafting aminocyclopentane (ACP) and aminocyclohexane (ACH) carboxylic acids onto the tetra-peptide RGDK. In practice, the ACP or ACH was grafted to the ring chain of the cyclic RGD peptides with a common bifunctional chelator (1,4,7,10-tetraazacyclododecane-tetraacetic acid [DOTA] or 1,4,7-triazacyclononane-triacetic acid [NOTA]) that was labeled with a radionuclide such as ^{68}Ga or ^{64}Cu. An improved PET imaging was reported in terms of the tumor-to-blood ratio of the ^{64}Cu(NOTA) complex relatively higher than that of the ^{64}Cu(DOTA) complex (Lee et al., 2016), which could be exploited for tumor-targeted radionuclide therapy in the brain.

The ligand-based method is constructed based on the conjugation of DDSs with specific ligands to improve the pharmacokinetic properties of nanocarrier. Further, such ligands may also play a critical role in direct interaction with receptors involved with receptor-mediated endocytosis/transcytosis (Newton, 2006; Vyas et al., 2001). Further, commonly used ligands for targeted endocytosis/transcytosis include Tf, OX26 antibody, thiamine, folate, and insulin and/or designated peptides.

2.9 MICROFLUIDICS: A ROBUST PLATFORM FOR EVALUATION OF BRAIN DRUG DELIVERY AND TARGETING

Microfluidic technology is multidisciplinary field—a combination of chemistry, physics and material science—and is based on the passage of nanoliter fluids through micrometric channels (Whitesides, 2006). In fact, the microfluidic systems have been designed in a way that any changes on passing fluids in channels to be monitored. At the microscale, the behaviors of fluids differ from the macrofluidic in several parameters such as surface tension and fluidic resistance start. Within a very small channel, at a size range about 100 nm–500 μm, very interesting phenomena can occur in terms of flow of fluid. Of note, the Reynolds number becomes very low, and hence there would be no mixing between two coflowing fluids mainly due to the laminar flow of fluid. This clearly implies that, in such small scale, efficient mixing and chemical reaction rates can be precisely controlled (Chen et al., 2012; Jahn et al., 2008). To date, microfluidics techniques have been applied in various dominions—from automated high-throughput screening to lab-on-a-chip and tissue-on-a-chip. It has indeed revolutionized the omics technologies in particular molecular marker detection for the immediate point-of-care monitoring of diseases. Microfluidics can be performed as closed channel continuous-flow using an external driving force. It has also been applied as microdroplet-flow or open system using techniques such as electrowetting.

Using traditional synthesis methods (e.g., dispersion, emulsion, precipitation, and suspension), one could not exactly control the particle size and morphology of engineered particulate systems. However, the synthesis of NPs by microfluidic method provides a number of advantages, including higher accuracy of the synthesis process, flexibility in multistep procedures, reproducibility, cost-effectiveness, and low usage of reagents (Kucuk and Edirisinghe, 2014; Yu et al., 2009). In a study, using microfluidics technique, three liposomal formulations encapsulated with a fluorescence dye were developed, including plain, PEGylated, and FA-armed liposomes. The size and surface properties of formulations were merely accurately controlled through tuning the flow rate ratio and the ratio of the lipids to other components. Having used a 3D tumoroid cultivation, significantly higher internalization was observed for the FA-armed liposomal formulation as compared to the PLs (Ran et al., 2016). In addition to application of microfluidics in production of NPs, this technique has been applied to establish BBB models (Alcendor et al., 2013; Prabhakarpandian et al., 2013). For example, the synthetic microvasculature model of BBB (SyM-BBB), which comprises a plastic, disposable, and a transparent microfluidic chip with a microcirculation-sized two-compartment chamber, has been developed, which allows experiments as apical-to-basolateral and basolateral-to-apical. This can significantly facilitate the BBB trafficking studies and the integration of such experiments with various analytical instrumentations (Prabhakarpandian et al., 2013). Perhaps, we have to wait until getting appropriate models for the brain-on-a-chip, but for now, excellent microfluidics models for BBB have been developed and put in practice (van der Helm et al., 2016).

2.10   MULTIFUNCTIONAL NANOMEDICINES AS ULTIMATE BRAIN DRUG DELIVERY AND TARGETING SYSTEMS

The current treatment modalities in practice solely fight the symptoms of brain diseases. Such palliative agents are not curative and can affect different biological entities of the brain. Nanoscaled DDSs, however, provide improved clinical outcomes, in large part through accelerating the entry of drug molecules into the brain, and hence improving the pharmacokinetics profile and inducing minimal side-effects. Targeted multimodal nanomedicines such as polymeric/lipidic biodegradable NSs and inorganic NSs provide great potential for surface functionalization, upon which several moieties can be conjugated onto these systems bestowing great potential for simultaneous targeted detection, monitoring, and therapy (Omidi, 2011). Such nanoscaled DDSs with targeting devices such as Tf or FA show great capability to cross BBB via active targeting mechanisms and hence are of great benefits to patients with brain tumor (Tzeng and Green, 2013). Of these systems, a dual-targeting DDS has been developed using bovine serum albumin NPs loaded with DOX and functionalized with mPEG2000 as well as lactoferrin (Lf). The engineered NPs showed longer circulation in rat, while the PEGylated Lf-decorated DOX-loaded NPs were shown to be substantially accumulated in the target glioma cells (Su et al., 2014). Likewise, PEGylated cationic solid lipid NPs (CSLN) loaded with Baicalin and functionalized with OX26 Ab (OX26-PEG-CSLN) with a size range of about 30–50 nm showed significantly higher area under curve (AUC) than that of the Baicalin solution alone (Liu et al., 2015). ANG-2 peptide decorated PLGA NPs loaded with DOX and anti-epidermal growth factor receptor (EGFR) siRNA were reported to prolong the life span of the orthotopic U87MG glioma xenograft mice (Wang et al., 2015). Gold NPs were armed with Tf or anti-TfR Ab, which displayed substantially high internalization across the BBB (Clark and Davis, 2015). Use of autocatalytic

BDDT system such as poly(amine-*co*-ester) terpolymer NPs was reported to provide prominently high accumulation in the brain tumors (Han et al., 2016). The brain entry of these systems is based on passive mechanism, and if functionalized with targeting and imaging devices, they can provide great benefits in BDDT. For improved drug delivery into the brain, DDSs can be further armed with CPPs such as TAT and its derivatives (Blumling III and Silva, 2012; Qin et al., 2012). Further, various biological potentials of BBB such as CMTs and RMTs may provide great possibilities for BDDT, upon which trafficking of small and large molecules to brain become applicable (Barar et al., 2016). However, after internalization, brain DDSs need to reach the target cells. This clearly means that multifunctional NSs need to be established as dual-targeting systems to safely cross the BBB and to efficiently interact with the diseased cells. If used for monitoring and imaging, these systems must become seamless DDSs.

2.11 CONCLUDING REMARKS

Specific delivery of therapeutic and diagnosis agents to the brain and explicit targeting of the diseased cells within the brain parenchyma is one of the most challenging issues in CNS diseases. BBB defends the integrity of brain parenchyma and hence CNS functions in a way to makes it almost as an impregnable fort to blood-circulating outsiders. Various DDSs have been engineered for efficient delivery of small molecules and macromolecules to the brain. Of these DDSs, multifunctional nanomedicines and seamless NSs and theranostics/diapeutics seem to provide improved clinical outcomes. For efficient delivery of drugs into the brain, CMTs and RMTs of the BCECs are deemed to be clinically important. These machineries can be exploited for crossing the BBB, whereas designated drug molecules to be able to reach the target cells in the brain parenchyma demand implementation of targeted therapy depending on the diseased cells. One may also capitalize on the development of nanoscaled multifunctional stimuli-responsive DDSs to selectively/specifically deliver the cargo molecules to the target site, simultaneously detect the diseased cells, and release cargo molecules based on monitoring outcomes of the diseased cells/tissue. It is envisioned that this approach to be tailored for each patient depending on the holistic pattern of genomics, proteomics, metabolomics—an approach so-called personalized medicine.

ACKNOWLEDGMENTS

The authors would like to acknowledge Tabriz University of Medical Sciences for the financial support of the BDDT project.

KEYWORDS

- blood–brain barrier
- brain diseases
- brain drug delivery
- brain drug targeting
- nanomedicines
- multifunctional nanosystems
- theranostics

REFERENCES

Abbott, N. J. Inflammatory Mediators and Modulation of Blood–Brain Barrier Permeability. *Cell Mol. Neurobiol.* **2000,** *20*, 131–147.

Abbott, N. J.; Ronnback, L.; Hansson, E. Astrocyte–Endothelial Interactions at the Blood–Brain Barrier. *Nat. Rev. Neurosci.* **2006,** *7*, 41–53.

Accardo, A.; Morelli, G. Review Peptide-Targeted Liposomes for Selective Drug Delivery: Advantages and Problematic Issues. *Biopolymers* **2015,** *104*, 462–479.

Alcendor, D. J.; Block, F. E., 3rd; Cliffel, D. E.; Daniels, J. S.; Ellacott, K. L.; Goodwin, C. R.; Hofmeister, L. H.; Li, D.; Markov, D. A.; May, J. C.; McCawley, L. J.; McLaughlin, B.; McLean, J. A.; Niswender, K. D.; Pensabene, V.; Seale, K. T.; Sherrod, S. D.; Sung, H. J.; Tabb, D. L.; Webb, D. J.; Wikswo, J. P. Neurovascular Unit on a Chip: Implications for Translational Applications. *Stem Cell Res. Ther.* **2013,** *4* (Suppl. 1), S18.

Attama, A. A.; Umeyor, C. E. The Use of Solid Lipid Nanoparticles for Sustained Drug Release. *Ther. Deliv.* **2015,** *6*, 669–684.

Bae, Y.; Nishiyama, N.; Fukushima, S.; Koyama, H.; Yasuhiro, M.; Kataoka, K. Preparation and Biological Characterization of Polymeric Micelle Drug Carriers with Intracellular pH-Triggered Drug Release Property: Tumor Permeability, Controlled Subcellular Drug Distribution, and Enhanced In Vivo Antitumor Efficacy. *Bioconjug. Chem.* **2005,** *16*, 122–130.

Barar, J.; Omidi, Y. Dysregulated pH in Tumor Microenvironment Checkmates Cancer Therapy. *Bioimpacts* **2013a,** *3*, 149–162.

Barar, J.; Omidi, Y. Intrinsic Bio-Signature of Gene Delivery Nanocarriers May Impair Gene Therapy Goals. *Bioimpacts* **2013b,** *3*, 105–109.

Barar, J.; Omidi, Y. Surface Modified Multifunctional Nanomedicines for Simultaneous Imaging and Therapy of Cancer. *Bioimpacts* **2014**, *4*, 3–14.

Barar, J.; Rafi, M. A.; Pourseif, M. M.; Omidi, Y. Blood–Brain Barrier Transport Machineries and Targeted Therapy of Brain Diseases. *Bioimpacts* **2016**, *6*, 225–248.

Blumling III, J. P.; Silva, G. A. Targeting the Brain: Advances in Drug Delivery. *Curr. Pharm. Biotechnol.* **2012**, *13*, 2417–2426.

Bondi, M. L.; Montana, G.; Craparo, E. F.; Picone, P.; Capuano, G.; Di Carlo, M.; Giammona, G. Ferulic Acid-Loaded Lipid Nanostructures as Drug Delivery Systems for Alzheimer's Disease: Preparation, Characterization and Cytotoxicity Studies. *Curr. Nanosci.* **2009**, *5*, 26–32.

Bragagni, M.; Mennini, N.; Furlanetto, S.; Orlandini, S.; Ghelardini, C.; Mura, P. Development and Characterization of Functionalized Niosomes for Brain Targeting of Dynorphin-B. *Eur. J. Pharm. Biopharm.* **2014**, *87*, 73–79.

Cacciatore, I.; Ciulla, M.; Fornasari, E.; Marinelli, L.; Di Stefano, A. Solid Lipid Nanoparticles as a Drug Delivery System for the Treatment of Neurodegenerative Diseases. *Expert Opin. Drug Deliv.* **2016**, *13*, 1121–1131.

Chen, D.; Love, K. T.; Chen, Y.; Eltoukhy, A. A.; Kastrup, C.; Sahay, G.; Jeon, A.; Dong, Y.; Whitehead, K. A.; Anderson, D. G. Rapid Discovery of Potent siRNA-Containing Lipid Nanoparticles Enabled by Controlled Microfluidic Formulation. *J. Am. Chem. Soc.* **2012**, *134*, 6948–6951.

Chen, Y. C.; Chiang, C. F.; Chen, L. F.; Liang, P. C.; Hsieh, W. Y.; Lin, W. L. Polymersomes Conjugated with Des-Octanoyl Ghrelin and Folate as a BBB-Penetrating Cancer Cell-Targeting Delivery System. *Biomaterials* **2014a**, *35*, 4066–4081.

Chen, Y. C.; Chiang, C. F.; Chen, L. F.; Liao, S. C.; Hsieh, W. Y.; Lin, W. L. Polymersomes Conjugated with Des-Octanoyl Ghrelin for the Delivery of Therapeutic and Imaging Agents into Brain Tissues. *Biomaterials* **2014b**, *35*, 2051–2065.

Cheng, C.; Wei, H.; Zhang, X. Z.; Cheng, S. X.; Zhuo, R. X. Thermo-Triggered and Biotinylated Biotin-P (NIPAAm-*co*-HMAAm)-*b*-PMMA Micelles for Controlled Drug Release. *J. Biomed. Mater. Res. A* **2009**, *88*, 814–822.

Chung, J.; Yokoyama, M.; Yamato, M.; Aoyagi, T.; Sakurai, Y.; Okano, T. Thermo-Responsive Drug Delivery from Polymeric Micelles Constructed Using Block Copolymers of Poly(*N*-Isopropylacrylamide) and Poly(Butylmethacrylate). *J. Control. Release* **1999**, *62*, 115–127.

Clark, A. J.; Davis, M. E. Increased Brain Uptake of Targeted Nanoparticles by Adding an Acid-Cleavable Linkage between Transferrin and the Nanoparticle Core. *Proc. Natl. Acad. Sci. U.S.A.* **2015**, *112*, 12486–12491.

Dai, J.; Nagai, T.; Wang, X.; Zhang, T.; Meng, M.; Zhang, Q. pH-Sensitive Nanoparticles for Improving the Oral Bioavailability of Cyclosporine A. *Int. J. Pharm.* **2004**, *280*, 229–240.

Danhier, F.; Ansorena, E.; Silva, J. M.; Coco, R.; Le Breton, A.; Preat, V. PLGA-Based Nanoparticles: An Overview of Biomedical Applications. *J. Control. Release* **2012**, *161*, 505–522.

de Boer, A. G.; Gaillard, P. J. Drug Targeting to the Brain. *Annu. Rev. Pharmacol. Toxicol.* **2007**, *47*, 323–55.

Dieu, L. H.; Wu, D.; Palivan, C. G.; Balasubramanian, V.; Huwyler, J. Polymersomes Conjugated to 83-14 Monoclonal Antibodies: In Vitro Targeting of Brain Capillary Endothelial Cells. *Eur. J. Pharm. Biopharm.* **2014**, *88*, 316–324.

Du, J.; Lu, W. L.; Ying, X.; Liu, Y.; Du, P.; Tian, W.; Men, Y.; Guo, J.; Zhang, Y.; Li, R. J.; Zhou, J.; Lou, J. N.; Wang, J. C.; Zhang, X.; Zhang, Q. Dual-Targeting Topotecan Liposomes Modified with Tamoxifen and Wheat Germ Agglutinin Significantly Improve Drug Transport across the Blood–Brain Barrier and Survival of Brain Tumor-Bearing Animals. *Mol. Pharm.* **2009**, *6*, 905–917.

Dufes, C.; Gaillard, F.; Uchegbu, I. F.; Schatzlein, A. G.; Olivier, J. C.; Muller, J. M. Glucose-Targeted Niosomes Deliver Vasoactive Intestinal Peptide (VIP) to the Brain. *Int. J. Pharm.* **2004**, *285*, 77–85.

Eskandani, M.; Nazemiyeh, H. Self-Reporter Shikonin-Act-Loaded Solid Lipid Nanoparticle: Formulation, Physicochemical Characterization and Geno/Cytotoxicity Evaluation. *Eur. J. Pharm. Sci.* **2014**, *59*, 49–57.

Esposito, E.; Fantin, M.; Marti, M.; Drechsler, M.; Paccamiccio, L.; Mariani, P.; Sivieri, E.; Lain, F.; Menegatti, E.; Morari, M.; Cortesi, R. Solid Lipid Nanoparticles as Delivery Systems for Bromocriptine. *Pharm. Res.* **2008**, *25*, 1521–1530.

Ezzati Nazhad Dolatabadi, J.; Omidi, Y. Solid Lipid-Based Nanocarriers as Efficient Targeted Drug and Gene Delivery Systems. *TrAC Trends Anal. Chem.* **2016**, *77*, 100–108.

Ezzati Nazhad Dolatabadi, J.; Valizadeh, H.; Hamishehkar, H. Solid Lipid Nanoparticles as Efficient Drug and Gene Delivery Systems: Recent Breakthroughs. *Adv. Pharm. Bull.* **2015**, *5*, 151–159.

Fathi, M.; Zangabad, P. S.; Aghanejad, A.; Barar, J.; Erfan-Niya, H.; Omidi, Y. Folate-Conjugated Thermosensitive *O*-Maleoyl Modified Chitosan Micellar Nanoparticles for Targeted Delivery of Erlotinib. *Carbohydr. Polym.* **2017**, *172*, 130–141.

Gao, J.-Q.; Lv, Q.; Li, L.-M.; Tang, X.-J.; Li, F.-Z.; Hu, Y.-L.; Han, M. Glioma Targeting and Blood–Brain Barrier Penetration by Dual-Targeting Doxorubincin Liposomes. *Biomaterials* **2013**, *34*, 5628–5639.

Gao, W.; Chan, J. M.; Farokhzad, O. C. pH-Responsive Nanoparticles for Drug Delivery. *Mol. Pharm.* **2010**, *7*, 1913–1920.

Ge, S.; Song, L.; Pachter, J. S. Where is the Blood–Brain Barrier ... Really? *J. Neurosci. Res.* **2005**, *79*, 421–427.

Geszke-Moritz, M.; Moritz, M. Solid Lipid Nanoparticles as Attractive Drug Vehicles: Composition, Properties and Therapeutic Strategies. *Mater. Sci. Eng. C* **2016**, *68*, 982–994.

Hamidi, M.; Azadi, A.; Rafiei, P. Pharmacokinetic Consequences of Pegylation. *Drug Deliv.* **2006**, *13*, 399–409.

Han, L.; Kong, D. K.; Zheng, M. Q.; Murikinati, S.; Ma, C.; Yuan, P.; Li, L.; Tian, D.; Cai, Q.; Ye, C.; Holden, D.; Park, J. H.; Gao, X.; Thomas, J. L.; Grutzendler, J.; Carson, R. E.; Huang, Y.; Piepmeier, J. M.; Zhou, J. Increased Nanoparticle Delivery to Brain Tumors by Autocatalytic Priming for Improved Treatment and Imaging. *ACS Nano* **2016**, *10*, 4209–4218.

Harris, J. M.; Martin, N. E.; Modi, M. Pegylation: A Novel Process for Modifying Pharmacokinetics. *Clin. Pharmacokinet.* **2001**, *40*, 539–551.

Hollins, A. J.; Omidi, Y.; Benter, I. F.; Akhtar, S. Toxicogenomics of Drug Delivery Systems: Exploiting Delivery System-Induced Changes in Target Gene Expression to Enhance siRNA Activity. *J. Drug Target.* **2007**, *15*, 83–88.

Jahn, A.; Reiner, J. E.; Vreeland, W.; DeVoe, D. L.; Locascio, L. E.; Gaitan, M. Preparation of Nanoparticles by Continuous-Flow Microfluidics. *J. Nanopart. Res.* **2008**, *10*, 925–934.

Jose, S.; Ansa, C.; Cinu, T.; Chacko, A.; Aleykutty, N.; Ferreira, S.; Souto, E. Thermo-Sensitive Gels Containing Lorazepam Microspheres for Intranasal Brain Targeting. *Int. J. Pharm.* **2013**, *441*, 516–526.

Kang, W.; Svirskis, D.; Sarojini, V.; McGregor, A. L.; Bevitt, J.; Wu, Z. Cyclic-RGDyC functionalized Liposomes for Dual-Targeting of Tumor Vasculature and Cancer Cells in Glioblastoma: An In Vitro Boron Neutron Capture Therapy Study. *Oncotarget* **2017**, *8*, 36614–36627.

Kharya, P.; Jain, A.; Gulbake, A.; Shilpi, S.; Jain, A.; Hurkat, P.; Majumdar, S.; Jain, S. K. Phenylalanine-Coupled Solid Lipid Nanoparticles for Brain Tumor Targeting. *J. Nanopart. Res.* **2013**, *15*.

Kneidl, B.; Peller, M.; Winter, G.; Lindner, L. H.; Hossann, M. Thermosensitive Liposomal Drug Delivery Systems: State of the Art Review. *Int. J. Nanomed.* **2014**, *9*, 4387–4398.

Kohori, F.; Sakai, K.; Aoyagi, T.; Yokoyama, M.; Sakurai, Y.; Okano, T. Preparation and Characterization of Thermally Responsive Block Copolymer Micelles Comprising Poly(*N*-Isopropylacrylamide-*b*-dl-Lactide). *J. Control. Release* **1998**, *55*, 87–98.

Kono, K. Thermosensitive Polymer-Modified Liposomes. *Adv. Drug Deliv. Rev.* **2001**, *53*, 307–319.

Kreuter, J. Drug Delivery to the Central Nervous System by Polymeric Nanoparticles: What Do We Know? *Adv. Drug Deliv. Rev.* **2014**, *71*, 2–14.

Kreuter, J. Mechanism of Polymeric Nanoparticle-Based Drug Transport across the Blood–Brain Barrier (BBB). *J. Microencapsul.* **2013**, *30*, 49–54.

Krizbai, I. A.; Deli, M. A. Signalling Pathways Regulating the Tight Junction Permeability in the Blood–Brain Barrier. *Cell. Mol. Biol. (Noisy-le-grand).* **2003**, *49*, 23–31.

Kucuk, I.; Edirisinghe, M. Microfluidic Preparation of Polymer Nanospheres. *J. Nanopart. Res.* **2014**, *16*, 2626.

Kuo, Y. C.; Lee, C. H. Dual Targeting of Solid Lipid Nanoparticles Grafted with 83-14 MAb and Anti-EGF Receptor for Malignant Brain Tumor Therapy. *Life Sci.* **2016**, *146*, 222–231.

Kuo, Y. C.; Wang, I. H. Enhanced Delivery of Etoposide across the Blood–Brain Barrier to Restrain Brain Tumor Growth Using Melanotransferrin Antibody- and Tamoxifen-Conjugated Solid Lipid Nanoparticles. *J. Drug Target.* **2016**, *24*, 645–654.

Lai, F.; Fadda, A. M.; Sinico, C. Liposomes for Brain Delivery. *Expert Opin. Drug Deliv.* **2013**, *10*, 1003–1022.

Lee, J. W.; Lee, Y. J.; Shin, U. C.; Kim, S. W.; Kim, B. I.; Lee, K. C.; Kim, J. Y.; Park, J. A. Improved Pharmacokinetics Following PEGylation and Dimerization of a c(RGD-ACH-K) Conjugate Used for Tumor Positron Emission Tomography Imaging. *Cancer Biother. Radiopharm.* **2016**, *31*, 295–301.

Li, G.; Meng, Y.; Guo, L.; Zhang, T.; Liu, J. Formation of Thermo-Sensitive Polyelectrolyte Complex Micelles from Two Biocompatible Graft Copolymers for Drug Delivery. *J. Biomed. Mater. Res. A* **2014**, *102*, 2163–2172.

Li, Y.; Gao, G. H.; Lee, D. S. Stimulus-Sensitive Polymeric Nanoparticles and Their Applications as Drug and Gene Carriers. *Adv. Healthc. Mater.* **2013**, *2*, 388–417.

Li, Y.; He, H.; Jia, X.; Lu, W. L.; Lou, J.; Wei, Y. A Dual-Targeting Nanocarrier Based on Poly(Amidoamine) Dendrimers Conjugated with Transferrin and Tamoxifen for Treating Brain Gliomas. *Biomaterials* **2012**, *33*, 3899–3908.

Liu, W. Y.; Wang, Z. B.; Zhang, L. C.; Wei, X.; Li, L. Tight Junction in Blood–Brain Barrier: An Overview of Structure, Regulation, and Regulator Substances. *CNS Neurosci. Ther.* **2012**, *18*, 609–615.

Liu, Z.; Zhao, H.; Shu, L.; Zhang, Y.; Okeke, C.; Zhang, L.; Li, J.; Li, N. Preparation and Evaluation of Baicalin-Loaded Cationic Solid Lipid Nanoparticles Conjugated with OX26 for Improved Delivery across the BBB. *Drug Dev. Ind. Pharm.* **2015**, *41*, 353–361.

Loureiro, J. A.; Andrade, S.; Duarte, A.; Neves, A. R.; Queiroz, J. F.; Nunes, C.; Sevin, E.; Fenart, L.; Gosselet, F.; Coelho, M. A. N.; Pereira, M. C.; Latruffe, N. Resveratrol and Grape Extract-Loaded Solid Lipid Nanoparticles for the Treatment of Alzheimer's Disease. *Molecules* **2017**, *22*, 1–16.

Mahringer, A.; Fricker, G. ABC Transporters at the Blood–Brain Barrier. *Expert Opin Drug Metab Toxicol.* **2016**, *12*, 499–508.

Matthaiou, E. I.; Barar, J.; Sandaltzopoulos, R.; Li, C.; Coukos, G.; Omidi, Y. Shikonin-Loaded Antibody-Armed Nanoparticles for Targeted Therapy of Ovarian Cancer. *Int. J. Nanomed.* **2014**, *9*, 1855–1870.

Moghassemi, S.; Hadjizadeh, A. Nano-Niosomes as Nanoscale Drug Delivery Systems: An Illustrated Review. *J. Control. Release* **2014**, *185*, 22–36.

Monaco, I.; Camorani, S.; Colecchia, D.; Locatelli, E.; Calandro, P.; Oudin, A.; Niclou, S.; Arra, C.; Chiariello, M.; Cerchia, L.; Comes Franchini, M. Aptamer Functionalization of Nanosystems for Glioblastoma Targeting through the Blood–Brain Barrier. *J. Med. Chem.* **2017**, *60*, 4510–4516.

Mooguee, M.; Omidi, Y.; Davaran, S. Synthesis and In Vitro Release of Adriamycin from Star-Shaped Poly(Lactide-*co*-Glycolide) Nano- and Microparticles. *J. Pharm. Sci.* **2010**, *99*, 3389–3397.

Müller, R. H.; Mäder, K.; Gohla, S. Solid Lipid Nanoparticles (SLN) for Controlled Drug delivery—A Review of the State of the Art. *Eur. J. Pharm. Biopharm.* **2000**, *50*, 161–177.

Muller, R. H.; Radtke, M.; Wissing, S. A. Solid Lipid Nanoparticles (SLN) and Nanostructured Lipid Carriers (NLC) in Cosmetic and Dermatological Preparations. *Adv. Drug Deliv. Rev.* **2002**, *54* (Suppl. 1), S131–S155.

Muntimadugu, E.; Dhommati, R.; Jain, A.; Challa, V. G. S.; Shaheen, M.; Khan, W. Intranasal Delivery of Nanoparticle Encapsulated Tarenflurbil: A Potential Brain Targeting Strategy for Alzheimer's Disease. *Eur. J. Pharm. Sci.* **2016**, *92*, 224–234.

Mura, S.; Nicolas, J.; Couvreur, P. Stimuli-Responsive Nanocarriers for Drug Delivery. *Nat. Mater.* **2013**, *12*, 991–1003.

Newton, H. B. Advances in Strategies to Improve Drug Delivery to Brain Tumors. *Expert Rev. Neurother.* **2006**, *6*, 1495–1509.

Ojeda, E.; Puras, G.; Agirre, M.; Zarate, J.; Grijalvo, S.; Eritja, R.; Martinez-Navarrete, G.; Soto-Sanchez, C.; Diaz-Tahoces, A.; Aviles-Trigueros, M.; Fernandez, E.; Pedraz, J. L. The Influence of the Polar Head-Group of Synthetic Cationic Lipids on the Transfection Efficiency Mediated by Niosomes in Rat Retina and Brain. *Biomaterials* **2016**, *77*, 267–279.

Omidi, Y. Smart Multifunctional Theranostics: Simultaneous Diagnosis and Therapy of Cancer. *Bioimpacts* **2011**, *1*, 145–147.

Omidi, Y.; Barar, J. Blood–Brain Barrier and Effectiveness of Therapy Against Brain Tumors In *Novel Therapeutic Concepts in Targeting Glioma*; Farassati, F., Ed.; Rijeka, Croatia: Intech, 2012a.

Omidi, Y.; Barar, J.; Ahmadian, S.; Heidari, H. R.; Gumbleton, M. Characterization and Astrocytic Modulation of System L Transporters in Brain Microvasculature Endothelial Cells. *Cell Biochem Funct.* **2008a**, *26*, 381–391.

Omidi, Y.; Barar, J. Impacts of Blood–Brain Barrier in Drug Delivery and Targeting of Brain Tumors. *Bioimpacts* **2012b**, *2*, 5–22.

Omidi, Y.; Barar, J.; Akhtar, S. Toxicogenomics of Cationic Lipid-Based Vectors for Gene Therapy: Impact of Microarray Technology. *Curr. Drug Deliv.* **2005a**, *2*, 429–441.

Omidi, Y.; Barar, J.; Heidari, H. R.; Ahmadian, S.; Yazdi, H. A.; Akhtar, S. Microarray Analysis of the Toxicogenomics and the Genotoxic Potential of a Cationic Lipid-Based Gene Delivery Nanosystem in Human Alveolar Epithelial a549 Cells. *Toxicol. Mech. Methods* **2008b**, *18*, 369–378.

Omidi, Y.; Hollins, A. J.; Benboubetra, M.; Drayton, R.; Benter, I. F.; Akhtar, S. Toxicogenomics of Non-Viral Vectors for Gene Therapy: A Microarray Study of Lipofectin- and Oligofectamine-Induced Gene Expression Changes in Human Epithelial Cells. *J. Drug Target.* **2003**, *11*, 311–323.

Omidi, Y.; Hollins, A. J.; Drayton, R. M.; Akhtar, S. Polypropylenimine Dendrimer-Induced Gene Expression Changes: The Effect of Complexation with DNA, Dendrimer Generation and Cell Type. *J. Drug Target.* **2005b**, *13*, 431–443.

Orlu-Gul, M.; Topcu, A. A.; Shams, T.; Mahalingam, S.; Edirisinghe, M. Novel Encapsulation Systems and Processes for Overcoming the Challenges of Polypharmacy. *Curr. Opin. Pharmacol.* **2014**, *18*, 28–34.

Paliwal, S. R.; Paliwal, R.; Vyas, S. P. A Review of Mechanistic Insight and Application of pH-Sensitive Liposomes in Drug Delivery. *Drug Deliv.* **2015**, *22*, 231–242.

Pardeshi, C. V.; Rajput, P. V.; Belgamwar, V. S.; Tekade, A. R.; Surana, S. J. Novel Surface Modified Solid Lipid Nanoparticles as Intranasal Carriers for Ropinirole Hydrochloride: Application of Factorial Design Approach. *Drug Deliv.* **2013**, *20*, 47–56.

Pardridge, W. M. Blood–Brain Barrier Endogenous Transporters as Therapeutic Targets: A New Model for Small Molecule CNS Drug Discovery. *Expert Opin. Ther. Targets* **2015**, *19*, 1059–1072.

Pardridge, W. M. Drug Targeting to the Brain. *Pharm. Res.* **2007**, *24*, 1733–1744.

Pardridge, W. M. Drug Transport Across the Blood–Brain Barrier. *J. Cereb. Blood Flow Metab.* **2012**, *32*, 1959–1972.

Patel, M.; Souto, E. B.; Singh, K. K. Advances in Brain Drug Targeting and Delivery: Limitations and Challenges of Solid Lipid Nanoparticles. *Expert Opin. Drug Deliv.* **2013**, *10*, 889–905.

Prabhakarpandian, B.; Shen, M. C.; Nichols, J. B.; Mills, I. R.; Sidoryk-Wegrzynowicz, M.; Aschner, M.; Pant, K. SyM-BBB: A Microfluidic Blood–Brain Barrier Model. *Lab Chip* **2013**, *13*, 1093–1101.

Preston, J. E.; Joan Abbott, N.; Begley, D. J. Transcytosis of Macromolecules at the Blood–Brain Barrier. *Adv. Pharmacol.* **2014**, *71*, 147–163.

Qin, Y.; Zhang, Q.; Chen, H.; Yuan, W.; Kuai, R.; Xie, F.; Zhang, L.; Wang, X.; Zhang, Z.; Liu, J.; He, Q. Comparison of Four Different Peptides to Enhance Accumulation of Liposomes into the Brain. *J. Drug Target.* **2012**, *20*, 235–245.

Rafi, M. A.; Omidi, Y. A Prospective Highlight on Exosomal Nanoshuttles and Cancer Immunotherapy and Vaccination. *Bioimpacts* **2015**, *5*, 117–122.

Ramos-Cabrer, P.; Campos, F. Liposomes and Nanotechnology in Drug Development: Focus on Neurological Targets. *Int. J. Nanomed.* **2013**, *8*, 951–960.

Ran, R.; Middelberg, A. P.; Zhao, C. X. Microfluidic Synthesis of Multifunctional Liposomes for Tumour Targeting. *Colloids Surf. B: Biointerfaces* **2016**, *148*, 402–410.

Rapoport, S. I. Advances in Osmotic Opening of the Blood–Brain Barrier to Enhance CNS Chemotherapy. *Expert Opin. Investig. Drugs* **2001**, *10*, 1809–1818.

Rassu, G.; Soddu, E.; Posadino, A. M.; Pintus, G.; Sarmento, B.; Giunchedi, P.; Gavini, E. Nose-to-Brain Delivery of BACE1 siRNA Loaded in Solid Lipid Nanoparticles for Alzheimer's Therapy. *Colloids Surf. B: Biointerfaces* **2017**, *152*, 296–301.

Ruozi, B.; Belletti, D.; Pederzoli, F.; Forni, F.; Vandelli, M. A.; Tosi, G. Potential Use of Nanomedicine for Drug Delivery Across the Blood–Brain Barrier in Healthy and Diseased Brain. *CNS Neurol. Disord. Drug Targets* **2016**, *15*, 1079–1091.

Russo, A.; DeGraff, W.; Friedman, N.; Mitchell, J. B. Selective Modulation of Glutathione Levels in Human Normal versus Tumor Cells and Subsequent Differential Response to Chemotherapy Drugs. *Cancer Res.* **1986**, *46*, 2845–2848.

Sahoo, B.; Devi, K. S. P.; Banerjee, R.; Maiti, T. K.; Pramanik, P.; Dhara, D. Thermal and pH Responsive Polymer-Tethered Multifunctional Magnetic Nanoparticles for Targeted Delivery of Anticancer Drug. *ACS Appl. Mater. Interfaces* **2013**, *5*, 3884–3893.

Samad, A.; Sultana, Y.; Aqil, M. Liposomal Drug Delivery Systems: An Update Review. *Curr. Drug Deliv.* **2007**, *4*, 297–305.

Schwerdt, A.; Zintchenko, A.; Concia, M.; Roesen, N.; Fisher, K.; Lindner, L. H.; Issels, R.; Wagner, E.; Ogris, M. Hyperthermia-Induced Targeting of Thermosensitive Gene Carriers to Tumors. *Hum. Gene Ther.* **2008**, *19*, 1283–1292.

Shanmukhi, P. Solid Lipid Nanoparticles—A Novel Solid Lipid Based Technology for Poorly Water Soluble Drugs: A Review. *Int. J. Pharm. Technol.* **2013**, *5*, 2645–2674.

Shui-ming, Z.; Dong-zhi, H.; Qi-neng, P.; Yan, X. Preparation and Entrapment Efficiency Determination of Solid Lipid Nanoparticles Loaded Levodopa. *Ch. J. Hosp. Pharm.* **2010**, *14*, 23–86.

Sinha, V. R.; Srivastava, S.; Goel, H.; Jindal, V. Solid Lipid Nanoparticles (SLN'S)—Trends and Implications in Drug Targeting. *Int. J. Adv. Pharm. Sci.* **2010**, *1*, 212–238.

Speiser, P. P. Nanoparticles and Liposomes: A State of the Art. *Methods Find Exp. Clin. Pharmacol.* **1991**, *13*, 337–342.

Stevenson, B. R.; Siliciano, J. D.; Mooseker, M. S.; Goodenough, D. A. Identification of ZO-1: A High Molecular Weight Polypeptide Associated with the Tight Junction (Zonula Occludens) in a Variety of Epithelia. *J. Cell Biol.* **1986**, *103*, 755–766.

Su, Z.; Xing, L.; Chen, Y.; Xu, Y.; Yang, F.; Zhang, C.; Ping, Q.; Xiao, Y. Lactoferrin-Modified Poly(Ethylene Glycol)-Grafted BSA Nanoparticles as a Dual-Targeting Carrier for Treating Brain Gliomas. *Mol. Pharm.* **2014**, *11*, 1823–1834.

Torchilin, V. Multifunctional and Stimuli-Sensitive Pharmaceutical Nanocarriers. *Eur. J. Pharm. Biopharm.* **2009**, *71*, 431–444.

Tsai, M. J.; Huang, Y. B.; Wu, P. C.; Fu, Y. S.; Kao, Y. R.; Fang, J. Y.; Tsai, Y. H. Oral Apomorphine Delivery from Solid Lipid Nanoparticles with Different Monostearate Emulsifiers: Pharmacokinetic and Behavioral Evaluations. *J. Pharm. Sci.* **2011**, *100*, 547–557.

Tzeng, S. Y.; Green, J. J. Therapeutic Nanomedicine for Brain Cancer. *Ther. Deliv.* **2013**, *4*, 687–704.

van der Helm, M. W.; van der Meer, A. D.; Eijkel, J. C.; van den Berg, A.; Segerink, L. I. Microfluidic Organ-on-chip Technology for Blood–Brain Barrier Research. *Tissue Barriers* **2016**, *4*, e1142493.

Vedagiri, A.; Thangarajan, S. Mitigating Effect of Chrysin Loaded Solid Lipid Nanoparticles against Amyloid β25–35 Induced Oxidative Stress in Rat Hippocampal Region: An Efficient Formulation Approach for Alzheimer's Disease. *Neuropeptides* **2016**, *58*, 111–125.

Vyas, S. P.; Singh, A.; Sihorkar, V. Ligand–Receptor-Mediated Drug Delivery: An Emerging Paradigm in Cellular Drug Targeting. *Crit. Rev. Ther. Drug Carrier* **2001**, *18*, 1–76.

Wang, A. Z.; Langer, R.; Farokhzad, O. C. Nanoparticle Delivery of Cancer Drugs. *Annu. Rev. Med.* **2012**, *63*, 185–198.

Wang, L.; Hao, Y.; Li, H.; Zhao, Y.; Meng, D.; Li, D.; Shi, J.; Zhang, H.; Zhang, Z.; Zhang, Y. Co-Delivery of Doxorubicin and siRNA for Glioma Therapy by a Brain Targeting System: Angiopep-2-Modified Poly(Lactic-*co*-Glycolic Acid) Nanoparticles. *J. Drug Target.* **2015**, *23*, 832–846.

Wang, Y. Y.; Lui, P. C.; Li, J. Y. Receptor-Mediated Therapeutic Transport across the Blood–Brain Barrier. *Immunotherapy* **2009**, *1*, 983–993.

Whitesides, G. M. The Origins and the Future of Microfluidics. *Nature* **2006**, *442*, 368–373.

Willis, C. L. Imaging In Vivo Astrocyte/Endothelial Cell Interactions at the Blood–Brain Barrier. *Methods Mol. Biol.* **2012**, *814*, 515–529.

Wood, M. J.; O'Loughlin, A. J.; Samira, L. Exosomes and the Blood–Brain Barrier: Implications for Neurological Diseases. *Ther. Deliv.* **2011**, *2*, 1095–1099.

Xin, H.; Jiang, X.; Gu, J.; Sha, X.; Chen, L.; Law, K.; Chen, Y.; Wang, X.; Jiang, Y.; Fang, X. Angiopep-Conjugated Poly(Ethylene Glycol)-*co*-Poly(ε-Caprolactone) Nanoparticles as Dual-Targeting Drug Delivery System for Brain Glioma. *Biomaterials* **2011**, *32*, 4293–4305.

Yang, M.; Ding, Y.; Zhang, L.; Qian, X.; Jiang, X.; Liu, B. Novel Thermosensitive Polymeric Micelles for Docetaxel Delivery. *J. Biomater. Mater. Res. A* **2007**, *81*, 847–857.

Yang, T.; Martin, P.; Fogarty, B.; Brown, A.; Schurman, K.; Phipps, R.; Yin, V. P.; Lockman, P.; Bai, S. Exosome Delivered Anticancer Drugs across the Blood–Brain Barrier for Brain Cancer Therapy in *Danio rerio. Pharm. Res.* **2015**, *32*, 2003–2014.

Yemisci, M.; Caban, S.; Gursoy-Ozdemir, Y.; Lule, S.; Novoa-Carballal, R.; Riguera, R.; Fernandez-Megia, E.; Andrieux, K.; Couvreur, P.; Capan, Y.; Dalkara, T. Systemically Administered Brain-Targeted Nanoparticles Transport Peptides across the Blood–Brain Barrier and Provide Neuroprotection. *J. Cereb. Blood Flow Metab.* **2015**, *35*, 469–475.

Yu, B.; Lee, R. J.; Lee, L. J. Microfluidic Methods for Production of Liposomes. *Methods Enzymol.* **2009**, *465*, 129–141.

Yu, Y.; Pang, Z.; Lu, W.; Yin, Q.; Gao, H.; Jiang, X. Self-Assembled Polymersomes Conjugated with Lactoferrin as Novel Drug Carrier for Brain Delivery. *Pharm. Res.* **2012**, *29*, 83–96.

Yue, P. J.; He, L.; Qiu, S. W.; Li, Y.; Liao, Y. J.; Li, X. P.; Xie, D.; Peng, Y. OX26/CTX-Conjugated PEGylated Liposome as a Dual-Targeting Gene Delivery System for Brain Glioma. *Mol. Cancer* **2014**, *13*, 191.

Yusuf, M.; Khan, M.; Khan, R. A.; Ahmed, B. Preparation, Characterization, In Vivo and Biochemical Evaluation of Brain Targeted Piperine Solid Lipid Nanoparticles in an Experimentally Induced Alzheimer's Disease Model. *J. Drug Target.* **2013**, *21*, 300–311.

Zhan, C.; Meng, Q.; Li, Q.; Feng, L.; Zhu, J.; Lu, W. Cyclic RGD-Polyethylene Glycol-Polyethylenimine for Intracranial Glioblastoma-Targeted Gene Delivery. *Chem. Asian J.* **2012**, *7*, 91–96.

Zhang, C.; Wan, X.; Zheng, X.; Shao, X.; Liu, Q.; Zhang, Q.; Qian, Y. Dual-Functional Nanoparticles Targeting Amyloid Plaques in the Brains of Alzheimer's Disease Mice. *Biomaterials* **2014**, *35*, 456–465.

Zhang, H.; Vinogradov, S. V. Short Biodegradable Polyamines for Gene Delivery and Transfection of Brain Capillary Endothelial Cells. *J. Control. Release* **2010**, *143*, 359–366.

Zhao, Y.; Ren, W.; Zhong, T.; Zhang, S.; Huang, D.; Guo, Y.; Yao, X.; Wang, C.; Zhang, W. Q.; Zhang, X.; Zhang, Q. Tumor-Specific pH-Responsive Peptide-Modified pH-Sensitive Liposomes Containing Doxorubicin for Enhancing Glioma Targeting and Anti-Tumor Activity. *J. Control. Release* **2016**, *222*, 56–66.

Zhu, L.; Torchilin, V. P. Stimulus-Responsive Nanopreparations for Tumor Targeting. *Integr.*
  Biol. **2013,** *5,* 96–107.
Zong, T.; Mei, L.; Gao, H.; Cai, W.; Zhu, P.; Shi, K.; Chen, J.; Wang, Y.; Gao, F.; He, Q.
  Synergistic Dual-Ligand Doxorubicin Liposomes Improve Targeting and Therapeutic Effi-
  cacy of Brain Glioma in Animals. *Mol. Pharm.* **2014,** *11,* 2346–2357.

CHAPTER 3

NANOTECHNOLOGY FOR BRAIN TARGETING

DEBJANI BANERJEE, HUMA SHAMSHAD, and NIDHI MISHRA*

Department of Applied Science, Indian Institute of Information Technology, Allahabad 211012, Uttar Pradesh, India

*Corresponding author. E-mail: nidhimishra@iiita.ac.in

ABSTRACT

The blood–brain barrier (BBB) which maintains the neural homeostasis remains as a major hindrance for many medical imaging and drug penetration procedures in case of tumors or cancers. Through intensive research works, many routes have been found to penetrate the BBB using micelles, polymeric nanoparticles, and liposomes as drug carriers. One of the most focused and emerging fields is using gold nanoparticles (AuNPs). Its tunable optoelectronic properties, along with high photostability, multifunctionality, and biocompatibility, have given it an upper hand in comparison with other existing techniques. In this chapter, we are going to focus on synthesis of AuNPs and utilizing them for brain targeting using its multifunctional properties.

3.1 INTRODUCTION

Selection and targeting of drugs for brain-related impairments has always been intricate due to the blood–brain barrier (BBB). BBB is a semipermeable complex network with tightly fitted endothelial cells, which do not allow substances to cross the bloodstream and enter the brain. Substances like glucose are transported through special means and transport facilities; therefore, the BBB restricts any kind of systemic drug-delivery methods. It has been estimated that more than 98% of low-molecular weight drugs and nearly 100% of high-molecular weight drugs developed for central nervous system (CNS) pathologies do not readily cross the BBB (Banks, 2009). Cancer is one of the major causes of death worldwide almost constituting one quarter of all deaths across the United States and United Kingdom (Dreaden, 2012). Amongst all, glioblastoma multiforme (GBM) forms the deadliest type of brain tumor with a life expectancy little above a year (Dreaden, 2012). The most common form of treatment involves surgical resection accompanied by chemotherapy sessions or radiotherapy sessions.

The brain tumors themselves can damage the normal functioning of BBB by secreting some soluble factors that dissolve the tight junctions with concomitant defective blood vessels formation. These are the results of subsequent loss of occludin and abnormal angiogenesis that cause aberrations in the endothelial permeability (Papadopoulos et al., 2001). Proper diagnosis and treatment of brain tumors like GBM becomes troublesome due to this network. The cumulative dose of radiation therapy that can be given safely (minimum loss to nearby tissues) along with the BBB penetration makes the treatment situation grim.

Recently, solid metal nanoparticles (NPs) have shown the therapeutic abilities in GBM researches. These abilities can be accredited due to their physical properties such as small size, enhanced physiological stability, and biocompatibility. Moreover, their surface functionalities help them target tumor-specific moieties such as antibodies, ligands, and peptides (Whitesides, 2003). NPs help to encapsulate many anticancer hydrophobic drugs and deliver them safely through the BBB without any change in their action. Among the various available NPs, gold nanoparticles (AuNPs) have proven to be very effective in treatment of GBM due to their various unique properties which will be discussed further in the chapter.

3.2   THE BLOOD-BRAIN BARRIER AND DRUGS

3.2.1   BLOOD-BRAIN BARRIER ANATOMY AND PHYSIOLOGY

The BBB is a highly selective structure composed of complex system of endothelial cells, pericytes, astrocytes, and perivascular mast cells that prevent most of the circulating cells, small ions, and even more restrictive movement of macromolecules (Wahl et al., 1988, Breimer and de Boer, 1998). Thus, this vasculature protects the brain from diseases as well as any noxious substance, thus preventing any kind of injury.

The vascular layer of brain capillary endothelial cells are attached to each other, side by side through tight and adherent junctions, thus giving it a dense form.The BBB allows the movement of water, some gases (oxygen, carbon dioxide, volatile anesthetics) through rapid diffusion, lipid soluble molecules by passive diffusion and crucialpolar molecules like amino acids, glucose, nucleosides are transported via carrier-mediated influx (Pardridge et al.,1985). Large molecules (proteins, peptides) which have growth and signaling function in the central nervous system (CNS) enter the brain via a restrictive manner by receptor mediated transcytosis.

These junctions perform two main functions: first, they prevent passage of most small molecules and ions from spaces between the cells, and second, they also prevent the movement of integral membrane proteins so that each cell membrane surface maintains its unique functions (Rabanel et al., 2012).

3.2.2   BLOOD–BRAIN BARRIER DYNAMICS

The BBB is not only a mechanical defense but also a dynamic entity in which active metabolism and carrier-mediated transport of nutrients such as glucose, amino acids, and ketones occur (Fig. 3.1). The BBB prevents uptake of most of the drugs, allowing only some specific hydrophilic molecules with mass less than 150 Da and highly hydrophobic compounds with a mass lower than 400–600 Da (Santaguida et al., 2012).

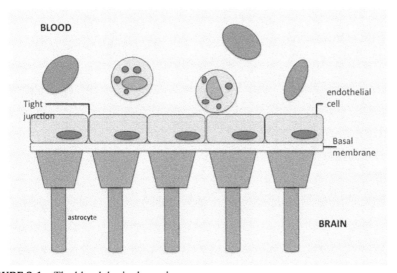

FIGURE 3.1   The blood–brain dynamics.

The vasculature of the BBB as compared to normal body organs strictly restricts the passage of most of the drugs from plasma to extracellular space. Therefore, knowing the fundamentals of BBB characteristics is very essential for proper drug delivery across the BBB. An important consideration is the change in the permeability of the BBB in case of pathogenicity, severe brain traumatic injury, and in malignant brain tumors as in case of GBM.

3.2.3   INTERACTION OF BLOOD–BRAIN BARRIER WITH DRUGS

Due to the given widely changing premises, most of the pharmacological therapies for neurological diseases render useless. As a consequence, many different types of strategies and methods are currently in quest for increasing the efficacy of drug delivery across the BBB. Most of the current strategies involve invasive techniques, for example, from osmotic and chemical opening of BBB, transporter/carrier, to chemical modifications of drugs (Kaur et al., 2008). By passing the BBB through alternative routes of delivery such as transnasal route may be considered. Alternative methods of drug delivery can also be used for crossing the cerebrospinal fluid also be considered or direct delivery of the drugs into the cerebrospinal fluid through lumbar puncture can also be adopted if targeted delivery is not required (Fig. 3.2). Invasive methods for bypassing the BBB involve direct introduction of drugs into the brain by surgical procedures (Sousa et al., 2010; He et al., 2011).

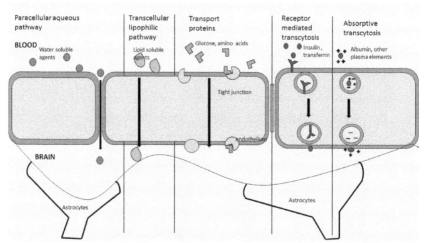

FIGURE 3.2   (See color insert.) Interaction of various kinds of drugs with the blood–brain barrier.

However, these methods are facing difficulties due to their minimal surface of absorption as well as in vivo toxicity, thus reducing the penetration possibilities of the respective drugs. The forcible opening of the BBB causes damage and immutable changes to BBB and uncontrolled passage of the drugs into the brain. The upper limit pore size of the BBB allows passage of particles of size less than 1 nm (Youns et al., 2011). However, particles of size up to several nanometers in diameter can also be passaged through carrier-mediated transport. Thus, nanocarriers are emerging as the most promising contenders for brain targeting and penetration of the BBB.

3.3 BRAIN TARGETING THROUGH NANOPARTICLES

3.3.1 *NANOPARTICLES AS NANOMEDICINE*

The general definition of nanotechnology is dealing with materials of dimensions ranging from 1 to 100 nm. Nanomedicine has recently gained much of its popularity due to ever-increasing demand of more efficacious treatments and health-care cost reductions. There is a belief that NPs are the future of medical industry.

Drug delivery is becoming the most important use of nanomaterials. About 75% share of total sales, 76% share of scientific papers on nanomedicine, and 23 clinically approved devices are based upon NP-mediated drug delivery. Biomedical nanotechnology shares in the global market are increasing enormously (Wagner et al., 2006; Zhang et al., 2007). Diagnosis and treatment of cancer is one of the most pivotal roles in the area of drug delivery. For last a few years, medical field has been focusing on treatment of brain cancers, but results are still inadequate. Most of the available treatments are restricted by the BBB, thus rendering most of them unfruitful. Therefore, significant advances in methods of treating as well as diagnosing brain cancers are necessary. Nanomedicine has thus become a huge market for brain-related ailments.

3.3.2 *PROPERTIES OF NANOPARTICLES*

Recent developments in the field of nanomedicine have provided a profitable help in therapeutics of brain cancers. These nanoscale products provide a wide range of fundamentally advanced properties, as follows:

(1) Nonimmunogenic
(2) Nontoxic, biocompatible
(3) Biodegradable
(4) Prolonged circulation time
(5) No aggregation or dissociation in blood
(6) BBB-targeted moiety (receptor absorptive uptake)
(7) Enhanced parent drug stability
(8) Can be conjugated with small molecules such as proteins, peptides, or nucleic acid

All of the given properties can be put into use for improving the present techniques of detecting, monitoring, and treating the disease at its various stages. Further, the novel interactions between these nanomaterials and their correspondingly sized physiological structures to DNA, organelles, and proteins, can also be used to provide a compliment to existing medical diagnostic/treatment strategies and to nurture the development of new and potentially more efficient approaches (Sarin et al., 2008). Among all the approaches, methods build on the ground of nanobiotechnology provide best aspects for achieving the required deal for effective drug delivery across the BBB.

3.3.3   MECHANISM FOR DRUG TRANSPORT BY NANOPARTICLES TO CROSS BLOOD–BRAIN BARRIER

The NPs provide a significant advantage over the present strategies. The NPs can be used to transport drugs across the BBB without forcibly opening it; thus, controllable drug delivery can be achieved. This helps in avoiding the risk of systemic toxicity. The transport of the NPs across the BBB depends on various factors such as type of surfactant used, size of NP, and type of drug loaded (Fig. 3.3).

There are several mechanisms through which NPs enhance the delivery of the drugs across the BBB, some of which are by opening the tight junctions between endothelial cells which allow the drug to creep across the BBB. Some NPs are transcytosed into the endothelial layers, endocytosis of the NPs by the endothelial cells, and release of the drugs. Coating agents of NPs prevent the exocytosis of the drugs by the efflux system of the BBB.

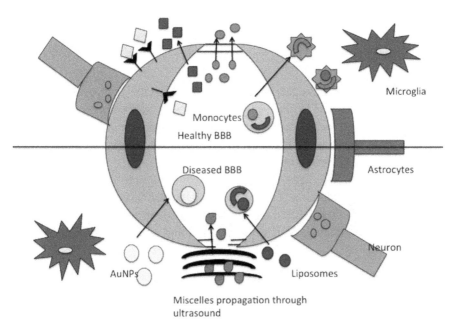

FIGURE 3.3 **(See color insert.)** Different mechanisms for drug penetration by nanoparticles across the blood–brain barrier.

3.3.4 NANOPARTICLES FOR BRAIN DRUG DELIVERY

In brain cancer researches, variety of NPs are being used, be it as an imaging agent or for drug-delivery purposes. These NPs include both solid-inorganic (magnetic Fe_3O_4 NPs, AuNPs, gadolinium NPs) and semiconductor quantum dots (QDs) and organic based (dendrimers, hydrogel, and polymers) (Koziara et al., 2005). The expected route for each of the systems is different from each other. The most commonly used NPs for brain targeting are as follows.

3.3.4.1 LIPOSOMES

Nanoliposomes size ranges from 25 to 50 nm in diameter. These liposomes are highly biocompatible and their properties vary largely upon lipid composition surface charge and size. Liposomes can be used to deliver antidrugs across the BBB. Functionalized liposomes with ApoE-derived peptides increase the cellular uptake of the drugs across the BBB (Koziara et al., 2005).

3.3.4.2 POLYMERIC NANOPARTICLES

- *Chitosan NPs*: They facilitate drug delivery as they can overcome any biological barrier. Surface-modified chitosan NPs with PEG (polyethylene glycol) have been synthesized to increase the retention time of the drug by escaping the capture by reticuloendothelial system (RES) of the BBB (Koziara et al., 2005).
- *Dendrimers*: They are novel 3D NPs having a core–shell with a wide range of applications. They are very efficient in crossing cell membranes, including BBB. Fourth-generation dendrimers have been synthesized to enhance BBB transportation and improve accumulation in glia cells.

3.3.4.3 INORGANIC NANOPARTICLES

Inorganic nanoparticles include a solid core material, which can be magnetic metallic oxide or semiconductor network. From recent surveys, it was seen that magnetic Fe_3O_4 is one of the most popular motifs for brain targeting. They can be used as contrasting agents in MRI for increasing the efficiency to detect brain tumors (Koziara et al., 2005).

3.3.4.4 QUANTUM DOTS

Many conjugated QDs have been successfully used to monitor the efficiency of the BBB transport for a number of molecules. They have very low toxicity levels, which allow them to be used as agents for treating brain cancers as well.

3.3.4.5 GOLD NANOPARTICLES

They have been recently being investigated as better drug-delivery agents as well as for cases of brain cancers. It has been found that AuNPs of size up to 50 nm easily disrupt the BBB, thus helping the therapeutics to cross the BBB. They are superior to the other existing NPs due to their easy synthesis methods, easy surface functionality, high biocompatibility, and high biodegradability.

3.4 GOLD NANOPARTICLES FOR BRAIN TARGETING

3.4.1 WHAT ARE GOLD NANOPARTICLES?

AuNPs are generally gold particles in various forms with dimensions less than a micrometer. They have been used extensively in the field of biology and photonics due to their size-based different optical properties. Recent studies and researches have led to their use in various fields, such as photovoltaics, therapeutic agents, drug delivery in medical applications, and catalysis. The optical and electronic properties can be changed according to its shape, size, aggregation state, and surface chemistry (Perrault et al., 2010).

There are different types of AuNPs available in different sizes and shapes, which are exploited for various purposes. Some of them are gold nanorods, gold bi-pyramids, gold nanorods with silver coating, "nanorice"-Fe_2O_3 nanorods coated with gold nanoshells, and gold nanocages (Perrault et al., 2010).

3.4.2 PROPERTIES OF GOLD NANOPARTICLES

3.4.2.1 PHYSICAL PROPERTIES

- Due to their small size, they have a very high surface area-to-volume ratio they prove useful for catalysis.
- They usually have anisotropic shapes.
- Due to their very high surface energies, they are often stabilized using some capping polymers, such as polyvinylpyrolidone and tannic acid, to prevent their aggregation.

3.4.2.2 OPTICAL PROPERTIES

Interactions of AuNP with light are depended on its environment and physical dimensions. Oscillating electric field of light when interacts with the electrons of colloidal AuNPs causes the electrons to oscillate with the same resonance frequency as the incident light. This phenomenon is termed as surface plasmon resonance due to the numerous plasmons created at the surface of the colloidal gold (Stuchinskaya et al., 2011).

For small AuNPs (30 nm), an absorption of light in the blue-green spectrum and reflection of red light is seen imparting them a rich red color. As the particle size increases, the absorption shifts toward longer wavelength and reflection of blue light takes place. As particle size increases to bulk, the absorption shifts toward infrared light, and most of the visible light are reflected giving a translucent and clear color.

3.4.2.3   THERMAL PROPERTIES

As we know that AuNPs have free electrons on its surface that are not bound to any atom, this property allows it to absorb different wavelengths of light. Depending upon their sizes, they absorb light waves, and at a particular wavelength, their electron cloud starts to oscillate at a resonant frequency, which gradually yields thermal energy. This property has been utilized for localized heat treatment in diseased body organ (Stuchinskaya et al., 2011).

3.4.2.4   PHARMACOKINETICS AND BIODISTRIBUTION PROPERTY

- Tumor targeting (EPR effect)
- Liver uptake (RES)
- Penetration of biological barriers (BBB)
- Renal clearance
- Metabolism

3.4.3   WHY GOLD NANOPARTICLES FOR BRAIN TARGETING?

AuNPs have efficiently been used as drug-delivery systems due to their following properties:

- Tunable sizes allow easy passage across the BBB.
- Very low toxicity levels.
- Stable attachment sites for ligands.
- Enhanced CT contrast capability helps in finding the tumor size as well as distance between tumors.
- Very high biocompatibility.

- Highly stable against oxidation and degradation.
- AuNP thermal properties provide a great alternative for radiation therapies without any kind of harmful effects to the nearby tissues.
- PEG-coated AuNPs have longer circulation time as well as help in encapsulating hydrophobic cancer drugs, thus helping a lot in cancer therapy.
- Ligand-conjugated AuNPs, which utilize cancer biomarkers over-expressed by the tumor cells, are becoming attractive for cancer diagnosis.
- Naked AuNPs have the capability of inhibiting angiogenesis, which can prove useful in brain tumor therapies as angiogenesis plays an important role in spread and growth of cancer.

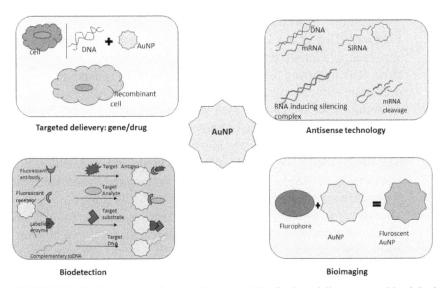

FIGURE 3.4   Befitting properties of gold nanoparticles for drug delivery across blood–brain barrier.

All these properties have led to an extensive research to utilize AuNPs for diagnosing as well as treating brain-related diseases. Recent studies have revealed that GBM cell lines show highest Tfr expressions, which results in iron accumulation in tumor-specific sites. Therefore, Tfr has become an attractive site for targeting brain tumors for drug delivery (Hainfeld et al., 2013). Tfr PEG-coated AuNPs have shown drastic improvement in drug uptake capacities of the BBB. Thus, AuNPs along with their carrier properties as well as therapeutic and diagnostic abilities have proved as a boon for brain-targeting strategies (Fig. 3.4).

3.5   SYNTHESIS OF GOLD NANOPARTICLES

In general, there are several ways to synthesize AuNPs. Some of the most contemporary methods are as follows:

    (a) *Physical methods*: The best method, which is known to synthesize AuNPs, is ϒ-irradiation method having the property of high purity and controllable size. Using this method, AuNPs of sizes ranging between 5 and 40 nm can be synthesized using alginate solution as stabilizer (Rezende et al., 2010). However, AuNPs with the size in between 2 and 7 nm are synthesized by the method of single-step ϒ-irradiation, in which bovine serum albumin protein is used as a stabilizer (Deb et al., 2011). Photochemical synthetic approach is also used to synthesize AuNPs, in which aqueous glycine solution along with hydrogen tetrachlorocuprate ($HAuCl_4$) is exposed to UV irradiation. Glycine is functionalized by capping the AuNPs by using an amino acid, which acts as a photochemical initiator (Ganeshkumar et al., 2012). Heating or photochemical reduction approaches are used to synthesize AuNPs, in which tartrate, malate, and citrate ligands are used to reduce $HAuCl_4$ (Guo et al., 2012). AuNPs with size ranging between 20 and 140 nm are synthesized by solvent-free photochemical method (Sawada and Takahashi, 2010).

        Microwave irradiation is another method to synthesize AuNPs, in which cetyltrimethyl ammonium bromide (CTAB) acts as a binding agent, whereas citric acid acts as a reducing agent, and sometimes aqueous extracts of *Cissus quadrangularis* are used as reducing or capping agent (Pérez et al., 2011). Silver and gold alloys can be used to synthesize porous AuNPs. In this method, the first step is to prepare microemulsion of NPs by using precursors such as $AgNO_3$ and $HAuCl_4$ and is then reduced by sodium borohydride ($NaBH_4$). The second step involves de-alloying by nitric acid (Kojima et al., 2010).

    (b) *Chemical methods*: Citrate thermo-reduction method is used to synthesize AuNPs having efficient surface-enhanced Raman spectroscopy in a short reaction time with the use of a low-cost reagent like inositol hexaphosphate which acts as a reduction agent for $HAuCl_4$ (Heidari et al., 2014). Partially functionalized AuNPs are synthesized by surfactant-assisted method, in which a bifunctional ligand such as hexadecyltrimethyl ammonium bromide is used which acts as a linker between AuNPs and solid substrate (Mishra et al., 2012). AuNPs synthesized by hot injection technique uses variety of surfactants such as 1-octadecanethiol, oleylamine, $AgNO_3$, and poly(N-vinyl pyrrolidone) in 1,5-pentanediol to stabilize colloidal solution (Guo et al., 2012).

AuNPs with a size of 2–20 nm is synthesize by the reduction of $HAuCl_4$ with $NaBH_4$ using isopropanol and ethanol in presence of tris(2-aminoethyl)amine (Khan et al., 2013). AuNP having a size of 6 nm is synthesized by the intramolecular reduction of $Na_3Au(SO_3)_2$ (Sawada and Takahashi, 2010). Solution phase method is also known to synthesize AuNPs with a size of 13 nm in which CTAB is used as capping agent or template to avoid agglomeration. Other method which is used to synthesize AuNPs is thermosensitive in nature, involves trisodium citrate which reduced the AuNPs later combined with chloroauric acid, and is modified with 11-mercaptoundecanoic acid by the formation of self-assembly monolayers (Mendoza et al., 2010).

(c) *Green methods*: The easy green biosynthesis method to prepare AuNPs having size of (25 + 7) nm is done by using eggshell membrane (ESM) which is natural biomaterial. In this method, ESM is completely immersed in aqueous solution of $HAuCl_4$ without the use of any reductant (Murawala et al., 2014). Sodium dehydrate and high-power ultrasounds are the green synthetic approach which is used to synthesize AuNPs of size ranging between 5 and 17 nm (Giljohann et al., 2010).

AuNPs are also synthesized by sunlight irradiation technique, in which gold salt was reduced by solar energy, and they were capped with 6-mercaptopurine and were modified by folic acid (Lan et al., 2013). AuNPs having size in the range of 15–80 nm are synthesize by other green synthetic method, in which citrus fruit juice extracts reduce the $HAuCl_4$ (Tarnawski and Ulbricht, 2011).

Another green method involves AuNP formation within the aqueous NaCl solution without the use of any external reductant and stabilizer from the bulk of gold substrate using natural chitosan (Fard and Namazi, 2011). AuNPs by another set of experiments were also synthesized by edible mushroom via sunlight exposure (Nalawade et al., 2012).

3.6 SYNTHESIS OF GOLD-NANOPARTICLE-CONJUGATED DRUG FOR BRAIN TARGETING

The following methods have been utilized to synthesize nanoconjugated drugs for effective brain targeting for diagnosis as well as therapy application:

(1) AuNPs synthesized by any of the common synthesis methods, which have already been discussed earlier in this chapter, can be conjugated with barbiturate. MDDA (12-mercaptododecanoic acid) acts as a linker, which conjugates between the barbiturate and the AuNP (Tosi et al., 2011). To every individual solution of different AuNP sizes, excessive amount of MDDA was added and then the mixture was continuously stirred for about another 4 h. After that, the solutions were centrifuged so that it can reach its higher concentration. Following this step, the activating agents 1-ethyl-3-(3-dimethyl-aminopropyl) carbodiimide, Hall, and N-hydroxysulfosuccinimide sodium salt (NHS) were added in the mixture along with the barbiturate and allowed to stir overnight (Tosi et al., 2011). This solution was again centrifuged to increase the concentrations. This leads to the formation barbiturate-conjugated AuNP. These conjugated AuNPs can easily penetrate the BBB and help in diagnosis and therapy.

(2) AuNPs of about 4 nm in diameter are synthesized by the reaction between tri-sodium citrate and gold chloride ($HAuCl_4$) by following the complete procedure described by Jana et al. (Fard and Namazi, 2011). Then, the pH of the solution is adjusted to 11 by using 0.1 M NaOH. Tri-sodium citrate is replaced by adding mercaptosuc-cinic acid (MSA), which acts as a capping ligand by establishing a strong Au–S bond (Moghimi et al., 2001). After continuous stirring (12 h), MSA-capped AuNPs are formed which is a final product. It was concentrated to 2.5 mg (gold)/mL by washing it with deionized water and ultracentrifugation. After that, pro-THP (pro-theophyl-line) is chemically introduced to MSA-capped AuNPs to form the nanoconjugate drug AuNP–pro-THP to allow in vivo drug release (Gref et al., 2000).

(3) Another method is using carbodiimide chemistry for coupling carboxyls and amines to form WGA (wheat germ agglutinin)–HRP (horseradish peroxidase)–AuNP–pro-THP nanoconjugate (Thorne et al., 1995). An amide bond is established between WGA–HRP and AuNP–pro-THP by diluting the AuNP–pro-THP solution by

deionized water and maintained at the temperature 4°C and pH 6.6, which is followed by adding WGA-HRP, *N*-hydroxysuccinimide (NHS), and EDC to establish amide bond between the MSA-capped AuNPs and WGA–HRP. The reaction is allowed to occur for 1 h and was followed by series of washing using deionized water. WGA has increasingly being used to enhance intranasal delivery of peptides to the brain through the olfactory route because of its affinity to the olfactory mucosa (Thorne et al., 1995). While the bioconjugation chemistry has been used now for a long time, specifically use of WGA as a targeting moiety exclusively in the case of phrenic motor-neurons, which are responsible for control of respiratory function, help in efficient crossing over of the BBB has been recently reported successful.

(4) Au and S both have sufficiently strong soft atom characteristics (Nuzzo et al., 1987). Thus, stabilizing polymers or biomolecules (containing an –SH (thiol) group) are used to enhance their conjugation on the gold surface. Kang et al. (2010) used this form of conjugation to attach poly(ethylene glycol) (PEG) to 30-nm AuNPs, resulting to increased particle stability in biological environments.

Stability in the physiological environments is increased since this type of cross-coupling undergoes ligand exchange. For biological applications, polymers such as PEG and oligoethylene glycol are used to lessen the ubiquitous adsorption of other bioconstructs on the surface of the particles and increase the stability of the complex in highly concentrated ionic environments, such as cell culture medium and biological fluids (Owens and Peppas, 2006). They are also not recognized by the wandering macrophages, therefore ensuring better uptake as they are not discarded by the body's immune system.

3.7 HOW DO GOLD NANOPARTICLES TARGET THE BRAIN?

Normally, transition across the BBB is carried out by carrier system, diffusion transport, and receptor-mediated endocytosis. Entrance of the AuNPs into these specialized cells shows penetration through BBB. It has been seen that various factors such as size, coating, and surface charge of AuNPs have a pivotal role in the intercellular uptake process. A group of scientists found that hydrophilic AuNPs were absorbed onto the particle surface but the hydrophobic AuNPs did not (Elia et al., 2014). These results showed that although the lipophilic nature of the BBB, hydrophilic AuNPs have greater portability to spontaneously penetrate the BBB.

AuNPs penetrate the BBB through various mechanisms such as diffusion, energy-mediated transport system, direct penetration, and endocytosis. Many studies have revealed that cationic particles have much greater tendency to cross the BBB as compared to anionic ones. So, it can be concluded that AuNP diffusion through the BBB is not predominantly by penetration. Another set of experiments showed that more specific site for transport of AuNPs can also be considered like the K^+, Na^+, and Ca^{2+} ion channels (Persidsky et al., 2006). These ion channels showed marked importance in the transport of AuNPs through the BBB giving the evidence that either the ions channels directly affect the passage of AuNPs transport or they alter the tight junctions of the BBB. This provides an alternate pathway for the AuNPs to target the brain. One more way through which the AuNPs can be targeted for BBB penetration is through Glut-1 receptor, which mediates the glucose transport (Kreuter et al., 1995). Glucose-conjugated AuNPs showed better BBB uptake than normal AuNPs. Another strategy that has been found to penetrate the BBB is through the olfactory route by using aerosolized AuNPs. As the sensory neurons are directly projected into the brain, the delivered AuNPs reach the brain crossing the BBB without much clearance by the barrier and get accumulated near the tumor site with due course of time (Barnes, 2013). AuNPs thus can prove very competent as theranostic agents for brain malignancies.

Gold's high atomic number as well as its capacity to be conjugated with MRI contrasting agents helps in multimodal imaging of the tumor tissue and its boundaries. AuNPs, which have extravasated into the tumor tissue due to their leaky vasculature, can spread free radicals, which can induce damage of the tumor cells (Kircher et al., 2012). Besides this, AuNPs can complement radiation therapy by localizing the radiation to tumor-associated endothelial cells. Thus, AuNPs along with radiation therapy can be used as a synergistic approach to treat brain malignancies such as GBM.

3.8   FUTURE ASPECTS

AuNPs have arisen as potential candidates for drug delivery as well as cancer therapy. Currently, most of its application is primarily focused on the detection of biological occurrences. However, more research is being done for making a multifaceted vector, which treat diseases like cancer. One such idea can be a nanosized scaffold which can hold a prodrug along with a targeting moiety as well as an imaging agent. This type of NPs would have multifunctional activities and prove more fruitful for clinical purposes. The future of these nanoconjugates is still questionable as the major factors like biodegradability and lowest possible systemic cytotoxicity in invivo models are still to be achieved. Clinical researches by applying basic sciences would also benefit long-term studies which are being continuously conducted to investigate the potential deleterious effects from chronic inflammation (Cho et al., 2009) associated with depreciated clear passage of the NPs, potential mutagenicity, and effects on reproductive health.

3.9 CONCLUSION

Nanotechnology has proven itself as a boon in individualistic medicine by increasing affinity of the NP for cancerous cells. Although extensive research is necessary, it has aptly given a great impact on our understanding of different aspects of biology by introducing some unique approaches to various fields of research, such as from bioanalysis to imaging. The term nanobiotechnology includes numerous interdisciplinary technologies based on the molecular-scale interaction utilizing nanoscale materials to enhance as well as affect biological components.

AuNPs have proven to be very effective to penetrate the BBB due to its various beneficial properties. They provide such provisions by which it can be used simultaneously for drug delivery as well as targeted agents and imaging agents.

This allows different drugs to be tailored for patients, making AuNPs essentially the delivery vehicles and cocktail of drugs can be added in it (Cho et al., 2009). The therapeutic efficiency of the drug would be improved to a great extent with the increasing use of AuNP targeting agents as they accumulate specifically at the tumor sites, rather at normal sites of brain tissues. Researchers are still finding many more innovative techniques to find a solution to these problems. So, there is lot more to come in our near future which will provide us with better strategies to decipher the complex BBB.

KEYWORDS

- **blood–brain barrier**
- **glioblastoma multiforme**
- **nanoparticles**
- **nanomedicines**
- **gold nanoparticle**
- **nanoconjugates**

REFERENCES

Akers, W. S.; Ferraris, S. P.; Koziara, J. M.; Mumper, R. J.; Oh, J. J. Blood Compatibility of Cetyl Alcohol/Polysorbate-Based Nanoparticles. *Pharm. Res.* **2005**, *22*, 1821–1828.

Ala, T. A.; Emory, C. R.; Frey, W. H.; Thorne, R. G., 2nd. Quantitative Analysis of the Olfactory Pathway for Drug Delivery to the Brain. *Brain Res.* **1995**, *692*, 278–282.

Almeida, L. E.; Andrade, G. R. S.; Barreto, L. S.; Costa, Jr., N. B.; Gimenez, I. F.; Rezende, T. S. Facile Preparation of Catalytically Active Gold Nanoparticles on a Thiolated Chitosan. *Mater. Lett.* **2010**, *64*, 882–884.

Alyautdin, R. N.; Ivanov, A. A.; Kharkevich, D. A.; Kreuter, J. Passage of Peptides through the Blood-Brain Barrier with Colloidal Polymer Particles (Nanoparticles). *Brain Res.* **1995**, *674*, 171–174.

Aoun, V.; Elkin, I.; Hildgen, P.; Mokhtar, M.; Rabanel, J. M. Drug-Loaded Nanocarriers: Passive Targeting and Crossing of Biological Barriers. *Curr. Med. Chem.* **2012**, *19* (19), 3070–3102.

Badiali, L.; Benassi, R.; Bondioli, L.; Fano, R. A.; Rivasi, F.; Tosi, G.; et al. Investigation on Mechanisms of Glycopeptide Nanoparticles for Drug Delivery across the Blood-Brain Barrier. *Nanomedicine (Lond.)* **2011**, *6*, 423–36.

Baethmann, A.; Schilling, L.; Unterberg, A.; Wahl, M. Mediators of Blood-Brain Barrier Dysfunction and Formation of Vasogenic Brain Edema. *J. Cereb. Blood Flow Metab.* **1988**, *8* (5), 621–634.

Banks, W. A. Characteristics of Compounds That Cross the Blood-Brain Barrier. *Neurol.* **2009**, *9*, 1471–2377.

Barnes, P. J. Theophylline. *J. Am. Respir. Crit. Care Med.* **2013**, *188*, 901–906.

Bock, A. K.; Dullaart, A.; Wagner, V.; Zweck, A. The Emerging Nanomedicine Landscape. *Nat. Biotechnol.* **2006**, *24* (10), 1211–1217.

Breimer, D. D.; de Boer, A. G. Cytokines and Blood-Brain Barrier Permeability. *Prog. Brain Res.* **1998**, *115*, 425–451.

Chakrabarti, K.; Chaudhuri, U.; Dasgupta, A. K.; Deb, S.; Lahiri, P.; Patra, H. K. Multistability in Platelets and Their Response to Gold Nanoparticles. *Nanomed.: Nanotechnol. Biol. Med.* **2011**, *7*, 376–384.

Chan, J.; Farokhzad, O.; Gu, F.; Langer, R.; Wang, A.; Zhang, L. Nanoparticles in Medicine: Therapeutic Applications and Developments. *Clin. Pharmacol. Ther.* **2007,** *83* (5), 761–769.

Chan, W. C. W.; Perrault, S. D. *Proc. Natl. Acad. Sci. U.S.A.* **2010,** *107*, 11194–11199.

Cho, M.; Cho, W. S.; Jeong, J.; et al. Acute Toxicity and Pharmacokinetics of 13-nm-sized PEG-Coated Gold Nanoparticles. *Toxicol. Appl. Pharmacol.* **2009,** *236* (1), 16–24.

Cook, M. J.; Edwards, D. R.; Moreno, M.; Russell, D. A.; Stuchinskaya, T. *Photochem. Photobiol. Sci.* **2011,** *10*, 822–831.

Costa-Martins, P.; Davies, D. C.; Papadopoulos, M. C.; Saadoun, S.; Woodrow, C. J.; et al. Occludin Expression in Microvessels of Neoplastic and Non-Neoplastic Human Brain. *Neuropathol. Appl. Neurobiol.* **2001,** *27*, 384–395.

Cucullo, L.; Hossain, M.; Janigro, D.; Oby, E.; Rapp, E.; Santaguida, S. Side by Side Comparison between Dynamic versus Static Models of Blood-Brain Barrier In Vitro: A Permeability Study. *Brain Res.* **2006,** *1109* (1), 1–13.

Daniel, W. L.; Giljohann, D. A.; Massich, M. D.; Mirkin, C. A.; Patel, P. C.; Seferos, D. S. Gold Nanoparticles for Biology and Medicine. Angew. Chem. **2010,** *49*, 3280–3294.

de la Zerda, A.; Jokerst, J. V.; Kircher, M. F.; et al. A Brain Tumor Molecular Imaging Strategy Using a New Triple-Modality MRI–Photoacoustic–Raman Nanoparticle. *Nat. Med.* **2012,** *18*, 829–834.

Dellacherie, E.; Gref, R.; Harnisch, S.; Lück, M.; Marchand, M.; Quellec, P.; et al. "Stealth" Corona-Core Nanoparticles Surface Modified by Polyethylene Glycol (PEG): Influences of the Corona (PEG Chain Length and Surface Density) and of the Core Composition on Phagocytic Uptake and Plasma Protein Adsorption. *Colloids Surf., B: Biointerfaces* **2000,** *18*, 301–313.

Dilmanian, F. A.; Hainfeld, J. F.; O'Connor, M. J.; Slatkin, D. N.; Smilowitz, H. M. Gold Nanoparticle Imaging and Radiotherapy of Brain Tumors in Mice. *Nanomedicine* **2013,** *8*, 1601–1609.

Dinesh, M. G.; Ganesh kumar, M.; Kannappan, S.; Sastry, T. P.; Sathish Kumar, M.; Suguna, L. Sun Light Mediated Synthesis of Gold Nanoparticles as Carrier for 6-Mercaptopurine: Preparation, Characterization and Toxicity Studies in Zebrafish Embryo Model. *Mater. Res. Bull.* **2012,** *47*, 2113–2119.

Dreaden, E. C. Size Matters: Gold Nanoparticles in Targeted Cancer Drug Delivery. *Ther. Deliv.* **2012,** *3* (4), 457–478.

Dubois, L. H.; Nuzzo, R. G.; Zegarski, B. R. Fundamental Studies of the Chemisorption of Organosulfur Compounds on Gold(III). Implications for Molecular Self-Assembly on Gold Surfaces. *J. Am. Chem. Soc.* **1987,** *109*, 733–740.

Efferth, T.; Hoheisel, J. D.; Youns, M. Therapeutic and Diagnostic Applications of Nanoparticles. *Curr. Drug Targets* **2011,** *12* (3), 357–365.

Eisenberg, J.; Pardridge, W. M.; Yang, J. Human Blood-Brain Barrier Insulin Receptor. J. Neurochem. **1985,** *44* (6), 1771–1778.

Elia, P.; Hazan, S.; Kolusheva, S.; Porat, Z.; Zach, R.; Zeiri, Y. Green Synthesis of Gold Nanoparticles Using Plant Extracts as Reducing Agents. *Int. J. Nanomed.* **2014,** *9*, 4007–4021.

El-Sayed, M. A.; Kang, B.; Mackey, M. A. Nuclear Targeting of Gold Nanoparticles in Cancer Cells Induces DNA Damage, Causing Cytokinesis Arrest and Apoptosis. *J. Am. Chem. Soc.* **2010,** *132* (5), 1517–1519.

Fard, A. M. P.; Namazi, H. Preparation of Gold Nanoparticles in the Presence of Citric Acid-Based Dendrimers Containing Periphery Hydroxyl Groups. *Mater. Chem. Phys.* **2011,** *129,* 189–194.

Garrovo, C.; Mandal, S.; Sousa, F.; et al. Functionalized Gold Nanoparticles: A Detailed *In Vivo* Multimodal Microscopic Brain Distribution Study. *Nanoscale* **2010,** *2* (12), 2826–2834.

Griffin, J. D.; Kim, S.; McLane, V. D.; Mendoza, K. C. In Vitro Application of Gold Nanoprobes in Live Neurons for Phenotypical Classification, Connectivity Assessment, and Electrophysiological Recording. *Brain Res.* **2010,** *1325,* 19–27.

Guo, W.; Pi, Y.; Song, H.; Sun, J.; Tang, W. Layer-by-Layer Assembled Gold Nanoparticles Modified Anode and Its Application in Microbial Fuel Cells. *Colloids Surf., A: Physicochem. Eng. Aspects* **2012,** *415,* 105–111.

Haorah, J.; Kanmogne, G. D.; Persidsky, Y.; Ramirez, S. H. Blood–Brain Barrier: Structural Components and Function under Physiologic and Pathologic Conditions. *J. Neuroimmune Pharmacol.* **2006,** *1* (3), 223–36.

Harada, A.; Kojima, C.; Kono, K.; Umeda, Y. Preparation of Near-Infrared Light Absorbing Gold Nanoparticles Using Polyethylene Glycol-Attached Dendrimers. *Colloids Surf., B: Biointerfaces* **2010,** *81,* 648–651.

He, H.; Jia, X. R.; Li, Y.; et al. PEGylated Poly(Amidoamine) Dendrimer-Based Dual-Targeting Carrier for Treating Brain Tumors. *Biomaterials* **2011,** *32* (2), 478–487.

Heidari, Z.; Salouti, M.; Sariri, R. Gold Nanorods–Bombesin Conjugate as a Potential Targeted Imaging Agent for Detection of Breast Cancer. *J. Photochem. Photobiol. B: Biol.* **2014,** *130,* 40–46.

Herradón, B.; Mann, E.; Pérez, Y. Preparation and Characterization of Gold Nanoparticles Capped by Peptide–Biphenyl Hybrids. *J. Colloid Interface Sci.* **2011,** *359,* 443–453.

Ho, C. Y.; Hsu, C. H.; Hsu, Y. B.; Lan, M. Y.; Lee, S. W.; Lin, J. C. Induction of Apoptosis by High-Dose Gold Nanoparticles in Nasopharyngeal Carcinoma Cells. *Auris Nasus Larynx* **2013,** *40,* 563–568.

Hunter, A. C.; Moghimi, S. M.; Murray, J. C. Long-Circulating and Target-Specific Nanoparticles: Theory to Practice. *Pharmacol. Rev.* **2001,** *53,* 283–318.

Kakkar, V.; Kaur, I. P.; Bhandari, R.; Bhandari, S. Potential of Solid Lipid Nanoparticles in Brain Targeting. *J. Control. Release* **2008,** *127* (2), 97–109.

Kanevsky, A. S.; Sarin, H.; Wu, H.; et al. Effective Transvascular Delivery of Nanoparticles across the Blood-Brain Tumor Barrier into Malignant Glioma Cells. *J. Transl. Med.* **2008,** *6,* 80.

Kapoor, S.; Mukherjee, T.; Nalawade, P. High-Yield Synthesis of Multispiked Gold Nanoparticles: Characterization and Catalytic Reactions. *Colloids Surf. A: Physicochem. Eng. Aspects* **2012,** *396,* 336–340.

Khan, M. S.; Siddaramaiah, H.; Vishakante, G. D. Gold Nanoparticles: A Paradigm Shift in Biomedical Applications. *Adv. Colloid Interface Sci.* **2013,** *199–200,* 44–58.

Mishra, A.; Tripathy, S. K.; Yun, S. I. Fungus Mediated Synthesis of Gold Nanoparticles and Their Conjugation with Genomic DNA Isolated from *Escherichia coli* and *Staphylococcus aureus*. *Process Biochem.* **2012,** *47,* 701–711.

Murawala, P.; Prasad, B. L. V.; Shiras, A.; Tirmale, A. In Situ Synthesized BSA Capped Gold Nanoparticles: Effective Carrier of Anticancer Drug Methotrexate to MCF-7 Breast Cancer Cells. *Mater. Sci. Eng. C* **2014,** *34,* 158–167.

Owens, I. D. E.; Peppas, N. A. Opsonization, Biodistribution, and Pharmacokinetics of Polymeric Nanoparticles. *Int. J. Pharm.* **2006,** *307* (1), 93–102.

Sawada, H.; Takahashi, K. Facile Preparation of Gold Nanoparticles through Autoreduction of Gold Ions in the Presence of Fluoroalkyl End-Capped Cooligomeric Aggregates: LCST-Triggered Sol–Gel Switching Behavior of Novel Thermo Responsive Fluoroalkyl End Capped Co-Oligomeric Nanocomposite-Encapsulated Gold Nanoparticles. *J. Colloid Interface Sci.* **2010,** *351,* 166–170.

Tarnawski, R.; Ulbricht, M. Amphiphilic Gold Nanoparticles: Synthesis, Characterization and Adsorption to PEGylated Polymer Surfaces. *Colloids Surf. A: Physicochem. Eng. Aspects* **2011,** *374,* 13–21.

Whitesides, G. M. The 'Right' Size in Nanobiotechnology. *Nat. Biotechnol.* **2003,** *21,* 1161–1165.

CHAPTER 4

NANOCARRIERS FOR BRAIN TARGETING: AN OVERVIEW

B. A. ADERIBIGBE[1,*], I. A. ADERIBIGBE[2], and A. P. I. POPOOLA[2]

[1]Department of Chemistry, University of Fort Hare, Alice Campus, Eastern Cape, South Africa

[2]Department of Chemical and Metallurgical Engineering, Tshwane University of Technology, Pretoria, South Africa

*Corresponding author. E-mail: blessingaderibigbe@gmail.com

ABSTRACT

The delivery of a medicinal agent to the brain in order to diagnose, treat, and prevent brain disease is hindered by the blood–brain barrier (BBB). Several methods of administration of bioactive agents to the brain have been developed over the years. However, they have been reported to be unsafe, expensive, and ineffective. Nano-based drug delivery systems that can permeate the BBB have been reported to be effective, safe with reduced side effects, and are suggested to be potential carriers for the delivery of bioactive to the brain. The physicochemical properties of the carriers play a crucial role in the brain uptake of the carriers and their interaction with the brain cells. This chapter provides a detailed information of an overview of nano-based carriers reported for the delivery of drugs to the brain resulting in enhanced therapeutic effects, their mechanisms, and biological efficacy (in vitro and in vivo). Some of the carriers this chapter will be focused on are inorganic nanoparticles, lipid nanoparticles, polymer-based nanoparticles, nanogels, nanoliposomes, nanoemulsions, quantum dots, micelles, dendrimers, exosomes, and polymersome.

4.1   INTRODUCTION

Diseases affecting the brain and the associated death rate are very significant. According to WHO, 50 million people are affected by brain diseases
worldwide (Mental Health). *Alzheimer's and dementia deaths amongst ages
55–74 years old was reported to be high in the* United States *when compared
to other countries between 2005 and 2007.* In a report by Pritchard et al.
(2013), *Alzheimer's and dementia deaths were estimated to be 186 men and
187 women rate per millions* in the United States. In other developed countries such as the United Kingdom, the estimated deaths have been reported to
be 6500 in which 32% are men and 48% are women (Pritchard et al., 2013).
In England, cases of brain diseases have reached 12.5 million with 1.3
million hospital admissions (Health and Social Care Information Centre).
Factors that contribute to brain diseases are environmental, aging, harmful
chemicals, stress, poor nutrition, genetics, etc. (Neurological Diseases and
Disorders Health Impacts of Climate Change). The impact of brain diseases
on an individual include mobility (Pearson et al., 2004), behavior (Levenson
et al., 2014; Marvel et al., 2004), the function of the bladder and bowel
(Dourado et al., 2012; Panicker et al., 2015), and communication (Miller et
al., 2012). Some of the diseases that affect the brain are tumor, Parkinson's
disease, schizophrenia, multiple sclerosis, Alzheimer's disease, epilepsy,
cancer, stroke, etc.

The administration of bioactive agents in order to diagnose, treat, and
prevent brain diseases is hampered by multicellular vascular structures
known as blood–brain barrier (BBB), which controls the passage of oxygen,
bioactive agents, molecules, and nutrients passage thereby protecting the
brain and maintaining a suitable environment for proper functioning of the
neurons (Obermeier et al., 2013). However, BBB is an obstacle to drug
delivery to the central nervous system. This chapter provides a detailed overview of nano-based carriers reported for the delivery of drugs to the brain,
their mechanisms, and biological efficacy (in vitro and in vivo).

4.2   A BRIEF ANATOMY OF THE BRAIN

The central nervous system is composed of the spinal cord and the brain
(Fig. 4.1) (Rughani, 2015). The brain is composed of the cerebrum, the
brainstem, and the cerebellum (Rughani, 2015). At the lower part of the
brain is the brainstem and the cerebellum lies posterior to the brainstem.
The brainstem extends from the cervical spinal cord to the diencephalon

of the cerebrum, the largest part of the brain (Rughani, 2015). The cerebrum is responsible for the decisions, behavior, perception, vision, emotion, speech, and memory (Anatomy of the brain). The brainstem is responsible for involuntary actions, such as blood pressure, digestion, hormone regulation, breathing, heartbeat, etc. The cerebellum is responsible for coordination, movement, and balance (Anatomy of the Brain). The brain is made of two types of nerves, namely, neuron and glia cells (Neurons: The Building Blocks of the Nervous System). The neurons are of varied shapes and sizes. They are composed of axon, cell body, and dendrites. They transmit information via signals that are chemical and electrical across a tiny gap known as a synapse. The dendrites act as antennae by receiving messages from the nerve cells that are transmitted to the cell body, which then accesses the message to determine its suitability for further transmission (Neurons: The Building Blocks of the Nervous System). Important messages are transmitted to the end of the axon, which contains neurotransmitters that open into the synapse. Glia provides the neurons with structural support, protection, and nourishment (Neurons: The Building Blocks of the Nervous System). Some of the glia cells are oligodendroglia, astroglia, ependymal, and microglia that act as an insulator to the neurons, transport nutrients to the neurons, secrete cerebrospinal fluid, digest dead neurons, respectively (Neurons: The Building Blocks of the Nervous System).

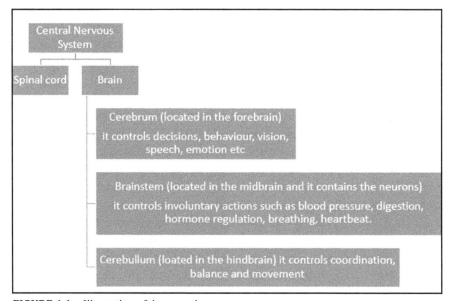

FIGURE 4.1   Illustration of the central nervous system.

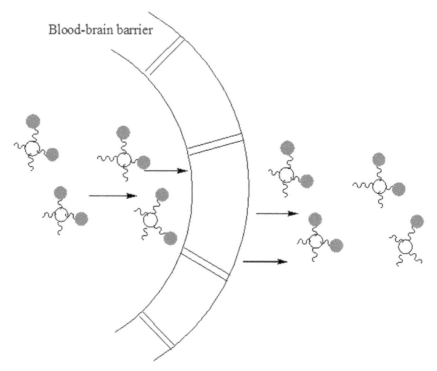

FIGURE 4.2   (See color insert.) Mechanism of nanocarriers via blood–brain barrier.

The brain is composed of BBB, a vascular system made up of capillary endothelial cells found on the cerebral microvessels, and perivascular tissues (Abbott et al., 2006; Demeule et al., 2008; Serlin et al., 2015; Shilo et al., 2014). It contains junctions adjacent to the endothelial cells. It removes waste products, allows diffusion of gases and molecules with a molecular weight between 400 and 500 Da, and maintains the ionic and fluid balance thereby protecting the brain (Abbott et al., 2006; Serlin et al., 2015). Nano-sized molecules penetrate the BBB via adsorptive-mediated transcytosis and receptor-mediated transcytosis (Demeule et al., 2008; Serlin et al., 2015) (Fig. 4.2). The incorporation of targeting moieties into nanocarriers is also useful for brain targeting moieties that are recognized by receptors (Demeule et al., 2008; Serlin et al., 2015). Nanosized molecules have been reported to be potential therapeutics for the treatment of brain diseases. Some of the brain diseases are Alzheimer's disease, Parkinson's disease, brain tumor, multiple sclerosis, epilepsy, cancer, stroke, schizophrenia, etc. (Fig. 4.3).

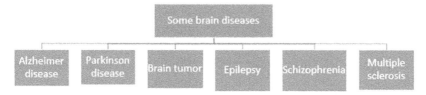

FIGURE 4.3   Brain diseases.

4.2.1   SOME BRAIN DISEASES

4.2.1.1   ALZHEIMER'S DISEASE

Alzheimer's disease is a neurodegenerative disorder that is incurable, progressive, and the cause of dementia. The disease progress by the formation of amyloid beta, a peptide that results in the formation of amyloid plaques on the outside surface of neurons of the brain resulting in the killing of neuron (Rukavina, 2004). The disease is caused by factors that are genetic and environmental. Mutations of *amyloid protein precursor* (*APP*), *presenilin* (*PS*)-*1*, and *PS-2* have been reported to be among the causes of few cases of the disease (Huang and Mucke, 2012; Bertram et al., 2012; Sanchez et al., 2012). Mutations that affect the processing of APP can alter the production Aβ peptides (Bertram et al., 2012; Huang and Mucke, 2012; Sanchez et al., 2012). Alzheimer's disease can be mild, moderate, and severe. The severe form is characterized by weight loss, lack of bladder and bowel control, seizures, infections, and increased sleeping (Lakhan, 2017). The decreased availability of oxygen can affect the function of neurons resulting in cell death (Kvq et al., 2013). Head injury has been reported to be a factor that can cause Alzheimer's disease (Fleminger et al., 2003). The deficiency of vitamins such as vitamin D (Evatt et al., 2008; Kvq et al., 2013) and *thiamine* (Molina et al., 2002; Kvq et al., 2013) increased concentration of copper and homocysteine levels result in damage to the neurons and constitute risk factor pathways to Alzheimer's disease (Huang et al., 2012; Sudduth et al., 2013).

4.2.1.2   BRAIN TUMOR

Brain tumors are classified as primary tumors that occur within the brain and metastatic that spreads into the brain from outside the central nervous system (Cerna et al., 2016). Brain tumor is life threatening and causes blockage of cerebrospinal fluid resulting in an increase in intracranial pressure and swelling (brain tumors). Majority of brain tumors originate from glial cells and are classified as grade I (pilocytic astrocytoma), grade II (diffuse astrocytoma), grade III (anaplastic astrocytoma), and grade IV (glioblastoma multiforme) (Cerna et al., 2016). Brain tumors, such as glioblastoma are life threatening due to their invasive nature and resistance to treatment (Cerna et al., 2016; Liu and Zong, 2012).

4.2.1.3   PARKINSON'S DISEASE

Parkinson's disease is identified by the presence of Lewy bodies, neuritis, and loss of pigmented dopaminergic neurons in the substantia nigra pars compacta (Hauser, 2017). The disease is characterized by resting tremor, rigidity, and bradykinesia (Hauser, 2017). The disease is caused by genetic and environmental factors. The environmental factors include exposure to weedicides, pesticides, industrial activity, smoking, and well water (Hauser, 2017; Liu et al., 2012; Pezzoli and Cereda, 2013; Wirdefeldt et al., 2011). Other factors have also been identified to be a cause of Parkinson's disease such as head injury (Fang et al., 2012; Kvq et al., 2013), vitamin D deficiency (Evatt et al., 2011; Kvq et al., 2013), and homocysteine (Shin et al., 2012; Kvq et al., 2013).

4.2.1.4   MULTIPLE SCLEROSIS

Multiple sclerosis is an inflammatory disease that is autoimmune. It affects the myelinated axons in the central nervous system and is characterized by the destruction of the myelin and axon (Luzzio 2017). It is believed to be caused by a combination of viral infections, vitamin D deficiency, and several genes play a crucial role (Luzzio, 2017).

4.2.1.5   EPILEPSY

Epilepsy is a brain disorder that is chronic and noninfectious. Symptoms associated with epilepsy include recurrent seizures that result in loss of consciousness and loss of control of bladder or bowel function (WHO, 2017; Yuan et al., 2013). Seizures are caused by electrical discharges that are excessive in the brain cells resulting from the abnormality of sodium–potassium

ATPase pump (WHO, 2017; Scharfman 2007). Seizure varies in frequency and causes social stigma in many countries thereby affecting the quality of life of the patients and their families (Thomas and Nair, 2011; WHO, 2017). It is caused by various factors such as head injury, brain tumor, brain infections, genetic conditions, etc. (WHO, 2017).

4.2.1.6 *SCHIZOPHRENIA*

Schizophrenia is a genetic brain disorder that involves episodes of psychosis and altered brain function (Frackenburg, 2017; Hannon et al., 2016). A new study revealed that the disease is linked to pruning away of the parts of the brain (Sekar et al., 2016). According to WHO, 21 million people are living with the disease (Schizophrenia). The disease is treatable.

4.3 DRUG ADMINISTRATION ROUTES TO THE BRAIN

There are different routes through which the drugs are administered for the treatment of brain diseases (Fig. 4.4). Some of the routes are intra-arterial, intranasal, intracerebral, intracerebroventricular, intrathecal, and intraventricular.

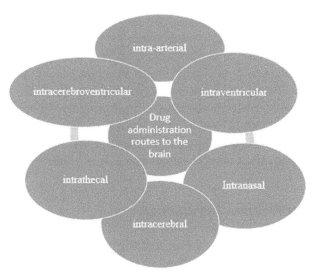

FIGURE 4.4 Drug administration routes to the brain.

4.3.1 INTRANASAL ADMINISTRATION OF DRUGS

In intranasal administration of bioactive agents, the agents are transported from the nasal cavity to the brain. It is used to deliver drugs that cannot be administered orally. The drugs delivered via this route are lipophilic molecules having a molecular weight that less than 600 Da (Illum, 2000; Van Woensel et al., 2013). Intranasal administration of drugs offers several advantages such as protection from systemic circulation, reduced systemic side effects, patient compliance, noninvasive with rapid onset of action (Van Woensel et al., 2013). Therapeutics administered intranasally reaches the central nervous system via the olfactory and trigeminal neural pathways (Hanson et al., 2008). The class of bioactive agents that are delivered intranasally are exendin (Hanson et al., 2008; Zhang 2016), genes (Aly et al., 2015; Hanson et al., 2008), polynucleotides (Han et al., 2007; Hanson et al., 2008), cytokines (Hanson et al., 2008; Ma et al., 2008), neuropeptides (Spetter et al., 2015), neutrophins (Chen et al., 1994), erythropoietin (Fletcher, 2009), chemotherapeutics (League-Pascual et al., 2017), and carbamazepine (Serralheiro et al., 2014).

4.3.2 INTRA-ARTERIAL

The pharmacokinetics of drugs delivered intra-arterially is very complex. Some therapeutics are delivered intra-arterially for the treatment of brain diseases. However, treating brain diseases by intra-arterial administration is restricted due to lack of understanding of the pharmacokinetics, lack of convenient high-speed drug concentration measurement methods, and poor rationalization of drug injection protocols (Ellis et al., 2015; Joshi et al., 2014). The aforementioned factors can result in treatment failure and adverse side effects. This route is used to treat diseases such as cerebral vasopasm (Mikeladze et al., 2012), brain cancer (Gobin et al., 2001), stroke (Abou-Chebl, 2011), and raised intracranial pressure (Kim, 2009; Yokota, 1993).

4.3.3 INTRACEREBROVENTRICULAR

Intracerebroventricular route of drug administration involves delivery of the drug into the cerebral ventricles through an implanted device on the scalp. This route is employed for treatment of diseases such as cancer (Meijer et al., 2009), Alzheimer's disease (Eriksdotter et al., 1998), meningitis (Li et al., 2007), etc. However, intracerebroventricular route of drug administration usually results in increased intracranial pressure (Cohen-Pfeffer et al., 2017; Li et al., 2007; Patel et al., 2009). It is useful for long-term administration of drugs (Cohen-Pfeffer et al., 2017).

4.4   CARRIERS FOR THE DELIVERY OF BIOACTIVE AGENTS TO THE BRAIN

Different types of carriers have been evaluated as a potential delivery system for transport of therapeutic agents to the brain for the treatment of neurological disorder. The uptake capacity of the carriers by the brain is dependent on the physicochemical properties, type of material used, and design of the carriers. Several types of carriers used for delivery of drugs to the brain include inorganic nanoparticles (magnetic), lipid nanoparticles, nanoemulsions, polymer-based nanoparticles, nanogels, dendrimers, nanoliposomes, quantum dots, exosomes and polymersome, nanocapsules, micelles, and nanospheres (Fig. 4.5).

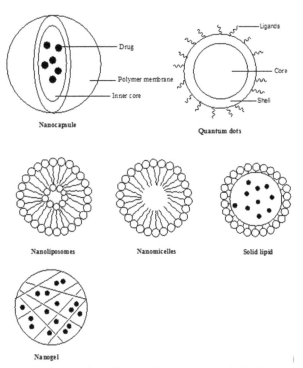

FIGURE 4.5   Nanocarriers for the delivery of therapeutics to the brain.

4.4.1 NANOGELS

Nanogels is a three-dimensional nanosized network formed by chemical or physical crosslink of polymers. Nanogels are biodegradable, biocompatible, and can be manipulated based on the nature of the application, exhibit controlled, and targeted drug release mechanism (Neamtu et al., 2017; Raemdonck et al., 2009; Sultana et al., 2013). Loading of therapeutic agents onto nanogels is by physical loading, covalent, and noncovalent interactions (Raemdonck et al., 2009). The mesh size of the nanogels influences drug diffusion from the network after swelling of the nanogels (Neamtu et al., 2017; Raemdonck et al., 2009; Sultana et al., 2013). In some cases, the drugs are entrapped covalently or noncovalently onto the nanogel network (Raemdonck et al., 2009), and the release is dependent on the swelling of the nanogel network. However, the swelling and drug release behavior of nanogels is dependent on factors such as pH, temperature, pressure, electricity, and selected molecular recognition (Kikuchi et al., 2002; Neamtu et al., 2017). Nanogels are soft and this influences their biodistribution and circulation time (Zhang et al., 2012). Nanogels size can be modified thereby enhancing their cellular interaction and uptake and enhances the permeability and retention effect (Ding et al., 2011; Gao et al., 2014; Molina et al., 2015). However, despite the unique good features of nanogels they are limited by their poor specificity of the target cell/tissue. Therefore, they require incorporation of targeting moieties in order to improve their targeted drug (Eckmann et al., 2014).

Nanogels have been employed and evaluated as potential carriers for delivery of therapeutic agents for the treatment of brain diseases (Table 4.1). Nanogels have been designed and injected into the brain to activate growth of neurons, which is suitable because damaged brain cells do not regenerate (Tamariz et al., 2011). The biocompatible nature of nanogels makes them a better option when compared to the application of stem cells, which are easily rejected by the brain immune system. Their ability to activate neurons makes them suitable for the treatment of Alzheimer's and Parkinson diseases (Hanson et al., 2016). Azadi et al. (2013) developed chitosan-based nanogels loaded with methotrexate for delivery to the brain. The nanogels were prepared by ionic gelation followed by modification of the surface with polysorbate 80. The cellular uptake of the formulations was high when compared to the free drug. However, the cellular uptake of the surface modified drug-loaded nanogels was not enhanced when compared to the unmodified drug loaded nanogels in vivo. The nanogels hydrophilic property, positive zeta potential, and the transcytosis delivery mechanisms of the drug-loaded nanogels enhanced the delivery across BBB (Azadi et al., 2013). Epsilon-polylysine-based nanogels have been designed to deliver nucleoside reverse transcriptase inhibitors to viral reservoir in the central nervous system with reduced toxicity (Warren et al., 2015). The nanogels were loaded with nucleoside reverse transcriptase inhibitors, zidovudine, lamivudine, and abacavir. Brain-targeting peptides were used resulting in enhanced accumulation in the central nervous system in vivo when compared to the free drug (Warren et al., 2015). Nanogels have been employed for the treatment of encephalitis caused by a viral infection (Madhav et al. 2015). Cefuroxime-loaded nanogel was prepared from a polymer, *Cucumis sativus*, by double emulsion solvent diffusion method. The formulation was stable and useful for administration intranasally (Madhav et al., 2015). Nanogels have been evaluated for the treatment of aggressive brain tumor. The prepared nanogels were loaded with cisplatin, a chemotherapeutic agent, conjugated with monoclonal antibodies, and brain targeting moeities (Nukolova, 2012). In vivo evaluation on tumor-bearing rats revealed enhanced reduction of tumor burden with reduced systemic and neurotoxicity when compared to the free drug. Nanogels with pullulan containing cholesteryl moieties were prepared and their interaction with beta-amyloids was reported (Boridy et al., 2009). Neutral and positively charged nanogels interacted with the beta-amyloids. The positively charged nanogels were toxic in vitro. Their interaction suggested that they are potential systems for antibody immunotherapy in Alzheimer's disease (Boridy et al., 2009). Beta-cyclodextrin-based nanogel prepared by Michael addition reaction have been reported to increase payload of hydrophobic drugs and enhanced transportation of the nanogels across the BBB in vitro (Wu, 2008).

TABLE 4.1 Nanogels for Treatment of Brain Diseases.

Polymer used	Drug encapsulated	Therapeutic outcomes	References
Polyethylene glycol	Semaphorin 3A	Enhanced axon outgrowth of dopaminergic neurons	Tamariz et al. (2011)
Chitosan	Methotrexate	Targeted drug delivery of the drug into the brain	Azadi et al. (2013)
Epsilon-polylysine	Zidovudine, lamivudine, and abacavir	Reduced neurotoxicity	Warren et al. (2015)
Cucumis sativus	Cefuroxime	Good stability and potential therapeutic for treatment of encephalitis	Madhav et al. (2015)
Dextrose	Cisplatin	Inhibited tumor growth	Nukolova (2012)

4.4.2   EXOSOMES

Exosomes are membrane secreted lipid vesicles that consist of an aqueous core and a lipid membrane with particle sizes between 40 and 100 nm (Kooijmans et al., 2012; Yang et al., 2015). They can be loaded with drugs that are either lipophilic or hydrophilic. They are more stable than synthetic-polymer-based nanoparticle because of their composition and they originate from the organism (Batrakova et al., 2016; Lakhal et al., 2011; Yang et al., 2015). The aqueous core is surrounded by lipid bilayer, and the surface is composed of proteins on the phospholipid bilayer (van den Boorn et al., 2013). They can be used to deliver small drug molecules, they have broad distribution with long circulation time, they can penetrate and cross physiological barrier and can reach target cell/tissue, and they have good selectivity and excellent safety profile (Yang et al., 2015). Exosomes obtained from brain endothelial cells were loaded with doxorubicin and paclitaxel and evaluated as potential therapeutics for treatment brain cancer in a zebrafish model (Yang et al., 2015). The intracellular uptake of the anticancer drugs enhanced significantly in vitro on human glioblastoma-astrocytoma U-87 MG cells. In vivo study on zebrafish further revealed that exosome transportation across BBB was high compared to the free drug (Yang et al., 2015). Exosomes loaded with RNA and proteins have been studied for the treatment of chronic neurodegenerative disorders (Alvarez-Erviti et al., 2011). In another study, exosomes were encapsulated with curcumin, a signal transducer and activator of transcription 3 inhibitor (Zhuang et al., 2011). *In vivo evaluation* was performed on lipopolysaccharide-induced brain inflammation model, experimental autoimmune encephalitis, and GL26 brain tumor model. The mice treated via intranasal administration of the formulation exhibited reduced brain inflammation. The delivery of exosome formulation to the microglia cells of the brain was selective, rapidly resulting in induced apoptosis of microglial cells. The transportation of the formulation was dependent on the particle sizes (Zhuang et al., 2011). No toxic side effects were visible. Exosome have also been reported to degrade β-amyloid (Lai et al., 2010; Yuyama et al., 2015). However, factors that must be addressed before exosomes can reach clinical use include the need to understand and select the right exosome donor cell, determine the right drug loading technique, and the selection of targeting peptides for the surface of exosome is crucial (Yang et al., 2015).

4.4.3   NANOCAPSULES

Nanocapsules are nanocarriers that are composed of an outer protective shell and an inner core where therapeutic agents are encapsulated (Bulloj et al., 2010). They are suitable for controlled and targeted drug release mechanism (Alvarez-Erviti et al., 2011). They are prepared by emulsion-coacervation, double emulsification, emulsion-diffusion, polymer-coating, layer-by-layer, and nanoprecipitation (Radtchenko et al., 2002; Mora-Huertas et al., 2010). Rodrigues et al. (2016) reported nanocapsules composed of poly(epsilon-caprolactone) and encapsulated with indomethacin. In vivo evaluation revealed that after administration intravenously on the animal model resulted in reduced glioblastoma (Table 4.2) (Rodrigues et al., 2016). Nanocapsule have been prepared from sasanquasaponin and natural phaeophorbide, as a photosensitizer with neroprotective effects in vivo (Ye et al., 2014). The formulation was able to pass through brain–blood barrier. Nanocapsules have been encapsulated with vitamin D3 (Ślusarczyk et al., 2016). The polymeric shell was composed of Poly(l-lysine hydrobromide)), Poly(l-glutamic acid) grafted, and polyethylene glycol. In vitro evaluation in hippocampal organotypic cultures revealed neuroprotective action of vitamin D3, which was characterized by an increase in suppressed release of nitric oxide (Ślusarczyk et al., 2016). Ghosh et al. (2013) prepared nanocapsules encapsulated with quercetin. The drug encapsulated nanocapsules exhibited protection against oxidative damage, which was evident by the loss of pyramidal neurons from the hippocampal CA1 and CA3 subfields in ischemia-reperfusion-induced young and aged rats (Ghosh et al., 2013). Bernardi et al. (2012) reported protective effect of nanocapsules against cell damage and neuroinflammation induced by amyloid beta (Aβ)1-42 in Alzheimer's disease models. The nanocapsules induced Aβ cell death and inhibited neuroinflammation caused by Aβ1-42 in vitro. Intracerebroventricular administration of the formulation resulted in reduction in synaptophysin levels and suppressed the activation of microglial and glial (Bernardi et al., 2012). Meloxicam-loaded nanocapsules have been reported to treat memory impairment induced by amyloid β-peptide in vivo (Ianiski et al., 2016). Carradori et al. (2016) prepared nanocapsules with adsorbed NFL-TBS.40-63 for selective delivery of bioactive molecules to brain neural stem cells. The uptake of the formulation by the neural stem cells from the brain was improved (Carradori et al., 2016).

4.4.4 NANOSPHERES

Nanospheres are the spherically shaped particles. Their sizes are in the range of 10–200 nm. The drug is encapsulated onto the polymer matrix. Their physicochemical properties influence how the drug is absorbed, circulated, and eliminated. They offers several advantages such as targeted drug release mechanisms, they can penetrate cells/tissue due to their sizes, exhibit extended circulation in the bloodstream, they can be administered orally, nasally, and parenterally (Mamo et al., 2015; Singh et al., 2010). However, their biomedical applications is hindered by susceptibility to particle aggregation resulting from their small size, reduced drug loading, and difficulty in handling them (Singh et al., 2010). There are several reports on their potential application for delivery of drugs to the brain (Table 4.2). Cai et al. (2016) prepared poly(vinylalcohol) modified with lactoferrin moiety for targeted brain delivery. Herran et al. (2015) prepared poly(lactic co-glycolic acid)-based nanospheres encapsulated with vascular endothelial growth factor. The nanocapsules were prepared by double emulsion solvent evaporation. Implantation of the nanospheres onto the cerebral cortex of APP/Ps1 mice revealed cellular proliferation (Herran et al., 2015).

Cerebrolsyin delivery using titanate nanospheres in an animal model of Parkinson disease was evaluated by Sharma et al. (2016). The administration of the formulation resulted in high level of dopamine, its metabolites and homovanillic acid suggesting good neuroprotective effects when compared to the free cerebrolsyin (Sharma et al., 2016). Al-Dhubiab et al. (2016) developed buccal film loaded with selegiline nanospheres. The nanospheres were prepared from poly(lactide-co-glycolide), and the buccal films were prepared from varied composition of hydroxypropyl methylcellulose and eudragit. In vivo studies on rabbits after oral administration resulted in increased time of absorption with enhanced absorption of the formulation into the systemic circulation than the free drug (Al-Dhubiab et al., 2016). Cannava et al. (2016) reported PLGA nanospheres for delivery of methylene blue. *In vitro* evaluation on human neuroblastoma SH-SY5Y cells revealed that the nanospheres had no effect on the cell viability. However, the nanospheres exhibited good neuroprotection against the metabolic effects of iodoacetic acid (Cannava et al., 2016). Arshad et al. (2014) encapsulated carboplatin onto PLGA-based nanospheres. In vivo studies showed that the formulations cytotoxic effect on the tumor was significant with reduced neuronal toxicity and prolonged tissue half-life when compared to the free drug (Arshad et al., 2014). Karatas et al. (2009) prepared chitosan-based nanospheres loaded with a caspase-3 inhibitor. Intravenous administration of the formulation in mice resulted in decreased infarct volume and neurological deficit (Karatas et al., 2009).

TABLE 4.2   Nanocapsules and Nanospheres Application for the Delivery of Drug to the Brain.

Polymer used	Drug encapsulated	Therapeutic outcomes	References
Polyethylene glycol	Semaphorin 3A	Enhanced axon outgrowth of dopaminergic neurons	Tamariz et al. (2011)
Chitosan	Methotrexate	Targeted drug delivery of the drug into the brain	Azadi et al. (2013)
Epsilon-polylysine	Zidovudine, Lamivudine, and Abacavir	Reduced neurotoxicity	Warren et al. (2015)
Cucumis sativus	Cefuroxime	Good stability and potential therapeutic for treatment of encephalitis	Madhav et al. (2015)
Dextrose	Cisplatin	Inhibited tumor growth	Nukolova (2012)

4.4.5   POLYMERSOMES

Polymersomes are drug delivery systems, which are hollow spheres with a core surrounded by bilayer membrane containing aqueous solution (Lee et al., 2012; Massignani et al., 2010; Meng and Zhong 2011). The aqueous core is used for the encapsulation of therapeutic agents (Lee et al., 2012; Massignani et al., 2010; Meng and Zhong 2011). In the applications of polymersomes for targeted drug delivery, drug-selected targeting moieties are incorporated in order to enhance their cellular interaction (Lee et al., 2012). Some of the selected targeting moieties employed are peptides, antibodies, antibody fragments, etc. (Lee et al., 2011, 2012; Lin et al., 2006). These targeting moieties enhance drug release ability of polymersomes at target cell/tissue by external stimuli thereby reducing toxic drug side effects and increasing the therapeutic efficacy (Lee et al., 2012). They also exhibit extended blood circulation time that is influenced by the presence of hydrophilic surface layers (Awasthi et al., 2004). Drug release from polymersomes is via the membrane by diffusion, and the rate of drug release is influenced by the particle distribution, composition of the membrane, and properties of polymers used, for example, polymers that are responsive to external stimuli such as light, magnetic field, pH, etc. (Lee et al., 2012).

Polymersomes have been studied for therapeutic delivery to the brain. Jia et al. (2016) developed polymersomes for dual drug targeting functionalized with transferrin and Tet-1 peptide and encapsulated with curcumin into the brain. Functionalizing the polymersomes with transferrin and Tet-1 peptide provided neuroprotection and enhanced targeted drug delivery to the brain (Jia et al., 2016). In vivo studies revealed significant cellular uptake of the formulation by the brain capillary endothelial cells. The formulation improved cognitive dysfunction in intrahippocampal amyloid-β_{1-42}-injected mice showing that polymersomes are suitable as noninvasive drug delivery systems for the brain (Jia et al., 2016). Pang et al. (2012) prepared polymersomes with conjugated analogue of cationic bovine serum albumin on the surface. The formulation was prepared from poly(ethylene glycol)-poly(ε-caprolactone) and was biodegradable, nontoxic, extended blood circulation time, and penetrated BBB significantly.

The aforementioned factors were influenced by the conjugation of cationic bovine serum albumin (Pang et al., 2012). Yu et al. (2012) developed polymersomes conjugated with varied densities of lactoferrin loaded with 6-coumarin and S14G-humanin, neuroprotective peptide. The permeability effect of the formulation through the BBB was significant. However, the brain uptake of the formulation was not time-dependent as a result of receptor saturation suggesting that selected range of lactoferrin densities influenced the brain uptake of the formulation (Yu et al., 2012). Pang et al. (2010) reported that polymersomes conjugated with lactoferrin or transferrin. The uptake of the formulation was by endocytosis and dependent on factors such as temperature, time, and concentration. In vivo studies in mice further revealed that the brain uptake of transferrin conjugates polymersomes was higher than lactoferrin conjugates polymersomes (Pang et al., 201). Georgieva et al. (2012) prepared polymersomes with dodecamer peptide for cellular interaction and targeted delivery at gangliosides GM1 and GT1b. Pang et al. (2011) prepared transferrin-conjugated biodegradable polymersomes, encapsulated with doxorubicin by nanoprecipitation method. In vitro evaluation on C6 glioma cells revealed high intracellular delivery and cytotoxic effects when compared to the free drug. In vivo evaluation on glioma rats suggested increased delivery of doxorubicin to the brain tumor cells and a significant reduction of tumor volume when compared to the free drug (Pang et al., 2011).

4.4.6 MICELLES

Polymeric micelles are composed of an inner core surrounded by a shell. The inner core is hydrophobic, which is used to encapsulate therapeutic agents. The outer shell is hydrophilic, which protects the therapeutic agent interaction with the environment and enhances stability (Kedar et al., 2010; Yokoyama et al., 1996). Micelles passive targeting mechanism is by enhanced permeability and retention effect. They are biodegradable, exhibit high loading capacity, enhanced water solubility of the encapsulated bioactive agents, and extended blood circulation time (Kedar et al., 2010; Yokoyama et al., 1996). Selected moieties can be incorporated onto the surface for active targeting (Savic et al., 2006; Shuai et al., 2004). Uptake of micelles into the cells is by fluid-state endocytosis (Kedar et al., 2010; Savic et al., 2006; Shuai et al., 2004).

Micelles have been reported to cross the BBB (Table 4.3). Xie et al. (2012) prepared micelles from chitosan modified with stearic acid. The micelles were encapsulated with doxorubicin (Xie et al., 2012). In vitro evaluation on glioma C6 cells showed that the drug-loaded micelles release profile was slow over a period of 48 h. The cytotoxic effect of DOX-loaded micelles was enhanced. In vivo studies further revealed that the micelles transportation of the drug across the BBB was rapid after administration (Xie et al., 2012). The modification of the micelles with stearic acid enhanced their cellular uptake. Li et al. (2015) prepared micelles by incorporating cyclic Arginine-GlycineAspartic acidDTyrosineLysine onto polylactic acidpolyethylene glycol. The micelle was loaded with docetaxel (Li et al., 2015). The intracellular accumulation of docetaxel was increased with good cytotoxic effect when compared to the free drug. Intravenous administration of the formulation resulted in improved accumulation in brain tumor in vivo with significant suppressed growth (Li et al., 2015). Muthu et al. (2012) reported similar findings whereby polyethylene-glycol-based micelles loaded with vitamin E and docetaxel resulted in extended blood circulation and increased cellular uptake into the brain. Zhang et al. (2010) prepared poly(ε-caprolactone)-block-poly(ethyl ethylene phosphate) micelles loaded with coumarin-6 for delivery to the brain. In vitro evaluation on BBB showed that cellular uptake of the formulation by the model was high with low toxicity. In vivo studies indicated the accumulation of the formulation in the hippocampus and striatum (Zhang et al., 2010). Liang et al. (2015) developed poly(ε-caprolactone urethanes)s-based micelles for drug delivery to the brain parenchyma. The micelles were loaded with doxorubicin and their accumulation in human brain microvascular endothelial cells was dependent on time and concentration (Liang et al., 2015). Ran et al. (2017) prepared DWSW (DWDSDWDGDPDYDS)-peptide-based micelles loaded with paclitaxel with antiglioma activity. The micelles accumulated significantly in glioma region *in* vivo with extended median survival time in glioma-bearing nude mice (Ran et al., 2017). Miura et al. (2013) loaded platinum-based anticancer drug onto polymeric micelle with cyclic Arg-Gly-Asp (cRGD) ligand molecules. The accumulation of the drug loaded micelles was high in the tumor parenchyma when compared to the micelles without targeting ligands. The accumulation of the drug into the tumor was via transcytosis (Miura et al., 2013). Zhan et al. (2011) prepared micelles with candoxin ligands for brain targeted drug delivery mechanism (Zhan et al., 2011). Micelles are potential drug delivery system for treatment of neurological disorder.

4.4.7 DENDRIMERS

Dendrimers nano-size macromolecules composed of a core, branches from the core with repeat units arranged in a geometrical progression, and terminal functional groups on the surface of dendritic architecture (Frechet et al., 1994). The drugs are encapsulated within the dendritic structure or covalently via functional groups on the surface of the dendrimers. The drug release from the dendrimer involves degradation of the covalent bonding by enzymes or environment, resulting in cleavage of the bonds and change in factors such as pH and temperature resulting in drug release (Frechet et al., 1994; Kesharwani et al., 2014; Tripathy et al., 2013). Their biodegradability and biocompatibility have made them useful as drug delivery systems.

Dendrimers have been used for delivery of drug to the brain with improved therapeutic outcomes (Table 4.3). Srinageshwar et al. (2017) prepared poly-amidoamine dendrimers with a mixed surface that can penetrate the BBB in mice, in which the formulation was administered via carotid artery. The dendrimers crossed the BBB and were localized within the neurons and glial cells when administered via carotid and migrated to the corpus callosum when administered intracranially over a period of 1 week (Srinageshwar et al., 2017). The dendrimers were not found in organs such the liver, lungs, and spleen tissue when administered via the carotid suggesting that the route resulted in targeted drug delivery (Srinageshwar et al., 2017).

Swami et al. (2015) prepared p-hydroxyl benzoic acid-dendrimer loaded with docetaxel for targeted delivery to brain tumors. Cell uptake of the formu-lation by U87MG human glioblastoma cells was significant when compared to the free drug (Swami et al., 2015). The formulation was characterized by sustained drug release, good physical stability, decreased hemolysis, and enhanced cytotoxic effects (Swami et al., 2015). Dai et al. (2010) demonstrated the targeting ability of dendrimers in the brain. In vivo evaluation of generation-4-polyamidoamine dendrimers by subarachnoid administration in a rabbit model of cerebral palsy showed that the dendrimers were localized in microglia and astrocytes. The targeted delivery of the dendrimers suggested the potential application of dendrimers for the treatment of cerebral palsy, Alzheimer's, and multiple sclerosis diseases (Dai et al., 2010). Bullen et al. (2010) reported the interaction of biotinylated G4 poly (amidoamine) dendrimer conjugates and G4 PAMAM dendrimers in endothelial cell culture models of the BBB s in vitro. Disruption of the composition of the liquid condensed and liquid expanded phases of the 1,2-dipalmitoyl-sn-glycero-3-phosphocholine lipid monolayer was influenced by the concentration of the formulation (Bullen et al., 2011). However, biotinylation of the dendrimers resulted in higher levels of toxicity when compared to the non-biotinylation dendrimers (Bullen et al., 2011). Kannan et al. (2012) prepared N-acetyl-L-cysteine-based dendrimers for the treatment of brain injury. The dendrimers contained glutathione disulphide linkers and suppressed neuro-inflammation. The dendrimers accumulated in the microglia and astrocytes (Kannan et al., 2012). Kim et al. (2012) encpsulated HMGB1 siRNA onto polyamidoamine dendrimers, and the administration of the formulation in postichemic rat brain model resulted in a decreased infarct volume (Kim et al., 2012). Other types of polymers have also been reported such as carbosilane dendrimer that transported siRNA into the brain (Serramia et al., 2015). Neelov et al. (2013) prepared poly-lysine dendrimers that were localized in the cytoplasm of neurons, periventricular zone and glial cells in the forebrain after intraventricular administration of the formulation (Neelov et al., 2013).

TABLE 4.3 Micelles and Dendrimers for Targeted Delivery to the Brain.

Nanocarrier	Materials used	Drug loaded	Therapeutic outcome	References
Micelles	Chitosan	Doxorubicin	The cellular uptake of the carriers was excellent in vivo. Drug release was slow with good cytotoxic effects	Xie et al. (2012)

TABLE 4.3 *(Continued)*

Micelles	polylactic acidpolyethylene glycol	Docetaxel	The accumulation of formulation in brain tumor tissues was enhanced with good antiglioma activity and no sign of toxicity	Li et al. (2015)
Micelles	Polyethylene glycol	Docetaxel, vitamin E	Enhanced cellular uptake and non-toxic	Muthu et al. (2012)
Micelles	Poly(ethyl ethylene phosphate) and poly(ε-caprolactone)	Coumarin-6	The accumulation of the formulation in the brain was high with low toxicity	Zhang et al. (2010)
Micelles	poly(ε-caprolactone urethanes)	Doxorubicin	The accumulation of the formulation in the subcortical area after administrated intravenously was high when compared to for doxorubicin which was not detected	Liang et al. (2015)
Micelles	Peptide	Paclitaxel	Good accumulation in the glioma region	Ran et al. (2017)
Micelles	Peptides	Platinum drug	antitumor effects in orthotopic mouse model of U87MG human glioblastoma was significant	Miura et al. (2013)
Micelles	Peptides	Candid candoxin	Good brain targeting	Zhan et al. (2011)
Dendrimers	Polyamido-amine	docetaxel	The formulation exhibited good physico-chemical stability and decrease hemolytic toxicity. Effective against U87MG human glioblastoma cells	Swami et al. (2015)
Dendrimers	Polyamido-amine	HMGB1 siRNA, siRNA	Formulation was localized in activated microglia and astrocytes	Bullen et al. (2011); Dai et al. (2010); Kannan et al. (2012); Kim et al. (2012); Neelov et al. (2013); Serramía et al. (2015)

4.4.8 NANOEMULSIONS

Nanoemulsions are a colloidal particulate system with particle sizes between 10 and 1000 nm (Singh et al., 2017). They are useful for enhancing the therapeutic efficacy of the loaded drug with reduced adverse and toxic effects. There are three types of nanoemulsion such as oil in water, bi-continuous, and water in oil nanoemulsions. They can be administered topically, intravenously, orally, intranasally, ocularly, and by pulmonary routes (Jaiswal et al., 2015; Singh et al., 2017; Solans et al., 2003). They are physically stable due to their small droplet size. The protect drugs form degrading environmental factors, exhibited targeted delivery mechanism, enhance bioavailability, prolong gastric retention (Jaiswal et al., 2015; Singh et al., 2017; Solans et al., 2003). Nanoemulsions undergo direct lymphatic absorption thereby enhancing its bioavailability. It is used to shield the after taste of drugs that can result in patient noncompliance (Jaiswal et al., 2015; Singh et al., 2017; Solans et al., 2003).

Mahajan et al. (2014) developed nanoemulsion loaded with saquinavir mesylate, an anti HIV drug for the treatment of neuro-AIDS (Table 4.4). Ex vivo studies on sheep nasal mucosa indicated an increase in drug permeation with no sign of toxicity when compared to the free drug. In vivo biodistribution studies showed a high drug uptake in the brain after intranasal administration of the formulation (Mahajan et al., 2014). A similar observation was reported by Prabhakar et al. (2013) in which indinavir, an anti-HIV, was loaded into oil in water nanoemulsion for targeted brain delivery. The uptake of the formulation into the brain was high when compared to other organs in vitro after Intravenous administration high (Prabhakar et al., 2013). Nanoemulsion have also been used to deliver other classes of drugs to the brain such as anti-migranes drugs, rizatriptan benzoate (Bhanushali et al., 2008), antispychotics, for example, Ziprasidone Hydrochloride, quetiapine (Bahadur et al., 2012; Boche et al., 2017; Setia et al., 2016), psychotropic agent, olanzapine (Shailaja et al., 2012), antidepressant, paroxetine (Pandey, 2015), mu-opioid peptide (Shah et al., 2014), antiamyloid. and anti-inflammatory (Sood et al., 2013). The cellular uptake of the nanoemulsion formulations by the brain was high.

4.4.9 CARBON-BASED NANOMATERIALS

Carbon-based nanomaterials such as carbon nanotubes, fullerenes, and graphene oxide have been utilized as drug delivery systems. They are characterized by large surface area, easily modified chemically, good chemical stability, good capacity to adsorb therapeutic agents, and can deliver molecule to target cell/tissue (Chung et al., 2013; Lamberti et al., 2015; Rasovic et al., 2017). Drug can be loaded onto the structure or via functionalities attached to the surface of the materials. Drug delivery from carbon-based materials can be either by diffusion of endocytosis mechanisms (Chung et al., 2013; Lamberti et al., 2015; Rasovic et al., 2017). The uptake of the loaded drug can be by internalization whereby the drug is released after degradation intracellularly and it can also be by degradation of the drug from the attachment in physiological environment resulting in the cellular uptake of the drug (Chung et al., 2013; Lamberti et al., 2015; Rasovic et al., 2017).

Carbon-based nanomaterials have been used for the targeted brain delivery (Table 4.4). Derivatives of fullerenes C60 containing different linkers between the fullerene cage and the solubilizing moieties were reported to induce neural stem cell proliferation in vitro in zebrafish (Hsieh et al., 2017). The solubilizing moieties influenced the biological activity of the derivatives. They were also able to inhibit glioblastoma cell proliferation. Derivatives with phenylbutiryc acids moieties significantly enhanced neural stem cell proliferation while derivatives with phenylalanine moieties inhibited glioblastoma growth significantly (Hsieh et al., 2017). Piotrovskiy et al. (2016) prepared complexes of fullerene C60. Their penetration into the central nervous system was limited thereby inhibiting the effects of nicotine at high doses. Intraperitoneal administration of fullerene complexes blocked effects of nicotine indicating fullerenes capacity to interact with central nicotine receptors (Piotrovskiy et al., 2016). However, other research reports showed that the formulation of fullerene did not cross the BBB but affected the neurotransmission in the brain (Yamada et al., 2008, 2010).

Functionalized carbon nanotubes have been utilized for drug delivery to the brain. Kafa et al. (2015) reported amino-functionalized multiwalled carbon nanotubes that were able to cross the BBB in vitro in primary porcine brain endothelial cells and primary rat astrocytes (Kafa et al., 2015). Carbon nanotubes loaded with caspase-3 siRNA have been utilized to silent of caspase-3 that is responsible for brain tissue loss after brain traumatic brain injury (Al-Jamal et al., 2011). Peri-lesional stereotactic administration of the formulation reduced neurodegeneration and after the focal ischemic damage of the rodent motor cortex (Al-Jamal et al., 2011). The interaction of functionalized multiwalled carbon nanotubes and the neural tissue after administration by cortical stereotactic have been reported (Bardi et al., 2013). The cellular uptake by neural tissue cells indicated the potential of carbon nanotubes for the drug delivery to the brain (Bardi et al., 2013). Amine-modified single-walled carbon nanotubes were able to protect the neurons in rats induced with stroke, indicating the potential of carbon nanotubes in protecting the brains in ischaemic injury (Lee et al., 2011). Functionalized single-walled carbon nanotubes conjugated with CpG have been reported for delivery to tumor-associated inflammatory cells (Zhao et al., 2011). They were not toxic with good cellular uptake after a single intracranial injection reduced the intracranial GL261 gliomas to half in tumor-bearing mice (Zhao et al., 2011). PEGylated oxidized multiwalled carbon nanotubes with angiopep-2 and loaded with doxorubicin were found suitable for the treatment of brain glioma (Ren et al., 2012). The anti-glioma effect was significant when compared to the free drug (Ren et al., 2012).

TABLE 4.4 Nanoemulsions and Carbon-Based Materials Used for Targeted Drug Delivery to the Brain.

Carriers	Drug loaded/ encapsulated	Therapeutic outcome	References
Nanoemulsion	Saquinavir mesylate	In vivo biodistribution studies revealed high drug uptake in brain after intra-nasal administration	Mahajan et al. (2014)
Nanoemulsion	Indinavir	High brain uptake in mice after intravenous administration	Prabhakar et al. (2015)
Nanoemulsion	Rizatriptan benzoate	Formulation was stable with high drug incorporation	Bhanushali et al. (2008)
Nanoemulsion	Quetiapine	High brain uptake in mice after Intravenous administration	Boche and Pokharkar (2017); Setia and Ahlawat (2016)
Nanoemulsion	Ziprasidone hydrochloride	The formulation was nontoxic	Bahadur and Pathak (2012)
Nanoemulsion	Olanzapine	Highly stable over a period of 3 months	Shailaja et al. (2012)
Nanoemulsion	Paroxetine	improved the behavioral activities in vivo	Pandey et al. (2015)
Nanoemulsion	Mu-opioid	Nontoxic	Shah et al. (2014)
Nanoemulsion	Curcumin, done-pezil (combination therapy)	Acetylcholine levels were increased significantly	Sood et al. (2013)

TABLE 4.4 *(Continued)*

Carriers	Drug loaded/ encapsulated	Therapeutic outcome	References
Fullerenes		Formulation inhibited glioblastoma cell proliferation in vitro and reduced glioblastoma formation in vivo	Hsieh et al. (2017)
Fullerenes	Nicotine	Formulation interacted with central nicotine receptors	Piotrovskiy et al. (2016)
Carbon nanotubes	Caspase-3 siRNA	Peri-lesional stereo-tactic administration of the formulation reduced neurodegeneration	Al-Jamal et al. (2013)
Carbon nanotubes	Doxorubicin	The cellular uptake by neural tissue cells was significant. Protected the neurons in rats induced with stroke	Bardi et al. (2013); Lee et al. (2013); Ren et al. (2012) Zhao et al. (2011)

4.4.10 *LIPIDS NANOPARTICLES*

Nanoparticles classified as nanolipid have been used to deliver drugs to the brain. Nanolipids have several properties making them useful for drug delivery such a targeted drug release mechanisms, enhance the stability of encapsulated drug, nonallergenic, biodegradable, biocompatible, nontoxic, nonirritating, and can be applied with water-based techniques and is affordable (Attama et al., 2012). Nanolipids have been employed for the delivery of therapeutic drugs to the brain with therapeutic effects such as intraconazole (Lim et al., 2014), baclofen (Ghasemian et al., 2017), and embelin (Sharma et al., 2017).

Polymer nanoparticles have also been evaluated for delivery of drugs for the treatment of neurological disorder. Different polymers have been utilized for the preparation of nanoparticles. Polylactic acid nanoparticles prepared by solvent evaporation exhibited an average diameter of 15 nm, which is smaller than blood cells (Marini e al. 2010). Das et al. (2005) prepared poly(butylcyanoacrylate) nanoparticulate loaded with hexapeptide dalargin, an antinociceptive peptide for delivery to the brain (Das et al., 2005). Poly-(D,L)-lactide-co-glycolide loaded with Bacoside-A was found to be a potential therapeutic for the treatment of neurodegenerative disorders (Jose et al., 2014). Poly-lactide-co-glycolide nanoparticle encapsulated with antituberculosis drugs were evaluated for cerebral drug delivery in a murine model. The formulation was characterized by enhanced bioavailability when compared to the free drug (Pandey et al., 2006).

4.4.11 NANOLIPOSOMES

Nanoliposome are nanoscale lipid vesicles formed by hydrophilic–hydrophobic interaction between phospholipids and water molecules (Khosravi-Darani and Mozafari, 2010). They are stable when compared to liposomes due to their size (Khosravi-Darani and Mozafari, 2010). They are also biocompatible, biodegradable, and useful for controlled and targeted drug delivery with reduced level of toxicity (Noble et al., 2014; Vieria and Gamarra, 2016).

Upadhyay et al. (2016) encapsulated quetiapine fumarate onto nanoliposomes for delivery to the brain via nasal route. Shaw et al. (2017) loaded docetaxel onto nanoliposomes that exhibited good cellular uptake by C6 glioma cells indicating their potential applicaton for the treatment of glioma. Guo et al. (2017) developed nanoliposomes formulation of ferric ammonium citrate and increased brain iron levels in rats after administration nasally. The drug loading efficiency of the drug onto the nanoliposomes was high and the formulation was stable. There was no sign of toxicity and indicated that nanoliposomes are potential therapeutics for the treatment of iron deficiency (Guo et al., 2017). Salem et al. (2015) **developed** glutathione-modulated nanoliposomes loaded with flucytosine for delivery to the brain. **The cellular** uptake of the drug-loaded nanoliposomes by brain cells of the rats was enhanced resulting in targeted and effective delivery of the drug to the brain in vivo (Salem et al., 2015). Krauze et al. (2007) prepared nanoliposomal formulation containing topoisomerase I inhibitor CPT-11. In vivo evaluation by the intracranial administration on rodent brain tumor xenografts indicated that the formulation prolonged the drug release resulting in good therapeutic outcomes and no sign of toxicity (Krauze et al., 2007). Hu et al. (2016) prepared nanoliposomes loaded with quercetin and temozolomide for dual drug delivery to the brain. In vitro evaluation on U87 glioma cells revealed minimal cellular uptake. In vivo accumulation of formulation in the brain was significant with delayed clearance rat in the model of glioma (Hu et al., 2016). Butowski et al. (2014) reported the formulation of nanoliposomal containing irinotecan, a potential therapeutic for the treatment of brain tumor. Nobel et al. (2006) combined convection-enhanced delivery nanoliposome containing CPT-11 for dual drug delivery for the treatment of brain tumor. The formulation was characterized by prolonged tissue retention and reduced toxicity when compared to free CPT-11 (Noble et al., 2006). Huang et al. (2013) loaded radioisotope complex, 99mTc labelled *N*,*N*-bis(2-mercaptoethyl)-*N'*,*N'*-diethylethylenediamine to PEGylated liposomes conjugates with lactoferrin suitable for delivery to the brain. The in vivo uptake of the formulation by the mouse brain was high when compared to the free liposomes (Huang et al., 2013).

4.4.12   QUANTUM DOTS

Quantum dots are nanocarriers with optical properties suitable for biological applications such as imaging and drug delivery. Drugs are loaded onto them using various methods such as by dispersing, coupling, dissolving, and adsorption of the drugs (Agarwal et al., 2015; Zhao and Zhu, 2016). Quantum dots application for drug delivery enhances drug efficacy, targeted drug delivery when targeted moieties are used, prevents premature drug degradation, reduce the loaded drug side effects, and they are characterized by high reactivity sites, large specific surface area, and strong adsorption capacity (Zhao and Zhu, 2016). Some researchers investigated the potential application of quantum dot for drug delivery to the brain. Agarwal et al. (2015) evaluated CdSe/ZnS quantum dots in which the surface was functionalized with a zwitterionic ligand and conjugates with the palmitoylated peptide. In vivo evaluation on chick embryo brain indicated that the formulation is nontoxic. The formulation was delivered to the brain and concentrated in the choroid plexus (Agarwal et al., 2015). Weng et al. (2013) prepared quantum dot-immunoliposome hybrid nanoparticles. The in vitro cellular uptake of the formulation was specific and efficient and in vivo evaluation further showed that the formulation can be used in combination with convection-enhanced delivery for treatment of glioblastoma (Weng et al., 2013). Paris-Robidas et al. (2016) conjugated quantum dots to monoclonal antibodies (Ri7). Intravenous administration in mice exhibited a high cellular uptake in brain tissues (Paris-Robidas et al., 2016). The formulation was internalized in the brain capillary endothelial cells 1 h after administration. However, the uptake of the formulation by the parenchymal was low. The uptake of the drug by the brain capillary endothelial cells was by endocytosis (Paris-Robidas et al., 2016). Getz et al. (2016) conjugated quantum dots containing (N-palmitate) and loaded with siRNAs for the treatment of neurological disorder.

4.5   CONCLUSION AND FUTURE PERSPECTIVE

The high rate of neurological disorders indicates that there is a serious need to develop therapeutics that can treat the disorder with reduced side effects. Some therapeutics used for the treatment of neurological disorder are limited by toxic drug side effects, nonspecificity, and inability to penetrate the BBB. Nanomaterials have been investigated and found to be potential materials for the development of brain delivery targeting systems and are able to deliver drugs to the brain by crossing the BBB, enhance drug bioavailability, reduce the toxic drug side effects. However, despite the potential application of these nanobiomaterials, most of them have only been evaluated in preclinical stage signifying the need for these carriers to reach the clinical trial stage. Some of the factors that must be addressed before these carriers can reach the clinical trial stage are the toxic effects of the carriers in long-term usage, the differences in species used in animal studies may limit the translation into clinical trial stage, improve the stability of the formulation over a long period of storage, there is a need to understand the interaction of the brain cells with the nanocarriers and the mechanism of the elimination of the carriers after cellular uptake of the drug loaded/encapsulated onto the nanocarriers. If the above factors are addressed, there is no doubt that in some years to come, nanocarriers will be the therapeutics that will be used to treat a neurological disorder.

ACKNOWLEDGMENTS

The financial assistance of the National Research Foundation (NRF) and Self-initiated Research Medical Research Council (MRC) South Africa are hereby acknowledged.

KEYWORDS

- brain
- nano-based drug delivery systems
- nanoparticles
- quantum dots
- nanogels
- brain disease
- micelles

REFERENCES

Abbott, N. J.; Ronnback, L.; Hansson, E. Astrocyte–Endothelial Interactions at the Blood-Brain Barrier. *Nat. Rev. Neurosci.* **2006,** *7,* 41–53.

Abou-Chebl, A. Intra-Arterial Therapy for Acute Ischemic Stroke. *Neurotherapeutics.* **2011,** *8,* 400–413.

Agarwal, R.; Domowicz, M. S.; Schwartz, N. B.; Henry, J.; Medintz, I.; Delehanty, J. B.; Stewart, M. H.; Susumu, K.; Huston, A. L.; Deschamps, J. R.; Dawson, P. E. Delivery and Tracking of Quantum Dot Peptide Bioconjugates in an Intact Developing Avian Brain. *ACS Chem. Neurosci.* **2015,** *6,* 494–504.

Al-Dhubiab, B. E.; Nair, A. B.; Kumria, R.; Attimarad, M.; Harsha, S. Development and Evaluation of Buccal Films Impregnated With Selegiline-Loaded Nanospheres. *Drug Deliv.* **2016,** *23,* 2154–2162.

Al-Jamal, K. T.; Gherardini, L.; Bardi, G.; Nunes, A.; Guo, C.; Bussy, C.; Herrero, M. A.; Bianco, A.; Prato, M.; Kostarelos, K.; Pizzorusso, T. Functional Motor Recovery from Brain Ischemic Insult by Carbon Nanotube-Mediated siRNA Silencing. *Proc. Natl. Acad. Sci.* **2011,** *108,* 10952–10957.

Alvarez-Erviti, L.; Seow, Y.; Yin, H.; Betts, C.; Lakhal, S.; Wood, M. J. Delivery of siRNA to the Mouse Brain by Systemic Injection of Targeted Exosomes. *Nat. Biotechnol.* **2011,** *29,* 341–345.

Aly, A. E.; Waszczak, B. L. Intranasal Gene Delivery for Treating Parkinson's Disease: Overcoming the Blood–Brain Barrier. *Expert Opin. Drug Deliv.* **2015,** *12,* 1923–1941.

Anatomy of the Brain. [Online] https://www.mayfieldclinic.com/PE-AnatBrain.htm (accessed May 28, 2017).

Arshad, A.; Yang, B.; Bienemann, A.; Barua, N.; Gill, S. P56convection-Enhanced Delivery (ced) of Carboplatin Nanospheres For The Treatment Of Glioblastoma. *Neuro Oncol.* **2014,** *16.* doi: 10.1093/neuonc/nou249

Attama, A. A.; Momoh, M. A.; Builders, P. F. *Lipid Nanoparticulate Drug Delivery Systems: A Revolution in Dosage form Design and Development.* In *Recent Advances in Novel Carrier System;* Sezer, A. D.; Ed.; INTECH Open Access Publisher; Croatia, 2012; pp 107–140

Awasthi, V. D.; Garcia, D.; Klipper, R.; Goins, B. A.; Phillips, W. T. Neutral and Anionic Liposome-Encapsulated Hemoglobin: Effect of Postinserted Poly(ethylene glycol)-distearoylphosphatidylethanolamine on Distribution and Circulation Kinetics. *J. Pharmacol. Exp. Ther.* **2004,** *309,* 241–248.

Azadi, A.; Hamidi, M.; Rouini, M. R. Methotrexate-Loaded Chitosan Nanogels as 'Trojan Horses' for Drug Delivery to Brain: Preparation and In Vitro/In Vivo Characterization. *Int. J. Biol. Macromol.* **2013,** *62,* 523–530.

Bahadur, S.; Pathak, K. Buffered Nanoemulsion for Nose to Brain Delivery of Ziprasidone Hydrochloride: Preformulation and Pharmacodynamic Evaluation. *Curr. Drug Deliv.* **2012,** *9,* 596–607.

Bardi, G.; Nunes, A.; Gherardini, L.; Bates, K.; Al-Jamal, K. T.; Gaillard, C.; Prato, M.; Bianco, A.; Pizzorusso, T.; Kostarelos, K. Functionalized Carbon Nanotubes in the Brain: Cellular Internalization and Neuroinflammatory Responses. *PLoS One.* **2013,** *8,* e80964.

Batrakova, E. V.; Kim, M. S. Development and Regulation of Exosome-Based Therapy Products. *Wiley Interdiscip. Rev. Nanomed. Nanobiotechnol.* **2016,** *8,* 744–757.

Bernardi, A.; Frozza, R. L.; Meneghetti, A.; Hoppe, J. B.; Battastini, A. M.; Pohlmann, A. R.; Guterres, S. S.; Salbego, C. G. Indomethacin-Loaded Lipid-Core Nanocapsules Reduce the Damage Triggered by Aβ1-42 in Alzheimer's Disease Models. *Int. J. Nanomed.* **2012,** *7*, 4927–4942.

Bertram, L.; Lill, C. M.; Tanzi, R. E. The Genetics of Alzheimer Disease: Back to the Future. *Neuron.* **2010,** *68*, 270–281.

Bhanushali, R. S.; Gatne, M. M.; Gaikwad, R. V.; Bajaj, A. N.; Morde, M. A. Nanoemulsion based Intranasal Delivery of Antimigraine Drugs for Nose to Brain Targeting. *Indian J. Pharm. Sci.* **2008,** *71*, 707–709.

Boche, M.; Pokharkar, V. Quetiapine Nanoemulsion for Intranasal Drug Delivery: Evaluation of Brain-Targeting Efficiency. *AAPS PharmSciTech.* **2017,** *19*, 686–696.

Boridy, S.; Takahashi, H.; Akiyoshi, K.; Maysinger, D. The Binding of Pullulan Modified Cholesteryl Nanogels to Aβ Oligomers and their Suppression of Cytotoxicity. *Biomaterials.* **2009,** *30*, 5583–5591.

Brain Tumours, an Introduction. [Online]. http://www.mayfieldclinic.com/PE-BrainTumor.htm (accessed May 29, 2017).

Bullen, H. A.; Hemmer, R.; Haskamp, A.; Cason, C.; Wall, S.; Spaulding, R.; Rossow, B.; Hester, M.; Caroway, M.; Haik, K. L. Evaluation of Biotinylated PAMAM Dendrimer Toxicity in Models of the Blood Brain Barrier: A Biophysical and Cellular Approach. *J. Biomater. Nanobiotechnol.* **2011,** *2*, 485.

Bulloj, A.; Leal, M. C.; Xu, H.; Castaño, E. M.; Morelli, L. Insulin-Degrading Enzyme Sorting in Exosomes: A Secretory Pathway for a Key Brain Amyloid-β Degrading Protease. *J. Alzheimers Dis.* **2010,** *19*, 79–95.

Butowski, N. A.; Han, S.; Taylor, J. W.; Aghi, M. K.; Prados, M.; Chang, S. M.; Clarke, J. L.; Bankiewicz, K.; Drummond, D. C.; Fitzgerald, J. A Phase I Study of Convection-Enhanced Delivery of Nanoliposomal Irinotecan Using Real-Time Imaging in Patients With Recurrent High Grade Glioma. J. Clin. Oncol. 2014. DOI: 10.1093/neuonc/nou206.46.

Cai, Q.; Ruan, C.; Jiang, L.; Ma, Y.; Pan, H. Preparation of Lactoferrin Modified Poly(vinyl alcohol) Nanospheres for Brain Drug Delivery. *Nanomed. Nanotechnol. Biol. Med.* **2016,** *12*, 542–543.

Cannava, C.; Stancanelli, R.; Marabeti, M. R.; Venuti, V.; Cascio, C.; Guarneri, P.; Bongiorno, C.; Sortino, G.; Majolino, D.; Mazzaglia, A.; Tommasini, S. Nanospheres Based on PLGA/Amphiphilic Cyclodextrin Assemblies as Potential Enhancers of Methylene Blue Neuroprotective Effect. *RSC Adv.* **2016,** *6*, 16720–16729.

Carradori, D.; Saulnier, P.; Préat, V.; Des Rieux, A.; Eyer, J. NFL-Lipid Nanocapsules for Brain Neural Stem Cell Targeting In Vitro and In Vivo. *J. Control Release.* **2016,** *238*, 253–262.

Cerna, T.; Stiborova, M.; Adam, V.; Kizek, R.; Eckschlager, T. Nanocarrier Drugs in the Treatment of Brain Tumors. *J. Cancer Metastasis Treat.* **2016,** *2*, 407–416.

Chen, X.-Q.; Fawcett, J. R.; Rahman. Y.-E.; Ala, T. A.; Frey, W. H. II. Delivery of Nerve Growth Factor to the Brain Via The Olfactory Pathway. *J. Alzheimer's Dis.* **1998,** *1*, 35–44.

Chung, C.; Kim, Y. K.; Shin, D.; Ryoo, S. R.; Hong, B. H.; Min, D. H. Biomedical Applications of Graphene and Graphene Oxide. *Acc. Chem. Res.* **2013,** *46*, 2211–2224.

Cohen-Pfeffer, J. L.; Gururangan, S.; Lester, T.; Lim, D. A.; Shaywitz, A. J.; Westphal, M.; Slavc, I. Intracerebroventricular Delivery as a Safe, Long-Term Route of Administration. *Pediatric Neurol.* **2017,** *67*, 23–35.

Dai, H.; Navath, R. S.; Balakrishnan, B.; Guru, B. R.; Mishra, M. K.; Romero, R.; Kannan, R. M.; Kannan, S. Intrinsic Targeting of Inflammatory Cells in the Brain by Polyamidoamine Dendrimers Upon Subarachnoid Administration. *Nanomedicine* **2010,** *5*, 1317–1329.

Das, D.; Lin, S. Double-Coated poly(butylcynanoacrylate) Nanoparticulate Delivery Systems for Brain Targeting of Dalargin Via Oral Administration. *J. Pharm. Sci.* **2005,** *94,* 1343–1353.

Demeule, M.; Currie, J. C.; Bertrand, Y.; Che, C.; Nguyen, T.; Regina, A.; Gabathuler, R.; Castaigne, J. P.; Beliveau, R. Involvement of the Low-Density Lipoprotein Receptor-Related Protein in the Transcytosis of the Brain Delivery Vector Angiopep-2. *J. Neurochem.* **2008,** *106,* 1534–1544.

Ding, J.; Shi, F.; Xiao, C. One-Step Preparation of Reduction-Responsive Poly(ethylene glycol)–poly(amino acid)s Nanogels as Efficient Intracellular Drug Delivery Platforms. *Polym. Chem.* **2011,** *2,* 2857–2864.

Dourado, C. C.; Engler, T. M.; Oliveira, S. B. Bowel Dysfunction in Patients with Brain Damage Resulting from Stroke and Traumatic Brain Injury: A Retrospective Study of a Case Series. *Texto Contexto Enfermagem.* **2012,** *21,* 905–911.

Eckmann, D. M.; Composto, R. J.; Tsourkas, A.; Muzykantov, V. R. Nanogel Carrier Design for Targeted Drug Delivery. *J. Mater. Chem. B Mater. Biol. Med.* **2014,** *2,* 8085–8097

Ellis, J. A.; Banu, M.; Hossain, S. S.; Singh-Moon, R.; Lavine, S. D.; Bruce, J. N.; Joshi, S. Reassessing the Role of Intra-Arterial Drug Delivery For Glioblastoma Multiforme Treatment. *J. Drug Deliv.* **2015,** 15 pages.

Epilepsy. WHO Facts Sheet February 2017. [Online]. http://www.who.int/mediacentre/factsheets/fs999/en/ (accessed May 29, 2017).

Eriksdotter, J. M.; Nordberg, A.; Amberla, K.; Bäckman, L.; Ebendal, T.; Meyerson, B. O.; Olson, L.; Seiger, Å.; Shigeta, M.; Theodorsson, E.; Viitanen, M. Intracerebroventricular Infusion of Nerve Growth Factor in Three Patients with Alzheimer's Disease. *Dement. Geriatr. Cogn. Disord.* **1998,** *9,* 246–257.

Evatt, M. L.; Delong, M. R.; Khazai, N.; Rosen, A.; Triche, S.; Tangpicha, V. Prevalence of Vitamin D Insufficiency in Patients with Parkinson Disease and Alzheimer Disease. *Arch. Neurol.* **2008,** *65,* 1348–135

Evatt, M. L.; DeLong, M. R.; Kumari, M.; Auinger, P.; McDermott, M. P.; Tangpricha, V. High Prevalence of Hypovitaminosis D Status in Patients with Early Parkinson Disease. *Arch. Neurol.* **2011,** *68,* 314–319.

Fang, F.; Chen, H.; Feldman, A.; Kamel, F.; Ye, W.; Wirdefeldt, K. Head Injury and Parkinson's Disease: A Population-Based Study. *Mov. Disord.* **2012,** *27,* 1632–1635.

Fleminger, S.; Oliver, D. L.; Lovestone, S.; Rabe-Hesketh, S.; Giora, A. Head Injury as a Risk Factor for Alzheimer's Disease: The Evidence 10 Years On; A Partial Replication. *J. Neurol. Neurosurg. Psychiatr.* **2003,** *74,* 857–862.

Fletcher, L.; Kohli, S.; Sprague, S. M.; Scranton, R. A.; Lipton, S. A.; Parra, A.; Jimenez, D. F.; Digicaylioglu, M. Intranasal Delivery of Erythropoietin Plus Insulin-Like Growth Factor–I for Acute Neuroprotection in Stroke: Laboratory Investigation. *J. Neurosurg.* **2009,** *111,* 164–170.

Frackenburg, F. R. Schizophrenia, 2017. [Online] http://emedicine.medscape.com/article/288259-overview. (accessed May 29, 2017).

Frechet, J. M. Functional Polymers and Dendrimers: Reactivity, Molecular Architecture, and Interfacial Energy. *Science* **1994,** *263,* 1710–1714.

Gao, Y.; Xie, J.; Chen, H. Nanotechnology-Based Intelligent Drug Design for Cancer Metastasis Treatment. *Biotechnol. Adv.* **2014,** *32,* 761–777.

Georgieva, J. V.; Brinkhuis, R. P.; Stojanov, K.; Weijers, C. A.; Zuilhof, H.; Rutjes, F. P.; Hoekstra, D.; van Hest, J.; Zuhorn, I. S. Peptide-Mediated Blood–Brain Barrier Transport of Polymersomes. *Angew. Chem. Int. Ed.* **2012**, *51*, 8339–8342.

Getz, T.; Qin, J.; Medintz, I. L.; Delehanty, J. B.; Susumu, K.; Dawson, P. E.; Dawson, G. Quantum Dot-Mediated Delivery of siRNA to Inhibit Sphingomyelinase Activities in Brain-Derived Cells. *J. Neurochem.* **2016**, *139*, 872–85.

Ghasemian, E.; Vatanara, A.; Navidi, N.; Rouini, M. R. Brain Delivery of Baclofen as a Hydrophilic Drug by Nanolipid Carriers: Characteristics and Pharmacokinetics Evaluation. *J. Drug Deliv. Sci. Technol.* **2017**, *37*, 67–73.

Ghosh, A.; Sarkar, S.; Mandal, A. K.; Das, N. Neuroprotective role of Nanoencapsulated Quercetin in Combating Ischemia-Reperfusion Induced Neuronal Damage in Young and Aged Rats. *PLoS One.* **2013**, *8*, e57735.

Gobin, Y. P.; Cloughesy, T. F.; Chow, K. L.; Duckwiler, G. R.; Sayre, J. W.; Milanese, K.; Vinuela, F. Intraarterial Chemotherapy for Brain Tumors by Using a Spatial Dose Fractionation Algorithm and Pulsatile Delivery 1. *Radiology.* **2001**, *218*, 724–32.

Guo, X.; Zheng, H.; Guo, Y.; Wang, Y.; Anderson, G. J.; Ci, Y.; Yu, P.; Geng, L.; Chang, Y. Z. Nasal Delivery of Nanoliposome-Encapsulated Ferric Ammonium Citrate Can Increase the Iron Content of Rat Brain. *J. Nanobiotechnol.* **2017**, *15*, 42.

Han, I. K.; Kim, M. Y.; Byun, H. M.; Hwang, T. S.; Kim, J. M.; Hwang, K. W.; Park, T. G.; Jung, W. W.; Chun, T.; Jeong, G. J.; Oh, Y. K. Enhanced Brain Targeting Efficiency of Intranasally Administered Plasmid DNA: An Alternative Route for Brain Gene Therapy. *J. Mol. Med.* **2007**, *85*, 75–83.

Hannon, E.; Dempster, E.; Viana, J.; Burrage, J.; Smith, A. R.; Macdonald, R.; Clair, D. S.; Mustard, C.; Breen, G.; Therman, S.; Kaprio, J.; Toulopoulou, T.; Hulshoff Pol, H. E.; Bohlken, M. M; Kahn, R. S.; Nenadic, I.; Hultman, C. M.; Murray, R. M.; Collier, D. A.; Bass, N.; Gurling, H.; McQuillin, A.; Schalkwyk, L.; Mill, J. An Integrated Genetic-Epigenetic Analysis of Schizophrenia: Evidence for Co-Localization of Genetic Associations and Differential DNA Methylation. *Genome Biol.* **2016**, *17*. DOI: 10.1186/s13059-016-1041-x.

Hanson, L. R.; Frey, W. H. Intranasal Delivery Bypasses the Blood-Brain Barrier to Target Therapeutic Agents to the Central Nervous System and Treat Neurodegenerative Disease. *BMC Neurosci.* **2008**, *9*.

Hanson, S. Will Nanogel Produce New Neurons to Fight Alzheimer's and Parkinson's? June 20, **2016**. [Online]. http://www.homecareassistancerockwall.com/nanogel-drug-delivery/ (accessed June 2, 2017).

Hauser, R. A. Parkinson Disease, 2017. [Online]. http://emedicine.medscape.com/article/1831191-overview (accessed May 29, 2017).

Health and Social Care Information Centre, Tables Showing Mean and Median Length of Stay for Patients with Primary and Secondary Diagnoses of a Neurological Condition, By Primary Care Trust (of Main Provider) for the Years 2007–2008 to 2011–2012. [Online]. http://www.parliament.uk/business/publications/business-papers/ commons/deposited-papers/?page=3&td=2014-01- 01&search_term=Department%20of%20Health&itemId=1 19014#toggle-1749 (accessed May 28, 2017).

Herran, E.; Perez-Gonzalez, R.; Igartua, M.; Pedraz, J. L.; Carro, E.; Hernandez, R. M. Enhanced Hippocampal Neurogenesis in APP/PS1 Mouse Model of Alzheimer's Disease after Implantation of VEGF-Loaded PLGA Nanospheres. *Curr. Alzheimer Res.* **2015**, *12*, 932–940.

Hsieh, F. Y.; Zhilenkov, A. V.; Voronov, I. I.; Khakina, E. A.; Mischenko, D. V.; Troshin, P. A.; Hsu, S. H. Water-Soluble Fullerene Derivatives as Brain Medicine: Surface Chemistry Determines if They are Neuroprotective and Antitumor. *ACS Appl. Mater. Interfaces.* **2017,** *9,* 11482–11492.

Hu, J.; Wang, J.; Wang, G.; Yao, Z.; Dang, X. Pharmacokinetics and Antitumor Efficacy of Dspe-Peg2000 Polymeric Liposomes Loaded with Quercetin and Temozolomide: Analysis of their Effectiveness in Enhancing the Chemosensitization of Drug-Resistant Glioma Cells. *Int. J. Mol. Med.* **2016,** *37,* 690–702.

Huang, F. Y.; Chen, W. J.; Lee, W. Y.; Lo, S. T.; Lee, T. W.; Lo, J. M. In Vitro and In Vivo Evaluation of Lactoferrin-Conjugated Liposomes as a Novel Carrier to Improve the Brain Delivery. *Int. J. Mol. Sci.* **2013,** *14,* 2862–2874.

Huang, Y.; Mucke, L. Alzheimer Mechanisms and Therapeutic Strategies. *Cell* **2012,** *148,* 1204–1222.

Ianiski, F. R.; Alves, C. B.; Ferreira, C. F.; Rech, V. C.; Savegnago, L.; Wilhelm, E. A.; Luchese, C. Meloxicam-Loaded Nanocapsules as an Alternative to Improve Memory Decline in an Alzheimer's Disease Model in Mice: Involvement Of Na$^+$. *Metab. Brain Dis.* **2016,** *31,* 793–802.

Illum, L. Transport of Drugs from the Nasal Cavity to the Central Nervous System. *Eur J Pharm Sci.* **2000,** *11,* 1–18.

Jaiswal, M.; Dudhe, R.; Sharma, P. K. Nanoemulsion: An Advanced Mode of Drug Delivery System. *3 Biotech.* **2015,** *5,* 123–127.

Jia, T.; Sun, Z.; Lu, Y.; Gao, J.; Zou, H.; Xie, F.; Zhang, G.; Xu, H.; Sun, D.; Yu, Y.; Zhong, Y. A Dual Brain-Targeting Curcumin-Loaded Polymersomes Ameliorated Cognitive Dysfunction in Intrahippocampal Amyloid-B1–42-Injected Mice. *Int. J. Nanomed.* **2016,** *11,* 3765–3775.

Jose, S.; Sowmya, S.; Cinu, T. A.; Aleykutty, N. A.; Thomas, S.; Souto, E. B. Surface Modified PLGA Nanoparticles for Brain Targeting of Bacoside-A. *Eur. J. Pharm. Sci.* **2014,** *63,* 29–35.

Joshi, S.; Ellis, J. A.; Emala, C. W. Revisiting Intra-Arterial Drug Delivery for Treating Brain Diseases or is it "Déjà-Vu, All Over Again"? *J. Neuroanaesth. Crit. Care.* **2014,** *1,* 108–115.

Kafa, H.; Wang, J. T.; Rubio, N.; Venner, K.; Anderson, G.; Pach, E.; Ballesteros, B.; Preston, J. E.; Abbott, N. J.; Al-Jamal, K. T. The Interaction of Carbon Nanotubes with an In Vitro Blood-Brain Barrier Model and Mouse Brain In Vivo. *Biomaterials* **2015,** *53,* 437–452.

Kannan, S.; Dai, H.; Navath, R. S.; Balakrishnan, B.; Jyoti, A.; Janisse, J.; Romero, R.; Kannan, R. M. Dendrimer-Based Postnatal Therapy for Neuroinflammation and Cerebral Palsy in a Rabbit Model. *Sci. Transl. Med.* **2012,** *4,* 130ra46.

Karatas, H.; Aktas, Y.; Gursoy-Ozdemir, Y.; Bodur, E.; Yemisci, M.; Caban, S.; Vural, A.; Pinarbasli, O.; Capan, Y.; Fernandez-Megia, E.; Novoa-Carballal, R. A Nanomedicine Transports a Peptide Caspase-3 Inhibitor Across the Blood–Brain Barrier and Provides Neuroprotection. *J. Neurosci.* **2009,** *29,* 13761–13769.

Kedar, U.; Phutane, P.; Shidhaye, S.; Kadam, V. Advances in Polymeric Micelles for Drug Delivery and Tumor Targeting. *Nanomed. Nanotechnol. Biol. Med.* **2010,** *6,* 714–729.

Kesharwani, P.; Jain, K.; Jain, N. K. Dendrimer as Nanocarrier for Drug Delivery. *Prog. Polym. Sci.* **2014,** *39,* 268–307.

Khosravi-Darani, K.; Mozafari, M. R. Nanoliposome Potentials in Nanotherapy: A Concise Overview. *Int. J. Nanosci. Nanotechnol.* **2010,** *6,* 3–13.

Kikuchi, A.; Okano, T. Pulsatile Drug Release Control Using Hydrogels. *Adv. Drug Deliv. Rev.* **2002**, *54*, 53–77.

Kim, I. D.; Shin, J. H.; Kim, S. W.; Choi, S.; Ahn, J.; Han, P. L.; Park, J. S.; Lee, J. K. Intranasal Delivery of HMGB1 siRNA Confers Target Gene Knockdown and Robust Neuroprotection in the Postischemic Brain. *Mol. Ther.* **2012**, *20*, 829–839.

Kim, J. H.; Park, I. S.; Park, K. B., Kang, D. H.; Hwang, S. H. Intraarterial Nimodipine Infusion to Treat Symptomatic Cerebral Vasospasm After Aneurysmal Subarachnoid Hemorrhage. *J. Korean Neurosurg. Soc.* **2009**, *46*, 239–244.

Kooijmans, S. A.; Vader, P.; van Dommelen, S. M.; van Solinge, W. W.; Schiffelers, R. M. Exosome Mimetics: A Novel Class of Drug Delivery Systems. *Int. J. Nanomed.* **2012**, *7*, 1525–1541.

Krauze, M. T.; Noble, C. O.; Kawaguchi, T.; Drummond, D.; Kirpotin, D. B.; Yamashita, Y.; Kullberg, E.; Forsayeth, J.; Park, J. W.; Bankiewicz, K. S. Convection-Enhanced Delivery of Nanoliposomal CPT-11 (Irinotecan) and PEGylated Liposomal Doxorubicin (Doxil) in Rodent Intracranial Brain Tumor Xenografts. *Neuro-Oncol.* **2007**, *9*, 393–403.

Kvq, L.; Nguyen, L. T. Environmental Factors in Alzheimer's and Parkinson's Diseases. *J Alzheimers Dis Parkinsonism.* **2013**, *3*, 12 pages.

Lai, R. C.; Arslan, F.; Lee, M. M.; Sze, N. S.; Choo, A.; Chen, T. S.; Salto-Tellez, M.; Timmers, L.; Lee, C. N.; El Oakley, R. M.; Pasterkamp, G. Exosome Secreted by MSC Reduces Myocardial Ischemia/Reperfusion Injury. *Stem Cell Res.* **2010**, *4*, 214–222.

Lakhal, S.; Wood, M. J. Exosome Nanotechnology: An Emerging Paradigm Shift in Drug Delivery: Exploitation of Exosome Nanovesicles for Systemic In Vivo Delivery of Rnai Heralds New Horizons for Drug Delivery Across Biological Barriers. *Bioessays.* **2011**, *33*, 737–741.

Lakhan, S. E. Alzheimer Disease, **2017**. [Online]. http://emedicine.medscape.com/article/1134817-overview. (accessed May 29, 2017).

Lamberti, M.; Pedata, P.; Sannolo, N.; Porto, S.; De Rosa, A.; Caraglia, M. Carbon Nanotubes: Properties, Biomedical Applications, Advantages and Risks in Patients and Occupationally-Exposed Workers. *Int. J. Immunopathol. Pharmacol.* **2015**, *28*, 4–13

League-Pascual, J. C.; Lester-McCully, C. M.; Shandilya, S.; Ronner, L.; Rodgers, L.; Cruz, R.; Peer, C. J.; Figg, W. D.; Warren, K. E. Plasma and Cerebrospinal Fluid Pharmacokinetics of Select Chemotherapeutic Agents Following Intranasal Delivery in a Non-Human Primate Model. *J. Neuro-Oncol.* **2017**, *13*, 1–7.

Lee, H. J.; Park, J.; Yoon, O. J.; Kim, H. W.; Kim, D. H.; Lee, W. B.; Lee, N. E.; Bonventre, J. V.; Kim, S. S. Amine-Modified Single-Walled Carbon Nanotubes Protect Neurons from Injury in a Rat Stroke Model. *Nat. Nanotechnol.* **2011**, *6*, 121–125.

Lee, J. S.; Feijen, J. Polymersomes for Drug Delivery: Design, Formation and Characterization. *J. Control Release.* **2012**, *161*, 473–478.

Lee, J. S.; Groothuis, T.; Cusan, C.; Mink, D.; Feijen, J. Lysosomally Cleavable Peptide-Containing Polymersomes Modified With Anti-EGFR Antibody for Systemic Cancer Chemotherapy. *Biomaterials* **2011**, *32*, 9144–9153.

Levenson, R. W.; Sturm, V. E.; Haase, C. M. Emotional and Behavioral Symptoms in Neurodegenerative Disease: A Model for Studying the Neural Bases of Psychopathology. *Ann. Rev. Clin. Psychol.* **2014**, *10*, 581–606.

Li, A. J.; Zheng, Y. H.; Liu, G. D.; Liu, W. S.; Cao, P. C.; Bu, Z. F. Efficient Delivery of Docetaxel for the Treatment of Brain Tumors by Cyclic RGD-Tagged Polymeric Micelles. *Mol. Med. Rep.* **2015**, *11*, 3078–3086.

Li, L.; Shui, Q. X.; Liang, K.; Ren, H. Brain-Derived Neurotrophic Factor Rescues Neurons From Bacterial Meningitis. *Pediatric Neurol.* **2007**, *36*, 324–329.

Liang, R. C.; Fang, F.; Wang, Y. C.; Song, N. J.; Li, J. H.; Zhao, C. J.; Peng, X. C.; Tong, A. P.; Fang, Y.; He, M.; You, C. Gemini Quaternary Ammonium-Incorporated Biodegradable Multiblock Polyurethane Micelles for Brain Drug Delivery. *RSC Adv.* **2015**, *5*, 6160–6171.

Lim, W. M.; Rajinikanth, P. S.; Mallikarjun, C.; Kang, Y. B. Formulation and Delivery of Itraconazole to the Brain Using a Nanolipid Carrier System. *Int. J. Nanomed.* **2014**, *9*, 2117.

Lin, J. J., Ghoroghchian, P. P., Zhang, Y., Hammer, D. A. Adhesion of Antibody-Functionalized Polymersomes. *Langmuir* **2006**, *22*, 3975–3979.

Liu, C.; Zong, H. Developmental Origins of Brain Tumors. *Curr. Opin. Neurobiol.* **2012**, *22*, 844–849.

Liu, R.; Guo, X.; Park, Y.; Huang, X.; Sinha, R.; Freedman, N. D.; Hollenbeck, A. R.; Blair, A.; Chen, H. Caffeine Intake, Smoking, and Risk of Parkinson Disease in Men and Women. *Am. J. Epidemiol.* **2012**, *175*, 1200–1207.

Luzzio, C. Multiple Sclerosis, 2017. [Online]. http://emedicine.medscape.com/article/1146199-overview#a4 (accessed May 29, 2017).

Ma, M.; Ma, Y.; Yi, X.; Guo, R.; Zhu, W.; Fan, X.; Xu, G.; Frey, W. H.; Liu, X. Intranasal Delivery of Transforming Growth Factor-Beta1 in Mice After Stroke Reduces Infarct Volume and Increases Neurogenesis in the Subventricular Zone. *BMC Neurosci.* **2008**, *9*, 117.

Madhav, N. V. S.; Raina, D.; Mala, M. Design of Cefuroxime Loaded Bio-Nano Gels For Brain Specificity via Ear. *J. Adv. Drug Deliv.* **2015**, *2*. 12 pages.

Mahajan, H. S.; Mahajan, M. S.; Nerkar, P. P.; Agrawal, A. Nanoemulsion-Based Intranasal Drug Delivery System of Saquinavir Mesylate For Brain Targeting. *Drug Deliv.* **2014**, *21*, 148–154.

Mamo, B.; Abebe, W.; Gabriel, T. D. Literature Review on Biodegradable Nanospheres for Oral and Targeted Drug Delivery. *Asian J. Biomed. Pharm. Sci.* **2015**, *5*, 1–12.

Marini, M. Preparation of Star-Shaped Polylactic Acid Drug Carrier Nanoparticles. *Mater. Sci. Appl.* **2010**, *1*, 36–38.

Marvel, C. L.; Paradiso, S. Cognitive and Neurological Impairment in Mood Disorders. *Psychiatr. Clin. North Am.* **2004**, *27*, 19–36.

Massignani, M.; Lomas, H.; Battaglia, G. *Polymersomes: A Synthetic Biological Approach to Encapsulation and Delivery.* In *Modern Techniques for Nano-and Microreactors/-Reactions*; Caruso, F., Ed.; Springer: Berlin, Germany, 2010; Vol. 223, pp 115–154.

Meijer, D. H.; Maguire, C. A.; LeRoy, S. G.; Sena-Esteves, M. Controlling Brain Tumor Growth Via Intraventricular Administration of an AAV Vector Encoding IFN-β. *Cancer Gene Ther.* **2009**, *16*, 664–671.

Meng, F.; Zhong, Z. Polymersomes Spanning from Nano-to-Microscales: Advanced Vehicles for Controlled Drug Delivery and Robust Vesicles for Virus and Cell Mimicking. *J. Phys. Chem. Lett.* **2011**, *2*, 1533–1539.

Mental Health. [Online]. http://www.who.int/mental_health/neurology/en/. (accessed May 28, 2017).

Mikeladze, K. G.; Éliava, S.; Shekhtman, O. D.; Lubnin, A.; Tabasaranskiĭ, T. F.; Iakovlev, S. B. Intra-Arterial Injection of Verapamil in Treatment of Cerebral Vasospasm in a Patient with Acute Subarachnoid Hemorrhage from an Aneurysm: Case Report. *Zh. Vopr. Neirokhir. Im. N N Burdenko.* **2012**, *77*, 57–60.

Miller, N.; Mshana, G.; Msuya, O.; Dotchin, C.; Walker, R.; Aris, E. Assessment of Speech in Neurological Disorders: Development of a Swahili Screening Test. *S. Afr. J. Commun. Disord.* **2012**, *59*, 27–33.

Miura, Y.; Takenaka, T.; Toh, K.; Wu, S.; Nishihara, H.; Kano, M. R.; Ino, Y.; Nomoto, T.; Matsumoto, Y.; Koyama, H.; Cabral, H. Cyclic RGD-Linked Polymeric Micelles for Targeted Delivery of Platinum Anticancer Drugs to Glioblastoma Through the Blood–Brain Tumor Barrier. *ACS Nano.* **2013**, *7*, 8583–8592.

Molina, J. A.; Jimenez-Jimenez, F. J.; Hernanz, A.; Fernandez-Vivancos, E.; Medina, S.; De Bustos, F.; Gomez-Escalonilla, C.; Sayed, Y. Cerebrospinal Fluid Levels of Thiamine in Patients with Alzheimer's Disease. *J. Neural Transm.* **2002**, *109*, 1035–1044.

Molina, M.; Asadian-Birjand, M.; Balach, J. Stimuli-Responsive Nanogel Composites and their Application in Nanomedicine. *Chem. Soc. Rev.* **2015**, *44*, 6161–6186.

Mora-Huertas, C. E.; Fessi, H.; Elaissari, A. Polymer-Based Nanocapsules for Drug Delivery. *Int. J. Pharm.* **2010**, *385*, 113–142.

Muthu, M. S.; Kulkarni, S. A.; Liu, Y.; Feng, S. S. Development of Docetaxel-Loaded Vitamin E TPGS Micelles: Formulation Optimization, Effects on Brain Cancer Cells and Biodistribution in Rats. *Nanomed.* **2012**, *7*, 353–364.

Neamtu, I.; Rusu, A. G.; Diaconu, A.; Nita, L. E.; Chiriac, A. P. Basic Concepts and Recent Advances in Nanogels as Carriers for Medical Applications. *Drug Deliv.* **2017**, *24*, 539–557.

Neelov, I. M.; Janaszewska, A.; Klajnert, B.; Bryszewska, M.; Makova, N. Z.; Hicks, D.; Pearson, H. A.; Vlasov, G. P.; Ilyash, M. Y.; Vasilev, D. S.; Dubrovskaya, N. M.; Tumanova, N. L.; Zhuravin, I. A.; Turner, A. J.; Nalivaeva, N. N. Molecular Properties of Lysine Dendrimers and their Interactions with Abeta-Peptides and Neuronal Cells. *Curr. Med. Chem.* **2013**, *20*, 134–143.

Neurological Diseases and Disorders Health Impacts of Climate Change. Environmental Health Perspectives and the National Institute of Environmental Health Sciences April 22, 2010. [Online]. https://www.niehs.nih.gov/health/materials/a_human_health_perspective_on_climate_change_full_report_508.pdf (accessed May 28, 2017).

Neurons: The Building Blocks of the Nervous System. [Online]. https://www.ck12.org/book/Human-Biology-Nervous-System/section/4.1/ (accessed May 29, 2017).

Noble, C. O.; Krauze, M. T.; Drummond, D. C.; Yamashita, Y.; Saito, R.; Berger, M. S.; Kirpotin, D. B.; Bankiewicz, K. S.; Park, J. W. Novel Nanoliposomal CPT-11 Infused by Convection-Enhanced Delivery in Intracranial Tumors: Pharmacology and Efficacy. *Cancer Res.* **2006**, *66*, 2801–2806.

Noble, G. T.; Stefanick, J. F.; Ashley, J. D.; Kiziltepe, T.; Bilgicer, B. Ligand-Targeted Liposome Design: Challenges and Fundamental Considerations. *Trends Biotechnol.* **2014**, *32*, 32–45.

Nukolova, N. Targeted Delivery of Chemotherapeutic Agents to Brain Tumors. International Conference on Central Nervous System-Drug Effects and Novel Drug Development. *J. Neurol Neurophysiol.* **2012**. DOI: 10.4172/2155-9562.S1.004.

Obermeier, B.; Daneman, R.; Ransohoff, R. M. Development, Maintenance and Disruption of the Blood-Brain Barrier. *Nat. Med.* **2013**, *19*, 1584–9156.

Pandey, R.; Khuller, G. K. Oral Nanoparticle-Based Antituberculosis Drug Delivery to the Brain in an Experimental Model. *J. Antimicrob. Chemother.* **2006**, *57*, 1146–1152.

Pandey, Y. R.; Kumar, S.; Gupta, B. K.; Ali J, Baboota S. Intranasal Delivery of Paroxetine Nanoemulsion via the Olfactory Region for the Management of Depression: Formulation, Behavioural and Biochemical Estimation. *Nanotechnol.* **2015**, *27*, 025102.

Pang, Z.; Gao, H.; Chen, J.; Shen, S.; Zhang, B.; Ren, J.; Guo, L.; Qian, Y.; Jiang, X.; Mei, H. Intracellular Delivery Mechanism and Brain Delivery Kinetics of Biodegradable Cationic Bovine Serum Albumin-Conjugated Polymersomes. *Int. J. Nanomed.* **2012,** *7,* 3421–3432.

Pang, Z.; Gao, H.; Yu, Y.; Guo, L.; Chen, J.; Pan, S.; Ren, J.; Wen, Z.; Jiang, X. Enhanced Intracellular Delivery and Chemotherapy for Glioma Rats by Transferrin-Conjugated Biodegradable Polymersomes Loaded with Doxorubicin. *Bioconjugate Chem.* **2011,** *22,* 1171–1180.

Pang, Z. Q.; Fan, L.; Hu, K. L.; Wu, B. X.; Jiang, X. G. Effect of Lactoferrin-and Transferrin-Conjugated Polymersomes in Brain Targeting: In Vitro and In Vivo Evaluations. *Acta Pharmacol. Sinica.* **2010,** *31,* 237–243.

Panicker, J. N.; Fowler, C. J.; Kessler, T. M. Lower Urinary Tract Dysfunction in the Neurological Patient: Clinical Assessment and Management. *Lancet Neurol.* **2015,** *14,* 720–732.

Paris-Robidas, S.; Brouard, D.; Emond, V.; Parent, M.; Calon, F. Internalization of Targeted Quantum Dots by Brain Capillary Endothelial Cells In Vivo. *J. Cerebral Blood Flow Metab.* **2016,** *36,* 731–742.

Patel, M. M.; Goyal, B. R.; Bhadada, S. V.; Bhatt, J. S.; Amin, A. F. Getting into the Brain. *CNS Drugs.* **2009,** *23,* 35–58.

Pearson, O. R.; Busse, M. E.; Van Deursen, R. W.; Wiles C. M. Quantification of Walking Mobility in Neurological Disorders. *QJM.* **2004,** *97,* 463–475.

Pezzoli, G.; Cereda, E. Exposure to Pesticides or Solvents and Risk of Parkinson Disease. *Neurol.* **2013,** *80,* 2035–2041.

Piotrovskiy, L. B.; Litasova, E. V.; Dumpis, M. A.; Nikolaev, D. N.; Yakovleva, E. E.; Dravolina, O. A.; Bespalov, A. Y. Enhanced Brain Penetration of Hexamethonium in Complexes with Derivatives of Fullerene C60. *Dokl. Biochem. Biophys.* **2016,** *468,* 173–175.

Prabhakar, K.; Afzal, S. M.; Surender, G.; Kishan, V. Tween 80 Containing Lipid Nanoemulsions for Delivery of Indinavir to Brain. *Acta Pharmaceutica Sinica B.* **2013,** *3,* 345–353.

Pritchard, C.; Mayers, A.; Baldwin, D. Changing Patterns of Neurological Mortality in the 10 Major Developed Countries 1979-2010. *Public Health,* **2013** DOI: 10.1016/j. puhe.2012.12.018

Radtchenko, I. L.; Sukhorukov, G. B.; Möhwald, H. A Novel Method for Encapsulation of Poorly Water-Soluble Drugs: Precipitation in Polyelectrolyte Multilayer Shells. *Int. J. Pharm.* **2002,** *242,* 219–223.

Raemdonck, K.; Demeester, J.; De Smedt, S. Advanced Nanogel Engineering for Drug Delivery. *Soft Matter.* **2009,** *5,* 707–715.

Ran, D.; Mao, J.; Zhan, C.; Xie, C.; Ruan, H.; Ying, M.; Zhou, J.; Lu, W.L.; Lu, W. D-Retro-Enantiomer of Quorum Sensing Peptides-Modified Polymeric Micelles for Brain Tumor Targeted Drug Delivery. *ACS Appl. Mater. Interfaces.* **2017.** DOI: 10.1021/acsami.7b03518

Rašović, I. Water-Soluble Fullerenes for Medical Applications. *Mater. Sci. Technol.* **2017,** *33,* 777–794.

Ren, J.; Shen, S.; Wang, D.; Xi, Z.; Guo, L.; Pang, Z.; Qian, Y.; Sun, X.; Jiang, X. The Targeted Delivery of Anticancer Drugs to Brain Glioma by PEGylated Oxidized Multi-Walled Carbon Nanotubes Modified With Angiopep-2. *Biomaterials.* **2012,** *33,* 3324–3333.

Rodrigues, S. F.; Fiel, L. A.; Shimada, A. L.; Pereira, N. R.; Guterres, S. S.; Pohlmann, A. R.; Farsky, S. H. Lipid-Core Nanocapsules Act as a Drug Shuttle Through the Blood Brain Barrier and Reduce Glioblastoma After Intravenous or Oral Administration. *J. Biomed. Nanotechnol.* **2016,** *12,* 986–1000.

Rughani, A. I. Brain Anatomy. August 24, 2015. [Online]. http://emedicine.medscape.com/article/1898830-overview (accessed May 28, 2017).

Rukavina A. S. A. Molecular Mechanisms in Alzheimer's Disease, 2004. [Online]. http://www.ifcc.org/ifccfiles/docs/150309200416.pdf (accessed May 29, 2017).

Salem, H. F.; Ahmed, S. M.; Hassaballah, A. E.; Omar, M. M. Targeting Brain Cells with Glutathione-Modulated Nanoliposomes: In Vitro and In Vivo Study. *Drug Design, Dev. Ther.* **2015,** *9*, 3705.

Sanchez, M. M.; Heyn, S. N.; Das, D.; Moghadam, S.; Martin, K. J.; Salehi, A. Neurobiological Elements of Cognitive Dysfunction in Down Syndrome: Exploring the Role of APP. *Biol. Psychiatr.* **2012,** *71*, 403–409.

Savić, R.; Eisenberg, A.; Maysinger, D. Block Copolymer Micelles as Delivery Vehicles of Hydrophobic Drugs: Micelle–Cell Interactions. *J. Drug Target.* **2006,** *14*, 343–355.

Scharfman, H. E. The Neurobiology of Epilepsy. *Curr. Neurol. Neurosci. Rep.* **2007,** *7*, 348–354.

Schizophrenia. [Online]. http://www.who.int/mental_health/management/schizophrenia/en/ (accessed May 30, 2017).

Sekar, A.; Bialas, A. R.; de Rivera, H.; Davis, A.; Hammond, T. R.; Kamitaki, N.; Tooley, K.; Presumey, J.; Baum, M.; Van Doren, V.; Genovese, G. Schizophrenia Risk from Complex Variation of Complement Component 4. *Nature* **2016,** *530*, 177–183.

Serlin, Y.; Shelef, I.; Knyazer, B.; Friedman, A. Anatomy and Physiology of the Blood–Brain Barrier. *Semin. Cell Dev. Biol.* **2015,** *38*, 2–6.

Serralheiro, A.; Alves, G.; Fortuna, A.; Falcão, A. Intranasal Administration of Carbamazepine to Mice: A Direct Delivery Pathway for Brain Targeting. *Eur. J. Pharm. Sci.* **2014,** *60*, 32–39.

Serramía, M. J.; Álvarez, S.; Fuentes-Paniagua, E.; Clemente, M. I.; Sánchez-Nieves, J.; Gómez, R.; de la Mata, J.; Muñoz-Fernández, M. Á. In Vivo Delivery of siRNA to the Brain by Carbosilane Dendrimer. *J. Control Release.* **2015,** *200*, 60–70.

Setia, A.; Ahlawat, S. Quetiapine Fumarate Loaded Mucoadhesive Nanoemulsion: Formulation, Optimization & Brain Delivery via Intranasal Administration. *Drug Deliv. Lett.* **2016,** *6*, 64–76.

Shah, L.; Kulkarni, P.; Ferris, C.; Amiji, M. M. Analgesic Efficacy and Safety of DALDA Peptide Analog Delivery to the Brain Using Oil-in-Water Nanoemulsion Formulation. *Pharm. Res.* **2014,** *31*, 2724–2734.

Shailaja, M.; Diwan, P. V.; Ramakrishna, S.; Ramesh, G.; Reddy, K. H.; Rao, Y. M. Development of Olanzapine Nano-Emulsion for Enhanced Brain Delivery. *Int. J. Pharm. Sci. Drug Res.* **2012,** *5*, 1648–1659.

Sharma, H. S.; Muresanu, D. F.; Lafuente, J. V.; Ozkizilcik, A.; Tian, Z. R.; Patnaik, R.; Mössler, H.; Sharma, A. *Timed Release of Cerebrolysin Using Titanate Nanospheres Induces Neuroprotection in Parkinson's Disease.* In *Biotech, Biomaterials and Biomedical TechConnect Briefs* 2016, Materials for Drug & Gene Delivery Chapter 3, 2016, pp 105–108.

Sharma, N.; Bhandari, S.; Deshmukh, R.; Yadav, A. K.; Mishra, N. Development and Characterization of Embelin-Loaded Nanolipid Carriers for Brain Targeting. *Artif. Cells, Nanomed. Biotechnol.* **2017,** *45*, 409–413.

Shaw, T. K.; Mandal, D.; Dey, G.; Pal, M. M.; Paul, P.; Chakraborty, S.; Ali, K. A.; Mukherjee, B.; Bandyopadhyay, A. K.; Mandal, M. Successful Delivery of Docetaxel to Rat Brain

Using Experimentally Developed Nanoliposome: A Treatment Strategy for Brain Tumor. *Drug Deliv.* **2017**, *24*, 346–357.

Shilo, M.; Motiei, M.; Hana, P.; Popovtzer, R. Transport of Nanoparticles Through the Blood-Brain Barrier for Imaging and Therapeutic Applications. *Nanoscale* **2014**, *6*, 2146–2152.

Shin, J. Y.; Ahn, Y. H.; Paik, M. J.; Park, H. J.; Sohn, Y. H.; Lee, P. H. Elevated Homocysteine by Levodopa is Detrimental to Neurogenesis in Parkinsonian Model. *PLoS One* **2012**, *7*, e50496.

Shuai, X.; Merdan, T.; Schaper, A. K.; Xi, F.; Kissel, T. Core-Cross-Linked Polymeric Micelles as Paclitaxel Carriers. *Bioconjugate Chem.* **2004**, *15*, 441–448.

Singh, A.; Garg, G.; Sharma, P. K. Nanospheres: A Novel Approach for Targeted Drug Delivery System. *Int. J. Pharm. Sci. Rev. Res.* **2010**, *5*, 84–88.

Singh, Y.; Meher, J. G.; Raval, K.; Khan, F. A.; Chaurasia, M.; Jain, N. K.; Chourasia, M. K. Nanoemulsion: Concepts, Development and Applications in Drug Delivery. *J. Control Release*. 2017. DOI: https://doi.org/10.1016/j.jconrel.2017.03.008.

Ślusarczyk, J.; Piotrowski, M.; Szczepanowicz, K.; Regulska, M.; Leśkiewicz, M.; Warszyński, P.; Budziszewska, B.; Lasoń, W.; Basta-Kaim, A. Nanocapsules with Poly-electrolyte Shell as a Platform for 1, 25-dihydroxyvitamin D3 Neuroprotection: Study in Organotypic Hippocampal Slices. *Neurotox. Res.* **2016**, *30*, 581–592.

Solans, C.; Esquena, J. O.; Forgiarini, A. M.; Uson, N. U.; Morales, D. A.; Izquierdo, P.; Azemar, N.; Garcia-Celma, M. J. Nano-Emulsions: Formation, Properties, and Applica-tions. *Surfactant Sci. Ser.* **2003**, *12*, 525–554.

Sood, S.; Jain, K.; Gowthamarajan, K. Intranasal Delivery of Curcumin–/INS; Donepezil Nanoemulsion for Brain Targeting in Alzheimer's Disease. *J. Neurol. Sci.* **2013**, *333*, 316–317.

Spetter, M. S.; Hallschmid, M. Intranasal Neuropeptide Administration to Target the Human Brain in Health and Disease. *Mol. Pharm.* **2015**, *12*, 2767–2780.

Srinageshwar, B.; Peruzzaro, S.; Andrews, M.; Johnson, K.; Hietpas, A.; Clark, B.; McGuire, C.; Petersen, E.; Kippe, J.; Stewart, A.; Lossia, O. PAMAM Dendrimers Cross the Blood-Brain Barrier When Administered through the Carotid Artery in C57BL/6J Mice. *Inter. J. Mol. Sci.* **2017**, *18*, 628.

Sudduth, T. L.; Powell, D. K.; Smith, C. D.; Greenstein, A.; Wilcock, D. M. Induction of Hyperhomocysteinemia Models Vascular Dementia by Induction of Cerebral Microhemor-rhages and Neuroinflammation. *J. Cereb. Blood Flow Metab.* **2013**, *33*, 708–715.

Sultana, F.; Imran-Ul-Haque, M.; Arafat, M.; Sharmin, S. An Overview of Nanogel Drug Delivery System. *J. Appl. Pharm. Sci.* **2013**, *3*, S95–S105.

Swami, R.; Singh, I.; Kulhari, H.; Jeengar, M. K.; Khan, W.; Sistla, R. p-Hydroxy Benzoic Acid-Conjugated Dendrimer Nanotherapeutics as Potential Carriers for Targeted Drug Delivery to Brain: An In Vitro and In Vivo Evaluation. *J. Nanopart. Res.* **2015**, *17*, 265.

Tamariz, E.; Wan, A. C.; Pek, Y. S.; Giordano, M.; Hernández-Padrón, G.; Varela-Echavarría, A.; Velasco, I.; Castaño, V. M. Delivery of Chemotropic Proteins and Improvement of Dopaminergic Neuron Outgrowth Through a Thixotropic Hybrid Nano-Gel. *J. Mater. Sci: Mater. Med.* **2011**, *22*, 2097–2109.

Thomas, S. V.; Nair, A. Confronting the Stigma of Epilepsy. *Ann. Indian Acad. Neurol.* **2011**, *14*, 158–163.

Tripathy, S.; Das, M. K. Dendrimers and Their Applications as Novel Drug Delivery Carriers. *J. Appl. Pharm. Sci.* **2013**, *3*, 142–149.

Upadhyay, P.; Trivedi, J.; Pundarikakshudu, K.; Sheth, N. Direct and Enhanced Delivery of Nanoliposomes of Anti-Schizophrenic Agent to the Brain Through Nasal Route. *Saudi Pharm. J.* **2016**, *25*, 346–358.

van den Boorn, J. G.; Dassler, J.; Coch, C.; Schlee, M.; Hartmann, G. Exosomes as Nucleic Acid Nanocarriers. *Adv. Drug Deliv. Rev.* **2013**, *65*, 331–335.

Van Woensel, M.; Wauthoz, N.; Rosière, R.; Amighi, K.; Mathieu, V.; Lefranc, F.; Van Gool, S. W.; De Vleeschouwer, S. Formulations for Intranasal Delivery of Pharmacological Agents to Combat Brain Disease: A New Opportunity to Tackle GBM? *Cancers* **2013**, *5*, 1020–1048.

Vieira, D. B.; Gamarra, L. F. Getting into the Brain: Liposome-Based Strategies for Effective Drug Delivery Across the Blood–Brain Barrier. *Int. J. Nanomed.* **2016**, *11*, 5381.

Warren, G.; Makarov, E.; Lu, Y.; Senanayake, T.; Rivera, K.; Gorantla, S.; Poluektova, L. Y.; Vinogradov, S. V. Amphiphilic Cationic Nanogels as Brain-Targeted Carriers for Activated Nucleoside Reverse Transcriptase Inhibitors. *J. Neuroimmune Pharmacol.* **2015**, *10*, 88–101.

Weng, K. C.; Hashizume, R.; Noble, C. O.; Serwer, L. P.; Drummond, D. C.; Kirpotin, D. B.; Kuwabara, A. M.; Chao, L. X.; Chen, F. F.; James, C. D.; Park, J. W. Convection-Enhanced Delivery of Targeted Quantum Dot–Immunoliposome Hybrid Nanoparticles to Intracranial Brain Tumor Models. *Nanomed.* **2013**, *8*, 1913–1925.

Wirdefeldt, K.; Adami, H. O.; Cole, P.; Trichopoulos, D.; Mandel, J. Epidemiology and Etiology of Parkinson's Disease: A Review of the Evidence. *Eur. J. Epidemiol.* **2011**, *26*, S1–58.

Wu, L. Synthesis and Evaluation of Cationic Beta-Cyclodextrin Nanogels for Drug Delivery across the Blood Brain Barrier. *In AIChE Annual Meeting,* **2008**.

Xie, Y. T.; Du, Y. Z.; Yuan, H.; Hu, F. Q. Brain-Targeting Study of Stearic Acid-Grafted Chitosan Micelle Drug-Delivery System. *Int. J. Nanomed.* **2012**, *7*, 3235–3244.

Yamada, T.; Jung, D. Y.; Sawada, R.; Matsuoka, A.; Nakaoka, R.; Tsuchiya, T. Effects Intracerebral Microinjection and Intraperitoneal Injection of [60] Fullerene on Brain Functions Differ in Rats. *J. Nanosci. Nanotechnol.* **2008**, *8*, 3973–3980.

Yamada, T.; Nakaoka, R.; Sawada, R.; Matsuoka, A.; Tsuchiya, T. Effects of Intracerebral Microinjection of Hydroxylated-[60] Fullerene on Brain Monoamine Concentrations and Locomotor Behavior in Rats. *J. Nanosci. Nanotechnol.* **2010**, *10*, 604–611.

Yang, T.; Martin, P.; Fogarty, B.; Brown, A.; Schurman, K.; Phipps, R.; Yin, V. P.; Lockman, P.; Bai, S. Exosome Delivered Anticancer Drugs Across the Blood-Brain Barrier for Brain Cancer Therapy in Danio Rerio. *Pharm. Res.* **2015**, *32*, 2003–2014.

Ye, Y.; Xing, H.; Li, Y. Nanoencapsulation of the Sasanquasaponin from Camellia Oleifera, its Photo Responsiveness and Neuroprotective Effects. *Int. J. Nanomed.* **2014**, *9*, 4475–4484.

Yokota, H.; Nakabayashi, M.; Fuse, A.; Mashiko, K.; Yamamoto, Y.; Henmi, H.; Otsuka, T.; Awaya, S.; Kobayashi, S.; Nakazawa, S. Continuous Intracarotid Infusion of Mannitol in Severe Head Injury. *No Shinkei Geka.* **1993**, *21*, 205–211.

Yokoyama, M.; Okano, T.; Sakurai, Y.; Suwa, S.; Kataoka, K. Introduction of Cisplatin Into Polymeric Micelle. *J. Control Release.* **1996**, *39*, 351–356.

Yu, Y.; Pang, Z.; Lu, W.; Yin, Q.; Gao, H.; Jiang, X. Self-Assembled Polymersomes Conjugated With Lactoferrin as Novel Drug Carrier for Brain Delivery. *Pharm. Res.* **2012**, *29*, 83–96.

Yuan, G. Q.; Gao, D. D.; Lin, J.; Han, S.; Lv, B. C. Treatment of Recurrent Epileptic Seizures in Patients with Neurological Disorders. *Exp. Ther. Med.* **2013**, *5*, 267–270.

Yuyama, K.; Sun, H.; Usuki, S.; Sakai, S.; Hanamatsu, H.; Mioka, T.; Kimura, N.; Okada, M.; Tahara, H.; Furukawa, J. I.; Fujitani, N. A Potential Function for Neuronal Exosomes: Sequestering Intracerebral Amyloid-β Peptide. *FEBS Lett.* **2015,** *589,* 84–88.

Zhan, C.; Li, B.; Hu, L.; Wei, X.; Feng, L.; Fu, W.; Lu, W. Micelle-Based Brain-Targeted Drug Delivery Enabled by a Nicotine Acetylcholine Receptor Ligand. *Angew. Chem. Int. Ed.* **2011,** *50,* 5482–5485.

Zhang, H.; Meng, J.; Zhou, S.; Liu, Y.; Qu, D.; Wang, L.; Li, X.; Wang, N.; Luo, X.; Ma, X. Intranasal Delivery of Exendin-4 Confers Neuroprotective Effect Against Cerebral Ischemia in Mice. *AAPS J.* **2016,** *18,* 385–394.

Zhang, P.; Hu, L.; Wang, Y.; Wang, J.; Feng, L.; Li, Y. Poly (ε-caprolactone)-block-poly (ethyl ethylene phosphate) Micelles for Brain-Targeting Drug Delivery: In Vitro and In Vivo Valuation. *Pharm. Res.* **2010,** *27,* 2657–2669.

Zhang, L.; Cao, Z.; Li, Y.; Ella-Menye, J. R.; Bai, T.; Jiang, S. Softer Zwitterionic Nanogels for Longer Circulation and Lower Splenic Accumulation. *ACS Nano.* **2012,** *6,* 6681–6686

Zhao, D.; Alizadeh, D.; Zhang, L.; Liu, W.; Farrukh O.; Manuel, E.; Diamond, D. J.; Badie, B. Carbon Nanotubes Enhance CpG Uptake and Potentiate Antiglioma Immunity. *Clin. Cancer Res.* **2011,** *17,* 71–82.

Zhao, M. X.; Zhu, B. J. The Research and Applications of Quantum Dots as Nano-Carriers for Targeted Drug Delivery and Cancer Therapy. *Nanoscale Res. Lett.* **2016,** *11,* 207.

Zhuang, X.; Xiang, X.; Grizzle, W.; Sun, D.; Zhang, S.; Axtell, R. C.; Ju, S.; Mu, J.; Zhang, L.; Steinman, L.; Miller, D. Treatment of Brain Inflammatory Diseases by Delivering Exosome Encapsulated Anti-Inflammatory Drugs from the Nasal Region to the Brain. *Mol. Ther.* **2011,** *19,* 1769–1779.

CHAPTER 5

NATURAL BIOPOLYMERIC NANOFORMULATIONS FOR BRAIN DRUG DELIVERY

JOSEF JAMPÍLEK[1,2,*] and KATARÍNA KRÁĽOVÁ[3]

[1]*Department of Analytical Chemistry, Faculty of Natural Sciences, Comenius University, Ilkovičova 6, 84215, Bratislava, Slovakia*

[2]*Regional Centre of Advanced Technologies and Materials, Faculty of Science, Palacký University Olomouc, Šlechtitelů 27, 78371 Olomouc, Czech Republic*

[3]*Institute of Chemistry, Faculty of Natural Sciences, Comenius University, Ilkovičova 6, 84215, Bratislava, Slovakia*

Corresponding author. E-mail: josef.jampilek@gmail.com

ABSTRACT

The blood–brain barrier (BBB) prevents the passage of many drugs that target the central nervous system (CNS). The BBB represents a structure with a complex cellular organization that separates the brain parenchyma from the systemic circulation. The BBB acts not only as a mechanical barrier but also as a metabolic barrier because of the presence of numerous enzymes that are able to inactivate xenobiotics potentially harmful to the CNS. On the other hand, many drugs, such as anti-infectious drugs, antineoplastics and a number of drugs for the treatment of neurological or psychiatric diseases, have to circumvent the BBB and reach their targets in the CNS. An application of nanoparticles as drug carriers can be considered as one of the possible strategies of drug delivery to the CNS. The application of nanomaterials as drug carriers have a number of advantages, especially targeted biodistribution and controlled release and thus, increased bioavailability and decreased toxicity. Currently, various polymer materials are preferred as nanocarriers because of their compatibility, inertness, and biodegradability. This contribution is focused on nanoformulations of various classes of drugs, the site of action of which is in the CNS, that were prepared using polymers of natural origin and their semisynthetic modifications representing a class of biocompatible and biodegradable carriers ensuring better therapeutic efficiency and toxicity profile of drugs.

5.1 INTRODUCTION

A central nervous system (CNS) disease can affect either the brain or the spinal cord. Brain disorders are among the most serious health problems facing our society, causing untold human suffering and enormous economic costs. They are also among the most mysterious of all diseases, and our ignorance of the underlying disease mechanisms is a major obstacle to the development of better treatments (McGovern, 2017). The brain is a complex, sophisticated system that regulates and coordinates body activities. Thus it is clear that it is vulnerable to various disorders. The brain can be damaged by the following: trauma, infections, degeneration, structural defects, tumors, blood flow disruption, and autoimmune disorders. Disorders of brain may involve (1) vascular disorders (such as stroke, ischemic attack, various hemorrhages); (2) infections (bacterial, fungal, viral, parasitic); (3) structural disorders (such as brain injury, various syndromes, brain tumors); (4) functional disorders (such as a headache, epilepsy, dizziness, and neuralgia); and (5) degeneration (such as Parkinson's disease (PD), Alzheimer disease (AD), multiple sclerosis, amyotrophic lateral sclerosis, and Huntington chorea) (Johns Hopkins University, 2017). From this listing, it results that many of different drugs from various classes for the treatment of or at least moderating symptoms of these disorders have to act in the brain tissue.

It can be stated that the administration of drugs to the brain refers to many difficulties. The most important is the necessity to overcome a series of barriers that protect CNS from invading pathogens, neurotoxic molecules, and circulating blood cells. These structures with diverse degrees of permeability include the blood–cerebrospinal fluid (CSF) barrier, the blood–brain barrier (BBB), the blood–retinal barrier, and the blood–spinal cord barrier (Cipolla, 2009; Saraiva et al., 2016). Among these, the BBB is the last, critical and serious obstacle for the permeation of drugs that require CNS action. It was first discovered by Ehrlich in 1885 (Gao, 2016). The BBB is a specialized substructure of the vascular system composed of the capillary endothelial cells that line the cerebral microvessels and the surrounding perivascular elements, including the basal lamina, pericytes, astrocyte endfeet, and interneurons. The endothelial cells are connected by tight junctions and adherens junctions (Cipolla, 2009) resulting in a very high transendothelial electrical resistance. On the other hand, in some brain pathologies (e.g., brain infections, neurodegenerative disorders and stroke) the BBB is altered and becomes more permeable allowing the entry of molecules that can induce inflammatory responses and neuronal damage (Abbott et al., 2010; Di and Kerns, 2015; Passeleu-Le Bourdonnec et al., 2013; Posadas et al., 2016; Obermeier et al., 2013).

The mechanism of passage between endothelial cells is named paracellular and utilized for ions and solutes that depend on a gradient of concentration. The passage occurring through endothelial cells is termed transcellular, and the balance between paracellular and transcellular transport is decisive to define the degree of permeability in a healthy BBB (Wolburg and Lippoldt, 2002). The transcellular pathway occurs in most cases with passive diffusion of lipophilic molecules, which takes place through specific receptors to transport molecules (Abbott et al., 2010). The permeation of hydrophilic and bulky molecules (carbohydrates, proteins, peptides) to the brain depends on a specific type of transport and specific receptors involved in the transport of such molecules, for example, glucose, insulin or transferrin (Gao, 2016; Persidsky et al., 2006). Other forms of transport occur via the formation of cellular invaginations known as transcytosis, pinocytosis, and endocytosis, that is, through the formation of vesicles (Gao, 2016; Di and Kerns, 2015; Saraiva et al., 2016; Simko and Mattson, 2014).

The BBB also represents a metabolic barrier, because it contains numerous enzymes. These enzymes can either metabolize potentially harmful drugs to CNS-inactive compounds or convert inactive drugs to their CNS-active metabolites or degrade them into metabolites or substrates of specific efflux transporters, such as P-glycoprotein (P-gp)/multidrug resistance (MDR) proteins belonging to ATP-binding cassette transporter (ABCs) family (Abbott et al., 2010; Choi and Kim, 2008; Di and Kerns, 2015; Passeleu-Le Bourdonnec et al., 2013; Saraiva et al., 2016).

To deliver drugs to the brain, the BBB is the first barrier. Many strategies were developed to overcome or bypass the BBB and allow active CNS drugs to reach their target. These various kinds of strategies can be classified with respect to the BBB as either invasive (direct injection into the cerebrospinal fluid or therapeutic or osmotic opening of the BBB) or non-invasive, such as the use of alternative routes of administration (e.g., intraventricular/intrathecal route, nose-to-brain route, olfactory and trigeminal pathways to brain), inhibition of efflux transporters, chemical modification of drugs (lipophilic analogues, prodrugs or bioprecursors, coupling with Trojan horses or BBB shuttles) and encapsulation of drugs into various carriers together with surface modification of these carriers using fragments of endogenous molecules recognized by influx transporters or cell penetrating peptides (Barar et al., 2016; Černíková and Jampílek, 2014; Crawford et al., 2016; Di and Kerns, 2015; Gao, 2016; Lu et al., 2014; Passeleu-Le Bourdonnec et al., 2013; Posadas et al., 2016; Wang et al., 2015a).

Besides above-mentioned strategies used for drug delivery to the brain, nanoparticles (NPs) or drug-delivery nanosystems as drug nanocarriers have been extensively studied for last few decades as a promising strategy of direct drug delivery to the CNS. NPs can be used to maintain drug levels in a therapeutically desirable range as well as to increase the half-lives, solubility, stability, and permeability of drugs (Barar et al., 2016; Khanbabaie and Jahanshahi, 2012; Posadas et al., 2016). It has been shown that nano delivery systems represent an excellent carrier system for effective drug delivery across blood-brain barrier to brain, and thus they can be used successfully to treat a wide variety of brain diseases (Jain, 2012; Nair, et al., 2016; Sarkar et al., 2017; Soni et al., 2016). Their uptake (passive or active targeting) into the brain is hypothesized to occur via adsorptive transcytosis and receptor-mediated endocytosis (Bazak et al., 2015; Gao, 2016; Wong et al., 2012; Yang et al., 2010a). Particle size, surface affinity, and stability in circulation are important factors influencing the brain distribution of colloidal particles. Currently, a tendency when multiple functionalities and moieties are incorporated as drug-delivery nanosystems to overcome limiting barriers and to achieve successful accumulation of nanotherapeutics specifically at diseased sites can be observed. Thus, site-specific delivery of therapeutics is one of the most important contributions of modern nanotherapeutic design (Bhaskar et al., 2010; Blanco et al., 2015; Comoglu et al., 2017; Jain, 2012; Meng, et al. 2017; Nair et al., 2016; Tosi et al., 2016; Zhang et al., 2017a).

As mentioned above, nose-to-brain delivery can be considered as an important alternative route for brain drug delivery (Muntimadugu et al., 2016; Rassu et al., 2016; Warnken et al., 2016; Yasir and Sara, 2014). The role of nanotherapeutics in treating brain disorders via nose-to-brain delivery, nanoformulation strategies for enhanced brain targeting via nasal route and neurotoxicity issues of NPs were discussed by Kozlovskaya et al., (2014), Lin et al. (2016a) and Shadab et al. (2015).

Critical evaluation of various forms of nanomaterials such as micelles, liposomes, nanoemulsions, nanogels, quantum dots, dendrimers or inorganic/ solid lipid/polymeric NPs and their mechanisms for drug transport across the

BBB were presented by Alyautdin et al. (2014), Li et al. (2017), Masserini (2013), and Posadas et al. (2016). The unique properties of NPs such as nanoscale size, large surface, higher solubility, and multi-functionality represent the capability to interact with composite cellular functions in new ways (Cacciatore et al., 2016; Saraiva et al., 2016). Nanonized drug substances are protected from aggregation by matrices. Organic-based biodegradable, frequently of natural or semisynthetic origin, polymer matrices have been more and more popular for preparation of formulations of nanotherapeutics (Eroglu et al., 2017; Gagliardi and Borri, 2017; Jain, 2012; James et al., 2016; Jampílek et al., 2015, 2019; Jampílek and Kráľová, 2017, 2018, 2019a, 2019b; Král et al., 2011; Lalatsa and Barbu, 2016; Lehto, et al., 2016; Nair, et al., 2016;). These materials are degradable chemically (non-enzymatically) or enzymatically in vivo to biocompatible, toxicologically safe products that are eliminated from the human body by standard metabolic pathways (Engineer et al., 2011). A number of such materials have been used as excipients/matrices in controlled drug delivery (Makadia and Siegel, 2011). The formulations based on these matrices are prepared by encapsulation technology that allows the design of controlled-release nanocarriers (Di and Kerns, 2015; Gao, 2016; Kawabata et al., 2011; Posadas et al., 2016).

The rapid development of nanomedicine and various nanosystems employed for nano-based drug delivery systems has offered new opportunities in designing and evaluating novel treatment strategies. Nanomedicines used for the treatment of CNS diseases were comprehensively described in some review papers (Fu et al., 2016; Reynolds and Mahato, 2017; Sun et al., 2016). Emerging advances in nanoengineering strategies for the treatment of neurodegenerative disorders connected with the AD, PD, amyotrophic lateral sclerosis and prion disease were analyzed by Nguyen et al. (2017) and Saraiva et al. (2016). DeMarino et al. (2017) presented an overview focused on the formulation and applications of NPs in improving the delivery efficiency of drugs with particular emphasis on the delivery of antiretroviral therapeutics to the brain to treat HIV, and the recent findings concerning nanocarriers for effective brain drug delivery in the light of current literature were summarized by Comoglu et al. (2017), and Jampílek and Kráľová 2018, 2019a. Nanosystems used for antitumor drugs are discussed, for example, by Gao (2016), Posadas et al. (2016), Shi et al. (2017a), Yang (2010b), and Jampílek and Kráľová 2019b, while nanocarriers for the therapy of fungal brain diseases are discussed by Gazzoni et al. (2012), Shao et al. (2012), Soliman (2017), and Voltan et al. (2016) and bacterial infections by Jampílek and Kráľová (2017), Masserini (2013), and Upadhyay (2014).

This chapter is focused on the effects and CNS targeting of nanoscale formulations containing a wide range of drugs, such as therapeutics of nervous and psychiatric diseases, antineoplastics and anti-infectives, that were prepared using polymers of natural origin and their semisynthetic modifications representing a class of biocompatible and biodegradable carriers ensuring better therapeutic efficiency and the improved toxicity profile of drugs.

5.2   LACTIC ACID-BASED MATRICES

5.2.1   POLY(LACTIC) ACID

Poly(lactic) acid (PLA) is a biodegradable (by hydrolysis and/or enzymatic activity) and bioactive aliphatic polyester rising by polymerization of lactic acid or the cyclic di-ester lactide. Lactic acid is manufactured by a fermentation process using lactic acid bacteria that convert simple carbohydrates such as glucose, sucrose, and galactose to lactic acid. Potatoes, starch, corn starch, cassava roots, and sugarcane are used as sources of carbohydrates. As PLA is biodegradable, it is suitable for applications such as medical implants or drug delivery matrices. PLA has been widely used for various biomedical applications because of its biodegradability, biocompatibility and nontoxic properties. In addition, it has low immunogenicity. Various PLA-based formulations are popular for preparation of controlled drug delivery systems, in spite of the fact that pure PLA-based nanoparticles possess low drug loading capacity and low encapsulation efficiency (Alsaheb et al., 2015; Auras et al., 2010; Gai et al., 2017; Lee et al., 2016; Tyler et al., 2016).

Coated breviscapine (BVP)-PLA-NPs with particle sizes of 177 and 319 nm and smooth spherical shapes administered intravenously as suspension in rats reached 9/4-fold increase in area under plasma concentration-time curve and 11/17-fold increase of elimination $t_{1/2}$ at application of 177 nm/319 nm NPs, respectively, compared with that of free BVP; they were mainly distributed in liver, spleen, heart, and brain, and the coated BVP-PLA-NPs could prolong the half-life of BVP because of their ability to penetrate the BBB, thus enhancing the accumulation of BVP in brain (Liu et al., 2008). PLA NPs encapsulating Endostar, a recombinant human endostatin (Ling et al., 2007), which was further conjugated with GX1 peptide, a cyclic 7-mer peptide, binding specifically to the human vascular endothelial cancer cells (Chen et al., 2009), and the near-infrared dye IRDye 800CW (IGPNE) represent multifunctional NPs that are able to facilitate the fluorescence imaging of tumor and the delivery of drug to the tumor region. In tumor-bearing mice, IGPNE enabled real-time imaging of U87MG tumors and enhanced antitumor treatment efficacy compared to free Endostar, suggesting improved antiangiogenic therapeutic efficacy in vivo (Li et al., 2015a). PLA NPs coupled with natural tripeptide glutathione (GSH) loaded with fluorescein sodium with the mean diameter of 258 nm showed a significant increase (ca. 5-folds) in fluorescein sodium uptake in the brain of Wistar rats compared to PLA NPs without conjugated GSH (Patel and Acharya, 2012). Li and Sabliov (2013) summarized recent advances on PLGA/PLA NPs enhanced neural delivery of drugs and focused attention on enhancing drug delivery across the BBB by surface modification of PLGA/PLA NPs using a coating with surfactants/polymers or by covalent conjugation of the NPs with targeting ligands. Nanoparticles of a block copolymer of PLA and hyperbranched polyglycerol, surface modified with adenosine providing a controlled release of camptothecin and killing U87 glioma cells in culture did not increase animal survival following i.v. administration to mice with intracranial U87 tumors, suggesting that enhanced NPs transport across the BBB may not be always reflected in pharmacological effects (Saucier-Sawyer et al., 2015). Bago et al. (2016) designed a PLA biocompatible electrospun nanofibrous scaffold (BENS) bearing drug-releasing human mesenchymal stem cells (hMSCs) and found that such a BENS-based implant was able to enhance hMSC retention in the surgical cavity 5-fold and prolonged persistence 3-fold compared to standard direct injection using a mouse model of brain cancer glioblastoma (GBM) surgical resection/recurrence. In vivo, the volume of established GBM xenografts was reduced 3-fold, following administration of BENS loaded with hMSCs releasing the antitumor protein TRAIL, and their application in the post-operative GBM surgical cavity resulted in the inhibition of the re-growth of residual GBM foci 2.3-fold and the prolongation of post-surgical median survival from 13.5 to 31 days in mice. Transferrin (Tf)-coated 3-bis(2-chloroethyl)-1-nitrosourea (BCNU)-loaded PLA NPs and BCNU-loaded PLA wafers prepared from BCNU-loaded PLA microspheres were found to release BCNU in two distinct phases, and combinational chemotherapy of BCNU-loaded wafer and BCNU-loaded PLA NPs showed more efficient inhibition of tumor growth in C6 glioma-bearing rats and prolonged the mean survival time of animals (164%) compared to the control group, and following the treatment, no tumors were visible by examination at 4 weeks, indicating potential benefits of the combination therapy of implantation of a BCNU-loaded wafer and intracarotid perfusion of BCNU-loaded NPs for glioma gene therapy (Han et al., 2012).

A formulation of the glial-derived neurotrophic factor (GDNF) blended into and/or covalently attached to the composite scaffolds consisting of electrospun PLA short nanofibers embedded within a thermo-responsive xyloglucan hydrogel was found to provide controlled and sustained delivery of GDNF showing eliciting effects on cell survival and dopaminergic axon growth. In Parkinsonian mice, the application of these composite scaffolds without deleterious impact on the host immune response suggested a notable improvement of the milieu of the injured brain connected with enhanced survival and integration of grafted neurons (Wang et al., 2016a). Using cholesterol, hydrophobic-modified dextran polyaldehyd polysaccharide coated PLA NPs with the particle size of about 160 nm were prepared and subsequently functionalized simultaneously with CD71 and EGFR antibodies and showed facilitated penetration of NPs across the BBB compared with monofuctionalized and native NPs as well as higher tumor accumulation in vivo (Yuan et al., 2009). The absolute bioavailability of chitosan (CS)-modified PLA NPs with encapsulated neurotoxin (NT) separated from the venom of *Naja atra*, which can block the transmission of the nerve impulse by binding to the acetylcholine receptor in the membrane, showing spherical particles with the mean size of 140 nm and zeta potential +33.71±3.24 mV intranasally administered using a microdialysis technique in free-moving rats was found to be ca. 151% compared to the corresponding NPs without CS functionalization, suggesting better brain targeting efficiency (Zhang et al., 2013a). Amphiphilic NPs based on CS and carboxy-enriched PLA with a diameter ranging from 150 to 180 nm were found to improve the stability of encapsulated temozolomide (TMZ), significantly delayed the hydrolysis of drug in the inactive metabolite in simulated physiological solution at pH 7.4 and showed pH-dependent TMZ release kinetics with the opportunity to increase or decrease the release rate (Di Martino et al., 2017).

The application of microbubble-enhanced unfocused ultrasound (MEUUS) technique developed by Yao et al. (2014) to mediate the extensive brain delivery of PEG-PLA NPs enabled the penetration of NPs through vascular walls, and their deep distribution in the parenchyma was found to be more than 250% of that estimated for the NPs without application of the above-mentioned MEUUS technique. Facilitated brain delivery of PEG-PLA NPs functionalized with Aβ-specific antibody 6E10 because of the application of MEUUS technique was observed as well.

Amphotericin B/PLA-b-PEG NPs coated with polysorbate 80 (Tween® 80) markedly increased drug concentration in the brain, following their transfer across the BBB, and these NPs also reduced the drug toxicity to liver, kidney and blood system with improved therapeutic effect (Ren et al., 2009). TMZ-loaded PLA NPs surface-coated by PEG-1000 and Tween® 80 showed enhanced permeation of the drug into the brain and simultaneously less deposition in the highly perfused organs like liver (Jain et al., 2016a). Biocompatible curcumin (CUR) magnetic NPs coated with PEG-PLA block copolymer and polyvinylpyrrolidone with mean diameter <100 nm exhibited no cytotoxicity in differentiated human neuroblastoma cells (SH-SY5Y) and after injection enabled visualization of amyloid plaques in mouse brains, and it was observed that they were co-localized with amyloid plaques suggesting the potential of these magnetic NPs to be used for non-invasive diagnostics of AD using MRI (Cheng et al., 2015). Amantadine (AMT)–loaded polymeric micelles of PLA-b-PEG diblock copolymer modified with a sialic acid derivative (methyl-β-neuraminic acid: mNA) with the PLA block at the core, PEG-mNA block on the shell and an anionic charge on the surface of the micelle at physiological pH could exhibit multiple therapeutic activities connected with the antiviral activity of mNA on the shell and the sustained release of the antiviral drug AMT from the micellar core (Ahn et al., 2010).

B6 peptide-modified PEG-PLA NPs exhibited a notable improvement of accumulation in brain capillary endothelial cells via lipid raft-mediated and clathrin-mediated endocytosis compared to NPs without B6 functionalization and could be considered as a drug delivery system allowing the brain delivery of neuropeptides, because the administration of neuroprotective peptide NAPVSIPQ (NAP) encapsulated in these B6 functionalized NPs to AD mouse models resulted in remarkable amelioration in learning impairments, cholinergic disruption and loss of hippocampal neurons even at a lower dose (Liu et al., 2013a).

The study of the interaction of nanoscale micelles of PEG-b-PLA/paclitexal (PTX) conjugate with rat brain glioma C6 cells showed that they enter the cell and are distributed, and the released drug reacts with the genes in the cell nuclei resulting in changes in the gene, which affects the growth cycle of the cells causing G2-M cell cycle arrest and inducing apoptosis of C6 cells (Wang et al., 2008). PTX incorporated together with avidin and biotin in spherical PEG-PLA NPs with particle size of about 110 nm and zeta potential of -10 mV with Tf as the targeting ligand showed increased antitumoral activity of PTX against a brain tumor (glioma) cells (BT4C) in vitro compared to commercial PTX formulation Taxol® and PTX-loaded non-targeted biotinylated and non-biotinylated PEG-PLA NPs (Pulkkinen et al., 2008). Lactoferrin (Lf)-conjugated PEG-PLA NPs with 55 Lf molecules conjugated to each NP and incorporating fluorescent probe, coumarin-6 were uptaken by bEnd.3 cells in a considerably higher extent than unconjugated NPs, and also in vivo following the i.v. administration of Lf-conjugated NPs, the coumarin-6 concentration in the mice brain reached about 3-fold values compared to NPs without Lf functionalization (Hu et al., 2009). According to IC_{50} values, the formulation of Tf-conjugated PLA-D-α-tocopheryl polyethylene glycol succinate (TPGS) NPs with docetaxel (DTX) was found to be 23.4%, 16.9%, and 229% more efficient than PLGA NPs, PLA-TPGS NPs formulations and Taxotere®, respectively, after 24 h treatment (Gan and Feng, 2010). Biodegradable PEG-PLA copolymer NPs with biotin groups on a surface coupled with the functional protein Tf showed targeting ability and using C6 glioma tumor-bearing rat model it was shown that they could penetrate into the tumor in vivo (Ren et al., 2010). The Tf-modified PEG-PLA NPs conjugated with resveratrol with particle size approx. 150 nm and 32 molecules of Tf on surface showed higher in vitro cytotoxicity against C6 and U87 glioma cells than the free drug, and their higher cytotoxicity was connected with increased cellular uptake of the drug-modified conjugates by glioma cells. Following i.p. administration to C6 glioma-bearing mice, they notably reduced tumor volume and accumulated in a brain tumor causing a prolongation of animal survival (Guo et al., 2013).

PEG-PLA NPs functionalized on the surface with Penetratin (a cell-penetrating peptide with relatively low content of basic amino acids) with the particle size of 100 nm and the zeta potential of -4.42 mV notably improved brain uptake and reduced accumulation in the non-target tissues compared with low-molecular-weight protamine (a cell-penetrating peptide with high arginine content)-functionalized NPs (Xia et al., 2012). Methoxy-PEG-PLA and maleimide-PEG-PLA NPs surface functionalized with CD-71/OX-26 antibodies with particle sizes of 200 ± 25 nm and zeta potentials of -18 ± 1 mV showed considerably higher concentrations of fluorescein isothiocyanate-dextran (FD4) in the brain of male Sprague Dawley rats receiving 0.4 mg/kg FD4 and equivalent nanoparticulate formulation through lateral tail vein than pure-FD4 treated animals, and immunopegylated NPs were able to sustain and enhance the FD4 concentration in the CNS for at least 3 days (Bommana et al., 2012). Cyclic Arginine-Glycine-Aspartic acid-D-Tyrosine-Lysine (c(RGDyK)/DTX/PLA/PEG micelles showed enhanced accumulation in U87MG and 9L glioblastoma cell lines with notable cytotoxic effects as well as G2/M phase arrest and stronger microtubule-stabilizing effects. They showed excellent growth inhibition of glioma spheroids compared to the non-targeted micelles and free DTX, and more intensive tumor regression in U87MG tumor-bearing mice suggested their superior anti-glioblastoma efficacy, which could be connected with the ability of these micelles to bind the αvβ3 integrin receptor (Li et al., 2015b).

By conjugating peptides TGN (12-amino-acid-peptide displayed by bacteriophage Clone 12-2) (Ireson et al., 2006; Li et al., 2011) and QSH (showing good affinity with $A\beta_{1-42}$ form the amyloid plaque) (Bartnik et al., 2010; van Groen et al., 2009) used as specifically targeting ligands at the BBB to the surface of PEGylated PLA NPs a dual-functional NPs drug delivery system was designed enabling enhanced and precise targeted delivery at the optimal maleimide/peptide molar ratio 3 for both TGN and QSH on the surface of NPs to amyloid plaque in the brains of AD model mice (Zhang et al., 2014a).

5.2.2 POLY(LACTIC-CO-GLYCOLIC) ACID

Biocompatible poly(lactic-co-glycolic acid) (PLGA) NPs showing reduced toxicity are frequently studied as vehicles suitable to deliver drugs to the site of action because of their increased drug loading capacity and versatile structure allowing surface functionalization (Jain et al., 2011a; Kapoor et al., 2015). As a component of various drug delivery systems for controlled and sustained-release properties, it has been approved by the US Food and Drug Administration and the European Medicine Agency (Danhier et al., 2012; Makadia and Siegel, 2011). The linear copolymer PLGA consists of different ratios of lactic and glycolic acids. Physicochemical properties, as well as crystallinity, can be changed depending on the ratio of lactic and glycolic acids. PLGA degrades by hydrolysis of its ester bonds in aqueous environments (Danhier et al., 2012; Engineer et al., 2011). PLGA NPs cannot distinguish between different cell types, however their surface functionalization and development of new nanosized dosage forms (e.g., core-shell-type lipid-PLGA hybrids, cell-PLGA hybrids, receptor-specific ligand-PLGA conjugates and theranostics) could contribute to overcoming these delivery challenges, improve the in vivo performance and achieve the desired therapeutic effects, whereby functional PLGA-based nanoparticulate systems are suitable for effective delivery of chemotherapeutic, diagnostic and imaging agents (Sah et al., 2013).

Because PLGA NPs are one of the most promising drug and gene delivery systems for crossing the BBB, multiple strategies have been developed for their use to deliver compounds across the BBB focused on (1) facilitation of the travel from the injection site (pre-transcytosis strategies), (2) enhancement of the passage across the brain endothelial cells (BBB transcytosis strategies), and (3) achievement of targeting of impaired nervous system cells (post-transcytosis strategies), and PLGA NPs modified according to these three strategies are denoted as first, second and third generation NPs, respectively (Cai et al., 2016).

Sempf et al. (2013) studied adsorption of plasma proteins on uncoated PLGA NPs after their incubation in human plasma using mass spectrometry and found that the best signal in terms of quality (high MASCOT score) in human plasma was estimated for apolipoprotein E, vitronectin, histidine-rich glycoprotein, and kininogen-1, that is, proteins being also constituents of high density lipoproteins. Zhang et al. (2017b) reported the feasibility of serving PEGylated PLGA-based phase shift nanodroplets as an effective alternative mediating agent for a focused ultrasound-induced BBB opening.

Zhang et al. (2017c) investigated the effect of CUR and the PLGA-encapsulated CUR NPs on BBB disruption and used an endovascular perfo-ration rat subarachnoid hemorrhage (SAH) model to evaluate a possible mechanism underlying BBB dysfunction in early brain injury. The CUR NPs improved neurological function, reduced brain water content and Evans blue dye extravasation after SAH over free CUR, they prevented the disruption of tight junction protein (ZO-1, occludin and claudin-5) and in this way weak-ened BBB dysfunction, up-regulated glutamate transporter-1 and attenuated the glutamate concentration of cerebrospinal fluid following SAH, signifi-cantly reduced SAH-mediated oxidative stress and eventually reversed SAH-induced cell apoptosis in rats. Protective effects of CUR-NPs resulted in inhibition of inflammatory response and microglia activation. TMZ-loaded PLGA NPs exhibited a sustained drug release and higher uptake by C6 glioma cells and were found to prolong the activity of the loaded drug while retaining the anti-metastatic activity (Jain et al., 2014). In vitro cell permeability experiment using MDCK cell line as a BBB model showed that metoclopramide loaded in spherical PLGA NPs with particle size about 150 nm exhibited greater drug crossing than free metoclopramide and PLGA NPs did not have destructive effects on the P-gp receptors (Nikandish et al., 2016).

PLGA NPs coated with poloxamer 188 were found to be able to deliver effectively the brain-derived neurotrophic factor (BDNF) into the brain and also to improve neurological and cognitive deficits in mice with traumatic brain injury, indicating their neuroprotective effect (Khalin et al., 2016). The transport of doxorubicin (DOX)-loaded PLGA NPs overcoated with polox-amer 188 across the BBB was assumed to be connected with the adsorption of blood apolipoproteins (ApoE or ApoA-I) on the nanoparticle surface caused by the poloxamer 188-coating, followed by receptor-mediated transcytosis of the NPs, and as the main mechanism clathrin-mediated endocytosis of PLGA NPs by U87 cells was reported (Malinovskaya et al., 2017). The treatment with indole-3-carbinol-loaded PLGA NPs stabilized by Tween® 80 resulted in higher survival rates of PC12 neuronal cells injured by glutamate excitotoxicity compared with those treated with free indole-3-carbinol as well as in decreased levels of reactive oxygen species (ROS) and apoptosis-related enzymes (caspase-3 and -8) in these cells, indicating neuroprotective effects of these NPs (Jeong et al., 2015). Bacoside A-loaded PLGA NPs coated with Tween® 80 with particle size 70–200 nm, polydispersity index 0.391±1.2, encapsulation efficiency 57.11±7.11% and drug loading capacity 20.5±1.98% showed a sustained release pattern with the maximum release of up to 83.04±2.55% in 48 h, and in the in vivo study, treatment with these NPs resulted in higher brain concentration of the drug (23.94±1.74 µg/g tissue) compared to pure Bacoside solution (Jose et al., 2014). Tween® 80 coated PLGA NPs loaded with methotrexate (MTX)–Tf conjugates showed successful migration and trans-BBB passage after their administration to albino rats through the intravenous route, and such nanoformulations were characterized with greater compatibility, less organ toxicity and higher anti-tumor activity connected with their targeting and sustained delivery potential at brain cancer treatment (Jain et al., 2015). PLGA NPs with or without Tween® 80 coating were tested for delivery of a protein-tissue inhibitor of matrix metalloproteinases 1 (TIMP-1) across the BBB, and binding and penetration on primary rat brain capillary endothelial cell cultures and the rat brain endothelial 4 cell line was estimated. The toxicity of the coated NPs to endothelial cells was not significant, however, the coating contributed to the enhanced delivery of the protein across endothelial cell barriers, both in vitro and in vivo, while the uncoated TIMP-1 NPs, as well as TIMP-1 alone, did not cross the endothelial monolayer (Chaturvedi et al., 2014).

GSH-coated PLGA NPs loaded with PTX with particle size ca. 200 nm were found to be not a substrate for P-gp and were not be effluxed by P-gp present in the BBB. They showed a sustained drug release, higher cytotoxicity in RG2 cells compared with uncoated NPs, and higher cell death caused by these NPs could be connected with increased microtubule stabilization. An in vivo brain uptake study in mice showed higher brain uptake of the NPs containing coumarin-6 compared with a solution (Geldenhuys et al., 2011). PLGA NPs with conjugated CRT peptide, an iron-mimicry moiety targeting the whole complex of transferrin/transferrin receptor (Tf/Tf-receptor), which were loaded with PTX, showed superior antiproliferation effect on C6 glioma cells and stronger inhibitory effect on glioma spheroids compared to Tf NPs and Taxol and much deeper distribution pattern in glioma parenchyma compared with unmodified NPs and Tf NPs and caused considerably prolonged median survival of mice than the application of Taxol and Tf NPs (Kang et al., 2015). CS coating of PLGA NPs conjugated with a novel anti-Aβ antibody contributed to higher stability and targeting of these immune-nano vehicles to cerebrovascular deposits of AD amyloid protein (Jaruszewski et al., 2012). High crossing efficiencies through the BBB in vivo of loperamide-loaded PLGA NPs surface functionalized with an active targeting moiety, a monoclonal antibody against the Tf-receptor, was reported also by Fornaguera et al. (2015). Mathew et al. (2012) prepared water-soluble PLGA coated CUR NPs coupled with a Tet-1 peptide having the affinity to neurons and possessing retrograde transportation properties that were able to destroy amyloid aggregates and exhibit anti-oxidative property without being cytotoxic. PLGA NPs surface functionalized with poly-γ-glutamic acid were reported to be efficient nanocarriers for delivery of the antiretroviral drug saquinavir across the BBB (Kuo and Yu, 2011). Tramadol-loaded PLGA NPs conjugated with Tf and Lf were found to be stable for 6 months and demonstrated enhanced efficacy and significantly higher pharmacological effect over a period of 24 h than the unconjugated NPs, whereby Lf-functionalized NPs showed better antinociceptive effect than the Tf-functionalized NPs (Lalani et al., 2013). After intravenous administration to mice of tramadol-loaded PLGA NPs surface-modified with Lf, the brain targeting was found to be 1.62-fold higher than that of Tf-modified NPs (Lalani et al., 2012). Cell-penetrating peptide-modified PLGA NPs were reported to be able to potentially deliver insulin into the brain via the nasal route, showing a total brain delivery efficiency of 6%, indicating that they have a potential to be applied in the treatment of neurodegenerative diseases (Yan et al., 2013). The Lf and folic acid (FA) grafted PLGA NPs evaluated for transporting etoposide across the BBB were tested on the permeation of the monolayer of human brain-microvascular endothelial cells regulated by human astrocytes and inhibition of the multiplication of U87MG cells. The permeability coefficient for etoposide across the BBB using Lf/FA/PLGA NPs increased about 2-fold compared to PLGA NPs, and the antiproliferative efficacy against the growth of U87MG cells decreased as follows: Lf/FA/PLGA NPs > FA/PLGA NPs > PLGA NPs > free etoposide solution (Kuo and Chen, 2015).

Tosi et al. (2014) studied in vitro and in vivo possible mechanisms of a cell-to-cell transport of PLGA NPs modified with a 7-aminoacid glycopeptide (g7) and eventually also with the antibody. They found that g7-modified NPs could be transported intra- and intercellularly within vesicles after vesicular internalization, and cell-to-cell transport is mediated by tunneling-nanotube (TNT)-like structures in cell lines and in glial as well as neuronal cells in vitro and depends on F-actin. The cell-to-cell transport, which occurs independently from NP surface modification with antibodies can be increased by induction of TNT-like structures overexpressing M-Sec, a central factor and an inducer of TNT formation. Using primary hippocampal cell cultures, it was shown that PLGA NPs modified with g7 for BBB crossing with encapsulated CUR significantly decreased Aβ aggregates (Ruozi et al., 2017). PLGA NPs modified with a mutated form of diphtheria toxin (CRM197) were found crossing the BBB at a similar extent as g7-modified PLGA NPs in vivo, and in vivo experiments using CRM197-modified PLGA NPs with loperamide confirmed also their ability to function as drug carriers to CNS (Tosi et al., 2015). Salvalaio et al. (2016) investigated the ability of PLGA NPs modified with a g7 ligand to transfer fluorescein isothiocyanate-tagged albumin (FITC-albumin) as a model drug with a high molecular weight, comparable to the therapeutic enzymes necessary to be delivered across the BBB for the therapy of lysosomal storage disorders in the brain. In vivo experiments, conducted on wild-type mice and knockout mouse models for mucopolysaccharidosis type I and II showed that in all injected mice, the BBB crossing of albumin was efficient, indicating that g7-modified PLGA NPs could deliver molecules of high molecular weight to the brain. Loperamide and rhodamine-123 PLGA NPs differently modified with glycopeptide g7 or with a random sequence of the same aminoamides (random-g7) were administered via the tail vein in rats and it was estimated that a high central analgesia, corresponding to 14% of the injected dose was caused only by loperamide delivered to the brain with g7-NPs. Based on the computational analysis, linear conformation of g7 and globular conformation of random-g7 were estimated, indicating their different interaction with the BBB, and it was suggested that BBB crossing by g7-NP utilizes multiple pathways, mainly membrane–membrane interaction and macropinocytosis-like mechanisms (Tosi et al., 2011).

Rabies virus-derived peptide (RDP) conjugated to DOX-loaded PLGA NPs of average diameter ca. 257 nm and zeta potential −5.51±0.73 mV was found to enhance DOX accumulation in an SH-SYSY neural cell line specifically as opposed to non-neural cell lines (e.g., MDA-MD-23), and nicotinic acetylcholine receptors were found to be responsible for cellular interaction with RDP. It could be assumed that RDP is suitable to be used in a non-toxic, non-invasive brain-targeting drug delivery system (Huey et al.,

2017). Multifunctional PLGA NPs modified by angiopep-2 encapsulating both DOX and epidermal growth factor receptor (EGFR) siRNA efficiently delivered DOX and siRNA into U87MG cells, resulting in considerable cell inhibition, apoptosis and EGFR silencing in vitro, and they were also able to penetrate the BBB in vivo and securely enhanced accumulation of drugs in the brain. In the animal study using the brain orthotopic U87MG glioma xenograft model they prolonged the lifespan of glioma-bearing mice and caused also an evident cell apoptosis in glioma tissue (Wang et al., 2015b). DOX-loaded PLGA NPs coated with GSH-PEG conjugate showing a sustained drug release for up to 96 h in vitro exhibited notably higher permeation of the drug-loaded NPs compared with the free DOX solution through the coculture of rat brain endothelial (RBE4) and C6 astrocytoma cells, which indicates their potential in delivering DOX to cancer cells (Geldenhuys et al., 2015). Sharma et al. (2014) investigated the effect of short-chain cell-penetrating peptides CPPs-TAT, Penetratin and Mastoparan on the transport of Tf-liposomes with encapsulated DOX across brain endothelial barrier in vitro and in vivo and found that dual-functionalized liposomes showed improved DOX delivery as compared to single-ligand or unmodified liposomes.

Fluorescent-tagged NPs were observed in the cerebral cortex parenchyma after i.v. administration of PLGA NPs functionalized with an apolipoprotein E modified peptide (pep-apoE) responsible for low-density lipoprotein receptor (LDLR) binding or with lipocalin-type prostaglandin-D-synthase (L-PGDS) to mice, whereby the NPs were mostly internalized by neurons and microglia; glial cells showed a weak activation. Because this observation was not obtained with non-functionalized NPs, it could be stated that the peptidic moieties enable BBB traversal of the NPs and could be considered as potential brain drug carriers (Portioli et al., 2017).

The Tf-receptor-binding peptide T7-modified magnetic PLGA NPs containing co-encapsulated hydrophobic magnetic NPs, PTX and CUR exhibited synergistic effects on inhibition of tumor growth via mechanisms of apoptosis induction and cell cycle arrest, displaying significantly increased efficacy relative to the single use of each drug, which was reflected in a more than 10-fold increase in cellular uptake studies and a more than 5-fold enhancement in brain delivery compared to nontargeting NPs. In vivo study with an orthotopic glioma model showed that the application of a magnetic field resulted in higher treatment efficiency and reduced adverse effects, and all mice bearing orthotopic glioma survived in comparison to 62.5% survival of animals treated with free drugs (Cui et al., 2016). The cytotoxicity of Tf-conjugated magnetic silica PLGA NPs loaded with DOX and PTX evaluated in U-87 cells was found to be higher than the cytotoxicity of DOX-PTX-NPs or DOX-PTX-NPs with free Tf and increased with application of magnetic field. Also the in vivo therapeutic efficacy testing of drug-loaded NPs, which was evaluated in intracranial U-87 MG-luc2 xenograft of BALB/c nude mice, confirmed their strongest anti-glioma activity indicating the potential of these NPs for the delivery of dual drugs for the effective treatment of brain glioma (Cui et al., 2013).

PLGA NPs (150–170 nm) functionalized with two different types of monoclonal antibodies, OX26 (against transferrin receptors) and DE2B4 antibody, a molecule able to recognize the Aβ peptide with encapsulated peptide iAβ$_5$, secured significantly better uptake of immune NPs with a controlled delivery of the peptide iAβ$_5$ by porcine brain capillary endothelial cells compared to NPs without monoclonal antibody functionalization (Loureiro et al., 2016a).

Micelles of I6P8 peptide conjugated to biodegradable PEG-PLGA and loaded with DOX could be efficiently transported across the BBB and subsequently target glioma cells and could introduce the highest glioma apoptosis and the longest survival of glioma-bearing mice in vivo, indicating that such an interleukin-6 receptor-mediated micelle system is a promising cascade-targeting therapeutic agent capable of overcoming the BBB for glioma-targeted therapy (Shi et al., 2017b). The intravenous administration of PEG-PLGA NPs modified with phage display peptide TGN and loaded with active peptide NAP, a highly active fragment of activity-dependent neuroprotective protein, resulted in better improvement in spatial learning than NAP solution and NAP-loaded NPs in a Morris water maze experiment, and following treatment with TGN-modified NPs, neither morphological damage nor detectable Aβ plaques were estimated in mice hippocampus and cortex (Li et al., 2013). The powerful brain selectivity of 12-amino-acid-peptide TGN covalently conjugated onto the surface of PEG-PLGA NPs was reported previously by Li et al. (2011). Lf-conjugated PEG-PLGA NPs showed considerably higher accumulation of coumarin-6 in bEnd.3 cells than unconjugated NPs, whereby their increased uptake occurred via additional clathrin-mediated endocytosis processes, and after i.v. administration of Lf-conjugated NPs, about 3-fold amount of coumarin-6 was found in the mice brain compared to NPs without LF functionalization. Moreover, i.v. injection of urocortin loaded Lf-conjugated NPs resulted in powerful weakening of the striatum lesion caused by 6-hydroxydopamine (6-OHDA) in rats (Hu et al., 2011). The T7 peptide-conjugated carmustine (BCNU)-loaded PEG-PLGA micelles were found to accumulate in a tumor more efficiently than in the unconjugated probe and showed the improved curative effect, which was reflected in the lower loss of mice body weight, smaller tumor size and prolonged survival time (Bi et al., 2015). Donepezil (DPZ) loaded PLGA-b-PEG NPs showed the destabilizing effect on the amyloid-beta fibril ($A\beta_{1-40}$ and $A\beta_{1-42}$) formation in vitro, the ability to cross the BBB model in vitro and a controlled release profile. In astrocytes incubated with amyloid fibrils in the in vitro BBB model, increased gene and protein expression levels of IL-1β, IL-6, GM-CSF, TGF-β, MCP-1 and TNF-α were estimated, indicating an increased inflammation, while the administration of free DPZ and DPZ-loaded NPs resulted in a considerable dose-dependent decrease in both gene and protein expression levels of IL-1β, IL-6, GM-CSF and TNF-α, whereby in TGF-β and MCP-1 levels, no changes were detected (Baysal et al., 2017). Methylene blue-loaded GSH coated PLGA-b-PEG NPs with the mean particle size of 136 nm showing a sustained drug release for up to 144 h caused a potent reduction in both endogenous and overexpressed tau protein levels in human neuroblastoma SHSY-5Y cells expressing endogenous tau and transfected HeLa cells over-expressing tau protein and considerably higher permeation through the coculture of rat brain endothelial 4 (RBE4) and C6 astrocytoma cells compared to the methylene blue solution (Jinwal et al., 2014).

The NPs consisting of loperamide and PLGA-PEG-PLGA triblock and coated with poloxamer 188 or Tween® 80 showed higher in vitro BBB penetration (14.4–21.2%) than uncoated PLGA-PEG-PLGA NPs (8.2%) and PLGA NPs (4.3%) and also higher accumulation in brain tissue, and at 150 min after i.v. administration the maximal possible antinociception effect for surfactant-coated NPs was 21–35%, that is, 2- or 3-fold higher than that of uncoated PLGA-PEG-PLGA NPs (Chen et al., 2013).

Rosmarinic acid-loaded polyacrylamide-CS-PLGA NPs grafted with cross-reacting material 197 (CRM197) and apolipoprotein E (ApoE) were used to treat human brain microvascular endothelial cells, RWA264.7 cells and Aβ-insulted SK-N-MC cells and an increase in the concentration of CRM197 and ApoE led to a decrease of the transendothelial electrical resistance and increased the ability of rosmarinic acid to cross the BBB (Kuo and Rajesh, 2017).

5.3 PROTEIN-BASED MATRICES

5.3.1 ALBUMIN

Albumin is one of the blood plasma proteins, and in contrast to other plasma proteins, it is not a glycoprotein. It is synthesized by the liver. Albumin constitutes about 60% of all plasma proteins. It is a water-soluble globular protein of an ellipsoidal shape, consisting of a single polypeptide chain of 585 amino acids with 17 disulfide bridges. It is cleaved by proteases into three domains with different functions. The molecular weight of albumin is 67 kDa. It helps to maintain the constant internal environment of the organism and is especially important for the transport of various compounds including drugs through blood. Thus, it can be widely used in different drug formulations (Kratz, 2008; Loureiro et al., 2016b; Naveen et al., 2016).

The albumin NPs can penetrate the BBB and target glioma cells via the mechanisms of biomimetic transport mediated by albumin-binding proteins (e.g., SPARC and gp60), which are overexpressed in many tumors for transport of albumin as an amino acid and an energy source for fast-growing cancer cells. Moreover, enhanced BBB penetration, intratumoral infiltration, and cellular uptake could be obtained by modification of albumin NPs with the cell-penetrating peptide LMWP, which was confirmed in a study using an intracranial glioma model, and therapeutic mechanisms were associated with induction of apoptosis, antiangiogenesis, and tumor immune micro-environment regulation (Lin et al., 2016b). Albumin corona as a protective coating for an NPs-based drug delivery system could inhibit plasma protein adsorption and decrease the complement activation and ultimately prolong the blood circulation time and reduce the toxicity of the polymeric (e.g., poly(3-hydroxybutyrate-co-3-hydroxyhexanoate) NPs (Peng et al., 2013). Different drug delivery systems using albumin as a drug carrier were already reviewed by Kratz (2008). Later, a review paper focused on albumin nano-structures as advanced drug delivery systems highlighting particular benefits of albumin use in drug delivery systems, such as ready availability, ease of chemical modification, good biocompatibility, and low immunogenicity, was presented by Karimi et al. (2016).

The targeting of different sites within the CNS by magnetic albumin nanospheres (MANs), consisting of maghemite NPs hosted by albumin-based nanosphere was evaluated in the range from 30 min up to 30 days after administration to mice. It was found that the erythrocytes internalized and transported the MANs across the BBB and transferred them to glial cells and neuropils before internalization by neurons, mainly in the cerebellum, and the fact that no pathological alterations because of crossing the BBB were detected could be connected with the synergistic effect of Fe-based nanoscale particles and hosting albumin-based nanospheres (Silva et al., 2012). The desirable in vivo biocompatibility of MANs denoting promising potential for the use as a magnetic drug delivery system, especially in CNS disease therapy, was reported also by Estevanato et al. (2011).

Rivastigmine-loaded human serum albumin (HSA) NPs with drug/polymer ratio 1:2 showed the particle size of 83.71±4.2 nm, drug entrapment of 81.46±0.76 % and drug release of 55.59±3.80% in 12 h in vitro (Avachat et al., 2014). Gabapentin-loaded albumin NPs coated with Tween® 80 increased the drug concentration in the brain about 3-fold compared to the free drug and caused considerable reduction of the duration of all phases of convulsion in both maximal electroshock-induced and pentylenetetrazole-induced convulsion models in comparison with the free drug or drug-loaded NPs without coating, suggesting that the coating could contribute to enhanced effectiveness in treating epilepsy (Wilson et al., 2014). Spherical 5-fluorouracil-loaded albumin NPs with the average particle size of 141.9 nm, the surface charge of −30.3 mV, the polydispersity index of 0.374 and the drug loading capacity in the range from 4.22% to 19.8% w/w showed a sustained release in pH 7.4 phosphate buffer after an initial burst effect and were found to have superior cytotoxicity compared to the free drug (Wilson et al., 2012).

Apolipoprotein E attached to the surface of HSA NPs was found to facilitate drug transport across the BBB, probably after interaction with lipoprotein receptors on the brain capillary endothelial cell membranes (Michaelis et al., 2006). Unlike non-functionalized HSA NPs, the HSA NPs with covalently bound Apo E injected intravenously into SV 129 mice were detected in brain capillary endothelial cells and neurons, and their uptake into mouse endothelial (b.End3) cells in vitro was observed as well, indicating that these NPs are taken up into the cerebral endothelium by an endocytic mechanism followed by transcytosis into the brain parenchyma (Zensi et al., 2009). HSA NPs with covalently bound apolipoprotein A-I (Apo A-I) administered to SV 129 mice and Wistar rats as i.v. injection was estimated inside the endothelial cells of brain capillaries as well as within parenchymal brain tissue of mice and rats, in contrast to NPs without Apo A-I functionalization, which did not cross the BBB during the experiments (Zensi et al., 2010). Loperamide-loaded HSA NPs with covalently attached apolipoproteins E3, A-I, and B-100 that were administered to mice using intravenous injection by the tail-flick test exhibited notable antinociceptive effects after 15 min, which lasted for over 1 h. The maximally possible effects of these nano-formulations differed from each other and were statistically different from the controls, while the loperamide solution achieved no effect, and it was suggested that more than one mechanism is involved in the interaction of NPs with the brain endothelial cells and the resulting delivery of drugs to the CNS (Kreuter et al., 2007). Long circulatory PEGylated albumin NPs encapsulating water-soluble antiviral drug azidothymidine, the surface of which was modified by anchoring Tf, exhibited higher uptake in the brain tissues than unmodified NPs, and in albino rats, their localization in the brain showed also considerable enhancement following i.v. administration (Mishra et al., 2006).

GSH-conjugated bovine serum albumin (BSA) NPs with approx. 750 units of GSH conjugated per BSA NP incorporating fluorescein sodium exhibited significantly higher uptake by neuro-glial cells than unconjugated NPs, and following an intravenous administration, they were able to transport about 3-fold higher drug concentration compared to unconjugated NPs (Patel et al., 2013). DOX-loaded solid lipid NPs (SLNPs) conjugated with cationic BSA exhibited approximately 6-fold higher uptake by HNGC-1 cells than a free DOX solution, showed higher cytotoxicity than free DOX or unconjugated SLNPs and maximum transcytosis ability across brain capillary endothelial cells and were found to be less immunogenic compared to plain formulations (Agarwal et al., 2011a). Similar results were obtained also with MTX-loaded SLNPs conjugated with cationic BSA (Agarwal et al., 2011b). A DOX-loaded dual-targeting drug delivery system prepared by absorbing Lf onto the surface of mPEG2000-modified BSA-NPs was investigated using the BBB model in vitro, BCECs/C6 glioma coculture model in vitro and C6 glioma-bearing rats in vivo. It was shown that high levels of both Lf and mPEG2000 in this formulation resulted in the strongest cytotoxicity, the best effectiveness in the uptake both in BCECs and C6 and improved dual-targeting effects, and considerable accumulation of DOX in the brain, especially at 2 h postinjection, was estimated (Su et al., 2014). DTX-incorporated albumin-lipid NPs with mean particle size 110 nm and zeta potential −2.95 mV showed effective inhibition of the proliferation of several cell lines and induced cell apoptosis, and in vivo imaging confirmed their accumulation at the glioma site, whereby these NPs inhibited tumor growth, prolonged the median survival time of mice with gliomas and induced higher levels of apoptosis than standard DTX preparations (Gao et al., 2015). The feasibility of targeting of PTX-loaded BSA NPs to the brain using mice as an in vivo model was tested by Bansal et al. (2011).

A fluorescein-labeled HSA can be used for visualizing the malignant borders of brain tumors for improved surgical resection (Elsadek and Kratz, 2012). Tanshinone IIA (TIIA) is a diterpenoid naphthoquinone found in the root of *Salvia miltiorrhiza* Bunge that exerts antioxidant and anti-inflammatory actions. Cationic BSA-conjugated tanshinone TIIA PEGylated NPs prolonged circulation time and increased plasma concentration compared with intravenously administrated TIIA solution, they secured better brain delivery efficacy with a high drug accumulation in rat brain, reduced neurological dysfunctions, neutrophils infiltration, and neuronal apoptosis. Significant suppression of the expression of pro-inflammatory cytokines TNF-α and IL-8, upregulation of the expression of anti-inflammatory cytokines IL-10, the increase of TGF-β 1 level in the ischemic brain as well as significant mRNA expressions of GFAP, MMP-9, COX-2, p38MAPK, ERK1/2 and JNK, down-regulation of the protein levels of GFAP, MMP-9 and COX-2, and decreased phosphorylation of ERK1/2, p38MAPK, and JNK caused by these NPs suggested that they have superior neuroprotective effects on ischemic stroke through modulation of mitogen-activated protein kinase (MAPK) signal pathways involved in the cascades of neuroinflammation (Liu et al., 2013b, 2013c). The NPs of BSA linked with apolipoprotein E3 incorporating sumatriptan succinate applied to male Wistar rats reached high brain/plasma drug ratio of 9.45 and 12.67, respectively, 2 h post oral drug administration and behavioral studies on male Swiss albino mice showed enhanced anti-migraine potential compared to Tween® 80 coated poly(butyl cyanoacrylate) (PBCA) NPs (Girotra and Singh, 2016).

Neuroprotective features in the presence of H_2O_2 neuronal damage exhibited by species of cationized BSA with conjugated iron-chelating agent, deferasirox, was studied using PC12 cell line, and it was found that they hindered apoptotic cell death while enhancing autophagic process and also attenuated amyloid β-induced learning deficits when administered peripherally to rats (Kamalinia et al., 2015).

Zuo et al. (2017) conjugated two brain tumor targeting ligands angiopep-2 and rabies virus glycoprotein and coated via BSA as an intermatrix protein moiety, coated onto the surface of layered double hydroxide (LDH) through electrostatic interaction, and the immobilization of these BSAs on the LDH surface was achieved by cross-linking using glutaraldehyde. Such active targeting NPs were more efficiently uptaken by two neutral cells (U87 and N2a) compared to unmodified LDHs.

5.3.2 COLLAGEN AND GELATIN

Collagen is an extracellular, water-insoluble scleroprotein, fibers of which mainly consist of amino acids glycine, proline, hydroxyproline, and hydroxylysine. The tertiary structure of collagen consists of three chains that twine each other and have a common axis. Collagen forms the primary building block of connective tissues. It forms about 25–30% of all proteins in the body of mammals. The collagen is decomposed by collagenase. Collagen is used for the production of gelatine, glue, as a material for the production of surgical fibers, for the treatment of vascular prostheses and for sausage covers. It is also found in applications in plastic surgery, organ engineering, in cosmetics and as drug delivery carrier in its native or modified form. (Abou Neel et al., 2013; Choi et al., 2016; Jao et al., 2017; Posadas et al., 2016).

The sustained release of neurotrophin-3 (NT-3) and chondroitinase ABC (ChABC) from the electrospun collagen nanofiber scaffold for spinal cord injury repair was reported by Liu et al. (2012). Biofunctional nanofiber constructs prepared by incorporating NT-3 and ChABC onto electrospun collagen nanofibers may find useful applications in spinal cord injuries treatment by providing topographical signals and multiple biochemical cues that can promote nerve regeneration while antagonizing axonal growth inhibition for CNS regeneration.

Gelatin also belongs to carriers enabling a controlled drug release (Khan et al., 2016; Murai et al., 2017; Sabet et al., 2017; Sahoo et al., 2015). As mentioned above, gelatin is a mixture of peptides and proteins produced by partial hydrolysis of collagen obtained from various animal body parts. In spite of gelatin is 98–99% protein by dry weight, it has little additional nutritional value. It is soluble in most polar solvents; it is easily soluble in hot water and sets to a gel on cooling. It is not dissolved in cold water. Thus the mechanical properties of gelatin gels are very sensitive to temperature variations. It is used as a gelling agent in food, pharmaceutical drugs, vitamin capsules, photography and cosmetic manufacturing (Djagnya et al., 2010; Foox and Zilberman, 2015).

Gelatin NPs could be used for i.v. administration or for drug delivery to the brain, while gelatin microparticles are a suitable vehicle for cell amplification and for delivery of large bioactive molecules. Gelatin hydrogels can trap molecules between the polymer's crosslink gaps from which they diffuse into the bloodstream and by the modification of gelatin and its combinations with other biomaterials carrier systems permitting a specific, targeted and controlled release could be designed (Foox and Zilberman, 2015). In addition, nanofibrous scaffolds prepared from CS and gelatin (50:50) using electrospinning were found to be the ideal option for neural tissue engineering. The study of the effects of T-cells obtained from peripheral blood of multiple sclerosis patients on cells number and morphology of oligodendroglia cells OLN-93 derived from spontaneously transformed cells in primary rat brain glial cultures performed onto this nanofibrous scaffold, and mRNA analysis of oligodendrocytes focused on the evaluation of B7-1(CD80), B7-2(CD86) and IL-12p40 expressions confirmed that the expressions of these molecules in OLN-93 cells increased (Salati et al., 2016).

The enhanced neuroprotection effect of nerve growth factor (NGF) loaded gelatin nanostructured lipid carriers in spinal cord injury rats compared to free NGF was connected with the effective inhibition of endoplasmic reticulum stress-induced cell death via the activation of downstream signals PI3K/Akt/ GSK-3β and ERK1/2 (Zhu et al., 2016). Buyukoz et al. (2018) fabricated a gelatine-based biomimetic scaffold with interconnected macropores and nanofibrous structure, to which NGF-loaded alginate (ALG) microspheres were integrated, in which the encapsulation efficiency of NGF into 0.1% and 1% ALG microspheres was 85% and 100%, respectively; the release rate of NGF was controlled, and a notably extended release period was achieved by the immobilization of microspheres in the scaffold. The released NGF was found to induce neurite extension of PC12 cells, indicating that the above-mentioned scaffolds showing similar topologic and mechanical properties to brain tissue and pore structure suitable for cell growth and differentiation have promising potential to be used in brain tissue engineering. Zhang et al. (2015) used several biocompatible natural polyanions including heparin, dextran sulfate, and gelatin to form layer-by-layer (LbL) assembly with positively charged neurotrophin NGF and its model protein lysozyme to investigate whether the nanoscale thin LbL coatings could secure the release of NGF and lysozyme. The highest efficacy in loading proteins into LbL films was observed with gelatin enabling a sustained release of NGF and lysozymes in the period of about 2 weeks, and NGF maintained the bioactivity to stimulate neurite outgrowth from PC12 cells suggesting that such biocompatible LbL coating could be used for implanted cortical neural prostheses with better long-term performance in human patients.

Kim et al. (2016) investigated the intranasal delivery of inducible nitric oxide synthase (iNOS) siRNA encapsulated in gelatin NPs with diameter about 188 nm that were cross-linked with glutaraldehyde to the postischemic rat brain and found that infarct volumes were efficiently reduced (up to 42.1±2.6%) at 2 days after 60 min of middle cerebral artery occlusion (MCAO), when iNOS siRNA/gelatin NPs were delivered at 6 h post-MCAO, and reductions in neurological and behavioral deficits over a period of 2 weeks was observed suggesting a robust neuroprotective effect of this nanoformulation.

Gelatin NPs capable to pass into the brain parenchyma following i.n. administration used as a carrier for i.n. delivery of an osteopontin peptide for the treatment of ischemic stroke caused a 72% reduction in mean infarct volume and extended the therapeutic window of i.n. administered osteopontin peptide to at least 6 h post-middle cerebral artery occlusion (Joachim et al., 2014).

Gelatin nanospheres of selegiline containing 98.41±1.22% of the drug and having a round external smooth surface with a continuous wall, although they were shriveled, were found to show an initial burst release followed by a slow release over approx. 10 h and could be regarded as a formulation suitable to deliver the drug by oral route (Al-Dhubiab, 2013).

Zhao et al. (2016a) investigated the effect of i.n. administration of phospholipid-based gelatin NPs on the neuro-recovery effects of neuropeptide Substance P (SP) on hemiparkinsonian rats and found that this treatment resulted in better behavioral improvement, higher levels of tyrosine hydroxylase in substantia nigra along with much lower extent of phosphorylated c-Jun protein and caspase-3 than i.v. administration of these NPs at treatment with intranasal SP solution. These results confirmed that intranasally administered SP-loaded gelatin NPs could effectively deliver SP into the damaged substantia nigra region and exhibit neuro-recovery function. Similar results concerning i.n. administration of SP-loaded gelatin NPs to rats with 6-OHDA-induced hemiparkinsonism (daily for 2 weeks) supporting the finding that delivery of SP-loaded gelatin NPs can protect against 6-OHDA-induced apoptosis both in vitro and in vivo were published by Lu et al. (2015). Phospholipid-based gelatin NPs with particle size 143 nm and zeta potential −38.2±1.2 mV encapsulating basic fibroblast growth factor (bFGF) administered intranasally to hemiparkinsonian rats caused efficient enrichment of exogenous bFGF in olfactory bulb and striatum without affecting negatively the integrity of nasal mucosa and showed obvious therapeutic effects (Zhao et al., 2014).

HIV-derived Tat peptide and PEG-grafted onto gelatin-siloxane NPs were found to escape the capture by the reticuloendothelial system, cross the BBB and reach the CNS of mice indicating that they could serve as a new type of non-viral vector for the delivery of a drug or therapeutic DNA to the brain (Tian et al., 2012a). An in vivo experiment using a double-hemorrhage rat model showed that the Tat peptide-decorated gelatin-siloxane NPs used for the delivery of calcitonin gene-related peptide (CGRP) transgene caused enhanced vasodilatory CGRP expression in cerebrospinal fluid and thus could be applied for treatment of cerebral vasospasm after subarachnoid hemorrhage (Tian et al., 2013). Also, spherical PEG-gelatin-siloxane NPs conjugated with the SynB peptide with an average diameter of 150–200 nm showed more efficient brain capillary endothelial cell uptake and improved crossing of the BBB compared to similar NPs without conjugated SynB peptide (Tian et al., 2012b). Pang et al. (2016) reported that neural precursor cells generated from induced pluripotent stem cells with gelatin sponge-electrospun PLGA/PEG nanofibers could promote functional recovery after spinal cord injury.

Cilengitide NPs fabricated using gelatin and Poloxamer 188-grafted heparin copolymer exhibited considerable apoptotic and cytotoxic effects in C6 GBM cells, and their administration of 2 mg/kg twice weekly to rats combined with ultrasound-targeted microbubble destruction resulted in >3-fold higher drug concentration in the tumor of rats, prolonged drug retention in tumor and a notable reduction of renal clearance compared to free cilengitide and prolonged the median survival period of rats to approximately 80 days compared to the control animals (<20 days) and animals treated with the free drug (ca. 30 days) indicating that such combined therapy could be beneficial at treatment of gliomas (Zhao et al., 2016b). The use of gelatin NPs cross-linked with polyethylenimine forming a stable complex with a zeta potential of 42.47 mV and showing high transfection efficiency and low cell toxicity as a potential gene vector in gene therapy was proposed by Kuo et al. (2011).

As a promising, novel injectable system for transplantation of stem cells to the brain, a hydrogel biomaterial tissue scaffold derived from oligomeric gelatin and copper-capillary ALG gel was reported, in which the multipotent astrocytic stem cells (MASCs) retained their multipotency up to 2 weeks in vitro, while in vivo, in a transplant study in which this biomaterial tissue scaffold mixed with MASCs was injected into the brain of a neonatal rat pup, significant reactive gliosis was not observed, viable cells were retained within the injected scaffolds, and some delivered cells migrated into the surrounding brain tissue (Willenberg et al., 2011).

Ruan et al. (2015) designed a tumor microenvironment sensitive size-shrinkable theranostic system consisting of small-sized AuNPs fabricated onto matrix metalloproteinase-2 (MMP-2) degradable gelatin NPs, with DOX and Cy5.5 decorated onto AuNPs and surface-modified with a tandem peptide of RGD and octarginine to improve its glioma targeting ability. In vitro incubation with MMP-2 resulted in the size shrinking from 188.2 nm to 55.9 nm, and DOX and Cy5.5 were released in a pH-dependent manner. This system showed excellent glioma targeting and accumulation efficiency in vivo with good colocalization with neovessels, and colocalization of Cy5.5 with DOX suggested that Cy5.5 could be used for imaging of DOX delivery.

5.4 CARBOHYDRATE-BASED MATRICES

5.4.1 CYCLODEXTRINS

Cyclodextrins (CDs) are naturally occurring cyclic oligosaccharides consisting of α-(1→4)-linked glucopyranose units. According to a number of these units they are divided into α-CDs (6-membered ring molecule), β-CDs (7-membered ring molecule) and γ-CDs (8-membered ring molecule). They have a unique basket-shaped topology with an "inner-outer" amphiphilic character. The primary and secondary hydroxyl groups of the native CDs can be modified/substituted. They have been used as safe and effective carriers for both hydrophilic and lipophilic drugs (Frömming and Szejtli, 1994; Kurkov and Loftsson. 2013; Loftsson and Stefansson. 2007). The recent progress and future perspectives of CD-based supramolecular systems for drug delivery were summarized by Dash and Konkimalla (2015), Kurkov and Loftsson (2013), Yameogo et al. (2014) and Zhang and Ma (2013). Because of the favorable profile of CDs to form hydrophilic inclusion complexes with poorly soluble active pharmaceutical ingredients, they could be used as excipients for a wide spectrum of centrally acting drugs for the treatment of neurological/neurodegenerative diseases, stroke, neuroinfections and brain tumors (Vecsernyes et al., 2014). CDs could be used in site-specific or site-enhanced delivery through sequential, multistep enzymatic and chemical transformations, whereby enzymatic transformation of a redox targeting drug incorporated in the brain-targeting CDs is connected with a serious change in physicochemical properties. The inclusion complexes of drugs with CDs could cross the BBB and subsequently secure a controlled release of the drug (Buchwald and Bodor, 2016).

CDs have been extensively studied as drug delivery carriers through host-guest interactions. The CD-based electrospun formulation of aripiprazole was reported to be suitable for fast drug delivery through the oral mucosa based on the ultrafast dissolution of the drug from this formulation and the intensified flux across membranes (Borbas et al., 2015). Shityakov et al. (2016) investigated the cytotoxic effects of unmodified α-CD and modified CDs, including trimethyl-β-CD and hydroxypropyl-β-CD (HPCD), on immortalized murine microvascular endothelial (cEND) cells of the BBB, and found that trimethyl-β-CD was the most cytotoxic, while HPCD was non-toxic. Based on molecular dynamics simulation of cholesterol binding to the CDs, it was assumed that the cytotoxicity of CDs is connected with phospholipids and not with cholesterol extraction. The inclusion complex of HPCD with β-caryophyllene (BCP) that was administered to vascular dementia rats markedly increased the bioavailability of BCP, stimulated the recovery of cerebral blood flow, increased the expression levels of the cannabinoid receptor type 2 in brain tissues and the expression levels of PI3K and Akt, suggesting its protective effect on cognitive deficits induced by chronic cerebral ischemia (Lou et al., 2017).

The α, β, and γ fatty ester CDs with decyl alkyl chains on the secondary face, which self-assembled and formed nanostructures suitable to entrap diazepam released the drug in biphasic profiles, whereby the corresponding NPs of γ-CD were characterized with a low initial burst effect and a markedly delayed release (Geze et al., 2009). Mendez-Ardoy et al. (2012) prepared NPs of amphiphilic CD derivatives with a "skirt-type" architecture incorporating long-chain fatty esters at the secondary hydroxyl rim and a variety of chemical functionalities (e.g., iodo, bromo, azido, cysteaminyl, and isothiocyanato) at the primary hydroxyls rim with mean hydrodynamic diameters 100–240 nm capable to incorporate poorly bioavailable drugs such as diazepam with high efficiency. Gil et al. (2012) designed biodegradable, polymeric NPs composed of β-CD and poly(β-amino ester) segments and tested their permeability across in vitro BBB models constructed using bovine and human brain microvascular endothelial cell monolayers. It was found that the integrity of in vitro BBB models was not affected by the NPs, and the permeability of the NPs across the in vitro BBB models was significantly higher than that of dextran. The release of DOX from DOX-loaded NPs lasted for at least 1 month.

Godinho et al. (2014) described the use of PEGylated CDs as novel siRNA nanosystems, whereby increasing both PEG length and PEG density at the surface of NPs resulted in improved NPs stability in vitro, and the application of CD formulations resulted in increased systemic exposure and reduced clearance compared to naked siRNAs. Lf-conjugated β-CD NPs (ca. 92 nm) loaded with poorly water-soluble near-infrared fluorescent dye IR-775 chloride (IRC) showed greatly improved BBB transport efficiency and achieved 7-fold higher AUC0→2h of IRC in brain than free IRC, suggesting that Lf-conjugated β-CD NPs are suitable as a potential brain-targeting drug delivery system for hydrophobic drugs and diagnostic reagents that could not cross the BBB (Ye et al., 2013). Methylene Blue loaded nanospheres based on PLGA and a non-ionic amphiphilic CD with particle size ca. 200 nm did not affect the viability of human neuroblastoma SH-SY5Y cells in vitro and were found to exhibit notable neuroprotection against the metabolic effects of iodoacetic acid, mainly in the presence of NADH electron donor (Cannava et al., 2016).

Phospholipids were found to play an essential role in memory and learning abilities and act as a source of choline in acetylcholine synthesis. The effect of CS/phospholipid/β-CD microspheres on the improvement of cognitive impairment was studied by Shan et al. (2016). Formation of hydrogen bonds between phospholipids and the amide group of CS as well as with OH group of β-CD was estimated, and the treatment with the above-mentioned microspheres notably increased the learning and memory abilities of rats compared to the control group, attenuated the expression of protein kinase C-δ and inhibited the activation of microglia, indicating that they would be suitable for treatment of AD. Glycol CS/sulfobutylether-β-CD based NPs were tested for intranasal delivery of dopamine to the striatum by Di Gioia et al. (2015). An inclusion complex was formed between sulfobutylether-β-CD and dopamine, whereby the drug was situated on the external surface of NPs, and the acute administration of this nanoformulation into the right nostril of rats did not modify the levels of the neurotransmitter in both right and left striatum, while its repeated intranasal administration into the right nostril resulted in a considerable increase of drug concentration in the ipsilateral striatum. Following the acute administration of glycol CS/sulfobutylether-β-CD NPs incorporating dopamine into the right nostril, the presence of NPs was estimated only in the right olfactory bulb and no morphological tissue damage was observed, indicating that this preparation could be applied for nose-to-brain delivery of dopamine for the PD treatment. Anand et al. (2012) performed a spectroscopic and photophysical study of the NPs consisting of the association of DOX and artemisinin to β-CD-epichlorohydrin cross-linked polymers (ca. 15 nm) and found that the complexes evidenced an alcohol-like environment for artemisinin and improved inherent emission ability for DOX in the nanoparticle frame.

For the formulation of controlled-release dry suspension for reconstitution, complexes of cyclodextrin-based nanosponges with gabapentin were coated with Espheres and then with ethyl cellulose and Eudragit® RS-100, in which the drug was partially entrapped in nanocavities. The controlled release profile of gabapentin for 12 h was observed, and only minimal drug leaching was estimated in reconstituted suspension during storage (7 days, 45°C, relative humidity of 75%). The nanosponges effectively masked the taste of the drug, and the in vivo bioavailability of the controlled release suspension exceeded that of the free drug by about 24% (Rao and Bhingole, 2015).

5.4.2 CHITOSAN

Chitosan (CS) could be considered as an attractive natural polymer for brain-targeted nanoformulations. CS is a high-molecular-weight linear polycationic heteropolysaccharide comprising copolymers of β-(1–4)-linked β-D-glucosamine and N-acetyl-β-D-glucosamine. It is produced by partial alkaline N-deacetylation of chitin that is commercially extracted from shrimp and crab shells and is also found in nature in cell walls of fungi of the class *Zygomycetes* and in insect cuticles (Bernkop-Schnürch and Dünnhaupt 2012; Thakur and Thakur 2015). Biological applications of chitosan, chitin and other oligosaccharides and their derivatives were summarized by Manivasagan et al. (2014). Direct and indirect anti-Alzheimer effects of naked and surface-modified CS NPs with ligands improving permeation across the BBB and brain targeting CS NPs loaded with anti-Alzheimer drugs were overviewed by Sarvaiya and Agrawal (2015).

Trapani et al. (2011) tested CS NPs for dopamine brain delivery and found that such nanoformulation showed less cytotoxicity and neurotoxicity after 3 h than the free neurotransmitter, and its intraperitoneal acute administration at the dose of 6–12 mg/kg to rats induced a dose-dependent increase in striatal dopamine output. Following i.n. administration, ropinirole hydrochloride-loaded CS NPs showed sustained release profiles for up to 18 h and assured notable higher drug concentration in the brain compared to intranasally administered drug solution and higher brain/blood ratios of the drug (0.386±0.57) than estimated for i.n. administration of ropinirole HCl (0.251±0.09) at 0.5 h, suggesting direct nose-to-brain transport bypassing the BBB (Jafarieh et al., 2015). Intranasally administered CS-coated intranasal ropinirole nanoemulsion was also reported to represent a promising approach to the effective treatment of PD (Mustafa et al. 2012, 2015). Rasagiline-loaded CS glutamate NPs with mean particle size 151 nm administered by i.n. route resulted in notably higher drug concentrations in the brain than i.n. administration of the free drug or i.v. administration of these NPs, and following i.n. administration they showed also higher values of the direct transport percentage (69%) than the drug solution (i.n.; 66%), suggesting the potential of these NPs for the direct nose-to-brain targeting in PD therapy (Mittal et al., 2016). An optimized formulation of CS SLNPs containing sumatriptan succinate with particle size ranging from 192 to 301.4 nm and zeta potential of 30.2–51.4 mV showed a 4.5-fold increase in brain/blood ratio of drug after 2 h of drug administration in male Wistar rats, and notable anti-migraine activity exhibited by this nanoformulation confirmed the successful transfer of NPs across the BBB (Hansraj et al., 2015). Venlafaxine (VLF)-loaded CS NPs administered intranasally reached the brain/blood ratio of 0.1612 at 0.5 h compared to corresponding values of 0.0293 and 0.0700 estimated with the i.v. and i.n. administration of the free drug, respectively, and exhibited the highest drug transport efficiency (508.59%) and direct transport percentage (80.34%) compared to the administration of the free drug, suggesting that VLP-loaded NPs could be effectively used in the treatment of depression (Haque et al., 2012). Quetiapine fumarate (QF) based microemulsion with CS with spherical globules having the mean size of 35 nm and pH value of 5.6 showed excellent ex vivo nasal diffusion and, following intranasal administration, ensured prolonged drug retention at the site of action and ca. 3-fold higher nasal bioavailability in brain than the corresponding microemulsion without CS and the free drug, suggesting preferential nose-to-brain transport (80.51±6.46%) bypassing the BBB (Shah et al., 2016a). Similar results were obtained with QF-loaded CS NPs with particle size ca. 131.08±7.45 nm showing a considerably higher brain/blood ratio and 2-fold higher nasal bioavailability in the brain compared to the drug solution after i.n. administration and confirming the superiority of CS as a penetration enhancer (Shah et al., 2016b). A stable efficacious buffered mucoadhesive nanoemulsion of ziprasidone hydrochloride containing 0.5% by weight of CS with the mean globule size of 145 nm was reported to be effective for drug delivery to the brain via the i.n. route, which is connected with the beneficial effect of CS, which could modify the phospholipids bilayer resulting in the changed permeability of epithelial layers of nasal mucosa and affect the reversible opening of tight junctions between epithelial cells, and thus it increases the residence time of the formulation on nasal mucosa and acts also as a penetration enhancer. The superiority of the intranasal mucoadhesive nanoemulsion was observed also in the locomotor activity test and the paw test, where the formulation did not exhibit nasal ciliotoxicity (Bahadur and Pathak, 2012). The i.n. administration of olanzapine-loaded CS NPs to conscious rabbits resulted in notably higher systemic absorption and 51±11.2% absolute bioavailability compared to that of free drug (28±6.7%) (Baltzley et al., 2014). Bromocriptine (BRC) CS NPs showed ca. 4-fold longer retention into the nostrils of animals than the drug solution, and the blood ratio of i.n. administered BRC-CS NPs (0.96±0.05) exceeded that of i.n. administered BRC solution (0.73±0.15) as well as i.v. administered BRC NPs (0.25±0.05) at 0.5 h. Moreover, i.n. administration of BRC NPs resulted in notably higher dopamine concentration (20.65±1.08 ng/mL) than estimated for haloperidol-treated mice

(10.94±2.16 ng/mL), and they were able to significantly revert the degeneration of dopaminergic neurons in haloperidol-treated mice (Md et al., 2014). The drug targeting index and direct transport percentage for BRC loaded CS NPs with average particle size 161.3±4.7 nm and zeta potential +40.3±2.7 mV, following i.n. administration were found to be 6.3±0.8 and 84.2±1.9%, respectively (Md et al., 2013). The i.n. administration of rivastigmine-loaded CS NPs or of the pure drug to Wistar rats resulted in the brain/blood ratio of 1.712 and 0.790, respectively, at 30 min, and following the i.v. administration of rivastigmine, the corresponding value was only 0.235, suggesting that direct nose-to-brain transport bypassed the BBB and brain drug concentration, drug transport efficiency and direct transport percentage were notably higher with the application of CS NPs compared to other rivastigmine formulations, thus confirming their convenient brain targeting efficiency (Fazil et al., 2012). Following the nasal administration of AD drug galantamine (GAL) hydrobromide/CS complex NPs, the NPs were estimated intracellularly in the brain neurons, however they were found to exhibit a considerable reduction of the AChE protein level and activity in rat brains compared to the oral and nasal drug solutions (Hanafy et al., 2016). Donepezil-loaded CS nanosuspension with the mean particle size of 150–200 nm administered at the dose of 0.5 mg/mL into the nostrils of rats resulted in brain and plasma concentrations of 147.54±25.08 and 183.451±13.45 ng/mL, respectively, while following i.n. administration of drug suspension, the corresponding values reached only 7.2±0.86 and 82.8±5.42 ng/mL, respectively, and CS nanosuspension did not cause mortality, hematological changes or body weight variations in animals and was not toxic (Bhavna et al., 2014a). Also, donepezil loaded CS NPs with particle size ranging from 100 to 200 nm intranasally administered in rats showed higher drug transport efficiency (191.398%) and direct transport percentage (1834.480%) compared to the drug solution, suggesting that such nanoformulation could be used for the treatment of AD (Bhavna et al., 2014b). Spherical piperine-loaded CS NPs with particle size ca. 248 nm and zeta potential +56.30 mV considerably improved cognitive functions, inhibited acetylcholin esterase, showed antioxidant effects, reduced drug nasal irritation at i.n. administration and were no toxic to the brain, being effective at 20-fold decreased dose compared to the effective oral dose (Elnaggar et al., 2015). Following i.v. administration of the MTX-loaded CS nanogels, in which the NPs showing particle size of 118 nm was applied at the dose of 25 mg MTX/kg, to rats, a 10–15-fold increase in brain concentrations of the drug was observed compared to the free drug; however, surfactant-modified NPs did not deliver a significantly higher MTX amount to brain than naked NPs (Azadi et al., 2013).

Spherical CS NPs conjugated with Tf and bradykinin B2 antibodies as potential targeting ligands prepared to be delivered across the BBB into astrocytes with the mean particle size of 235 nm and the zeta potential of 22.88±1.78 mV showed notably improved cellular accumulation and gene silencing efficiency in astrocytes than non-functionalized CS NPs and single-antibody-modified CS NPs and considerably enhanced the knock-down effect of siRNA-loaded NPs, and these dual targeting NPs could be considered as a suitable strategy for inhibiting HIV replication in astrocytes (Gu et al., 2017). CS NPs loaded with intramembranous fragments of Aβ with the uniform particle size of 15.23 nm and the peptide association efficiency of 78.4% showed high brain uptake efficiency of nanoantigen of 81%, which was ca. 4-fold higher than that of antigen alone (20.6%), and advantageous immunogenicity and such a targeted nanovaccine delivery system could be used as a carrier for Aβ (Zhang and Wu, 2009). The delivery of small interfering RNA (siRNA) into the CNS tumor within hours could be obtained by concentrated suspensions of CS NPs, which are able to complex siRNA targeting galectin-1 (Gal-1) to a high percentage and protect them from RNAse degradation, and because of intracellular delivery of anti-Gal-1 siRNA, the expression of Gal-1 in both murine and human GBM cells was reduced, and >50% Gal-1 reduction in a tumor bearing mice was observed because of sequence-specific RNA interference (Van Woensel et al., 2016). Reduced activation of microglial cells in vitro and in rat models of spinal cord injury by CS polyplex mediated delivery of miRNA-124 reported by Louw et al., (2016) suggested that such treatment technique could be used to reduce inflammation for a multitude of CNS neurodegenerative conditions.

CS NPs surface modified with PEG and conjugated with monoclonal anti-transferrin receptor-1 antibody via streptavidin-biotin binding with particle size 100–800 nm and positive zeta potential, exhibiting an initial burst effect followed by a controlled release, were reported as a promising carrier system to transport nucleic acid-based drugs to brain parenchyma (Kozlu et al., 2014). CS NPs functionalized with PEG and modified with the monoclonal antibody OX26 (Cs-PEG-OX26) administered intraperitoneally in mice showed much higher brain uptake than corresponding unmodified NPs, which was connected with longer circulation and interaction between positively charged CS and negative charges of brain endothelium as well as with OX26 Tf-receptor affinity (Monsalve et al., 2015). N-Benzyloxycarbonyl-Asp(OMe)-Glu(OMe)-Val-Asp(OMe)-fluoromethyl ketone-loaded CS NPs surface-modified with PEG, to which monoclonal antibodies against the Tf receptor were conjugated via biotin-streptavidin bonds, exhibited a fast release of the effective compound within brain paren-chyma and inhibited ischemia-induced caspase-3 activity, thus providing neuroprotection (Yemisci et al., 2012). Nanocapsules coated with cationic CS or PEG with clozapine (CZP) as the drug model caused increased action duration in pseudo-psychosis induced by D,L-amphetamine in rats compared to corresponding Tween® 80-coated nanocapsules, the activity of CS-coated nanocapsules being superior, and these nanocapsules ensured the prolonged half-life of the drug in rats as well as a decrease in total clearance, suggesting their suitability as a delivery system for CZP in the treatment of schizo-phrenia (Vieira et al., 2016).

Riluzole encapsulated in CS conjugated N-isopropylacrylamide NPs coated with Tween® 80 with particle size about 50 nm intraperitoneally administered in doses 10, 20, and 40 µg drug/kg body weight of male Wistar rats after 1 h of middle cerebral artery occlusion caused an essential reduc-tion in infarct size, a reduction in the expression of immunological param-eters like NOS-2, NF-kB and COX-2 and was able to ensure considerable neuroprotection even at a very low concentration, showing therapeutic potential against cerebral ischemia (stroke) (Verma et al., 2016). Coating of gallic acid (GA)-loaded CS NPs with Tween® 80 was reported to improve brain targeting in treated male Swiss albino mice compared to free GA, which resulted in stronger effects connected with the reversal of scopol-amine-induced amnesia caused by GA showing antioxidant properties and thus improving the cholinergic functions that are impaired by ROS. On the other hand, the corresponding NPs without surfactant coating did not show a considerable change in pharmacodynamic or biochemical parameters (Nagpal et al., 2013).

Following i.v injection, the CS-conjugated Pluronic®-based nano-carrier with a specific target peptide for the brain (rabies virus glycopro-tein; RVG29) excellently accumulated in mice brain tissue exceeding the accumulation of the nanocarrier conjugated with the peptide only, and the effective delivery of β-galactosidase used as a model protein, which was loaded in the nanocarrier, in the brain was estimated as well, whereby the bioactivity of the delivered protein in the brain was maintained (Kim et al., 2013). CS-functionalized Pluronic® P123/F68 micelles loaded with myric-etin showed better cellular uptake and antitumor activity than the free drug in vitro, notably higher anticancer effect in vivo and altered the expression of apoptotic proteins, such as Bcl-2, BAD and BAX, in mice suggesting that they could be used as a drug delivery system for glioblastoma treatment (Wang et al., 2016b).

Following the i.n. administration of negative and positive CS-PLGA NPs labeled with rhodamine B to rats, it was found that both nanocarriers reached the brain and persisted in the brain up to 48 h after i.n. administration; however, the brain reaching was slowed down by positively charged NPs which used the trigeminal pathway, while the transport of negatively charged ones occurred by the olfactory pathway, and not even systemic pathways could be excluded (Bonaccorso et al., 2017).

CS-coated levodopa liposomes were found to notably reduce the scores of abnormal involuntary movement in rats compared to the free drug, and they could be applied in reducing dyskinesias inducing for PD. The mechanism might involve the pathway of signaling molecular phospho-ERK1/2, phospho-Thr34 DARPP-32 and ΔFosB in the striatum (Cao et al., 2016). Thermo-gelling injectable nanogels prepared by combining self-assembled nanocapsules of amphiphile-modified CS with glycerophosphate di-sodium salt and glycerol loaded with ethosuximide (ESM) exhibited an extended, mostly Fickian drug release in vitro, and in vivo, they were able to suppress spike-wave discharges in Long Evan rat model. Using human retinal pigmented epithelium cells, the excellent cytotoxicity of this nanoformulation was estimated, suggesting its suitability for being used as injectable depot gel for drug delivery (Hsiao et al., 2012). The ropinirole hydrochloride-loaded PLGA/dipalmitoylphospatidylcholine/trimethyl chitosan spray-dried spheroid microparticles with the size ranging from 2.09 μm to 2.41 μm showed a 235-fold enhancement of drug permeation across sheep nasal mucosa in an ex vivo study compared to the control, when the nanoformulation was prepared with trimethyl chitosan of low molecular weight (Karavasili et al., 2016).

Stearic acid-grafted CS micelles with the diameter of 22 nm and the surface potential of 36.4±0.71 mV in aqueous solution were efficiently uptaken by bEnd.3 cells (IC_{50} = 237.6±6.61 μg/mL), and micelles loaded with DOX showed slow drug-release behavior in vitro with a cumulative release up to 72% being observed within 48 h. The cytotoxicity of DOX-loaded micelles against glioma C6 cells was estimated as 2.664±0.036 μg/mL, while that of DOX.HCl was 0.181±0.066 μg/mL, and in vivo imaging confirmed the rapid transport of stearic acid-grafted CS micelles across the BBB and into the brain (Xie et al., 2012).

Biocompatible CS-coated nanostructured lipid carriers (CS NLCs) with the particle size of 114 nm and the surface charge of +28 mV were found to be uptaken by 16HBE14o-cells, did not cause haemagglutination or hemolysis processes during incubation with erythrocytes, and following i.n. administration did not affect adversely the nasal mucosa of mice, while securing efficient brain delivery of the NPs (Gartziandia et al., 2015). In addition, Gartziandia et al. (2017) investigated the in vivo neuroprotective effect of glial cell-derived neurotrophic factor (GDNF) encapsulated in a CS-coated nanostructured lipid carrier (CS-NLC-GDNF) with particle size of about 130 nm administered intranasally in a 6-OHDA partially lesioned rat model and found that daily intranasal administration of nanoformulation during 2 weeks resulted in behavioral improvement in rats and also notable improvement in both the density of tyrosine hydroxylase positive (TH$^+$) fibers in the striatum and the TH$^+$ neuronal density in the substantia nigra, and an in vitro experiment showed that PC-12 cells were protected by encapsulated GDNF against 6-OHDA toxin.

CS-coated and uncoated SLNPs as a nasal delivery system for the transport of β-secretase 1 (BACE1) siRNA potentially useful in the treatment of AD using RVG-9R to increase the transcellular pathway in neuronal cells were designed by Rassu et al. (2017). Such nanoformulations were characterized with satisfactory mucoadhesiveness and prolonged residence time in the nasal cavity and using Caco-2 as a model of epithelial-like phenotypes, it was found that siRNA released from CS-coated SLNPs permeated the monolayer to a greater extent than from other formulations.

The IC$_{50}$ values estimated for Tf functionalized DTX loaded TPGS conjugated CS NPs, NPs without Tf functionalization and Docel™ in C6 glioma cells were 0.41±0.04, 2.21±0.06 and 60.98±2.45 µg/mL, respectively, after 24 h incubation, suggesting that the drug formulated in Tf-functionalized TPGS CS NPs could be 149-fold more efficient than the commercial preparation Docel™. Similar results were obtained also in an in vivo study with i.v. administration of the above-mentioned DTX formulations to rats at the same drug concentration 7.5 mg/kg, and the total AUC for Tf-functionalized NPs, which showed also prolonged circulation in blood, was 4-fold higher than that for Docel™ (Agrawal et al., 2017a). Similar results were obtained also with Tf-receptor targeting bioadhesive micelles using TPGS conjugated CS, which, based on IC$_{50}$ values, were found to be 248-fold more effective than Docel™ after 24 h treatment of the C6 glioma cells and the relative bioavailability of these targeted micelles in vivo was 4-fold higher than that of Docel™ after 48 h of treatment (Agrawal et al., 2017b).

Thiolated trimethyl CS NPs grafted with a non-toxic carboxylic fragment of the tetanus neurotoxin tested to deliver a plasmid DNA encoding for the brain-derived neurotrophic factor (BDNF) in a peripheral nerve injury model were found to stimulate the release, and notable expression of BDNF in neural tissues causing better functional recovery after injury as compared to control treatments and increased expression of neurofilament and growth-associated protein GAP-43 in the injured nerves was observed, indicating that this nanoformulation could be applied in an effective thera-peutic intervention for peripheral neuropathies (Lopes et al., 2017). Also thiolated CS NPs with entrapped buspirone hydrochloride with the particle size of 226 nm evaluated for the drug for brain delivery through the intra-nasal route exhibited high cumulative percentage drug permeation through nasal mucosa (76.21%) and bioadhesion efficiency (90.218±0.134%) on porcine mucin and significantly higher brain drug concentration than drug solution after i.n. or i.v. administration (Bari et al., 2015). Thiolated CS NPs with the particle size of ca. 215 nm and zeta potential +17.06 mV were also reported to be suitable for nose-to-brain delivery of antidepressant selegiline hydrochloride (Singh et al., 2016).

AuNPs with the mean particle size of 18 nm coated on the surface with poly(ε-caprolactone diol) based polyurethane nanofibers with incorporated temozolomide-loaded CS NPs exhibiting a sustained drug release were found to effectively inhibit the growth of U-87 human glioblastoma cells and could be considered as a potential candidate for glioblastoma cancer treatment (Irani et al., 2017).

Ceria-containing hydroxyapatite NPs with attached GAL showing the prevailing existence of dispersed negatively charged rod-like particles conjugated with Ce nanodots exhibiting an optimizable in vitro release for GAL and Ce, which were i.p. injected into ovariectomized AD albino-rats during a period of 1 month at the dose of 2.5 mg/kg/day upregulated oxida-tive stress markers, degenerated neurons in hippocampal and cerebral tissues were wholly recovered, and Aβ-plaques were vanished. On the other hand, Ce-containing carboxymethyl CS-coated hydroxyapatite nanocomposites with well-structured developing aggregates of uncharged tetragonal-shaped particles laden with the accession of ceria quantum dots were found to delay the in vitro release of GAL and Ce, respectively (Wahba et al., 2016).

5.4.3 STARCH AND CELLULOSE

Starch is a mixture of amylose and amylopectin. Amylose is a linear polymer of α-(1→4)-linked D-glucopyranosyl units and represents about 20–30 wt% of standard starch. Amylopectin is composed of linear segments connected by α-(1→6)-branched D-glucopyranosyl units (5–6 wt% of standard starch) and non-branched α-(1→4)-linked D-glucopyranosyl units (70–80 wt% of standard starch) (Kim et al., 2015; Lancuški et al., 2015; Pérez and Bertoft, 2010). Natural starches can be chemically modified by substitutions of hydroxyl moieties (Chen et al., 2015; Engelberth et al., 2015).

Biocompatible starch-coated superparamagnetic iron oxide NPs (SPION) in buffered artificial cerebrospinal fluid injected into the brain parenchyma of anaesthetized rats were found to form a concentration gradient from the center of the injection site toward the periphery suggesting that they might be transported in the extracellular space as well as being internalized in nerve cells (Kim et al., 2003). PEG-modified, cross-linked starch-coated Fe_2O_3 NPs suitable for magnetic targeting showed improved plasma stability and enhanced tumor exposure of NPs in 9L-glioma rat model (Cole et al., 2011). The aminated, starch-coated, magnetic Fe_2O_3 NPs conjugated by β-glucosidase showed ca. 85% relative activity of the free enzyme but much better temperature stability and the application of an additional field on the surface of a tumor led to effective delivery of NPs into a subcutaneous tumor of a glioma-bearing mouse and the β-glucosidase activity in tumor lesions was about one order higher, with 2.14 of tumor/non-tumor enzyme activity (Zhou et al., 2013). Cationic protein protamine-loaded crosslinked and aminated starch-coated magnetic iron oxide NPs that were simultaneously PEGylated and heparinized were reported to be promising for simultaneous tumor targeting and imaging (Zhang et al., 2014b). They displayed 37-fold longer half-life (9.37 h) than heparin (0.15 h) and enabled extended exposure to tumor lesions in a flank 9L-glioma mouse model and reached 7-fold improvement of the tumor targeting ability over that of heparin (Zhang et al., 2013b).

Cellulose (CLS), the main structural polysaccharide in higher plants, is formed by β-(1→4)-linked D-glucopyranose chains that are laterally bound by hydrogen bonds to form microfibrils with a diameter in the nanoscale range, which are further organized in microfibril bundles (Klemm et al., 1998). Bacterial cellulose (BC) produced by aerobic bacteria of genera *Acetobacter*, *Agrobacterium*, and *Sarcina ventriculi* differs from plant CLS. Naturally, BC is characterized by high purity, strength, moldability, and increased water holding ability (Esa et al., 2014). Hydroxy moieties of cellulose can be modified by various substituents, thus materials based on CLS are characterized by low toxicity, endogenous and/or dietary decomposition products, stability, high water permeability, film strength, compatibility with a wide range of drugs and ability to form micro- and nanoparticles and, therefore, represent an appropriate material for drug delivery systems (Edgar, 2007; George and Sabapathi, 2015; Postek et al., 2013; Qiu and Hu, 2013; Sunasee et al., 2016; Tang et al., 2017).

Roman et al. (2009) tested the cytotoxicity and cellular uptake of cellulose nanocrystals (CNCs) and found that they were non-toxic to cells, and the cellular uptake of untargeted CNCs was minimal, suggesting their potential as carriers in targeted drug delivery applications. Kempaiah et al. (2013) reported about hydrogel made of bacterial nanocellulose that has been magnetically functionalized and was biocompatible for the purpose of rapid cellular localization and proliferation, and the application of such magnetic hydrogels integrated on intracranial stents in the treatment of brain aneurysms could be connected with the accelerating of the endothelialization process and reduced healing time.

Rod-like, glucose-based NPs cellulose nanocrystals are considered to be effective nanocarriers for targeted drug delivery to brain tumor cells. Dong et al. (2014) designed FA-conjugated CNCs for cancer targeting and found that cellular binding/uptake of the conjugate by human (DBTRG-05MG, H4) and rat (C6) brain tumor cells was 1452, 975 and 46-fold higher, respectively, than that of CNCs without conjugated FA, and the conjugate was internalized by DBTRG-05MG and C6 cells primarily via caveolae-mediated endocytosis, while its internalization by H4 cells occurred primarily via clathrin-mediated endocytosis. Similar results related to significantly increased cellular binding/uptake of FA-conjugated CNCs in DBTRG-05MG and U-87 MG cells because of FA conjugation enabling selective targeting of folate receptor-positive brain tumor cells was also reported, and DOX-loaded FA-conjugated CNCs exhibited a considerable cytotoxic effect on above-mentioned cells (Cho et al., 2011).

The stiffness of cellulose acetate nanofibers controlled by heat treatment was shown to regulate astrocyte activity. At stiffer substrate, enhanced adhesion and viability in culture were observed with saturation at >2 MPa of tensile strength, and higher activity of astrocytes was estimated in terms of increasing intermediate filament glial fibrillary acidic protein (Min et al., 2015). Pillay et al. (2009) designed an intracranial nano-enabled scaffold device (NESD) for the site-specific delivery of dopamine (DA) and, using biometric simulation and computational prototyping, prepared a binary crosslinked ALG scaffold embedding stable DA-loaded cellulose acetate phthalate (CAP) NPs with the optimum particle size of 197 nm and the zeta potential of −34.00 mV, enabling a modulated drug release due to polymorphic transitions of the crosslinked scaffold that further controlled the outward diffusion of the CAP NPs. Following implantation in the parenchyma of the frontal lobe of the Sprague-Dawley rat model, NESD provided high levels and controlled delivery of DA in the cerebrospinal fluid.

Oridonin suspended in carboxymethylcellulose or loaded with a nano-structured emulsion administered orally or injected to transgenic APP/PS1 mice, an animal model of cerebral amyloidosis for AD, notably reduced β-amyloid deposition, plaque-associated APP expression and microglial activation in brain of the animals following treatments of 10 days, and injected nanoemulsion of the drug resulted in the improvement of deficits in nesting, an important affiliative behavior and in social interaction (Zhang et al., 2013c).

Aripiprazole compositions prepared by nano milling using hydroxy-propyl cellulose (HPC) showed higher solubility and in vitro dissolution rate than those prepared by coprecipitation, which was connected with the disruption of HPC crystallinity in nanomilled compositions, and the NPs retained their in vitro dissolution rate also after compression into tablets (Abdelbary et al., 2014). Mucoadhesive temperature-mediated in situ gel formulations prepared using CS and hydroxypropyl methylcellulose (hypro-mellose, HPMC) used for i.n. delivery of ropinirole to the brain of albino rats showed absolute drug bioavailability of 82% and, following its nasal admin-istration, the AUC (0–480 min) in the brain was 8.5-fold of that estimated after i.v. administration and about 3-fold higher than the corresponding value obtained with intranasal ropinirole solution (Khan et al., 2010). The nasal formulations of thermoreversible nasal gels prepared using Pluronic® F-127 and HPMC K4M as gelling agents containing anti-Parkinson drug ropinirole exhibited an ex vivo drug release between 56–100% in 5 h; the gel had a protective effect on the mucosa and drug bioavailability in the brain following i.n. the administration was 5-fold higher than that obtained with i.v. administration (Rao et al., 2017).

5.4.4   PULLULAN

Pullulan is a copolymer with the chemical structure {(→6)-α-D-g-lucopyranosyl-(1→4)-α-D-glucopyranosyl-(1→4)-α-D-glucopyranosyl (1→)}$_n$ containing both α-(1→4)- and α-(1→6)-linkages in its structure. It is produced by fungus *Aureobasidium pullulans* from starch. Pullulan is highly soluble in water and shows no toxicity and immunogenicity. In addi-tion, because of the presence of a number of hydroxyl moieties can be highly functionalized, and thus its physicochemical properties can be modified to be used as a drug carrier (Chassot et al., 2016; Grenha and Rodrigues, 2013; Rekha and Sharma, 2007; Singh et al., 2008, 2015; Zarekar et al., 2017).

Pullulan-drug conjugates can target infected cells/tissues and exhibit high bioactivity with the release of cytotoxic molecules and, therefore, can be used for targeted drug delivery as well as targeted gene delivery for the treatment of a variety of diseases (Singh et al., 2015). Kawasaki et al. (2017) investigated carborane-bearing pullulan nanogels for effective delivery of boron to tumor cells in vitro and in vivo using murine fibrosarcoma cells (CMS5a) and a mouse fibrosarcoma xenograft model. They found that the nanogels were taken up by CMS5a cells, and the internalized nanogels were localized around the nuclear membrane and, after i.v. administration to mice bearing fibrosarcoma xenografts, accumulated in tumors in vivo and were found to reduce the cytotoxicity of carborane and to be internalized into tumor cells, acting as a dual-delivery therapeutic.

Boridy et al. (2009) studied the binding of pullulan modified choles-teryl (CHP) nanogels forming stable NPs of 20–30 nm to Aβ oligomers and their suppression of cytotoxicity and found that both neutral and posi-tively charged CHP nanoparticles interact with $A\beta_{1-42}$ monomers and oligo-mers; however, only positively charged derivatives are toxic, particularly in primary cortical cultures, and a notable reduction of $A\beta_{1-42}$ toxicity in both the primary cortical and microglial cells is due to the binding of both monomeric and oligomeric $A\beta_{1-42}$ to CHP. These results suggest that such nanogels could be promising in the antibody immunotherapy in neurological disorders connected with the formation of soluble toxic aggregates, such as an AD.

5.4.5 ALGINATE

Alginate (alginic acid, ALG) is an anionic polysaccharide distributed widely in the cell walls of brown seaweeds (*Phaeophyceae*), from which it can be also obtained. Also, it is a component of biofilms produced by *Pseudomonas aeru-ginosa*. ALG is a linear copolymer with homopolymeric blocks of β-(1→4)-linked D-mannopyranuronic acid and its $C_{(5)}$ epimer α-L-gulopyranuronic acid covalently linked together in different sequences or blocks. Its color ranges from white to yellowish-brown. ALG has a number of free hydroxyl moieties and carboxyl groups, and thus its physicochemical properties can be modified by means of varied substitutions or formation salts resulting in ALG-based drug carries/matrices. Moreover, alginates are biocompatible, biodegradable and non-toxic. The ALG-based drug delivery systems can be formulated as gels, matrices, membranes, nanospheres, microspheres, etc. (Bhatia, 2016; Jain and Bar-Shalom, 2014; Lee and Mooney, 2012; Sarei et al., 2013; Sosnik, 2014; Tonnesen and Karlsen, 2002; Yang et al., 2011; Zia et al., 2015).

ALG-CUR nanocomposite supplemented with a diet for 24 days at a dose of 10^{-5}, 10^{-3}, and 10^{-1} g/mL to PD *Drosophila* flies caused a considerable delay in the loss of climbing ability and a reduction in the oxidative stress and apoptosis in the brain of PD model flies (Siddique et al., 2013). Using PD model flies, it was shown that following the treatment of PD flies with bromocriptine ALG nanocomposite, a notable reduction in lipid peroxidation and glutathione-*S*-transferase activity and an increase in GSH content were observed without gross morphological changes in their brains compared with controls (Siddique et al., 2016). The i.n. administration of venlafaxine-loaded ALG NPs to albino Wistar rats resulted in notable improvement of the behavioral analysis parameters, i.e. swimming, climbing and immobility and locomotor activity in comparison with the drug solution (i.n.) and drug tablet (oral), and also a greater brain/blood ratio at 30 min was observed for VLF ALG NPs (i.n.) compared to VLF solution administered by i.n. route (Haque et al., 2014). Compared to pure VLF hydrochloride, biopolymeric composites of montmorillonite ALG fabricated as microbeads with 97% EE for an oral extended release of VLF exhibited a notably less burst release with the cumulative release of 20% (over a period of 26 h) and 22% (over a period of 29 h) in the gastric and intestinal fluid respectively, and the applied clay caused a reduction of the burst effect as well as the extended release of drug, suggesting that such formulation could eliminate the repeated intake (every 3–4 h) of drug (Jain and Datta, 2016).

Ni et al. (2014) reported about the fabrication of biodegradable fiber suitable for efficient simultaneous delivery of PTX and TMZ to glioma C6 cancer cells in vitro. They added polypropylene carbonate and TMZ to the emulsion of PTX-loaded Ca-ALG microparticles and used an electrospinning process to prepare fibers showing prolonged release time compared to that of formulations with only one drug as well as higher cytotoxicity to glioma C6 cells. Huang et al. (2016) developed novel device by integrating TMZ-doped polycaprolactone (PCL) nanofiber (TP) membrane prepared by direct electrospinning of TMZ/PCL and NGF-coated PCL (NGFP) membrane using sodium ALG hydrogel fabricated by a layer-by-layer assembly technology which enabled a sustained release of both TMZ and NGF, namely an efficient release of TMZ necessary for inhibition of the growth of C6 glioma cells and an adequate release of NGF to induce the differentiation of PC12 neuron cells over 4 weeks. According to researchers, such TP-NGFP-TP membrane device could be used "like a tampon to fill up the surgical residual cavity and afford residual glioma removal, structural support, hemostasis and local neural tissue reconstruction in the surgical treatment of glioma."

A study of ALG hydrogels functionalized with D- or L-penicillamine (D-, L-PEN) used as new 3D scaffolds for cell adhesion studies showed that the affinity of C-6-glioma and endothelial cells to D-PEN functionalized 3D ALG was higher than that to L-PEN functionalized one (Benson et al., 2014). The MP@Alg-CUR AuNPs conjugate consisting of covalently conjugated MTX to bis(aminopropyl) terminated PEG subsequently conjugated onto ALG–CUR AuNPs, representing a hybrid nanostructured drug delivery system, was found to be highly hemocompatible and showed improved cytotoxic activity potential against C6 glioma and MCF-7 cancer cell lines, suggesting its possible application in targeted combination chemotherapy for the treatment of cancer (Dey et al., 2016).

Biocompatible nanocomposites of ALG and Fe_2O_3 NPs encapsulating DOX have been reported to have a great safety and potential for brain tumor therapy (Su and Cheng, 2015). DOX-loaded into molecularly imprinted carboxymethyl cellulose-$(CS-ALG)_5$ biocompatible hollow microcapsules showed a sustained drug release for a period >168 h, which was connected with the fact that the imprinted sites were blocked by DOX molecules that entered the imprinted sites through electrostatic interaction. Encapsulated DOX was found to be effective in the induction of the apoptosis of U373 malignant glioma cells, and its efficacy related to tumor inhibition within 144 h exceeded that of free DOX (Wang et al., 2014). DOX-loaded ALG stabilized perfluorohexane nanodroplets with the particle diameter of 39 nm and the EE of 92% released 7–13% of DOX after 24 h incubation in phosphate-buffered saline with pH 7.4; however, after exposure to ultrasound for 10 min, a triggered release of 85.95% of DOX from this nanoformulation was observed (Baghbani et al., 2016).

5.4.6 HYALURONIC ACID

Hyaluronic acid (HA) is a linear polysaccharide composed of disaccharide units of N-acetyl-β-D-glucosamine and β-D-glucuronic acid, linked via (1→3) and (1→4) glycosidic bonds. HA is present in almost all biological fluids and tissues (Kogan et al., 2007; Ossipov, 2010; Yadav et al., 2008, 2010). Hyaluronic acid can be functionalized by various substitutions and can be used in drug delivery systems (Han et al., 2016; Mero et al., 2016; Rao et al., 2016).

The release rate of cisplatin (CDDP) from spherical CDDP-incorporated HA NPs with particle size about 100–200 nm was found to increase when hyaluronidase (HAse) was added to the release medium. This was shown by the addition of the U343MG cell line secreting HAse to the release medium, and because glioma tumor cell lines are able to secrete HAse, this nanoformulation could be used as a drug targeting system (Jeong et al., 2008).

Yin et al. (2016) designed a pH-sensitive prodrug by conjugating HA with DOX that was modified with Lf suitable for glioma dual-targeted treatment, showing a much quicker drug release at lower pH (5.0 and 6.0) than at pH 7.4 and higher cellular uptake in C6 cells than free DOX, and following administration of Lf-modified prodrug to BALB/C mice bearing C6 glioma, its enhanced accumulation in glioma and 2-fold longer survival time compared to that of the saline group were observed as well. The HA-conjugated liposomes specifically targeted GBM cells over other brain cells, which was connected with higher expression of CD44 in tumor cells, and because of the uptake of these liposomes into GBM cells, lyso-somal evasion and the increased efficacy of DOX were observed. On the other hand, astrocytes and microglia cells exhibited extensive HA-conju-gated liposomes-lysosome co-localization and reduced the antineoplastic potency (Hayward et al., 2016). Binding of hyaluronan-grafted lipid-based NPs (HA-LNPs) to glioblastoma multiforme (GBM) cell lines and primary neurospheres of GBM patients was estimated, and these NPs loaded with Polo-Like Kinase 1 (PLK1) siRNAs strongly reduced the expression of PLK1 mRNA and were found to cumulate in cell death also under shear flow that simulates the flow of the cerebrospinal fluid compared with the control groups. In a human GBM U87MG orthotopic xenograft model obtained by intracranial injection of U87MG cells into nude mice, treat-ment with PLK1 siRNA loaded HA-LNPs resulted in a decrease (>80%) of mRNA levels and prolonged the survival of animals (Cohen et al., 2015). Nasr (2016) developed a mucoadhesive optimized HA-based lipidic nano-emulsion co-encapsulating two polyphenols (CUR and resveratrol) with the particle size of 115 nm and the zeta potential of -23.9 ± 1.7 V for the nose-to-brain delivery, which was found to preserve the antioxidant ability of the two polyphenols and protect them from degradation, and its admin-istration to rats resulted in approx. a 9- and a 7-fold increase in $AUC0\rightarrow7h$ for CUR and resveratrol, respectively, suggesting increased amounts of these polyphenols in the brain.

Jiang et al. (2016) reported about the multifunctionality of nanogels of CUR-HA conjugates on inhibiting amyloid β-protein fibrillation and cytotoxicity, because CUR encapsulation into nanogels can protect cells from the toxicity of the free drug, the hydrogel network can hinder interactions between Aβ molecules, and the hydrophobic binding between Aβ and the conjugated CUR can counteract against the electrostatic repulsion between like-charged Aβ and HA. Spherical CUR-loaded polyelectrolyte complex NPs based on HA/CS NPs with the average diameter of 207 nm, the positive charge of 25.37 mV and the EE of 89.9% showed a sustained release pattern following initial burst lasting for 4 h as well as stronger cytotoxicity and uptake by C6 glioma cells than CUR solution, and the mechanism of their uptake by C6 cells involved active endocytosis, macropinocytosis, clathrin-, caveolae- and CD44-mediated endocytosis (Yang et al., 2015). The i.v. administration of CUR-loaded CS/HA/PEG NPs with the particle size of 246 nm and the zeta potential of -27.2 mV resulted in 4-fold greater area under the curve than that of CUR solution, suggesting the suitability of this nanoformulation for the therapy of brain tumors (Xu et al., 2017a). CUR-loaded Lf-coated NPs based on CS/HA/PEG for treating brain glioma were found more efficiently taken up by brain capillary endothelial cells than the corresponding NPs without Lf coating; after crossing the BBB they targeted the glioma C6 cells more effectively; and the treatment of mice in vivo resulted in 2.39 higher accumulation of these NPs in the brain compared to NPs without Lf coating (Xu et al., 2017b).

The tumor-homing penetrating peptide tLyP-1-functionalized NPs (tLPTS/HATS NPs), composed of two modularized amphiphilic conjugates of tLyP-1-PEG-D-α-tocopheryl succinate (tLPTS) and D-α-tocopheryl succinate-grafted hyaluronic acid (HATS), designed for tumor-targeted delivery of DTX exhibited significant penetrability and inhibitory effect on both PC-3 and MDA-MB-231 multicellular tumor spheroids, and in vivo they showed much more lasting accumulation and extensive distribution throughout tumor regions than HATS NPs (Liang et al., 2015).

5.4.7   GUMS

Gums are polysaccharides of natural origin that are able to increase the viscosity of solutions, thus they are used as gelling, emulsifying, encapsulating and stabilizing agents. Natural gums can be classified according to the origin (they are obtained from bacteria (e.g., gellan, xanthan gums), seaweeds (e.g., agar, carrageenan), woody elements of plants and seed coatings (e.g.,

Arabic, tragacanth gums, or beta-glucan)) or when polysaccharides are uncharged or ionic. (Choudhary and Pawar, 2014; Dewick 2009; Granzotto et al,. 2017). Diverse structures of these natural gums ensure different physicochemical properties, which resulted in their applications for drug nanoformulations as drug delivery systems (Goswami and Nai, 2014; Hamman et al., 2015; Prajapati et al., 2013). Various gums that have been investigated as drug carriers for BBB targeting of drugs are mentioned bellow.

Gellan gum (GG) is a linear water-soluble anionic deacetylated exopolysaccharide that is prepared by aerobic fermentation of *Sphingomonas elodea*. It consists of a tetrasaccharide repeating unit of α-L-rhamnopyranose, β-D-glucuronic acid and two β-D-glucopyranoses (Jansson et al., 1983). Rivastigmine-loaded in situ gelling nanostructured lipid carriers (NLCs) with mean particle size ca. 123 nm that were prepared using glyceryl monostearate, Capmul MCM C8, Lecithin and Tween® 80 and incorporated into an in situ gelling system using 0.8% GG and 15% Lutrol F showed a 2-fold increase in nasal permeation of the drug over plain drug solution and caused a 3-fold increase in enzyme inhibition efficacy (Wavikar and Vavia, 2015). Hao et al. (2016) designed an in situ gelling formulation by combining nanoparticles and an ionic-triggered deacetylated gellan gum (DGG) matrix for challenging intranasal drug (resveratrol) delivery. The formulation showed an about 3-fold increase of bioavailability in the brain following its intranasal application, and direct delivery via the nose-brain pathway was confirmed by higher drug targeting efficiency (458.2%) and direct transport percentage (78.18%). Koivisto et al. (2017) designed GG hydrogels crosslinked by bioamines spermidine and spermine for neural tissue engineering that exhibited biomimicking properties comparable to naive rabbit brain tissue under physiologically relevant stress and strain; the human pluripotent stem cell-derived neuronal cells showed good cytocompatibility in these hydrogels; and by functionalization of hydrogels with laminin cell type-specific behavior, neuronal cell maturation and neurite migration were obtained. Sophorolipid-conjugated GG reduced AuNPs effectively killed human glioma cell line LN-229 as well as human glioma stem cell line HNGC-2, and conjugation of DOX.HCl to these NPs resulted in enhanced cytotoxic effects (Dhar et al., 2011). The cellular uptake by C6 glioma cells of superparamagnetic Au-coated magnetite (Fe_3O_4@Au) NPs with a negatively charged shell consisting of GG, onto which positively charged DOX molecules were loaded via electrostatic attraction, could be enhanced by the application of magnetic fields, which results in greater tumor cell death (Venugopal et al., 2016).

Gum arabic (GA) is non-digestible branched neutral or weakly acid polysaccharide with protein residues obtained from trees *Acacia senegal* and *Vachellia seyal*. The backbone of GA is composed of $(1{\rightarrow}3)$-linked β-D-galactopyranose bearing in $C_{(6)}$ side chains consisting of L-arabinose, L-rhamnopyranose, D-galactopyranose and D-glucuronic acid (Belitz et al., 2009; Dewick, 2009). GA-capped AgNPs caused decreased adduct formation of glycated protein and, at higher concentrations, strongly inhibited the advanced glycation end products formation, which are major contributors to the pathology of diabetes, Alzheimer's disease and atherosclerosis (Ashraf et al., 2014). Magnetic Fe_3O_4 NPs coated with GA with the mean particle size of ca. 100 nm and the GA content of 15.6% by dry weight were effectively uptaken by 9L glioma cells in vitro, and after their i.v. administration to rats harboring 9L glioma and the application of an external magnetic field, they accumulated at the tumor site, and the accumulation of these NPs in excised tumors showed a 12-fold increase compared to contralateral normal brain, suggesting their potential use for simultaneous tumor imaging and targeted intra-tumoral drug delivery (Zhang et al., 2009).

Xanthan gum (XG) is an anionic non-toxic mucoadhesive polysaccharide produced by the bacterium *Xanthomonas campestris*. $(1{\rightarrow}4)$-Linked β-D-glucopyranose creates the backbone of XG. Every 2nd glucopyranose unit is substituted in $C_{(3)}$ by a trisaccharide side chain consisting of β-D-mannopyranose-$(1{\rightarrow}4)$-β-D-glucuronic acid-$(1{\rightarrow}2)$-α-D-mannopyranose that is acetylated and pyruvylated (Kobori and Nakao, 2010; Muddineti et al., 2015; Palaniraja and Jayaraman, 2011). Carbamazepine (CBZ) formulated in a nanoemulgel system containing oleic acid/labrasol in the ratio of 1:5 as oil/surfactant and 0.1% XG showed a very low in vitro release of the drug from mucoadhesive o/w nanoemulgel, while CBZ uptake via the liposomal membrane represented 65% within 1 h, and treatment of animals with this nanoemulgel resulted in considerably prolonged onset times for convulsion of chemically convulsive mice and protected the animals from two electric shocks (Samia et al., 2012).

Ondansetron hydrochloride loaded mucoadhesive nanostructured lipid carriers prepared by the high-pressure homogenization technique, in which *Delonix regia* gum (flamboyant seed gum) isolated from seeds of *D. regia* was used as a mucoadhesive polymer, with particle size 92.28–135 nm and zeta potential ranging from −11.5 to −36.2 mV administered by i.n. route showed also the higher drug targeting efficiency of 506% and the direct transport percentage of 197.14% (Devkar et al., 2014). This type of gum consists of a $(1{\rightarrow}4)$-α-D-mannopyranose-$(1{\rightarrow}6)$-α-D-galactopyranose, whereas ratio mannose–galactose is 2:1 (Pacheco-Aguirre et al., 2010). Spherical donepezil-loaded mango gum polymeric NPs, obtained from the mango tree *Mangifera indica* L., with particle size about 90–130 nm showed brain targeting ability in vivo, suggesting that the extracted water-soluble fraction of mango gum could be used as a nanocarrier for brain delivery of drugs (Jakki et al., 2016).

5.5   CONCLUSION

Drug targeting to the brain can be considered as a challenge because of the necessity to overcome the BBB that prevents the passage of many drugs, the site of action of which is the CNS. A number of drugs for the treatment of neurological or psychiatric diseases, antineoplastics, and anti-infectious (antiviral, antibacterial, antifungal, and antiparasitic) drugs are medicaments that constitute substrates of various degradation causing enzymes and/or efflux pumps. On the other hand, targeted delivery of drugs is a promising branch of nanotechnology, thus an application of nanoparticles as drug carriers can be considered as one of the possible strategies for drug delivery to the CNS. The benefits include non-invasive administration and reduced systemic side effects resulting in decreased toxicity. Currently, a shift from inorganic materials as drug matrices to various polymer biodegradable materials, especially of natural origin and their semisynthetic modifications, can be noted because of their compatibility, inertness, insignificant human toxicity and price acceptability. In addition, these matrices can be functionalized by other molecules, ligands or fragments, which results in higher affinity to the BBB and increased permeation to the brain. However, their application depends on the detailed knowledge of pharmacokinetic aspects of permeation over the BBB to the brain and advances in material science. Because the R&D process of drugs with a new/innovative mode of action is relatively long and risky, the preparation of drug NPs from clinically used drugs has been a perspective approach both for "drug life-cycle management," that is, economic use of a drug molecule, and for the treatment of various disorders because of less adverse effects based on the administration of lower drug doses with the same effect as the bulk drug, possibility of targeted biodistribution, etc. Thus, it seems that the applications of nanomaterials as drug nanocarriers/delivery systems have a number of advantages, especially in terms of facilitating the delivery of drugs to the brain through the BBB without forcible opening or damaging it. With respect to the progress in nanomedicine, both in research and clinical applications, the development of safe and effective methods of nanosystem-based delivery of drugs for CNS disorders is anticipated. It could be assumed that in the near future some of these above-discussed nanoformulations will move from in vivo evaluation to clinical trials and subsequently to the clinical practice. Such nanocarriers have the potential for the development of personalized nanomedicine to achieve targeted or on-demand release of drugs independent of physiological conditions. Therefore, these types of drug nanoformulations can be supposed to start a revolution in the treatment of CNS diseases.

ACKNOWLEDGMENTS

This study was supported by the Slovak Research and Development Agency (projects APVV-17-0373 and APVV-17-0318) and by the Ministry of Education of the Czech Republic (LO1305).

KEYWORDS

- **blood–brain barrier**
- **carbohydrates**
- **drug carriers**
- **nanoformulations**
- **lactic acid derivatives**
- **targeted delivery**
- **proteins**

REFERENCES

Abbott, N. J; Patabendige, A. A. K.; Dolman, D. E. M.; Yusof, S. R.; Begley, D. J. Structure and Function of the Blood-Brain Barrier. *Neurobiol. Dis.* **2010,** *37,* 13–25.

Abdelbary, A. A.; Li, X. L.; El-Nabarawi, M.; Elassasy, A.; Jasti, B. Comparison of Nano milling and Coprecipitation on the Enhancement of In Vitro Dissolution Rate of Poorly Water-Soluble Model Drug Aripiprazole. *Pharm. Dev. Technol.* **2014,** *19,* 491–500.

Abou Neel, E. A.; Bozec, L.; Knowles, J. C.; Syed, O.; Mudera, V.; Day, R.; Hyun, J. K. Collagen-Emerging Collagen-Based Therapies Hit the Patient. *Adv. Drug Deliv. Rev.* **2013,** *65,* 429–456.

Agarwal, A.; Agrawal, H.; Tiwari, S.; Jain, S.; Agrawal, G. P. Cationic Ligand Appended Nanoconstructs: a Prospective Strategy for Brain Targeting. *Int. J. Pharm.* **2011a,** *421,* 189–201.

Agarwal, A.; Majumder, S.; Agrawal, H.; Majumdar, S.; Agrawal, G. P. Cationized Albumin Conjugated Solid Lipid Nanoparticles as Vectors for Brain Delivery of an Anti-Cancer Drug. *Curr. Nanosci.* **2011b,** *7,* 71–80.

Agrawal, P.; Singh, R. P.; Sonali; Kumari, L.; Sharma, G.; Koch, B.; Rajesh, C. V.; Mehata, A. K.; Singh, S.; Pandey, B. L.; Muthu, M. S. TPGS-Chitosan Cross-Linked Targeted Nanoparticles for Effective Brain Cancer Therapy. *Mater. Sci. Eng. C Mater. Biol. Appl.* **2017a,** *74,* 167–176.

Agrawal, P.; Sonali; Singh, R. P.; Sharma, G.; Mehata, A. K.; Singh, S.; Rajesh, C. V.; Pandey, B. L.; Koch, B.; Muthu, M. S. Bioadhesive Micelles of d-α-Tocopherol Polyethylene Glycol Succinate 1000: Synergism of Chitosan and Transferrin in Targeted Drug Delivery. *Colloids Surf. B Biointerfaces* **2017b,** *152,* 277–288.

Ahn, Y. S.; Baik, H. J.; Lee, B. R.; Lee, E. S.; Oh, K. T.; Lee, D. H.; Youn, Y. S. Preparation of Multifunctional Polymeric Micelles for Antiviral Treatment. *Macromol. Res.* **2010,** *18,* 747–752.

Al-Dhubiab, B. E. Formulation and In Vitro Evaluation of Gelatin Nanospheres for the Oral Delivery of Selegiline. *Curr. Nanosci.* **2013,** *9,* 21–25.

Alsaheb, R. A. A.; Aladdin, A.; Othman, N. Z.; Malek, R. A.; Leng, O. M.; Aziz, R.; El Enshasy, H. A. Recent Applications of Polylactic Acid in Pharmaceutical and Medical Industries. *J. Chem. Pharm. Res.* **2015,** *7,* 51–63.

Alyautdin, R.; Khalin, I.; Nafeeza, M. I.; Haron, M. H.; Kuznetsov, D. Nanoscale Drug Delivery Systems and the Blood-Brain Barrier. *Int. J. Nanomedicine.* **2014,** *9,* 795–811.

Anand, R.; Manoli, F.; Manet, I.; Daoud-Mahammed, S.; Agostoni, V.; Gref, R.; Monti, S. β-Cyclodextrin Polymer Nanoparticles as Carriers for Doxorubicin and Artemisinin: a Spectroscopic and Photophysical Study. *Photochem. Photobiol.* Sci. **2012,** *11,* 1285–1292.

Ashraf, J. M.; Ansari, M. A.; Choi, I.; Khan, H. M.; Alzohairy, M. A. Antiglycating Potential of Gum Arabic Capped-Silver Nanoparticles. *Appl. Biochem. Biotechnol.* **2014,** *174,* 398–410.

Auras, R.; Lim, L. T.; Selke, S. E. M.; Tsuji, H. *Poly(lactic acid): Synthesis, Structures, Properties, Processing, and Applications.* John Wiley & Sons: Hoboken, **2010.**

Avachat, A. M.; Oswal, Y. M.; Gujar, K. N.; Shah, R. D. Preparation and Characterization of Rivastigmine Loaded Human Serum Albumin (HSA) Nanoparticles. *Curr. Drug Deliv.* **2014,** *11,* 359–370.

Azadi, A.; Hamidi, M.; Rouini, M. R. Methotrexate-Loaded Chitosan Nanogels as 'Trojan Horses' for Drug Delivery to Brain: Preparation and In Vitro/In Vivo Characterization. *Int. J. Biol. Macromol.* **2013,** *62,* 523–530.

Baghbani, F.; Mortarzadeh, F.; Mohandesi, J. A.; Yazdian, F.; Mokhtari-Dizaji, M.; Hamedi, S. Formulation Design, Preparation and Characterization of Multifunctional Alginate Stabilized Nanodroplets. *Int. J. Biol. Macromol.* **2016,** *89,* 550–558.

Bago, J. R.; Pegna, G. J.; Okolie, O.; Mohiti-Asli, M.; Loboa, E. G.; Hingtgen, S. D. Electrospun Nanofibrous Scaffolds Increase the Efficacy of Stem Cell-Mediated Therapy of Surgically Resected Glioblastoma. *Biomaterials.* **2016,** *90,* 116–125.

Bahadur, S.; Pathak, K. Buffered Nanoemulsion for Nose to Brain Delivery of Ziprasidone Hydrochloride: Preformulation and Pharmacodynamic Evaluation. *Curr. Drug Deliv.* **2012,** *9,* 596–607.

Baltzley, S.; Mohammad, A.; Malkawi, A. H.; Al-Ghananeem, A. M. Intranasal Drug Delivery of Olanzapine-Loaded Chitosan Nanoparticles. *AAPS PharmSciTech*. **2014**, *15*, 1598−1602.

Bansal, A.; Kapoor, D. N.; Kapil, R.; Chhabra, N.; Dhawan, S. Design and Development of Paclitaxel-Loaded Bovine Serum Albumin Nanoparticles for Brain Targeting. *Acta Pharm.* **2011**, *61*, 141−156.

Barar, J.; Rafi, M. A.; Pourseif, M. M.; Omidi, Y. Blood-Brain Barrier Transport Machineries and Targeted Therapy of Brain Diseases. *Bioimpacts*. **2016**, *6*, 225–248.

Bari, N. K.; Fazil, M.; Hassan, M. Q.; Haider, M. R.; Gaba, B.; Narang, J. K.; Baboota, S.; Ali, J. Brain Delivery of Buspirone Hydrochloride Chitosan Nanoparticles for the Treatment of General Anxiety Disorder. *Int. J. Biol. Macromol.* **2015**, *81*, 49−59.

Bartnik, D.; Funke, S. A.; Andrei-Selmer, L. C.; Bacher, M.; Dodel, R.; Willbold, D. Differently Selected d-Enantiomeric Peptides Act on Different Aβ Species. *Rejuvenation Res.* **2010**, *13*, 202−205.

Baysal, I.; Ucar, G.; Gultekinoglu, M.; Ulubayram, K.; Yabanoglu-Ciftci, S. Donepezil Loaded PLGA-b-PEG Nanoparticles: their Ability to Induce Destabilization of Amyloid Fibrils and to Cross Blood-Brain Barrier In Vitro. *J. Neural. Transm.* **2017**, *124*, 33−45.

Bazak, R.; Houri, M.; El Achy, S.; Kamel, S.; Refaat, T. Cancer Active Targeting by Nanoparticles: a Comprehensive Review of Literature. *J. Cancer Res. Clin. Oncol.* **2015**, *141*, 769–784.

Belitz, H. D.; Grosch, W.; Schieberle, P. *Food Chemistry, 4th Revised and Extended*; Springer-Verlag: Berlin-Heidelberg, **2009**.

Benson, K.; Galla, H. J.; Kehr, N. S. Cell Adhesion Behavior in 3D Hydrogel Scaffolds Functionalized with d- or l-Aminoacids. *Macromol. Biosci.* **2014**, *14*, 793−798.

Bernkop-Schnürch, A.; Dünnhaupt S. Chitosan-Based Drug Delivery Systems. *Eur. J. Pharm. Biopharm.* **2012**, *81*, 463−469.

Bhatia, S. Marine Polysaccharides Based Nano-Materials and Its Applications. *Natural Polymer Drug Delivery Systems: Nanoparticles, Plants, and Algae*. Springer: Berlin-Heidelberg, **2016**, 185–225.

Bhavna; M. D. S.; Ali, M.; Ali, R.; Bhatnagar, A.; Baboota, S.; Ali, J. Donepezil Nanosuspension Intended for Nose to Brain Targeting: In Vitro and In Vivo Safety Evaluation. *Int. J. Biol. Macromol.* **2014a**, *67*, 418−425.

Bhavna; M. D. S.; Ali, M.; Bhatnagar, A.; Baboota, S.; Sahni, J. K.; Ali, J. Design, Development, Optimization and Characterization of Donepezil Loaded Chitosan Nanoparticles for Brain Targeting to Treat Alzheimer's Disease. *Sci. Adv. Mater.* **2014b**, *6*, 720−735.

Bi, Y. K.; Liu, L. S.; Lu, Y. F.; Sun, T.; Shen, C.; Chen, X. L.; Chen, Q. J.; An, S.; He, X.; Ruan, C. H.; Wu, Y.; Zhang, Y.; Guo, Q.; Zheng, Z.; Liu, Y.; Lou, M.; Zhao, S.; Jiang, C. T7 Peptide-Functionalized PEG-PLGA Micelles Loaded with Carmustine for Targeting Therapy of Glioma. *ACS Appl. Mater. Interfaces*. **2016**, *8*, 27465−27473.

Bhaskar, S.; Tian, F.; Stoeger, T.; Kreyling, W.; de la Fuente, J. M.; Grazú, V.; Borm, P.; Estrada, G.; Ntziachristos, V.; Razansky, D. Multifunctional Nanocarriers for Diagnostics, Drug Delivery and Targeted Treatment Across Blood–Brain Barrier: Perspectives on Tracking and Neuroimaging. *Part Fibre Toxicol.* **2010**, *7*, 3.

Blanco, E.; Shen, H.; Ferrari, M. Principles of Nanoparticle Design for Overcoming Biological Barriers to Drug Delivery. *Nat Biotechnol.* **2015**, *33*, 941−951.

Bommana, M. M.; Kirthivasan, B.; Squillante, E. In Vivo Brain Microdialysis to Evaluate FITC-dextran Encapsulated Immunopegylated Nanoparticles. *Drug Deliv.* **2012**, *19*, 298−306.

Bonaccorso, A.; Musumeci, T.; Serapide, M. F.; Pellitteri, R.; Uchegbu, I. F.; Puglisi, G. Nose to Brain Delivery in Rats: Effect of Surface Charge of Rhodamine B Labeled Nanocarriers on Brain Subregion Localization. *Colloids Surf. B Biointerfaces* **2017**, *154*, 297−306.

Borbas, E; Balogh, A; Bocz, K; Mueller, J; Kiserdei, E; Vigh, T; Sinko, B; Marosi, A; Halasz, A; Dohanyos, Z; Szente, L.; Balogh, G. T.; Nagy Z. K. In Vitro Dissolution-Permeation Evaluation of an Electrospun Cyclodextrin-Based Formulation of Aripiprazole Using μFlux™. *Int. J. Pharm.* **2015**, *491*, 180−189.

Boridy, S.; Takahashi, H.; Akiyoshi, K.; Maysinger, D. The Binding of Pullulan Modified Cholesteryl Nanogels to A β Oligomers and Their Suppression of Cytotoxicity. *Biomaterials*. **2009**, *30*, 5583−5591.

Buchwald, P; Bodor, N. Brain-Targeting Chemical Delivery Systems and Their Cyclodextrin-Based Formulations in Light of the Contributions of Marcus E. Brewster. *J. Pharm. Sci.* **2016**, *105*, 2589−2600.

Buyukoz, M.; Erdal, E.; Sacide A. A. Nanofibrous Gelatine Scaffolds Integrated with Nerve Growth Factor-Loaded Alginate Microspheres for Brain Tissue Engineering. *J. Tissue Eng. Regen. Med.* **2018**, *12*, 707–719.

Cacciatore, I.; Ciulla, M.; Fornasari, E.; Marinelli, L.; Di Stefano, A. Solid Lipid Nanoparticles as a Drug Delivery System for the Treatment of Neurodegenerative Diseases. *Expert Opin. Drug. Deliv.* **2016**, *13*, 1121−1131.

Cai, Q.; Wang, L.; Deng, G.; Liu, J. H.; Chen, Q. X.; Chen, Z. B. Systemic Delivery to Central Nervous System by Engineered PLGA nanoparticles. *Am. J. Transl. Res.* **2016**, *8*, 749−764.

Cao, X. B.; Hou, D. Z.; Wang, L.; Li, S.; Sun, S. G.; Ping, Q. N.; Xu, Y. Effects and Molecular Mechanism of Chitosan-Coated Levodopa Nanoliposomes on Behavior of Dyskinesia Rats. *Biol. Res.* **2016**, *49*, 32.

Cannava, C.; Stancanelli, R.; Marabeti, M. R.; Venuti, V.; Cascio, C.; Guarneri, P.; Bongiorno, C.; Sortino, G.; Majolino, D.; Mazzaglia, A.; Tommasini, S.; Ventura, C. A. Nanospheres Based on PLGA/Amphiphilic Cyclodextrin Assemblies as Potential Enhancers of Methylene Blue Neuroprotective Effect. *RSC Adv.* **2016**, *6*, 16720−16729.

Černíková, A.; Jampílek, J. Structure Modification of Drugs Influencing Their Bioavailability and Therapeutic Effect. *Chem. Listy.* **2014**, *108*, 7−16.

Chassot, J. M.; Ferreira, L. M.; Gomes, F. P.; Silva, C. B.; Tasso, L.; Cruz, L. Pullulan as a Stabilizer Agent of Polymeric Nanocapsules for Drug Delivery. *Braz. J. Pharm. Sci.* **2016**, *52*, 735−740.

Chaturvedi, M.; Molino, Y.; Sreedhar, B.; Khrestchatisky, M.; Kaczmarek, L. Tissue Inhibitor of Matrix Metalloproteinases-l Loaded Poly(Lactic-Co-Glycolic Acid) Nanoparticles for Delivery Across the Blood-Brain Barrier. *Int. J. Nanomedicine.* **2014**, *9*, 575−588.

Chen, B.; Cao, S.; Zhang, Y.; Wang, X.; Liu, J.; Hui, X.; Wan, Y.; Du, W.; Wang, L.; Wu, K.; Fan, D. A Novel Peptide (GX1) Homing to Gastric Cancer Vasculature Inhibits Angiogenesis and Cooperates with TNF Alpha in Anti-Tumor Therapy. *BMC Cell Biol.* **2009**, *10*, 63.

Chen, Y. C.; Hsieh, W. Y.; Lee, W. F.; Zeng, D. T. Effects of Surface Modification of PLGA-PEG-PLGA Nanoparticles on Loperamide Delivery Efficiency Across the Blood-Brain Barrier. *J. Biomater. Appl.* **2013**, *27*, 909−922.

Chen, Q.; Yu, H. J.; Wang, L.; ul Abdin, Z.; Chen, Y. S.; Wang, J. H., Zhou, W. D.; Yang, X. P.; Khan, R. U.; Zhang, H. T.; Chen, X. Recent Progress in Chemical Modification of Starch and its Applications. *RSC Adv.* **2015,** *5,* 67459–67474.

Cheng, K. K.; Chan, P. S.; Fan, S. J.; Kwan, S. M.; Yeung, K. L.; Wang, Y. X. J.; Chow, A. H. L.; Wu, E. X.; Baum, L. Curcumin-Conjugated Magnetic Nanoparticles for Detecting Amyloid Plaques in Alzheimer's Disease Mice Using Magnetic Resonance Imaging (MRI). *Biomaterials.* **2015,** *44,* 155–172.

Cho, H. J.; Lee, S.; Dong, S. P.; Roman, M.; Lee, Y. W. Cellulose Nanocrystals as a Novel Nanocarrier for Targeted Drug Delivery to Brain Tumor Cells. *FASEB J.* **2011,** *25,* 762.2.

Choi, Y. K.; Kim, K.-W. Blood–Neural Barrier: Its Diversity and Coordinated Cell-To-Cell Communication. *BMB Rep.* **2008,** *41,* 345–352.

Choi, D. H.; Heo, J. W.; Park, J. H.; Jo. Y. H.; Jeong, H. J.; Chang, M. W.; Choi, J. H.; Hong, J. K. Nano-Film Coatings Onto Collagen Hydrogels with Desired Drug Release. *J. Ind. Eng. Chem.* **2016,** *36,* 326–333.

Choudhary, P. D.; Pawar, H. A. Recently Investigated Natural Gums and Mucilages as Pharmaceutical Excipients: An Overview. *J. Pharm.* **2014,** *2014,* 204849.

Cipolla, M. J. Barriers of the CNS. *The Cerebral Circulation*; Granger, D. N.; Granger, J. Eds.; Morgan & Claypool Life Sciences: San Rafael, 2009; pp 1–59.

Cohen, Z. R.; Ramishetti, S.; Peshes-Y. N.; Goldsmith, M.; Wohl, A.; Zibly, Z.; Peer, D. Localized RNAi Therapeutics of Chemoresistant Grade IV Glioma Using Hyaluronan-Grafted Lipid-Based Nanoparticles. *ACS Nano.* **2015,** *9,* 1581–1591.

Cole, A. J.; David, A. E.; Wang, J. X.; Galban, C. J.; Hill, H. L.; Yang, V. C. Polyethylene Glycol Modified, Cross-Linked Starch-Coated Iron Oxide Nanoparticles for Enhanced Magnetic Tumor Targeting. *Biomaterials* **2011,** *32,* 2183–2193.

Comoglu, T.; Arisoy, S.; Akkus, Z. B. Nanocarriers for Effective Brain Drug Delivery. *Curr. Top. Med. Chem.* **2017,** *17,* 1490–1506.

Crawford, L.; Rosch, J.; Putnam, D. Concepts, Technologies, and Practices for Drug Delivery Past the Blood-Brain Barrier to the Central Nervous System. *J. Control. Release.* **2016,** *240,* 251–266.

Cui, Y.; Xu, Q. X.; Chow, P. K. H.; Wang, D.; Wang, C. H. Transferrin-Conjugated Magnetic Silica PLGA Nanoparticles Loaded with Doxorubicin and Paclitaxel for Brain Glioma Treatment. *Biomaterials.* **2013,** *34,* 8511–8520.

Cui, Y.; Zhang, M.; Zeng, F.; Jin, H. Y.; Xu, Q.; Huang, Y. Z. Dual-Targeting Magnetic PLGA Nanoparticles for Codelivery of Paclitaxel and Curcumin for Brain Tumor Therapy. *ACS Appl. Mater. Interfaces* **2016,** *8,* 32159–32169.

Danhier, F.; Ansorena, E.; Silva, J. M.; Coco, R.; Le Breton, A.; Preat, V. PLGA-Based Nanoparticles: An Overview of Biomedical Applications. *J. Control. Release.* **2012,** *161,* 505–522.

Dash, T. K.; Konkimalla, V. B. Modification of Cyclodextrin for Improvement of Complexation and Formulation Properties. *Handbook of Polymers for Pharmaceutical Technologies–Biodegradable Polymers*; Thakur, V. K.; Thakur, M. K., Eds.; Scrivener Publishing & J. Wiley and Sons: Hoboken, **2015;** pp 205–224.

DeMarino, C.; Schwab, A.; Pleet, M.; Mathiesen, A.; Friedman, J.; El-Hage, N.; Kashanchi, F. Biodegradable Nanoparticles for Delivery of Therapeutics in CNS Infection. *J. Neuroimun. Pharmacol.* 2017, *12,* 31–50.

Devkar, T. B.; Tekade, A. R.; Khandelwal, K. R. Surface Engineered Nanostructured Lipid Carriers for Efficient Nose to Brain Delivery of Ondansetron HCl Using *Delonix regia* Gum as a Natural Mucoadhesive Polymer. *Colloids Surf. B Biointerfaces*. **2014,** *122,* 143–150.

Dewick P. M. *Medicinal Natural Products: A Biosynthetic Approach*, 3rd Edition. John Wiley & Sons: Chichester, **2009.**

Dey, S.; Sherly, M. C. D.; Rekha, M. R.; Sreenivasan, K. Alginate Stabilized Gold Nanoparticle as Multidrug Carrier: Evaluation of Cellular Interactions and Hemolytic Potential. *Carbohydr. Polym.* **2016,** *136,* 71–80.

Dhar, S.; Reddy, E. M.; Prabhune, A.; Pokharkar, V.; Shiras, A.; Prasad, B. L. V. Cytotoxicity of Sophorolipid-Gellan Gum-Gold Nanoparticle Conjugates and Their Doxorubicin Loaded Derivatives Towards Human Glioma and Human Glioma Stem Cell Lines. *Nanoscale* **2011,** *3,* 575–580.

Di Gioia, S.; Trapani, A.; Mandracchia, D.; De Giglio, E.; Cometa, S.; Mangini, V.; Arnesano, F.; Belgiovine, G.; Castellani, S.; Pace, L.; Lavecchia, M. A.; Trapani, G.; Conese, M.; Puglisi, G.; Cassano, T. Intranasal Delivery of Dopamine to the Striatum Using Glycol Chitosan/Sulfobutylether-β-Cyclodextrin-Based Nanoparticles. *Eur. J. Pharm. Biopharm.* **2015,** *94,* 180–193.

Di Martino, A.; Kucharczyk, P.; Capakova, Z.; Humpolicek, P.; Sedlarik, V. Enhancement of Temozolomide Stability by Loading in Chitosan-Carboxylated Polylactide-Based Nanoparticles. *J. Nanopart. Res.* **2017,** *19,* 71.

Djagnya, K. B.; Wang, Z.; Xu, S. Gelatin: A Valuable Protein for Food and Pharmaceutical Industries: Review. *Crit. Rev. Food Sci. Nutr.* **2010,** *41,* 481–492.

Dong, S. P.; Cho, H. J.; Lee, Y. W.; Roman, M. Synthesis and Cellular Uptake of Folic Acid-Conjugated Cellulose Nanocrystals for Cancer Targeting. *Biomacromolecules*. **2014,** *15,* 1560–1567.

Edgar, K. J. Cellulose Esters in Drug Delivery. *Cellulose*. **2014,** *14,* 49–64.

Elnaggar, Y. S. R.; Etman, S. M.; Abdelmonsif, D. A.; Abdallah, O. Y. Intranasal Piperine-Loaded Chitosan Nanoparticles as Brain-Targeted Therapy in Alzheimer's Disease: Optimization, Biological Efficacy, and Potential Toxicity. *J. Pharm. Sci.* **2015,** *104,* 3544–3556.

Elsadek, B.; Kratz, F. Impact of Albumin on Drug Delivery: New Applications on the Horizon. *J. Control. Release.* **2012,** *157,* 4–28.

Engelberth, S. A.; Hempel, N.; Bergkvist, M. Chemically Modified Dendritic Starch: A Novel Nanomaterial for siRNA Delivery. *Bioconjug. Chem.* **2015,** *26,* 1766–1774.

Engineer, C.; Parikh, J.; Raval, A. Review on Hydrolytic Degradation Behavior of Biodegradable Polymers from Controlled Drug Delivery System. *Trends Biomater. Artif. Organs.* **2011,** *25,* 79–85.

Eroglu, M. S.; Oner, E. T.; Mutlu, E. C.; Bostan, M. S. Sugar Based Biopolymers in Nanomedicine. New Emerging Era for Cancer Imaging and Therapy. *Curr. Top. Med. Chem.* **2017,** *17,* 1507–1520.

Esa, F.; Tasirin, S. M.; Rahman, N. A. Overview of Bacterial Cellulose Production and Application. *Agric. Agric. Sci. Proc.* **2014,** *2,* 113–119.

Estevanato, L.; Cintra, D.; Baldini, N.; Portilho, F.; Barbosa, L.; Martins, O.; Lacava, B.; Miranda-Vilela, A. L.; Tedesco, A. C.; Bao, S.; Morais, P. C.; Lacava, Z. C. Preliminary Biocompatibility Investigation of Magnetic Albumin Nanosphere Designed as a Potential Versatile Drug Delivery System. *Int. J. Nanomedicine.* **2011,** *6,* 1709–1717.

Fazil, M.; Md, S.; Hague, S.; Kumar, M.; Baboota, S.; Sahni, J. K.; Ali, J. Development and Evaluation of Rivastigmine Loaded Chitosan Nanoparticles for Brain Targeting. *Eur. J. Pharm. Sci.* **2012,** *47,* 6–15.

Foox, M.; Zilberman, M. Drug Delivery from Gelatin-Based Systems. *Expert Opin. Drug Deliv.* **2015,** *12,* 1547–1563.

Fornaguera, C.; Dols-Perez, A.; Caldero, G.; Garcia-Celma, M. J.; Camarasa, J.; Solans, C. PLGA Nanoparticles Prepared by Nano-Emulsion Templating Using Low-Energy Methods as Efficient Nanocarriers for Drug Delivery Across the Blood-Brain Barrier. *J. Control. Release.* **2015,** *211,* 134–143.

Frömming, K. H.; Szejtli, J. Cyclodextrins in pharmacy. Kluwer Academic Publishers: Dordrecht, **1994.**

Fu, S. Z.; Xia, J. Y.; Wu, J. B. Functional Chitosan Nanoparticles in Cancer Treatment. *J. Biomed. Nanotechnol.* **2016,** *12,* 1585–1603.

Gagliardi, M.; Borri, C. Polymer Nanoparticles as Smart Carriers for the Enhanced Release of Therapeutic Agents to the CNS. *Curr. Pharm. Des.* **2017,** *23,* 393–410.

Gai, M.; Frueh, J.; Tao, T.; Petrov, A. V.; Petrov, V. V.; Shesterikov, E. V.; Tverdokhlebov, S. I.; Sukhorukov, G. B. Polylactic Acid Nano- and Microchamber Arrays for Encapsulation of Small Hydrophilic Molecules Featuring Drug Release Via High Intensity Focused Ultrasound. *Nanoscale.* **2017,** *9,* 7063–7070.

Gao, H. L.; Cao, S. J.; Yang, Z.; Zhang, S.; Zhang, Q. Z.; Jiang, X. G. Preparation, Characterization and Anti-glioma Effects of Docetaxel-Incorporated Albumin-Lipid Nanoparticles. *J. Biomed. Nanotechnol.* **2015,** *11,* 2137–2147.

Gao, H. L. Progress and Perspectives on Targeting Nanoparticles for Brain Drug Delivery. *Acta Pharm. Sin. B* **2016,** *6,* 268–286.

Gan, C. W.; Feng, S. S. Transferrin-Conjugated Nanoparticles of Poly(Lactide)-D-Alpha-Tocopheryl Polyethylene Glycol Succinate Diblock Copolymer for Targeted Drug Delivery Across the Blood-Brain Barrier. *Biomaterials* **2010,** *31,* 7748–7757.

Gartziandia, O.; Herran, E.; Pedraz, J. L.; Carro, E.; Igartua, M.; Hernandez, R. M. Chitosan Coated Nanostructured Lipid Carriers for Brain Delivery of Proteins by Intranasal Administration. *Colloids Surf. B Biointerfaces* **2015,** *134,* 304–313.

Gartziandia, O.; Herran, E.; Ruiz-Ortega, J. A.; Miguelez, C.; Igartua, M.; Lafuente, J. V.; Pedraz, J. L.; Ugedo, L.; Hernandez, R. M. Intranasal Administration of Chitosan-Coated Nanostructured Lipid Carriers Loaded with GDNF Improves Behavioral and Histological Recovery in a Partial Lesion Model of Parkinson's Disease. *J. Biomed. Nanotechnol.* **2016,** *12,* 2220–2230.

Gazzoni, A. F.; Capilla, J.; Mayayo, E.; Guarro, J. Efficacy of Intrathecal Administration of Liposomal Amphotericin B Combined with Voriconazole in a Murine Model of Cryptococcal Meningitis. *Int. J. Antimicrob. Agents.* **2012,** *39,* 223–227.

Geldenhuys, W.; Mbimba, T.; Bui, T.; Harrison, K.; Sutariya, V. Brain-Targeted Delivery of Paclitaxel Using Glutathione-Coated Nanoparticles for Brain Cancers. *J. Drug Target.* **2011,** *19,* 837–845.

Geldenhuys, W.; Wehrung, D.; Groshev, A.; Hirani, A.; Sutariya, V. Brain-Targeted Delivery of Doxorubicin Using Glutathione-Coated Nanoparticles for Brain Cancers. *Pharm. Dev. Technol.* **2015,** *20,* 497–506.

George, J.; Sabapathi, S. N. Cellulose Nanocrystals: Synthesis, Functional Properties, and Applications. *Nanotechnol. Sci. Appl.* **2015,** *8,* 45–54.

Geze, A.; Choisnard, L.; Putaux, J. L.; Wouessidjewe, D. Colloidal Systems Made of Biotransesterified Alpha, Beta and Gamma Cyclodextrins Grafted with C10 Alkyl Chains. *Mat. Sci. Eng. C.* **2009,** *29,* 458–462.

Gil, E. S.; Wu, L. F.; Xu, L. C.; Lowe, T. L. β-Cyclodextrin-Poly(β-Amino Ester) Nanoparticles for Sustained Drug Delivery Across the Blood-Brain Barrier. *Biomacromolecules* **2012,** *13,* 3533–3541.

Girotra, P.; Singh, S. K. A Comparative Study of Orally Delivered PBCA and ApoE Coupled BSA Nanoparticles for Brain Targeting of Sumatriptan Succinate in Therapeutic Management of Migraine. *Pharm. Res.* **2016,** *33,* 1682–1695.

Godinho, B. M. D. C.; Ogier, J. R.; Quinlan, A.; Darcy, R.; Griffin, B. T.; Cryan, J. F.; Driscoll, C. M. PEGylated Cyclodextrins as Novel siRNA Nanosystems: Correlations Between Polyethylene Glycol Length and Nanoparticle Stability. *Int. J. Pharm.* **2014,** *473,* 105–112.

Goswami, S.; Naik, S. Natural Gums and Its Pharmaceutical Application. *J. Sci. Innov. Res.* **2014,** *3,* 112–121.

Granzotto, C.; Arslanoglu, J.; Rolando, C.; Tokarski, C. Plant gum identification in historic artworks. *Sci. Rep.* **2017,** *7,* 44538.

Grenha, A., Rodrigues, S. Pullulan-Based Nanoparticles: Future Therapeutic Applications in Transmucosal Protein Delivery. *Ther. Deliv.* **2013,** *4,* 1339–1341.

Gu, J. J.; Al-Bayati, K.; Ho, E. A. Development of Antibody-Modified Chitosan Nanoparticles for the Targeted Delivery of siRNA Across the Blood-Brain Barrier as a Strategy for Inhibiting HIV Replication in Astrocytes. *Drug Deliv. Transl. Res.* **2017,** *7,* 497–506.

Guo, W. H.; Li, A. M.; Jia, Z. J.; Yuan, Y.; Dai, H. F.; Li, H. X. Transferrin Modified PEG-PLA-Resveratrol Conjugates: In vitro and In Vivo Studies for Glioma. *Eur. J. Pharmacol.* **2013,** *718,* 41–47.

Hamman, H.; Steenekamp, J.; Hamman, J. Use of Natural Gums and Mucilages as Pharmaceutical Excipients. *Curr. Pharm. Des.* **2015,** *21,* 4775–4797.

Han, L.; Ren, Y.; Long, L. X.; Zhong, Y.; Shen, C. H.; Pu, P. Y.; Yuan, X. B.; Kang, C. S. Inhibition of C6 Glioma In Vivo by Combination Chemotherapy of Implantation of Polymer Wafer and Intracarotid Perfusion of Transferrin-Decorated Nanoparticles. *Oncol. Rep.* **2012,** *27,* 121–128.

Han, N. K.; Shin, D. H.; Kim, J. S.; Weon, K. Y.; Jang, C. Y.; Kim, J. S. Hyaluronan-Conjugated Liposomes Encapsulating Gemcitabine for Breast Cancer Stem Cells. *Int. J. Nanomedicine* **2016,** *11,* 1413–1425.

Hanafy, A. S.; Farid, R. M.; Helmy, M. W.; ElGamal, S. S. Pharmacological, Toxicological and Neuronal Localization Assessment of Galantamine/Chitosan Complex Nanoparticles in Rats: Future Potential Contribution in Alzheimer's Disease Management. *Drug Deliv.* **2016,** *23,* 3111–3122.

Hansraj, G. P.; Singh, S. K.; Kumar, P. Sumatriptan Succinate Loaded Chitosan Solid Lipid Nanoparticles for Enhanced Anti-Migraine Potential. *Int. J. Biol. Macromol.* **2015,** *81,* 467–476.

Hao, J. F.; Zhao, J.; Zhang, S. P.; Tong, T. T.; Zhuang, Q. N.; Jin, K.; Chen, W.; Tang, H. Fabrication of an Ionic-Sensitive In Situ Gel Loaded with Resveratrol Nanosuspensions Intended for Direct Nose-To-Brain Delivery. *Coll. Surf. B Biointerfaces.* **2016,** *147,* 376–386.

Haque, S.; Md, S.; Fazil, M.; Kumar, M.; Sahni, J. K.; Ali, J.; Baboota, S. Venlafaxine Loaded Chitosan NPs for Brain Targeting: Pharmacokinetic and Pharmacodynamic Evaluation. *Carbohydr. Polym.* **2012,** *89,* 72–79.

Haque, S.; Md, S.; Sahni, J. K.; Ali, J.; Baboota, S. Development and Evaluation of Brain Targeted Intranasal Alginate Nanoparticles for Treatment of Depression. *J. Psychiatr. Res.* **2014**, *48*, 1−12.

Hayward, S. L.; Wilson, C. L.; Kidambi, S. Hyaluronic Acid-Conjugated Liposome Nanoparticles for Targeted Delivery to CD44 Overexpressing Glioblastoma Cells. *Oncotargetics* **2016**, *7*, 34158−34171.

Hsiao, M. H.; Larsson, M.; Larsson, A.; Evenbratt, H.; Chen, Y. Y.; Chen, Y. Y.; Liu, D. M. Design and Characterization of a Novel Amphiphilic Chitosan Nanocapsule-Based Thermo-Gelling Biogel with sustained In Vivo Release of the Hydrophilic Anti-Epilepsy Drug Ethosuximide. *J. Control. Release* **2012**, *161*, 942−948.

Hu, K. L.; Li, J. W.; Shen, Y. H.; Lu, W.; Gao, X. L.; Zhang, Q. Z.; Jiang, X. G. Lactoferrin-Conjugated PEG-PLA Nanoparticles with Improved Brain Delivery: In Vitro and In Vivo Evaluations. *J. Control. Release* **2009**, *134*, 55−61.

Hu, K. L.; Shi, Y. B.; Jiang, W. M.; Han, J. Y.; Huang, S. X.; Jiang, X. G. Lactoferrin-Conjugated PEG-PLGA Nanoparticles for Brain Delivery: Preparation, Characterization and Efficacy in Parkinson's Disease. *Int. J. Pharm.* **2011**, *415*, 273−283.

Huang, D. X.; Lin, C.; Wen, X. J.; Gu, S. Y.; Zhao, P. A Potential Nanofiber Membrane Device for Filling Surgical Residual Cavity to Prevent Glioma Recurrence and Improve Local Neural Tissue Reconstruction. *PLoS One* **2016**, 11, e0161435.

Huey, R.; O'Hagan, B.; McCarron, P.; Hawthorne, S. Targeted Drug Delivery System to Neural Cells Utilizes the Nicotinic Acetylcholine Receptor. *Int. J. Pharm.* **2017**, *525*, 12−20.

Irani, M.; Sadeghi, G. M. M.; Haririan, I. A Novel Biocompatible Drug Delivery System of Chitosan/Temozolomide Nanoparticles Loaded PCL-PU Nanofibers for Sustained Delivery of Temozolomide. *Int. J. Biol. Macromol.* **2017**, *97*, 744−751.

Ireson, C. R.; Kelland, L. R. Discovery and Development of Anticancer Aptamers. *Mol. Cancer Ther.* **2006**, *5*, 2957–2962.

Jafarieh, O.; Md, S.; Ali, M.; Baboota, S.; Sahni, J. K.; Kumari, B.; Bhatnagar, A.; Ali, J. Design, Characterization, and Evaluation of Intranasal Delivery of Ropinirole-Loaded Mucoadhesive Nanoparticles for Brain Targeting. *Drug Dev. Ind. Pharm.* **2015**, *41*, 1674−1681.

Jain, A. K.; Das, M.; Swarnakar, N. K.; Jain, S. Engineered PLGA Nanoparticles: An Emerging Delivery Tool in Cancer Therapeutics. *Crit. Rev. Ther. Drug Carrier Syst.* **2011**, *28*, 1–45.

Jain, K. K. Nanobiotechnology-Based Strategies for Crossing the Blood–Brain Barrier. *Nanomedicine* **2012**, *7*, 1225–1233.

Jain, D.; Bar-Shalom, D. Alginate Drug Delivery Systems: Application in Context of Pharmaceutical and Biomedical Research. *Drug Dev. Ind. Pharm.* **2014**, *40*, 1576–1584.

Jain, D. S.; Athawale, R. B.; Bajaj, A. N.; Shrikhande, S. S.; Goel, P. N.; Nikam, Y.; Gude, R. P. Unraveling the Cytotoxic Potential of Temozolomide Loaded into PLGA Nanoparticles. *DARU-J. Pharm. Sci.* **2014**, *22*, 18.

Jain, A; Jain, A; Garg, N. K.; Tyagi, R. K.; Singh, B.; Katare, O. P.; Webster, T. J.; Soni, V. Surface Engineered Polymeric Nanocarriers Mediate the Delivery of Transferrin-Methotrexate Conjugates for an Improved Understanding of Brain Cancer. *Acta Biomater.* **2015**, *24*, 140−151.

Jain, D.; Bajaj, A.; Athawale, R.; Shrikhande, S.; Goel, P. N.; Nikam, Y.; Gude, R.; Patil, S.; Raut, P. P. Surface-Coated PLA Nanoparticles Loaded with Temozolomide for Improved

Brain Deposition and Potential Treatment of Gliomas: Development, Characterization and In Vivo Studies. *Drug Deliv.* **2016,** *23,* 999–1016.

Jain, S.; Datta, M. Montmorillonite-Alginate Microspheres as a Delivery Vehicle for Oral Extended Release of Venlafaxine Hydrochloride. *J. Drug Deliv. Sci. Technol.* **2016,** *33,* 149–156.

Jakki, S. L.; Ramesh, Y. V.; Gowthamarajan, K.; Senthil, V.; Jain, K.; Sood, S.; Pathak, D. Novel Anionic Polymer as a Carrier for CNS Delivery of Anti-Alzheimer Drug. *Drug Deliv.* **2016,** *23,* 3471–3479.

James, R.; Manoukian, O. S.; Kumbar, S. G. Poly(Lactic Acid) for Delivery of Bioactive Macromolecules. *Adv. Drug Deliv. Rev.* **2016,** *107,* 277–288.

Jampílek, J.; Záruba, K.; Oravec, M.; Kuneš, M.; Babula, P.; Ulbrich, P.; Brezaniová, I.; Opatřilová, R.; Tříska, J.; Suchý, P. Preparation of Silica Nanoparticles Loaded with Nootropics and Their In Vivo Permeation Through Blood-Brain Barrier. *Biomed. Res. Int.* **2015,** *2015,* 812673.

Jampílek, J.; Kráľová, K. Nano-Antimicrobials: Activity, Benefits and Weaknesses. *Nanostructures in Therapeutic Medicine,* Vol. 2. *Nanostructures for Antimicrobial Therapy.* Ficai, A.; Grumezescu, A. M. Eds.; Elsevier: Amsterdam, **2017,** 23–54.

Jampílek, J.; Kráľová, K. Application of Nanobioformulations for Controlled Release and Targeted Biodistribution of Drugs. *Nanobiomaterials: Applications in Drug Delivery.* Sharma, A. K.; Keservani, R. K.; Kesharwani, R. K. Eds.; Apple Academic Press & CRC Press: Oakville, **2018,** 131–208.

Jampílek, J.; Kráľová, K. Nanotechnology Based Formulations for Drug Targeting to Central Nervous System. *Nanoparticulate Drug Delivery Systems.* Keservani, R. K.; Sharma, A. K. Eds.; Apple Academic Press & CRC Press: Warentown, **2019a,** 151–220.

Jampílek, J.; Kos, J.; Kráľová, K. Potential of Nanomaterial Applications in Dietary Supplements and Foods for Special Medical Purposes. *Nanomaterials* **2019,** *9,* 296.

Jansson, P. E.; Lindberg, B.; Sandford, P. A. Structural Studies of Gellan Gum, an Extracellular Polysaccharide Elaborated by *Pseudomonas elodea. Carbohydr. Res.* **1983,** *124,* 135–139.

Jao, D.; Xue, Y.; Medina, J.; Hu, X. Protein-Based Drug-Delivery Materials. *Materials* **2017,** *10,* 517.

Jaruszewski, K. M.; Ramakrishnan, S.; Poduslo, J. F.; Kandimalla, K. K. Chitosan Enhances the Stability and Targeting of Immuno-Nanovehicles to Cerebro-Vascular Deposits of Alzheimer's Disease Amyloid Protein. *Nanomedicine* **2012,** *8,* 250–260.

Jeong, Y. I.; Kim, S. T.; Jin, S. G.; Ryu, H. H.; Jin, Y. H.; Jung, T. Y.; Kim, I. Y.; Jung, S. Cisplatin-Incorporated Hyaluronic Acid Nanoparticles Based on Ion-Complex Formation. *J. Pharm. Sci.* **2008,** *97,* 1268–1276.

Jeong, J. H.; Kim, J. J.; Bak, D. H.; Yu, K. S.; Lee, J. H.; Lee, N. S.; Jeong, Y. G.; Kim, D. K.; Kim, D. K.; Han, S. Y. Nanoparticles Against Glutamate-Induced Neurotoxicity. *J. Nanosci. Nanotechnol.* **2015,** *15,* 7922–7928.

Jiang, Z. Q.; Dong, X. Y.; Liu, H.; Wang, Y. J.; Zhang, L.; Sun, Y. Multifunctionality of Self-Assembled Nanogels of Curcumin-Hyaluronic Acid Conjugates on Inhibiting Amyloid β-Protein Fibrillation and Cytotoxicity. *React. Funct. Polym.* **2016,** *104,* 22–29.

Jinwal, U. K.; Groshev, A.; Zhang, J.; Grover, A.; Sutariya, V. B. Preparation and Characterization of Methylene Blue Nanoparticles for Alzheimer's Disease and Other Tauopathies. *Curr. Drug Deliv.* **2014,** *11,* 541–550.

Joachim, E.; Kim, I. D.; Jin, Y.; Kim, K.; Lee, J. K.; Choi, H. Gelatin Nanoparticles Enhance the Neuroprotective Effects of Intranasally Administered Osteopontin in Rat Ischemic Stroke Model. *Drug Deliv. Transl. Res.* **2014**, *4*, 395−399.

Johns Hopkins University. The Johns Hopkins Hospital and Johns Hopkins Health System – Overview of Nervous System Disorders, **2017**. http://www.hopkinsmedicine.org/ healthlibrary/conditions/nervous_system_disorders/overview_of_nervous_system_ disorders_85,P00799 (accessed May 12, 2019).

Jose, S.; Sowmya, S.; Cinu, T. A.; Aleykutty, N. A.; Thomas, S.; Souto, E. B. Surface-Modified PLGA Nanoparticles for Brain Targeting of Bacoside-A. *Eur. J. Pharm. Sci.* **2014**, *63*, 29−35.

Kamalinia, G.; Khodagholi, F.; Shaerzadeh, F.; Tavssolian, F.; Chaharband, F.; Atyabi, F.; Sharifzadeh, M.; Amini, M.; Dinarvand, R. Cationic Albumin-Conjugated Chelating Agent as a Novel Brain Drug Delivery System in Neurodegeneration. *Chem. Biol. Drug Des.* **2015**, *86*, 1203−1214.

Kang, T.; Jiang, M. Y.; Jiang, D.; Feng, X. Y.; Yao, J. H.; Song, Q. X.; Chen, H. Z.; Gao, X. L.; Chen, J. Enhancing Glioblastoma-Specific Penetration by Functionalization of Nanoparticles with an Iron-Mimic Peptide Targeting Transferrin/Transferrin Receptor Complex. *Mol. Pharm.* **2015**, *12*, 2947−2961.

Kapoor, D. N.; Bhatia, A.; Kaur, R.; Sharma, R.; Kaur, G.; Dhawan, S. PLGA: A Unique Polymer for Drug Delivery. *Ther. Deliv.* **2015**, *6*, 41–58.

Karavasili, C.; Bouropoulos, N.; Sygellou, L.; Amanatiadou, E. P.; Vizirianakis, I. S.; Fatouros, D. G. PLGA/DPPC/Trimethylchitosan Spray-Dried Microparticles for the Nasal Delivery of Ropinirole Hydrochloride: In Vitro, Ex Vivo and Cytocompatibility Assessment. *Mater. Sci. Eng. C Mater. Biol. Appl.* **2016**, *59*, 1053−1062.

Karimi, M.; Bahrami, S.; Ravari, S. B.; Zangabad, P. S.; Mirshekari, H.; Bozorgomid, M.; Shahreza, S.; Sori, M.; Hamblin, M. R. Albumin Nanostructures as Advanced Drug Delivery Systems. *Expert Opin. Drug Deliv.* **2016**, *13*, 1609−1623.

Kawabata, Y.; Wada, K.; Nakatani, M.; Yamada, S.; Onoue, S. Formulation Design for Poorly Water-Soluble Drugs Based on Biopharmaceutics Classification System: Basic Approaches and Practical Applications. *Int. J. Pharm.* **2011**, *420*, 1–10.

Kawasaki, R.; Sasaki, Y.; Akiyoshi, K. Intracellular Delivery and Passive Tumor Targeting of a Self-Assembled Nanogel Containing Carborane Clusters for Boron Neutron Capture Therapy. *Biochem. Biophys. Res. Commun.* **2017**, *483*, 147−152.

Kempaiah, R.; Arias, S. L.; Pastrana, F.; Alucozai, M.; Reece, L. M.; Pavon, J.; Allain, J. P. A New Nanostructured Material for Regenerative Vascular Treatments: Magnetic Bacterial Nanocellulose (MBNC). Pan American Health Care Exchanges: Medellin, 2013; pp 1−11.

Khalin, I.; Alyautdin, R.; Wong, T. W.; Gnanou, J.; Kocherga, G.; Kreuter, J. Brain-Derived Neurotrophic Factor Delivered to the Brain Using Poly(Lactide-Co-Glycolide) Nanoparticles Improves Neurological and Cognitive Outcome in Mice With Traumatic Brain Injury. *Drug Deliv.* **2016**, *23*, 3520−3528.

Khan, S.; Patil, K.; Bobade, N.; Yeole, P.; Gaikwad, R. Formulation of Intranasal Mucoadhesive Temperature-Mediated In Situ Gel Containing Ropinirole and Evaluation of Brain Targeting Efficiency in Rats. *J. Drug Target.* **2010**, *18*, 223−234.

Khan, H.; Shukla, R. N.; Bajpai, A. K. Genipin-Modified Gelatin Nanocarriers as Swelling Controlled Drug Delivery System for in vitro Release of Cytarabine. *Mater. Sci. Eng. C Mater. Biol. Appl.* **2016**, *61*, 457–465.

Khanbabaie, R.; Jahanshahi, M. Revolutionary Impact of Nanodrug Delivery on Neuroscience. *Curr. Neuropharmacol.* **2012**, *10*, 370–392.

Kim, D. K.; Mikhaylova, M.; Wang, F. H.; Kehr, J.; Bjelke, B.; Zhang, Y.; Tsakalakos, T.; Muhammed, M. Starch-Coated Superparamagnetic Nanoparticles as MR Contrast Agents. *Chem. Mater.* **2003**, *15*, 4343–4351.

Kim, J. Y.; Choi, W. I.; Kim, Y. H.; Tae, G. Brain-targeted Delivery of Protein Using Chitosan- and RVG Peptide-Conjugated, Pluronic-Based Nano-Carrier. *Biomaterials* **2013**, *34*, 1170–1178.

Kim, H. Y.; Park, S. S.; Lim, S. T. Preparation, Characterization and Utilization of Starch Nanoparticles. *Colloids Surf. B Biointerfaces.* **2015**, *126*, 607–620.

Kim, I. D.; Sawicki, E.; Lee, H. K.; Lee, E. H.; Park, H. J.; Han, P. L.; Kim, K.; Choi, H.; Lee, J. K. Robust Neuroprotective Effects of Intranasally Delivered iNOS siRNA Encapsulated in Gelatin Nanoparticles in the Postischemic Brain. *Nanomedicine* **2016**, *12*, 1219–1229.

Klemm, D.; Heinze, T.; Wagenknecht, W.; Phillip, B.; Heinze, U. Comprehensive Cellulose Chemistry, Wiley-VCH: Weinheim, **1998**.

Kobori, T.; Nakao, H. Xanthan Gum – Basic Properties, Applications, and Future Perspective in Nanotechnology. *Polysaccharides: Development, Properties and Applications*; Tiwari, A., Ed.; Nova Science Publishers: Hauppauge, 2010; pp 379–393.

Kogan, G.; Soltes, L.; Stern, R.; Gemeiner, P. Hyaluronic Acid: A Natural Biopolymer with a Broad Range of Biomedical and Industrial Applications. *Biotechnol. Lett.* **2007**, *29*, 17–25.

Koivisto, J. T.; Joki, T.; Parraga, J. E.; Paakkonen, R.; Yla-Outinen, L.; Salonen, L.; Jonkkari, I.; Peltola, M.; Ihalainen, T. O.; Narkilahti, S.; Kellomäki, M. Bioamine-Crosslinked Gellan Gum Hydrogel for Neural Tissue Engineering. *Biomed. Mater.* **2017**, *12*, 025014.

Kozlovskaya, L.; Abou-Kaoud, M.; Stepensky, D. Quantitative Analysis of Drug Delivery to the Brain Via Nasal Route. *J. Control. Release* **2014**, *189*, 133–140.

Kozlu, S.; Caban, S.; Yerlikaya, F.; Fernandez, M. E.; Novoa-Carballal, R.; Riguera, R.; Yemisci, M.; Gursoy-Ozdemir, Y.; Dalkara, T.; Couvreur, P.; Capan, Y. An Aquaporin 4 Antisense Oligonucleotide Loaded, Brain Targeted Nanoparticulate System Design. *Pharmazie* **2014**, *69*, 340–345.

Král, V.; Králová, J.; Flieger, M.; Jampílek, J.; Řezáčová, A.; Dohnal, J.; Oktábec, Z.; Záruba, K.; Grünwaldová, V.; Poučková, P.; Martásek, P. *Route of Drug Administration in Nanoparticle form to Enable Penetration through the Brain Blood Barrier*. PV 2011-366, June 21, **2011**.

Kratz, F. Albumin as a Drug Carrier: Design of Prodrugs, Drug Conjugates and Nanoparticles. *J. Control. Release* **2008**, *132*, 171–183.

Kreuter, J.; Hekmatara, T.; Dreis, S.; Vogel, T.; Gelperina, S.; Langer, K. Covalent Attachment of Apolipoprotein A-I and Apolipoprotein B-100 to Albumin Nanoparticles Enables Drug Transport into the Brain. *J. Control. Release* **2007**, *118*, 54–58.

Kuo, Y. C.; Yu, H. W. Transport of Saquinavir Across Human Brain-Microvascular Endothelial Cells by Poly(Lactide-Co-Glycolide) Nanoparticles with Surface Poly-(γ-Glutamic Acid). *Int. J. Pharm.* **2011**, *416*, 365–375.

Kuo, W. T.; Huang, H. Y.; Chou, M. J.; Wu, M. C.; Huang, Y. Y. Surface Modification of Gelatin Nanoparticles with Polyethylenimine as Gene Vector. *J. Nanomater.* **2011**, *2011*, 646538.

Kuo, Y. C.; Chen, Y. C. Targeting Delivery of Etoposide to Inhibit the Growth of Human Glioblastoma Multiforme Using Lactoferrin- and Folic Acid-Grafted Poly(Lactide-Co-Glycolide) Nanoparticles. *Int. J. Pharm.* **2015**, *479*, 138–149.

Kuo, Y. C.; Rajesh, R. Targeted Delivery of Rosmarinic Acid Across the Blood-Brain Barrier for Neuronal Rescue Using Polyacrylamide-Chitosan-Poly(Lactide-Co-Glycolide) Nanoparticles with Surface Cross-Reacting Material 197 and Apolipoprotein E. *Int. J. Pharm.* **2017,** *528*, 228–241.

Kurkov, S. V.; Loftsson, T. Cyclodextrins. *Int. J. Pharm.* **2013,** *453*, 167–180.

Lalani, J.; Raichandani, Y.; Mathur, R.; Lalan, M.; Chutani, K.; Mishra, A. K.; Misra, A. Comparative Receptor Based Brain Delivery of Tramadol-Loaded Poly(Lactic-Co-Glycolic Acid) Nanoparticles. *J. Biomed. Nanotechnol.* **2012,** *8*, 918–927.

Lalani, J.; Rathi, M.; Lalan, M.; Misra, A. Protein Functionalized Tramadol-Loaded PLGA Nanoparticles: Preparation, Optimization, Stability and Pharmacodynamic Studies. *Drug Dev. Ind. Pharm.* **2013,** *39*, 854–864.

Lalatsa, A; Barbu, E. Carbohydrate Nanoparticles for Brain Delivery. *Int. Rev. Neurobiol.* **2016,** *130*, 115–153.

Lancuški, A.; Vasilyev, G.; Putaux, J. L.; Zussman, E. Rheological Properties and Electrospinnability of High-Amylose Starch in Formic Acid. *Biomacromolecules* **2015,** *16*, 2529–2536.

Lee, K. Y.; Mooney, D. J. Alginate: Properties and Biomedical Applications. *Progr. Polym. Sci.* **2012,** *37*, 106–126.

Lee, B. K.; Yun, Y.; Park, K. PLA Micro- and Nano-Particles. *Adv Drug Deliv Rev.* **2016,** *107*, 176–191.

Lehto, T.; Ezzat, K.; Wood, M. J. A.; Andaloussi, S. E. Peptides for Nucleic Acid Delivery. *Adv. Drug Deliv. Rev.* **2016,** *106* A, 172–182.

Li, J. W.; Feng, L.; Fan, L.; Zha, Y.; Guo, L. R.; Zhang, Q. Z.; Chen, J.; Pang, Z. Q.; Wang, Y. C.; Jiang, X. G.; Yang, V. C.; Wen, L. Targeting the Brain with PEG-PLGA Nanoparticles Modified with Phage-Displayed Peptides. *Biomaterials* **2011,** *32*, 4943–4950.

Li, J. Y.; Sabliov, C. PLA/PLGA Nanoparticles for Delivery of Drugs Across the Blood-Brain Barrier. *Nanotechnol. Rev.* **2013,** *2*, 241–257.

Li, J. W.; Zhang, C.; Li, J.; Fan, L.; Jiang, X. G.; Chen, J.; Pang, Z. Q.; Zhang, Q. Z. Brain Delivery of NAP with PEG-PLGA Nanoparticles Modified with Phage Display Peptides. *Pharm. Res.* **2013,** *30*, 1813–1823.

Li, D.; Kerns, E. H. *Blood-Brain Barrier in Drug Discovery: Optimizing Brain Exposure of CNS Drugs and Minimizing Brain Side Effects for Peripheral Drugs.* John Wiley & Sons: Hoboken, 2015.

Li, Y. Q.; Du, Y.; Liu, X.; Zhang, Q.; Jing, L. J.; Liang, X. L.; Chi, C. W.; Dai, Z. F.; Tian, J. Conjugated Polylactic Acid Nanoparticles Encapsulating Endostar on Glioma by Optical Molecular Imaging. *Mol. Imagining.* **2015a,** *14*, 356–365.

Li, A. J.; Zheng, Y. H.; Liu, G. D.; Liu, W. S.; Cao, P. C.; Bu, Z. F. Efficient Delivery of Docetaxel for the Treatment of Brain Tumors by Cyclic RGD-Tagged Polymeric Micelles. *Mol. Med. Rep.* **2015b,** *11*, 3078–3086.

Li, X. M.; Tsiouklis, J.; Weng, T. T.; Zhang, B. N.; Yin, G. Q.; Feng, G. Z.; Cui, Y. D.; Savina, I. N.; Mikhalovska, L. I.; Sandeman, S. R.; Howel, C. A.; Mikhalovsky S. V. Nano Carriers for Drug Transport Across the Blood-Brain Barrier. *J. Drug Target.* **2017,** *25*, 17–28.

Liang, D. S.; Su, H. T.; Liu, Y. J.; Wang, A. T.; Qi, X. R. Tumor-Specific Penetrating Peptides-Functionalized Hyaluronic Acid-d-α-Tocopheryl Succinate-Based Nanoparticles for Multi-Task Delivery to Invasive Cancers. *Biomaterials* **2015,** *71*, 11–23.

Lin, T.; Liu, E.; He, H.; Shin, M. C.; Moon, C.; Yang, V. C.; Huang, Y. Nose-To-Brain Delivery of Macromolecules Mediated by Cell-Penetrating Peptides. *Acta Pharm. Sin. B* **2016a,** *6*, 352–358.

Lin, T. T.; Zhao, P. F.; Jiang, Y. F.; Tang, Y. S.; Jin, H. Y.; Pan, Z. Z.; He, H. N.; Yang, V. C.; Huang, Y. Z. Blood-Brain-Barrier-Penetrating Albumin Nanoparticles for Biomimetic Drug Delivery Via Albumin-Binding Protein Pathways for Antiglioma Therapy. *ACS Nano* **2016b,** *10*, 9999–10012.

Ling, Y.; Yang, Y.; Lu, N.; You, Q. D.; Wang, S.; Gao, Y.; Chen, Y.; Guo, Q. L. Endostar, A Novel Recombinant Human Endostatin, Exerts Antiangiogenic Effect Via Blocking VEGF-Induced Tyrosine Phosphorylation of KDR/Flk-1 of Endothelial Cells. *Biochem. Biophys. Res. Commun.* **2007,** *361*, 79–84.

Liu, M. X.; Li, H. F.; Luo, G.; Liu, Q. F.; Wang, Y. M. Pharmacokinetics and Biodistribution of Surface Modification Polymeric Nanoparticles. *Arch. Pharm. Res.* **2008,** *31*, 547–554.

Liu, T.; Xu, J.; Chan, B. P.; Chew, S. Y. Sustained Release of Neurotrophin-3 and Chondroitinase ABC from Electrospun Collagen Nanofiber Scaffold for Spinal Cord Injury Repair. *J. Biomed. Mater. Res. A* **2012,** *100*, 236–242.

Liu, Z. Y.; Gao, X. L.; Kang, T.; Jiang, M. Y.; Miao, D. Y.; Gu, G. Z.; Hu, Q. Y.; Song, Q. X.; Yao, L.; Tu, Y. F.; Chen, H.; Jiang, X.; Chen, J. B6 Peptide-Modified PEG-PLA Nanoparticles for Enhanced Brain Delivery of Neuroprotective Peptide. *Bioconjug. Chem.* **2013a,** *24*, 997–1007.

Liu, X.; An, C. Y.; Jin, P.; Liu, X. S.; Wang, L. H. Protective Effects of Cationic Bovine Serum Albumin-Conjugated PEGylated Tanshinone IIA Nanoparticles on Cerebral Ischemia. *Biomaterials* **2013b,** *34*, 817–830.

Liu, X.; Ye, M.; An, C. Y.; Pan, L. Q.; Ji, L. T. The Effect of Cationic Albumin-Conjugated PEGylated Tanshinone IIA Nanoparticles on Neuronal Signal Pathways and Neuroprotection in Cerebral Ischemia. *Biomaterials* **2013c,** *34*, 6893–6905.

Loftsson, T.; Stefansson, E. Cyclodextrins in Ocular Drug Delivery: Theoretical Basis with Dexamethasone as a Sample Drug. *J. Drug Deliv. Sci. Technol.* **2007,** *17*, 3–9.

Lopes, C. D. F.; Goncalves, N. P.; Gomes, C. P.; Saraiva, M. J.; Pego, A. P. BDNF gene Delivery Mediated by Neuron-Targeted Nanoparticles is Neuroprotective in Peripheral Nerve Injury. *Biomaterials* **2017,** *121*, 83–96.

Lou, J; Teng, Z. P.; Zhang, L. K.; Yang, J. D.; Ma, L. J.; Wang, F; Tian, X. C.; An, R; Yang, M; Zhang, Q; Xu, L.; Dong, Z. β-Caryophyllene/Hydroxypropyl-β-Cyclodextrin Inclusion Complex Improves Cognitive Deficits in Rats with Vascular Dementia Through the Cannabinoid Receptor Type 2-Mediated Pathway. *Front. Pharmacol.* **2017,** *8*, 2.

Loureiro, J. A.; Gomes, B.; Fricker, G.; Coelho, M. A. N.; Rocha, S.; Pereira, M. C. Cellular uptake of PLGA Nanoparticles Targeted with Anti-Amyloid and Anti-Transferrin Receptor Antibodies for Alzheimer's Disease Treatment. *Colloids Surf. B. Biointerfaces.* **2016a,** *145*, 8–13.

Loureiro, A.; Azoia, N. G.; Gomes, A. C.; Cavaco-Paulo, A. Albumin-Based Nanodevices as Drug Carriers. *Curr. Pharm. Des.* **2016b,** *22*, 1371–1390.

Louw, A. M.; Kolar, M. K.; Novikova, L. N.; Kingham, P. J.; Wiberg, M.; Kjems, J.; Novikov, L. N. Chitosan Polyplex Mediated Delivery of miRNA-124 Reduces Activation of Microglial Cells In Vitro and in Rat Models of Spinal Cord Injury. *Nanomedicine.* **2016,** *12*, 643–653.

Lu, C. T.; Zhao, Y. Z.; Wong, H. L.; Cai, J.; Peng, L.; Tian, X. Q. Current Approaches to Enhance CNS Delivery of Drugs Across the Brain Barriers. *Int. J. Nanomedicine.* **2014,** *9,* 2241–2257.

Lu, C. T.; Jin, R. R.; Jiang, Y. N.; Lin, Q.; Yu, W. Z.; Mao, K. L.; Tian, F. R.; Zhao, Y. P.; Zhao, Y. Z. Gelatin Nanoparticle-Mediated Intranasal Delivery of Substance P Protects Against 6-Hydroxydopamine-Induced Apoptosis: an In Vitro and In Vivo Study. *Drug Des. Devel. Ther.* **2015,** *9,* 1955–1962.

Makadia, H. K.; Siegel, S. J. Polylactic-Co-Glycolic Acid (PLGA) as Biodegradable Controlled Drug Delivery Carrier. *Polymers.* **2011,** *3,* 1377–1397.

Malinovskaya, Y; Melnikov, P; Baklaushev, V; Gabashvili, A; Osipova, N; Mantrov, S; Ermolenko, Y; Maksimenko, O; Gorshkova, M; Balabanyan, V.; Kreuter J.; Gelperina S. Delivery of Doxorubicin-Loaded PLGA Nanoparticles into U87 Human Glioblastoma Cells *Int. J. Pharm.* **2017,** *524,* 77–90.

Manivasagan, P.; Senthilkumar, K.; Venkatesan, J.; Kim, S. K. Biological Applications of Chitin, Chitosan, Oligosaccharides, and Their Derivatives. *Chitin and Chitosan Derivatives: Advances in Drug Discovery and Developments*; Kim, S. K. Ed., CRC Press: Boca Raton, **2014;** pp 223–242.

Masserini, M. Nanoparticles for Brain Drug Delivery. *ISRN Biochemistry.* **2013,** *2013,* 238428.

Mathew, A.; Fukuda, T.; Nagaoka, Y.; Hasumura, T.; Morimoto, H.; Yoshida, Y.; Maekawa, T.; Venugopal, K.; Kumar, D. S. Curcumin Loaded-PLGA Nanoparticles Conjugated with Tet-1 Peptide for Potential Use in Alzheimer's Disease. *PLoS One.* **2012,** 7, e32616.

McGovern Institute for Brain Research Brain Disorders. Massachusetts Institute of Technology: Cambridge, **2017.** https://mcgovern.mit.edu/brain-disorders (accessed May 12, 2019).

Md, S.; Khan, R. A.; Mustafa, G.; Chuttani, K.; Baboota, S.; Sahni, J. K.; Ali, J. Bromocriptine loaded Chitosan Nanoparticles Intended for Direct Nose to Brain Delivery: Pharmacodynamic, Pharmacokinetic and Scintigraphy Study in Mice Model. *Eur. J. Pharm. Sci.* **2013,** *48,* 393–405.

Md, S.; Haque, S.; Fazil, M.; Kumar, M.; Baboota, S.; Sahni, J. K.; Ali, J. Optimised Nanoformulation of Bromocriptine for Direct Nose-To-Brain Delivery: Biodistribution, Pharmacokinetic and Dopamine Estimation by Ultra-HPLC/Mass Spectrometry Method. *Expert Opin. Drug Deliv.* **2014,** *11,* 827–842.

Mendez-Ardoy, A.; Gomez-Garcia, M.; Geze, A.; Putaux, J. L.; Wouessidjewe, D; Mellet, C. O.; Defaye, J.; Fernandez, J. M. G.; Benito, J. M. Monodisperse Nanoparticles from Self-Assembling Amphiphilic Cyclodextrins: Modulable Tools for the Encapsulation and Controlled Release of Pharmaceuticals. *Med. Chem.* **2012,** *8,* 524–532.

Meng, J.; Agrahari, V.; Youm, I. Advances in Targeted Drug Delivery Approaches for the Central Nervous System Tumors: The Inspiration of Nanobiotechnology. *J Neuroimmune Pharmacol.* **2017,** *12,* 84–98.

Mero, A.; Grigoletto, A.; Martinez, G.; Pasut, G. Recent Developments in Hyaluronic Acid-Based Nanomedicine. *Recent Advances in Biotechnology*, Vol. 3, *Recent progress in glycotherapy*; Zhou, Q., Ed; Bentham Science Publishers – Bentham eBooks: Sharjah, 2016; pp 102–129.

Michaelis, K.; Hoffmann, M. M.; Dreis, S.; Herbert, E.; Alyautdin, R. N.; Michaelis, M.; Kreuter, J.; Langer, K. Covalent Linkage of Apolipoprotein E to Albumin Nanoparticles

Strongly Enhances Drug Transport into the Brain. *J. Pharmacol. Exp. Ther.* **2006,** *317,* 1246–1253.

Min, S. K.; Jung, S. M.; Ju, J. H.; Kwon, Y. S.; Yoon, G. H.; Shin, H. S. Regulation of Astrocyte Activity Via Control Over Stiffness of Cellulose Acetate Electrospun Nanofiber. *In vitro Cell. Dev. Biol. Anim.* **2015,** *51,* 933–940.

Mishra, V.; Mahor, S.; Rawat, A.; Gupta, P. N.; Dubey, P.; Khatri, K.; Vyas, S. P. Targeted Brain Delivery of AZT Via Transferrin Anchored Pegylated Albumin Nanoparticles. *J. Drug Target.* **2006,** *14,* 45–53.

Mittal, D.; Md, S.; Hasan, Q.; Fazil, M.; Ali, A.; Baboota, S.; Ali, J. Brain Targeted Nanoparticulate Drug Delivery System of Rasagiline Via Intranasal Route. *Drug Deliv.* **2016,** *23,* 130–139.

Monsalve, Y.; Tosi, G.; Ruozi, B.; Belletti, D.; Vilella, A.; Zoli, M.; Vandelli, M. A.; Forni, F.; Lopez, B. L.; Sierra, L. PEG-g-Chitosan Nanoparticles Functionalized with the Monoclonal Antibody OX26 for Brain Drug Targeting. *Nanomedicine* **2015,** *10,* 1735–1750.

Muddineti, O. S.; Ghosh, B.; Biswas, S. Current Trends in Using Polymer Coated Gold Nanoparticles for Cancer Therapy. *Int. J. Pharm.* **2015,** *484,* 252–267.

Muntimadugu, E.; Dhommati, R.; Jain, A.; Challa, V. G. S.; Shaheen, M.; Khan, W. Intranasal Delivery of Nanoparticle Encapsulated Tarenflurbil: A Potential Brain Targeting Strategy for Alzheimer's Disease. *Eur. J. Pharm. Sci.* **2016,** *92,* 224–234.

Murai, K.; Kurumisawa, K.; Nomura, Y.; Matsumoto, M. Regulated Drug Release Abilities of Calcium Carbonate-Gelatin Hybrid Nanocarriers Fabricated Via a Self-Organizational Process. *Chem Med Chem* **2017,** *2,* 1595–1599.

Mustafa, G.; Baboota, S.; Ahuja, A.; Ali, J. Formulation Development of Chitosan-Coated Intra Nasal Ropinirole Nanoemulsion for Better Management Option of Parkinson: An In Vitro Ex Vivo Evaluation. *Curr. Nanosci.* **2012,** *8,* 348–360.

Mustafa, G.; Ahuja, A.; Al Rohaimi, A. H.; Muslim, S.; Hassan, A. A.; Baboota, S.; Ali, J. Nano-Ropinirole for the Management of Parkinsonism: Blood-Brain Pharmacokinetics and Carrier Localization. *Exp. Rev. Neurother.* **2015,** *15,* 695–710.

Nagpal, K.; Singh, S. K.; Mishra, D. N. Nanoparticle Mediated Brain Targeted Delivery of Gallic Acid: In Vivo Behavioral and Biochemical Studies for Protection Against Scopolamine-Induced Amnesia. *Drug Deliv.* **2013,** *20,* 112–119.

Nair, M.; Jayant, R. D.; Kaushik, A.; Sagar, V. Getting into the Brain: Potential of Nanotechnology in the Management of Neuroaids. *Adv. Drug Deliv. Rev.* **2016,** *103,* 202–217.

Nasr, M. Development of an Optimized Hyaluronic Acid-Based Lipidic Nanoemulsion Co-Encapsulating two Polyphenols for Nose-to-brain Delivery. *Drug Deliv.* **2016,** *23,* 1444–1452.

Naveen, R.; Akshata, K.; Pimple, S.; Chaudhari, P. A Review on Albumin as Drug Carrier in Treating Different Diseases and Disorders. *Der Pharmacia Sinica* **2016,** *7,* 11–15.

Nguyen, K. T.; Pham, M. N.; Vo, T. V.; Duan, W.; Tran, P. H. L.; Thao, T. T. D. Strategies of Engineering Nanoparticles for Treating Neurodegenerative Disorders. *Curr. Drug Metab.* **2017,** *18,* 786–789.

Ni, S. L.; Fan, X. Z.; Wang, J. G.; Qi, H. X.; Li, X. G. Biodegradable Implants Efficiently Deliver Combination of Paclitaxel and Temozolomide to Glioma C6 Cancer Cells In Vitro. *Ann. Biomed. Eng.* **2014,** *42,* 214–221.

Nikandish, N; Hosseinzadeh, L; Azandaryani, A. H.; Derakhshandeh, K. The Role of Nanoparticle in Brain Permeability: An In Vitro BBB Model. *Iran. J. Pharm. Res.* **2016,** *15,* 403–413.

Obermeier, B.; Daneman, R.; Ransohoff, R. M. Development, Maintenance and Disruption of the Blood-Brain Barrier. *Nat. Med.* **2013**, *19*, 1584–1596.

Ossipov, D. A. Nanostructured Hyaluronic Acid-Based Materials for Active Delivery to Cancer. *Exp, Opin. Drug Deliv.* **2010**, *7*, 681–703.

Pacheco-Aguirre, J.; Rosado-Rubio, G..; Betancur-Ancona, D., Chel-Guerrero, L. Physicochemical Properties of Carboxymethylated Flamboyant (*Delonix regia*) Seed Gum. *CyTA-J. Food.* **2010**, *8*, 169–176.

Palaniraja, A.; Jayaraman, V. Production, Recovery and Applications of Xanthan Gum by *Xanthomonas campestris. J. Food Eng.* **2011**, *106*, 1–12.

Pang, M.; Shu, T.; Chen, R. Q.; Liu, C.; He, L.; Yang, Y.; Bardeesi, A. S. A.; Lin, C. K.; Zhang, L. M.; Wang, X.; Liu, B.; Rong, L. M. Neural Precursor Cells Generated from Induced Pluripotent Stem Cells with Gelatin Sponge-Electrospun PLGA/PEG Nanofibers for Spinal Cord Injury Repair. *Int. J. Clin. Exp. Med.* **2016**, *9*, 17985–17994.

Passeleu-Le Bourdonnec, C.; Carrupt, P. A.; Scherrmann, J. M.; Martel, S. Methodologies to Assess Drug Permeation through the Blood-Brain Barrier for Pharmaceutical Research. *Pharm. Res.* **2013**, *30*, 2729–2756.

Patel, P. J.; Acharya, S. R. Design and Development of Glutathione Conjugated Poly (d, l) Lactide Nanocarriers for Delivery of Hydrophilic Fluorescent Marker Across Blood-Brain Barrier. *Curr. Nanosci.* **2012**, *8*, 847–857.

Patel, P. J.; Acharya, N. S.; Acharya, S. R. Development and Characterization of Glutathione-Conjugated Albumin Nanoparticles for Improved Brain Delivery of Hydrophilic Fluorescent Marker. *Drug Deliv.* **2013**, *20*, 143–155.

Peng, Q.; Zhang, S.; Yang, Q.; Zhang, T.; Wei, X. Q.; Jiang, L.; Zhang, C. L.; Chen, Q. M.; Zhang, Z. R.; Lin, Y. F. Preformed Albumin Corona, a Protective Coating for Nanoparticles-Based Drug Delivery System. *Biomaterials.* **2013**, *34*, 8521–8530.

Pérez, S.; Bertoft, E. The Molecular Structures of Starch Components and Their Contribution to the Architecture of Starch Granules: A Comprehensive Review. *Starch – Starke.* **2010**, *62*, 389–420.

Persidsky, Y.; Ramirez, S. H.; Haorah, J.; Kanmogne, G. D. Blood–Brain Barrier: Structural Components and Function Under Physiologic and Pathologic Conditions, *J. NeuroImmune Pharmacol.* **2006**, *1*, 223–236.

Pillay, S.; Pillay, V.; Choonara, Y. E.; Naidoo, D.; Khan, R. A.; du Toit, L. C.; Ndesendo, V. M. K.; Modi, G.; Danckwerts, M. P.; Iyukee, S. E. Design, Biometric Simulation and Optimization of a Nano-Enabled Scaffold Device for Enhanced Delivery of Dopamine to the Brain. *Int. J. Pharm.* **2009**, *382*, 277–290.

Portioli, C.; Bovi, M.; Benati, D.; Donini, M.; Perduca, M.; Romeo, A.; Dusi, S.; Monaco, H. L.; Bentivoglio, M. Novel Functionalization Strategies of Polymeric Nanoparticles as Carriers for Brain Medications. *J. Biomed. Mater. Res. A.* **2017**, *105*, 847–858.

Posadas, I.; Monteagudo, S.; Ceña, V. Nanoparticles for Brain-Specific Drug and Genetic Material Delivery, Imaging and Diagnosis. *Nanomedicine* **2016**, *11*, 833–849.

Postek, M. T.; Moon, R. J.; Rudie, A. W. Bilodeau, M. A. *Production and Applications of Cellulose Nanomaterials.* Tappi Press: Peachtree Corners, **2013**. https://umaine.edu/pdc/wp-content/uploads/sites/398/2015/02/Nanocellulose-Book_Preview.pdf (accessed May 12, 2019).

Prajapati, V. D.; Jani, G. K.; Moradiya, N. G.; Randeria, N. P. Pharmaceutical Applications of Various Natural Gums, Mucilages and Their Modified Forms. *Carbohydr. Polym.* **2013**, *92*, 1685–1699.

Pulkkinen, M.; Pikkarainen, J.; Wirth, T.; Tarvainen, T.; Haapa-Acho, V.; Korhonen, H.; Seppala, J.; Jarvinen, K. Three-step Tumor Targeting of Paclitaxel Using Biotinylated PLA-PEG Nanoparticles and Avidin-Biotin Technology: Formulation Development and In Vitro Anticancer Activity. *Eur. J. Pharm. Biopharm.* **2008**, *70*, 66–74.

Qiu, X. Y.; Hu, S. W. "Smart" Materials Based on Cellulose: A Review of the Preparations, Properties, and Applications. *Materials* **2013**, *6*, 738–781.

Rao, M. R. P.; Bhingole, R. C. Nanosponge-Based Pediatric-Controlled Release Dry Suspension of Gabapentin for Reconstitution. *Drug Dev. Ind. Pharm.* **2015**, *41*, 2029–2036.

Rao, N. V.; Yoon, H. Y.; Han, H. S.; Ko, H.; Son, S.; Lee, M.; Lee, H.; Jo, D. G.; Kang, Y. M.; Park, J. H. Recent Developments in Hyaluronic Acid-Based Nanomedicine for Targeted Cancer Treatment. *Expert Opin. Drug Deliv.* **2016**, *13*, 239–252.

Rao, M.; Agrawal, D. K.; Shirsath, C. Thermoreversible Mucoadhesive In Situ Nasal Gel for Treatment of Parkinson's Disease. *Drug Dev. Ind. Pharm.* **2017**, *43*, 142–150.

Rassu, G.; Soddu, E.; Cossu, M.; Gavini, E.; Giunchedi, P.; Dalpiaz, A. Particulate Formulations Based on Chitosan for Nose-To-Brain Delivery of Drugs: A review. *J. Drug Deliv. Sci. Technol.* **2016**, *32* B, 77–87.

Rassu, G.; Soddu, E.; Posadino, A. M.; Pintus, G.; Sarmento, B.; Giunchedi, P.; Gavini, E. Nose-to-Brain Delivery of BACE1 siRNA Loaded in Solid Lipid Nanoparticles for Alzheimer's Therapy. *Colloids Surf. B Biointerfaces.* **2017**, *152*, 296–301.

Rekha, M.; Sharma, C. Pullulan as a Promising Biomaterial for Biomedical Applications: A Perspective. *Trends Biomater Artif. Organs.* **2007**, *20*, 116–121.

Ren, T. B.; Xu, N.; Cao, C. H.; Yuan, W. Z.; Yu, X.; Chen, J. H.; Ren, J. Preparation and Therapeutic Efficacy of Polysorbate-80-Coated Amphotericin B/PLA-b-PEG Nanoparticles. *J. Biomater. Sci. Polym. Ed.* **2009**, *20*, 1369–1380.

Ren, W. H.; Chang, J. A.; Yan, C. H.; Qian, X. M.; Long, L. X.; He, B.; Yuan, X. B.; Kang, C. S.; Betbeder, D.; Sheng, J.; Pu, P. Y. Development of Transferrin Functionalized Poly(Ethylene Glycol)/Poly(Lactic Acid) Amphiphilic Block Copolymeric Micelles as a Potential Delivery System Targeting Brain Glioma. *J. Mater. Sci. Mater. Med.* **2010**, *21*, 2673–2681.

Reynolds, J. K.; Mahato, R. I. Nanomedicines for the Treatment of CNS Diseases. *J. Neuroimun. Pharmacol.* **2017**, *12*, 1–5.

Roman, M.; Dong, S. P.; Hirani, A.; Lee, Y. W. Cellulose Nanocrystals for Drug Delivery. *Polysaccharide Materials: Performance by Design*; Edgar, K. J.; Heinze, T.; Buchanan, C. M. Eds.; American Chemical Society: Washington, DC, **2009**; pp 81–91.

Ruan, S. B.; He, Q.; Gao, H. L. Matrix Metalloproteinase Triggered Size-Shrinkable Gelatin-Gold Fabricated Nanoparticles for Tumor Microenvironment Sensitive Penetration and Diagnosis of Glioma. *Nanoscale.* **2015**, *7*, 9487–9496.

Ruozi, B.; Belletti, D.; Pederzoli, F.; Masoni, M.; Keller, K.; Ballestrazzi, A.; Vandelli, M. A.; Tosi, G.; Grabrucker, A. M. Novel Curcumin Loaded Nanoparticles Engineered for Blood-Brain Barrier Crossing and Able to Disrupt Abeta Aggregates. *Int. J. Pharm.* **2017**, *526*, 413–424.

Sabet, S.; George, M. A.; El-Shorbagy, H. M.; Bassiony, H.; Farroh, K. Y.; Youssef, T.; Salaheldin, T. A. Gelatin Nanoparticles Enhance Delivery of Hepatitis C Virus Recombinant NS2 gene. *PLoS One.* **2017**, *12*, e0181723.

Sah, H.; Thoma, L. A.; Desu, H. R.; Sah, E.; Wood, G. C. Concepts and Practices Used to Develop Functional PLGA-Based Nanoparticulate Systems. *Int. J. Nanomedicine.* **2013**, *8*, 747–765.

Sahoo, N.; Sahoo, R. K.; Biswas, N.; Guha, A.; Kuotsu, K. Recent Advancement of Gelatin Nanoparticles in Drug and Vaccine Delivery. *Int. J. Biol. Macromol.* **2015,** *81,* 317–331.

Salati, A.; Ahangari, G.; Keshvari, H.; Sanati, M. H. Modeling the Effect of Autoreactive T-cells on Oligodendrocytes in Mutiple Sclerosis Patients Using Chitosan/Gelatin Nanofibrous Scaffolds. *Biointerface Res. Appl. Chem.* **2016,** *6,* 1214–1221.

Salvalaio, M.; Rigon, L.; Belletti, D.; D'Avanzo, F.; Pederzoli, F.; Ruozi, B.; Marin, O.; Vandelli, M. A.; Forni, F.; Scarpa, M.; Tomanin, R.; Tosi, G. Targeted Polymeric Nanoparticles for Brain Delivery of High Molecular Weight Molecules in Lysosomal Storage Disorders. *PLoS One.* **2016,** *11,* e0156452.

Samia, O.; Hanan, R.; Kamal, E. Carbamazepine Mucoadhesive Nanoemulgel (MNEG) as Brain Targeting Delivery System Via the Olfactory Mucosa. *Drug Deliv.* **2012,** *19,* 58–67.

Saraiva, C.; Praça C.; Ferreira R.; Santos T.; Ferreira L.; Bernardino L. Nanoparticle-Mediated Brain Drug Delivery: Overcoming Blood-Brain Barrier to Treat Neurodegenerative Diseases. *J. Control. Release* **2016,** *235,* 34–47.

Sarei, F.; Dounighi, N. M.; Zolfagharian, H.; Khaki, P.; Moradi Bidhendi, S.. Alginate Nanoparticles as a Promising Adjuvant and Vaccine Delivery System. *Indian J. Pharm. Sci.* **2013,** *75,* 442–449.

Sarkar, A.; Fatima, I.; Jamal, Q. M. S.; Sayeed, U.; Khan, M. K. A.; Akhtar, S.; Kamal, M. A.; Farooqui, A.; Siddiqui, M. H. Nanoparticles as a Carrier System for Drug Delivery Across Blood-Brain Barrier. *Curr. Drug Metab.* **2017,** *18,* 129–137.

Sarvaiya, J.; Agrawal, Y. K. Chitosan as a Suitable Nanocarrier Material for Anti-Alzheimer Drug Delivery. *Int. J. Biol. Macromol.* **2015,** *72,* 454–465.

Saucier-Sawyer, J. K.; Deng, Y.; Seo, Y. E.; Cheng, C. J.; Zhang, J. W.; Quijano, E.; Saltzman, W. M. Systemic Delivery of Blood-Brain Barrier-Targeted Polymeric Nanoparticles Enhances Delivery to Brain Tissue. *J. Drug Target.* **2015,** *23,* 736–749.

Sempf, K.; Arrey, T.; Gelperina, S.; Schorge, T.; Meyer, B.; Karas, M.; Kreuter, J. Adsorption of Plasma Proteins on Uncoated PLGA Nanoparticles. *Eur. J. Pharm. Biopharm.* **2013,** *85,* 53–60.

Shadab, M.; Mustafa, G.; Baboota, S.; Ali, J. Nanoneurotherapeutics Approach Intended for Direct Nose to Brain Delivery. *Drug Dev. Ind. Pharm.* **2015,** *41,* 1922–1934.

Shah, B.; Khunt, D.; Misra, M.; Padh, H. Non-Invasive Intranasal Delivery of Quetiapine Fumarate Loaded Microemulsion for Brain Targeting: Formulation, Physicochemical and Pharmacokinetic Consideration. *Eur. J. Pharm. Sci.* **2016a,** *91,* 196–207.

Shah, B.; Khunt, D.; Misra, M.; Padh, H. Application of Box-Behnken Design for Optimization and Development of Quetiapine Fumarate Loaded Chitosan Nanoparticles for Brain Delivery Via Intranasal Route. *Int. J. Biol. Macromol.* **2016b,** *89,* 206–218.

Shan, L.; Tao, E. X.; Meng, Q. H.; Hou, W. X.; Liu, K.; Shang, H. C.; Tang, J. B.; Zhang, W. F. Formulation, Optimization, and Pharmacodynamic Evaluation of Chitosan/Phospholipid/β-Cyclodextrin Microspheres. *Drug Des Dev Ther.* **2016,** *10,* 417–429.

Shao, K.; Wu, J.; Chen, Z.; Huang, S.; Li, J.; Ye, L.; Lou, J.; Zhu, L.; Jiang, C. A Brain-Vectored Angiopep-2 Based Polymeric Micelles for the Treatment of Intracranial Fungal Infection. *Biomaterials* **2012,** *33,* 6898–6907.

Sharma, G.; Modgil, A.; Zhong, T. C.; Sun, C. W.; Singh, J. Influence of Short-Chain Cell-Penetrating Peptides on Transport of Doxorubicin Encapsulating Receptor-Targeted Liposomes Across Brain Endothelial Barrier. *Pharm. Res.* **2014,** *31,* 1194–1209.

Shi, J. J.; Kantoff, P. W.; Wooster, R.; Farokhzad, O. C. Cancer Nanomedicine: Progress, Challenges and Opportunities. *Nat. Rev. Cancer* **2017a,** *17,* 20–37.

Shi, W.; Cui, X. X.; Shi, J. L.; Chen, J.; Wang, Y. Overcoming the Blood-Brain Barrier for Glioma-Targeted Therapy Based on an Interleukin-6 Receptor-Mediated Micelle System. *RSC Adv.* **2017b**, *7*, 27162–27169.

Shityakov, S.; Salmas, R. E.; Salvador, E.; Roewer, N.; Broscheit, J.; Foerster, C. Evaluation of the Potential Toxicity of Unmodified and Modified Cyclodextrins on Murine Blood-Brain Barrier Endothelial Cells. *J. Toxicol. Sci.* **2016**, *41*, 175–184.

Siddique, Y. H.; Khan, W.; Singh, B. R.; Naqvi, A. H. Synthesis of Alginate-Curcumin Nano-composite and Its Protective Role in Transgenic *Drosophila* Model of Parkinson's Disease. *ISRN Pharmacol.* **2013**, *2013*, 794582.

Siddique, Y. H.; Khan, W.; Fatima, A.; Jyoti, S.; Khanam, S.; Naz, F.; Rahul; Ali, F.; Singh, B. R.; Naqvi, A. H. Effect of Bromocriptine Alginate Nanocomposite (BANC) on a Trans-genic *Drosophila* Model of Parkinson's Disease. *Dis. Model. Mech.* **2016**, *9*, 63–68.

Silva, D. D. C. E.; Estevanato, L. K. C.; Simioni, A. R.; Rodrigues, M. M. D.; Lacava, B. M.; Lacava, Z. G. M.; Tedesco, A. C.; Morais, P. C.; Bao, S. N. Successful Strategy for Targeting the Central Nervous System Using Magnetic Albumin Nanospheres. *J. Biomed. Nanotechnol.* **2012**, *8*, 182–189.

Singh, R. S.; Saini, G. K.; Kennedy, J. F. Pullulan: Microbial Sources, Production and Appli-cations. *Carbohydr. Polym.* **2008**, *73*, 515–531.

Singh, R. S.; Kaur, N.; Kennedy, J. F. Pullulan and Pullulan Derivatives as Promising Biomol-ecules for Drug and Gene Targeting. *Carbohydr. Polym.* **2015**, *123*, 190–207.

Singh, D.; Rashid, M.; Hallan, S. S.; Mehra, N. K.; Prakash, A.; Mishra, N. Pharmacolog-ical Evaluation of Nasal Delivery of Selegiline Hydrochloride-Loaded Thiolated Chitosan Nanoparticles for the Treatment of Depression. *Artif. Cells Nanomed. Biotechnol.* **2016**, *44*, 865–877.

Simko, M.; Mattson, M. O. Interactions between Nanonized Materials and the Brain. *Curr. Med. Chem.* **2014**, *21*, 4200–4214.

Soliman, G. M. Nanoparticles as Safe and Effective Delivery Systems of Antifungal Agents: Achievements and challenges. *Int. J. Pharm.* **2017**, *523*, 15–32.

Soni, S.; Ruhela, R. K.; Medhi, B. Nanomedicine in Central Nervous System (CNS) Disor-ders: A Present and Future Prospective. *Adv. Pharm. Bull.* **2016**, *6*, 319–335.

Sosnik, A. Alginate Particles as Platform for Drug Delivery by the Oral Route: State-of-the-Art. *ISRN Pharm.* **2014**, *17*, 926157.

Su, Z. G.; Xing, L.; Chen, Y. N.; Xu, Y. R.; Yang, F. F.; Zhang, C.; Ping, Q. N.; Xiao, Y. Y. Lactoferrin-Modified Poly(Ethylene Glycol)-Grafted BSA Nanoparticles as a Dual-Targeting Carrier for Treating Brain Gliomas. *Mol. Pharm.* **2014**, *11*, 1823–1834.

Su, C. H.; Cheng, F. Y. In vitro and In Vivo Applications of Alginate/Iron Oxide Nanocom-posites for Theranostic Molecular Imaging in a Brain Tumor Model. *RSC Adv.* **2015**, *5*, 90061–90064.

Sun, Y.; Kang, C.; Liu, F.; Song, L. Delivery of Antipsychotics with Nanoparticles. *Drug Dev. Res.* **2016**, *77*, 393–399.

Sunasee, R.; Hemraz, U. D.; Ckless, K. Cellulose Nanocrystals: A Versatile Nanoplatform for Emerging Biomedical Applications. *Expert Opin. Drug Deliv.* **2016**, *13*, 1243–1256.

Tang, J.; Sisler, J.; Grishkewich, N.; Tam, K. C. Functionalization of Cellulose Nanocrystals for Advanced Applications. *J. Colloid Interface Sci.* **2017**, *494*, 397–409.

Thakur, V. K.; Thakur, M. K. *Handbook of Polymers for Pharmaceutical Technologies – Biodegradable Polymers.* Scrivener Publishing & J. Wiley & Sons: Hoboken, **2015**, 33–60, 105–126, 275–298.

Tian, X. H.; Wei, F.; Wang, T. X.; Wang, D.; Wang, J; Lin, X. N.; Wang, P.; Ren, L. Blood-Brain Barrier Transport of Tat Peptide and Polyethylene Glycol Decorated Gelatin-Siloxane Nanoparticle. *Mater. Lett.* **2012a**, *68*, 94−96.

Tian, X. H.; Wei, F.; Wang, T. X.; Wang, P.; Lin, X. N.; Wang, J.; Wang, D.; Ren, L. In Vitro and In Vivo Studies on Gelatin-Siloxane Nanoparticles Conjugated with SynB Peptide to Increase Drug Delivery to the Brain. *Int. J. Nanomed.* **2012b**, *7*, 1031−1041.

Tian, X. H.; Wang, Z. G.; Meng, H.; Wang, Y. H.; Feng, W.; Wei, F.; Huang, Z. C.; Lin, X. N.; Ren, L. Tat Peptide-Decorated Gelatin-Siloxane Nanoparticles for Delivery of CGRP Transgene in Treatment of Cerebral Vasospasm. *Int. J. Nanomed.* **2013**, *8*, 865−876.

Tonnesen, H. H.; Karlsen, J. (2002). Alginate in Drug Delivery Systems. *Drug Dev. Ind. Pharm.* **2002**, *28*, 621−630.

Tosi, G.; Fano, R. A.; Bondioli, L.; Badiali, L.; Benassi, R.; Rivasi, F.; Ruozi, B.; Forni, F.; Vandelli, M. A. Investigation on Mechanisms of Glycopeptide Nanoparticles for Drug Delivery Across the Blood-Brain Barrier. *Nanomedicine.* **2011**, *6*, 423−436.

Tosi, G.; Vilella, A.; Chhabra, R.; Schmeisser, M. J.; Boeckers, T. M.; Ruozi, B.; Vandelli, M. A.; Forni, F.; Zoli, M.; Grabrucker, A. M. Insight on the Fate of CNS-Targeted Nanoparticles. Part II: Intercellular Neuronal Cell-to-Cell Transport. *J. Control. Release* **2014**, *177*, 96−107.

Tosi, G.; Vilella, A.; Veratti, P.; Belletti, D.; Pederzoli, F.; Ruozi, B.; Vandelli, M. A.; Zoli, M.; Forni, F. Exploiting Bacterial Pathways for BBB Crossing with PLGA Nanoparticles Modified with a Mutated form of Diphtheria Toxin (CRM197): In Vivo Experiments. *Mol. Pharm.* **2015**, *12*, 3672−3684.

Tosi, G.; Musumeci, T.; Ruozi, B.; Carbone, C.; Belletti, D.; Pignatello, R.; Vandelli, M. A.; Puglisi, G. The "fate" of Polymeric and Lipid Nanoparticles for Brain Delivery and Targeting: Strategies and Mechanism of Blood–Brain Barrier Crossing and Trafficking into the Central Nervous System. *J. Drug Deliv. Sci. Technol.* **2016**, *32* B, 66−76.

Trapani, A.; De Giglio, E.; Cafagna, D.; Denora, N.; Agrimi, G.; Cassano, T.; Gaetani, S.; Cuomo, V.; Trapani, G. Characterization and Evaluation of Chitosan Nanoparticles for Dopamine Brain Delivery. *Int. J. Pharm.* **2011**, *419*, 296−307.

Tyler, B.; Gullotti, D.; Mangraviti, A.; Utsuki, T.; Brem, H. Polylactic Acid (PLA) Controlled Delivery Carriers for Biomedical Applications. *Adv. Drug Deliv. Rev.* **2016**, *107*, 163−175.

Upadhyay R. K. Drug Delivery Systems, CNS Protection, and the Blood-Brain Barrier. *BioMed Res. Int.* **2014**, *2014*, 869269.

van Groen, T.; Kadish, I.; Wiesehan, K.; Funke, S. A.; Willbold, D. In Vitro and In Vivo Staining Characteristics of Small, Fluorescent, Aβ42-Binding d-Enantiomeric Peptides in Transgenic AD Mouse Models. *ChemMedChem.* **2009**, *4*, 276−282.

Van Woensel, M.; Wauthoz, N.; Rosiere, R.; Mathieu, V.; Kiss, R.; Lefranc, F.; Steelant, B.; Dilissen, E.; Van Gool, S. W.; Mathivet, T.; Gerhardt, H; Amighi, K; De Vleeschouwer, S. Development of siRNA-Loaded Chitosan Nanoparticles Targeting Galectin-1 for the Treatment of Glioblastoma Multiforme via Intranasal Administration. *J. Control. Release* **2016**, *227*, 71−81.

Vecsernyes, M.; Fenyvesi, F.; Bacskay, I.; Deli, M. A.; Szente, L.; Fenyvesi, E. Cyclodextrins, Blood-Brain Barrier, and Treatment of Neurological Diseases. *Arch. Med. Res.* **2014**, *45*, 711−729.

Venugopal, I.; Pernal, S.; Duproz, A.; Bentley, J.; Engelhard, H.; Linninger, A. Magnetic Field-Enhanced Cellular Uptake of Doxorubicin Loaded Magnetic Nanoparticles for Tumor Treatment. *Mater. Res. Express.* **2016**, *3*, 095010.

Verma, S. K.; Arora, I.; Javed, K.; Akhtar, M.; Samim, M. Enhancement in the Neuroprotective Power of Riluzole Against Cerebral Ischemia Using a Brain-Targeted Drug Delivery Vehicle. *ACS Appl. Mater. Interfaces.* **2016**, *8*, 19716–19723.

Vieira, S. M.; Michels, L. R.; Roversi, K.; Metz, V. G.; Moraes, B. K. S.; Piegas, E. M.; Freddo, R. J.; Gundel, A.; Costa, T. D.; Burger, M. E.; Colomé, K. M.; Haas, S. E. A Surface Modification of Clozapine-Loaded Nanocapsules Improves their Efficacy: A Study of Formulation Development and Biological Assessment. *Colloids Surf. B Biointerfaces* **2016**, *145*, 748–756.

Voltan, A. R.; Quindós, G.; Alarcón, K. P.; Fusco-Almeida A. M.; Mendes-Giannini, M. J.; Chorilli, M. Fungal Diseases: Could Nanostructured Drug Delivery Systems be a Novel Paradigm for Therapy? *Int. J. Nanomedicine.* **2016**, *11*, 3715–3730.

Wahba, S. M. R.; Darwish, A. S.; Kamal, S. M. Ceria-Containing Uncoated and Coated Hydroxyapatite-Based Galantamine Nanocomposites for Formidable Treatment of Alzheimer's Disease in Ovariectomized Albino-Rat Model. *Mater. Sci. Eng. C Mater. Biol. Appl.* **2016**, *65*, 151–163.

Wang Z. F.; Luo Y. N.; Zheng Y. H.; Han H. L.; Hong X. Y.; Jing X. B. Interaction of Copolymer-Paclitexal Conjugate Micelles with C-6 Glioma Cells. *Chem. J. Chinese Univ.* **2008**, *29*, 1671–1676.

Wang, P.; Zhang, A. X.; Jin, Y.; Zhang, Q.; Zhang, L. Y.; Peng, Y.; Du, S. H. Molecularly Imprinted Layer-Coated Hollow Polysaccharide Microcapsules Toward Gate-Controlled Release of Water-Soluble Drugs. *RSC Adv.* **2014**, *4*, 26063–26073.

Wang, X. Q.; Yu, X. W.; Vaughan, W.; Liu, M. Y.; Guan, J. T. Novel Drug-Delivery Approaches to the Blood-Brain Barrier. *Neurosci. Bull.* **2015a**, *31*, 257–264.

Wang, L.; Hao, Y. W.; Li, H. X.; Zhao, Y. L.; Meng, D. H.; Li, D.; Shi, J. J.; Zhang, H. L.; Zhang, Z. Z.; Zhang, Y. Co-Delivery of Doxorubicin and siRNA for Glioma Therapy by a Brain Targeting System: Angiopep-2-Modified Poly(Lactic-co-Glycolic Acid) Nanoparticles. *J. Drug Target.* **2015b**, *23*, 832–846.

Wang, T. Y.; Bruggeman, K. F.; Kauhausen, J. A.; Rodriguez, A. L.; Nisbet, D. R.; Parish, C. L. Functionalized Composite Scaffolds Improve the Engraftment of Transplanted Dopaminergic Progenitors in a Mouse Model of Parkinson's Disease. *Biomaterials* **2016a** *74*, 89–98.

Wang, G.; Wang, J. J.; Tang, X. J.; Du, L.; Li, F. In Vitro and In Vivo Evaluation of Functionalized Chitosan-Pluronic Micelles Loaded with Myricetin on Glioblastoma Cancer. *Nanomedicine.* **2016b** *12*, 1263–1278.

Warnken, Z. N.; Smyth, H. D. C.; Watts, A. B.; Weitman, S.; Kuhn, J. G.; Williams, R. O. Formulation and Device Design to Increase Nose to Brain Drug Delivery. *J. Drug Deliv. Sci. Technol.* **2016**, *35*, 213–222.

Wavikar, P. R.; Vavia, P. R. Rivastigmine-Loaded In Situ Gelling Nanostructured Lipid Carriers for Nose to Brain Delivery. *J. Liposome Res.* **2015**, *25*, 141–149.

Willenberg, B. J.; Zheng, T.; Meng, F. W.; Meneses, J. C.; Rossignol, C.; Batich, C. D.; Terada, N.; Steindler, D. A.; Weiss, M. D. Gelatinized Copper-Capillary Alginate Gel Functions as an Injectable Tissue Scaffolding System for Stem Cell Transplants. *J. Biomater. Sci. Polym. Ed.* **2011**, *22*, 1621–1637.

Wilson, B.; Ambika, T. V.; Patel, R. D. K.; Jenita, J. L.; Priyadarshini, S. R. B. Nanoparticles Based on Albumin: Preparation, Characterization and the Use for 5-Flurouracil Delivery. *Int. J. Biol. Macromol.* **2012**, *51*, 874–878.

Wilson, B.; Lavanya, Y.; Priyadarshini, S. R. B.; Ramasamy, M.; Jenita, J. L. Albumin Nanoparticles for the Delivery of Gabapentin: Preparation, Characterization and Pharmacodynamic Studies. *Int. J. Pharm.* **2014,** *473,* 73−79.

Wolburg, H.; Lippoldt, A. Tight Junctions of the Blood-Brain Barrier: Development, Composition and Regulation. *Vasc. Pharmacol.* **2002,** *38,* 323–337.

Wong, H. L.; Wu, X. Y.; Bendayan, R. Nanotechnological Advances for the Delivery of CNS Therapeutics. *Adv. Drug Deliv. Rev.* **2012,** *64,* 686–700.

Xia, H. M.; Gao, X. L.; Gu, G. Z.; Liu, Z. Y.; Hu, Q. Y.; Tu, Y. F.; Song, Q. X.; Yao, L.; Pang, Z. Q.; Jiang, X. G.; Chen, J.; Chen, H. Penetratin-Functionalized PEG-PLA Nanoparticles for Brain Drug Delivery. *Int. J. Pharm.* **2012,** *436,* 840–850.

Xie, Y. T.; Du, Y. Z.; Yuan, H.; Hu, F. Q. Brain-Targeting Study of Stearic Acid-Grafted Chitosan Micelle Drug-Delivery System. *Int. J. Nanomed.* **2012,** *7,* 3235−3244.

Xu, Y. R.; Asghar, S.; Yang, L.; Chen, Z. P.; Li, H. Y.; Shi, W. W.; Li, Y. B.; Shi, Q. Q.; Ping, Q. N.; Xiao, Y. Y. Nanoparticles Based on Chitosan Hydrochloride/Hyaluronic Acid/PEG Containing Curcumin: In vitro Evaluation and Pharmacokinetics in Rats. *Int. J. Biol. Macromol.* **2017a,** *102,* 1083−1091.

Xu, Y. R.; Asghar, S.; Yang, L.; Li, H. Y.; Wang, Z. L.; Ping, Q. N.; Xiao, Y. Y. Lactoferrin-Coated Polysaccharide Nanoparticles Based on Chitosan Hydrochloride/Hyaluronic acid/PEG for Treating Brain Glioma. *Carbohydr. Polym.* **2017b,** *157,* 419−428.

Yadav, A. K.; Mishra, P.; Agrawal, G. P. An Insight on Hyaluronic Acid in Drug Targeting and Drug Delivery. *J. Drug Target.* **2008,** *16,* 91−107.

Yadav, A. K.; Agarwal, A.; Rai, G.; Mishra, P.; Jain, S.; Mishra, A. K.; Agrawal, H.; Agrawal, G. P. Development and Characterization of Hyaluronic Acid Decorated PLGA Nanoparticles for Delivery of 5-Fluorouracil. *Drug Deliv.* **2010,** *17,* 561−572.

Yameogo, J. B. G.; Geze, A.; Choisnard, L.; Putaux, J. L.; Semde, R.; Wouessidjewe, D. Progress in Developing Amphiphilic Cyclodextrin-Based Nanodevices for Drug Delivery. *Curr. Topics Med. Chem.* **2014,** *14,* 526−541.

Yan, L.; Wang, H. Y.; Jiang, Y. F.; Liu, J. H.; Wang, Z.; Yang, Y. X.; Huang, S. W.; Huang, Y. Z. Cell-Penetrating Peptide-Modified PLGA Nanoparticles for Enhanced Nose-to-Brain Macromolecular Delivery. *Macromol. Res.* **2013,** *21,* 435−441.

Yang, Z.; Liu, Z. W.; Allaker, R. P.; Reip, P.; Oxford, J.; Ahmad, Z.; Ren, G. A Review of Nanoparticle Functionality and Toxicity on the Central Nervous System. *J. Royal Soc. Interface* **2010a,** *7,* S411–S422.

Yang, H. Nanoparticle-Mediated Brain-Specific Drug Delivery, Imaging, and Diagnosis. *Pharm. Res.* **2010b,** *27,* 1759–1771.

Yang, J. S.; Xie, Y. J.; He, W. Research Progress on Chemical Modification of Alginate: A Review. *Carbohydr. Polym.* **2011,** *84,* 33–39.

Yang, L.; Gao, S. Y.; Asghar, S.; Liu, G. H.; Song, J.; Wang, X.; Ping, Q. N.; Zhang, C.; Xiao, Y. Y. Hyaluronic Acid/Chitosan Nanoparticles for Delivery of Curcuminoid and Its In vitro Evaluation in Glioma Cells. *Int. J. Biol. Macromol.* **2015,** *72,* 1391−1401.

Yao, L.; Song, Q. X.; Bai, W. K.; Zhang, J. Z.; Miao, D. Y.; Jiang, M. Y.; Wang, Y.; Shen, Z. H.; Hu, Q. Y.; Gu, X.; Huang, M.; Zheng, G.; Gao, X. L.; Hu, B.; Chen J.;Chen K. Z. Facilitated Brain Delivery of Poly (Ethylene Glycol)-Poly (Lactic Acid) Nanoparticles by Microbubble-Enhanced Unfocused Ultrasound. *Biomaterials* **2014,** *35,* 3384−3395.

Yasir, M.; Sara, U. V. S. Solid Lipid Nanoparticles for Nose-to-Brain Delivery of Haloperidol: In Vitro Drug Release and Pharmacokinetics Evaluation. *Acta Pharm. Sin. B* **2014,** *4,* 454–463.

Ye, Y. J.; Sun, Y.; Zhao, H. L.; Lan, M. B.; Gao, F.; Song, C.; Lou, K. Y.; Li, H.; Wang, W. A Novel Lactoferrin-Modified β-Cyclodextrin Nanocarrier for Brain-Targeting Drug Delivery. *Int. J. Pharm.* **2013**, *458*, 110−117.

Yemisci, M.; Gursoy-Ozdemir, Y.; Caban, S.; Bodur, E.; Capan, Y.; Dalkara, T. Transport of a Caspase Inhibitor Across the Blood Brain Barrier by Chitosan Nanoparticles. *Methods in Enzymology*, vol. 508: *Nanomedicine: Cancer, Diabetes, and Cardiovascular, Central Nervous System, Pulmonary and Inflammatory Diseases*; Duzgunes, N., Ed.; Academic Press: San Diego, **2012**, 253−269.

Yin, Y. T.; Fu, C. P.; Li, M.; Li, X. P.; Wang, M. Y.; He, L.; Zhang, L. M.; Peng, Y. A pH-Sensitive Hyaluronic Acid Prodrug Modified with Lactoferrin for Glioma Dual-Targeted Treatment. *Mater. Sci. Eng. C Mater. Biol. Appl.* **2016**, *67*, 159−169.

Yuan, X. B.; Kang, C. S.; Zhao, Y. H.; Gu, M. Q.; Pu, P. Y.; Tian, N. J.; Sheng, J. Surface Multi-Functionalization of Poly(lactic acid) Nanoparticles and C6 Glioma Cell Trageting In *Vivo. Chin. J. Polym. Sci.* **2009**, *27*, 231−239.

Zarekar, N. S.; Lingayat, V. J.; Pande, V. V. Nanogel as a Novel Platform for Smart Drug Delivery System. *Nanosci. Nanotechnol. Res.* **2017**, *4*, 25−31.

Zensi, A.; Begley, D.; Pontikis, C.; Legros, C.; Mihoreanu, L.; Wagner, S.; Buechel, C.; von Briesen, H.; Kreuter, J. Albumin Nanoparticles Targeted with Apo E Enter the CNS by Transcytosis and are Delivered to Neurones. *J. Control. Release* **2009**, *137*, 78−86.

Zensi, A.; Begley, D.; Pontikis, C.; Legros, C.; Mihoreanu, L.; Buechel, C.; Kreuter, J. Human Serum Albumin Nanoparticles Modified with Apolipoprotein A-I Cross the Blood-Brain Barrier and Enter the Rodent Brain. *J. Drug Target.* **2010**, *18*, 842−848.

Zhang, S. J.; Wu, L. X. Amyloid-β Associated with Chitosan Nano-Carrier has Favorable Immunogenicity and Permeates the BBB. *AAPS PharmSciTech.* **2009**, *10*, 900−905.

Zhang, L.; Yu, F. Q.; Cole, A. J.; Chertok, B.; David, A. E.; Wang, J. K.; Yang, V. C. Gum Arabic-Coated Magnetic Nanoparticles for Potential Application in Simultaneous Magnetic Targeting and Tumor Imaging. *AAPS J.* **2009**, *11*, 693−699.

Zhang, J. X.; Ma, P. X. Cyclodextrin-Based Supramolecular Systems for Drug Delivery: Recent Progress and Future Perspective. *Adv. Drug Deliv. Rev.* **2013**, *65*, 1215−1233.

Zhang, X. G.; Liu, L.; Chai, G. B.; Zhang, X. Y.; Li, F. Z. Brain Pharmacokinetics of Neurotoxin-Loaded PLA Nanoparticles Modified with Chitosan after Intranasal Administration in Awake Rats. *Drug Dev. Ind. Pharm.* **2013a**, *39*, 1618−1624.

Zhang, J; Shin, M. C.; David, A. E.; Zhou, J.; Lee, K.; He, H.; Yang, V. C. Long-Circulating Heparin-Functionalized Magnetic Nanoparticles for Potential Application as a Protein Drug Delivery Platform. *Mol. Pharm.* **2013b**, *10*, 3892−3902.

Zhang, Z. Y.; Daniels, R.; Schluesener, H. J. Oridonin Ameliorates Neuropathological Changes and Behavioural Deficits in a Mouse Model of Cerebral Amyloidosis. *J. Cell. Mol. Med.* **2013c**, *17*, 1566−1576.

Zhang, C.; Wan, X.; Zheng, X. Y.; Shao, X. Y.; Liu, Q. F.; Zhang, Q. Z.; Qian, Y. Dual-Functional Nanoparticles Targeting Amyloid Plaques in the Brains of Alzheimer's Disease Mice. *Biomaterials* **2014a**, *35*, 456−465.

Zhang, J.; Shin, M. C.; Yang, V. C. Magnetic Targeting of Novel Heparinized Iron Oxide Nanoparticles Evaluated in a 9L-Glioma Mouse Model. *Pharm. Res.* **2014b**, *31*, 579−592.

Zhang, Z.; Li. Q.; Han, L.; Zhong, Y. Layer-by-Layer Films Assembled from Natural Polymers for Sustained Release of Neurotrophin. *Biomed. Mater.* **2015**, *11*, 055006.

Zhang, G. L.; Chen, L. K.; Guo, X. Y.; Khan, A. A.; Gu, Y. C.; Gu, N. Nanoparticle-Mediated Drug Delivery Systems (DDS) in the Central Nervous System. **2017a,** *Curr. Org. Chem. 21,* 272–283.

Zhang, X.; Hu, J. G.; Zhao, G. J.; Huang, N.; Tan, Y.; Pi, L.; Huang, Q.; Wang, F.; Wang, Z. G.; Wang, Z. B.; Cheng, Y. PEGylated PLGA-Based Phase Shift Nanodroplets Combined with Focused Ultrasound for Blood-Brain Barrier Opening in Rats. *Oncotarget* **2017b,** *8,* 38927–38936.

Zhang, Z. Y.; Jiang, M.; Fang, J.; Yang, M. F.; Zhang, S.; Yin, Y. X.; Li, D. W.; Mao, L. L.; Fu, X. Y.; Hou, Y. J.; Fu, X. T.; Fan, C. D.; Sun, B. L. Hemorrhage-Induced Blood-Brain Barrier Disruption Through Inhibition of Inflammatory Response and Oxidative Stress. *Mol. Neurobiol.* **2017c,** *54,* 1–14.

Zhao, Y. Z.; Li, X.; Lu, C. T.; Lin, M.; Chen, L. J.; Xiang, Q.; Zhang, M.; Jin, R. R.; Jiang, X.; Shen, X. T.; Li, XK; Cai, J. Gelatin Nanostructured Lipid Carriers-Mediated Intranasal Delivery of Basic Fibroblast Growth Factor Enhances Functional Recovery in Hemiparkinsonian Rats. *Nanomedicine* **2014,** *10,* 755–764.

Zhao, Y. Z.; Jin, R. R.; Yang, W.; Xiang, Q.; Yu, W. Z.; Lin, Q.; Tian, F. R.; Mao, K. L.; Lv, C. Z.; Wang, Y. X. J.; Lu, C. T. Using Gelatin Nanoparticle-Mediated Intranasal Delivery of Neuropeptide Substance P to Enhance Neurorecovery in Hemiparkinsonian Rats. *PLoS One* **2016a,** *11,* e0148848.

Zhao, Y. Z.; Lin, Q.; Wong, H. L.; Shen, X. T.; Yang, W.; Xu, H. L.; Mao, K. L.; Tian, F. R.; Yang, J. J.; Xu, J.; Wei, X.; Fu, X. B.; Li, X. K.; Xu, H. Z.; Xiao, J. Glioma-Targeted Therapy Using Cilengitide Nanoparticles Combined with UTMD-Enhanced Delivery. *J. Control. Release* **2016b,** *224,* 112–125.

Zhou, J.; Zhang, J.; David, A. E.; Yang, V. C. Magnetic Tumor Targeting of β-Glucosidase Immobilized Iron Oxide Nanoparticles. *Nanotechnology.* **2013,** *24,* 375102.

Zhu, S. P.; Wang, Z. G.; Zhao, Y. Z.; Wu, J.; Shi, H. X.; Ye, L. B.; Wu, F. Z.; Cheng, Y.; Zhang, H. Y.; He, S. B.; Wei, X. J.; Fu, X. B.; Li, X. K.; Xu, H. Z.; Xiao, J. Gelatin Nanostructured Lipid Carriers Incorporating Nerve Growth Factor Inhibit Endoplasmic Reticulum Stress-Induced Apoptosis and Improve Recovery in Spinal Cord Injury. *Mol. Neurobiol.* **2016,** *53,* 4375–4386.

Zia, K. M.; Zia, F.; Zuber, M.; Rehman, S.; Ahmad, M. N. Alginate-Based Polyurethanes: A Review of Recent Advances and Perspective. *Int. J. Biol. Macromol.* **2015,** *79,* 377–387.

Zuo, H. L.; Chen, W. Y.; Cooper, H. M.; Xu, Z. P. A Facile Way of Modifying Layered Double Hydroxide Nanoparticles with Targeting Ligand Conjugated Albumin for Enhanced Delivery to Brain Tumor Cells. *ACS Appl. Mater. Interfaces* **2017,** *9,* 20444–20453.

SECTION II
PARTICULATE CARRIERS

CHAPTER 6

NANOPARTICULATE DRUG-DELIVERY SYSTEMS FOR BRAIN TARGETING

EMIL JOSEPH, GAUTAM SINGHVI*, and SASWATA BANERJEE

Department of Pharmacy, Birla Institute of Technology and Science (BITS), Pilani 333031, Rajasthan, India

Corresponding author. E-mail: singhvigautam@gmail.com

ABSTRACT

The prevalence of neurological disorders and its untreatable nature owing to the omnipotent presence of the blood–brain barrier has posed a serious problem in the medical fraternity owing to the lack of efficient drug-delivery systems for efficient treatment of these ailments. Thus, with a view to mitigate these problems, increased research is going on to target the potent molecules at the site of action, that is, the brain and central nervous system. The advancements of nanotechnology have provided a huge respite to these problem owing to their nanodimensions, allowing to treat these ailments with much more efficiency than previously possible by conventional routes. In our chapter, we have highlighted with of those facts as well as the functional advantages that each of these nanoparticles, namely polymeric and lipid nanoparticles, provide owing to their chemical structure and composition. We also discussed various functional properties and synthesis techniques for each of these nanoparticles and possible improvements with regards to the current state of research in this and allied fields of treatment. We also highlighted the possible physiological hurdles that we as formulation scientist face while designing a particular formulation for targeted delivery of drugs to the brain, so that the reader can understand and comprehend the need for such drastic and complicated technological applications.

6.1 INTRODUCTION

With the advancement of the last century, the increased prevalence of neurological disorders with its untreatable nature have haunted people in healthcare system for long as efficient treatment remained inaccessible to patients and healthcare staff. With growing population and increased observance of neurological disorders, fear of the same taking the shape of an endemic have led to the growing concern among the general mass and forced the scientific fraternity to take notice of this situation and take corrective action. In recent years, a lot of effort regarding the same has been directed to treat individual symptoms associated with these disorders at local level, that is, at the site of abnormality in the central nervous system (CNS), through targeted delivery of therapeutic agents. But the complex physiological system providing immunity to the CNS against harmful/deleterious agents has also been the fundamental reason of failure for proper delivery of these highly efficient therapeutic agents and attains therapeutic concentration at the site of action. Thus, we can say that owing to practical obligations of providing better healthcare, development of a specialized field of expertise took place with formulation/pharmaceutical scientists concerned solely with formulation of efficient carrier systems for targeted delivery of therapeutic agents at the site of action, that is, brain or in a broader sense, the CNS.

The most challenging task for targeting therapeutic drug molecules to brain is the omnipotent presence of the blood–brain barrier (BBB). The fact that our BBB cannot discriminate between essential therapeutic molecules and other nonessential/harmful entities makes it a huge hurdle to our objective of targeted brain delivery (Pardridge et al., 2009). Thus, we can see that majority of small drug molecules (>98%) and almost all of the large molecule drugs, including biotechnology-based products, fail to cross this supportive barrier. In general, the motivation behind targeted delivery of drugs through the BBB stems from the very need to treat several CNS disorders. Traditional approaches like prodrugs, disruption of BBB, intracerebral injection/use of implants, etc. have been investigated for this very purpose by pharmaceutical scientists. As per the "prodrug" approach, the drug molecules have been converted to their bioreversible lipophilic form, thereby improving their passage across the BBB and subsequent biotransformation thereafter (Rautio et al., 2008). More sophisticated prodrug approaches have been developed in recent times, namely, receptor-mediated macromolecular delivery, via carrier-mediated endogenous transporters as well as gene-mediated enzyme prodrug therapy (Ratio et al., 2008) but prodrug approach gets restricted due to the premature conversion of prodrug by the plasma–enzymes in the plasma itself.

The BBB, made up of tight junctions of endothelial cells, can be disrupted by techniques, such as osmotic disruption, ultrasound disruption, and disruption by bradykinin-analog, thereby causing leakage in the tight junctions and subsequent passage through the BBB. A classic example of osmotic disruption is the administration of hypertonic mannitol solution along with subsequent administration of drugs through intracarotid artery, thereby increasing the drug concentration in brain and tumor tissues, owing to the osmotic shock to the endothelial cells leading to their shrinkage and subsequent leaky behavior due to disruption of the tight junctions (Gabathuler, 2010). Magnetic resonance imaging-guided focused ultrasound technique is another useful approach for BBB disruption. Feng et al., in his study, demonstrated that extravasation of Evans Blue was enhanced by nearly two folds in groups with second sonication in comparison to that of groups with a single sonication where ultrasound frequency was set at 1 MHz and a repetition frequency of 1 Hz. The selective B_2 bradykinin receptor agonist, Cereport (also called RMP-7), did not only show an increased transient permeability of the BBB but also enhanced CNS delivery of carboplatin, loperamide, and cyclosporin-A, when administered along with RMP-7 (Borlongan and Emerich, 2003). The major limitation associated with BBB disruption technique is the increased brain uptake of plasma albumin and other protein components of blood, which are toxic to brain cells (Vykhodtseva et al., 2008).

Another useful traditional approach is through intracerebral injection/use of implants whereby diffusion forms the basis of drug delivery into the infiltrated brain through successful injection of bolus active pharmaceutical ingredients (APIs) or through placement of impregnated biodegradable wafer. Various forms of lipidic implants and poly(D,L-lactide-co-glycolide) (PLGA)-based microparticles containing paclitaxel with potential of controlled release kinetics for several weeks have been formulated and characterized in vitro by Elkharraz et al. (2006). These devices when injected via the intracranial route into the brain tissue overcome the limitations associated with paclitaxel's inability to cross the BBB to an extent significant to have therapeutic levels on systemic administration. Thus, it can be inferred from their works that controlled drug-delivery system is helpful to improve the treatment of operable and inoperable brain tumors at a local level.

But these traditional approaches are still unable to achieve the desired outcome for better brain delivery with most of them being associated with numerous adverse effects owing to their highly destructive nature, which may prove fatal in long term. Thus, it is a challenge to effectively deliver therapeutic agents at their site of action, that is, brain and CNS with none or limited side-effects thereby promising better patient compliance.

6.1.1   STRUCTURE OF BLOOD BRAIN BARRIER AND TRANSPORT MECHANISMS

The BBB is regarded as a dynamic and complex barrier separating blood and the CNS and strictly controls the exchanges between the compartments of blood and brain thereby restricting entry of untoward substances, such as pathogens, toxic molecules, pathogens, along with other external molecules thereby maintaining brain homeostasis. An illustration of BBB is shown in Figure 6.1.

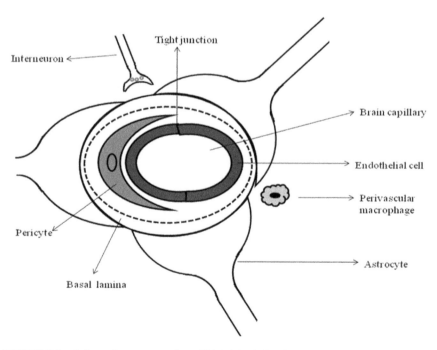

FIGURE 6.1   Schematic representation of blood–brain barrier.

Endothelial cells present in the brain and other parts of the body vary physiologically specifically due to the presence of intracellular tight junctions, resulting in lesser degree of paracellular diffusion of hydrophilic molecules. Also, the presence of relatively high number of mitochondrial cells results in higher degree of metabolic activity. But at the same time, there is a relatively higher number of active transporters present (for transport/efflux of nutrients/toxic compounds) (Weiss et al., 2009). The basal lamina is also composed chiefly of glycoproteins, collagen, and proteoglycans, which is involved in the dynamic regulation of BBB with the aid of multiple basal lamina proteins, matrix metalloproteases, their inhibitors, and the tissue inhibitor of metalloproteases. The astrocytes and glial cells present in the BBB also contribute largely to the barrier integrity through glial-derived neurotrophic factor, angiopoietin-1, and angiotensin II. Brain microvessels have numerous pericytes and ratio of pericytes to endothelial cells was linked with the barrier capacity. Endothelium, pericytes, and perivascular astrocytes are intimately associated with neuronal projections.

6.1.1.1 GENERAL TRANSPORT MECHANISMS ACROSS BLOOD–BRAIN BARRIER

The major classes/types of transport mechanisms to the BBB are depicted in Figure 6.2.

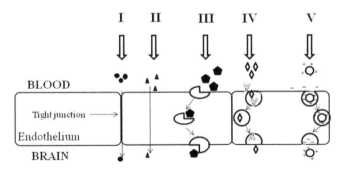

BLOOD-BRAIN BARRIER TRANSPORT MECHANISMS

I.   Paracellular aqueous pathway
II. Transcellular lipophilic pathway
III. Carrier mediated pathway
IV. Receptor mediated transcytosis
V.   Adsorptive transcytosis

FIGURE 6.2   General transport mechanisms across blood–brain barrier.

Paracellular aqueous pathway is a rare pathway which allows small water-soluble molecules to cross BBB into the brain. Lipophilic molecules such as steroid hormones, alcohols, etc. penetrate transcellularly by dissolving in the lipidic plasma membrane. A protein transporter is involved in transporting of glucose or amino acids across the BBB, which is triggered by a conformational change in the protein postbinding to the specific transporter and thereby aids in the transport of the molecule across. The receptor-mediated transcytosis, similar to protein transport, is associated with the selective uptake of macromolecules. These include receptors for lipoprotein, transferring, insulin, etc. and have also been explored for the delivery of drug at targeted sites through ligand-based nanoformulations. Adsorptive-mediated transcytosis, another class of transport mechanism present in BBB, is stimulated by electrostatic interactions between positively and negatively charged substances, for example, between positively charged moiety of a peptide and negatively charged plasma membrane surface of heparin sulfate proteoglycans.

The complex structure and specific transport mechanism of BBB, as can be seen through our discussion, restrict entry of various drug molecules owing to its highly dynamic nature. Since traditional approaches are unable to provide the desired solution, pharmaceutical scientists looked for other alternatives which led to further development and leading to acceptance of nanotechnology as the other alternative for drug delivery across BBB successfully.

6.1.2   NANOTECHNOLOGY FOR DELIVERY OF CENTRAL NERVOUS SYSTEM ACTIVE DRUGS

The application of nanotechnology in the field of drug delivery broadened the scope for formulation scientists for development of better brain delivery systems over the existing ones and accentuates potential of molecules with probable or proven CNS activity. It is widely accepted among the scientific fraternity that application of nanotechnology in the field of brain-targeted drug delivery, especially for CNS active drug which have inability to cross BBB, could be entrapped in nanocarriers to achieve better therapeutic activity. The same may also bring about the rebirth of many drugs, which had been discontinued due to their failure to gain therapeutic concentration in the brain.

6.1.2.1   MECHANISMS OF NANOFORMULATIONS FACILITATING BRAIN DELIVERY

Kreuter (2012) described a number of possibilities to explain the mechanisms of delivery of nanoformulations across the BBB, as discussed below:

(1) As compared to the pure drugs, there is an increased retention of the nanoformulations in the brain blood capillaries combined with more adsorption to the capillary walls. These retention and adsorption create a higher concentration gradient than possible for pure drugs and thereby enhance its transport through and across the endothelial cell layer resulting in better delivery to the brain.

(2) Opening of the tight endothelial junctions owing to nanoparticles (NPs) could promote permeation through these loose junctions in either free form or together with the NPs in bound form.

(3) The surfactants used in preparation of nanoformulations can have a characteristic effect of solubilization of the membrane lipids of endothelial cells in BBB, which would eventually lead to membrane fluidization and thereby enhanced drug permeability through BBB.

(4) Endocytosis of drug containing nanoformulation and subsequent release in the endothelial cells of the brain capillaries postlysis of the NPs would result in the release of the drugs within these cells and delivery to the brain.

(5) Surfactants used as coating agent of NPs have the potential to inhibit the efflux system, especially P-glycoprotein. Endocytosis mediated by the adsorption of apolipoprotein B and/or E from the blood through the low-density lipoprotein receptor is also a suggested mechanism for increased uptake/transport of the nanoformulations coated with various grades of polysorbate and poloxamers such as Pluronic F68 (Kreuter, 2004; 2012).

6.1.3   NANOFORMULATIONS INVESTIGATED

A wide variety of nanocarriers have been investigated in detail for improving brain delivery, which primarily involve various nanoparticulate systems of both polymeric and lipid nature, namely, dendrimers, nanosuspensions, liposomes, nanoemulsions, and ligand-mediated nanosystems out of which polymeric and solid lipid NPs (SLNs) have been discussed in detail below.

Summary of the various forms of nanosystems investigated for drug delivery with examples is listed in Table 6.1.

TABLE 6.1   List of Nanoformulations Investigated for Brain Distribution.

Formulation	Active drug	Component	Advantages
Polymeric NP	Etoposide	PLGA and PCL	Selective distribution with higher brain permeability (Snehalatha et al., 2008)
	Coenzyme Q10	Trimethylated chitosan surface-modified PLGA	The resulting nanoparticle system exhibited negligible cytotoxicity and enhanced the brain uptake of Trimethylated Chitosan (TMC)/ PLGA–NP (Wang et al., 2010)
	Venlafaxine	Chitosan	Better brain uptake, higher DTP% (Haque et al., 2012)
	Amphotericin B	PLA– Polyethylene glycol (PEG)–tween 80	Drug concentration in mice brain greatly enhanced, reduced the toxicity of Amphotericin B (AmB) to liver, kidney, etc. (Goldberg et al., 2009)
	Doxorubicin	PBCA–tween 80	Augmented accumulation of NP in the tumor site and in the contralateral hemisphere (Ambruosi et al., 2006)
	Doxorubicin	PLGA nanoparticles coated with poloxamer 188	The efficacy of brain delivery by nanoparticles not only is influenced by the coating surfactants but also by other formulation parameters (Gelperina et al., 2010)
Solid lipid nanoparticles/nanostructured lipid carriers	Camptothecin	Cetyl palmitate, dynasan and witepsol	Higher affinity to the porcine brain capillary endothelial cells as compared to macrophages (Martins et al., 2012)
	Clozapine	Trimyristin, tripalmitin, and tristearin	The AUC and MRT of clozapine SLN were significantly higher in brain (Manjunath and Venkateswarlu, 2005)

TABLE 6.1 *(Continued)*

Formulation	Active drug	Component	Advantages
	Olanzapine	Glyceryl mono-stearate	The AUC and MRT of olan-zapine SLN were consider-ably higher in brain (Joseph and Saha, 2017)
	Baicalein	Tripalmitin, gelucire, and vitamin E	Brain-targeting efficiency of baicalein was greatly improved by NLC (Tsai et al., 2012)
Liposomes	Oregon Green	1,2-dipalmitoyl-sn-glycero-3-phosphocholine, 3β-[N-(N′,N′-dimethylaminoethane)-carbamoyl], cholesterol , 1,2-dioleoyl-sn-glycero-3-phosphoethanolamine, 1,2-dihexadecanoyl-sn-glycero-3-phosphoethanol-amine	Liposomes were strongly internalized in cultured cell lines within 6 h (Bellavance et al., 2012)
	Citicoline	1,2-Distearoyl-sn-glycero-3-phosphocholine, cholesterol, 1,2-Distearoyl-sn-glycero-3-phosphoetha-nolamine	Considerable increase (10-fold) in the bioavail-ability of the drug in the brain parenchyma (Ramos-Cabrer et al., 2011)
	Interferon–gene plasmid	Cationic liposomes	Inhibition of growth tumor of resistant malignant glioma and aggressive tumor (Yoshida et al., 2004)
Dendrimers	Paclitaxel	Polyamidoamine	12-Fold greater perme-ability across porcine brain endothelial cells (Teow et al., 2013)
	Docetaxel	Polypropyleneimine	Higher targeting efficiency and biodistribution to the brain (Gajbhiye and Jain, 2011)
Micelles	Olanzapine	Block copolymers of ethylene oxide/propylene oxide	Demonstrated higher drug-targeting index (5.20), DTE% (520.26%), and DTP% (80.76%) (Abdelbary and Tadros, 2013)
	Digoxin	Pluronic P85 micelles	Pluronic P85 enhanced the delivery of digoxin to the brain (Batrakova et al., 2001)

TABLE 6.1   *(Continued)*

Formulation	Active drug	Component	Advantages
Nanoemul-sion	Risperidone	Glyceryl monocaprylate	Higher DTE) and increased direct nose to brain drug transport (DTP%) (Kumar et al., 2008)
	Rizatriptan	Labrafil, Cremophor RH, PF127, and PF68	Brain targeting of nano-emulsions (AUC = 302.52 µg min/g) was higher as compared to IV admin-istration (AUC = 109.63 µg min/g) of the drug (Bhanushali et al., 2009)
	Saquinavir	Flax-seed and safflower oil	Improved brain uptake (Vyas et al., 2008)

AUC, area under the curve; DTE%, drug transport efficiency; DTP%, direct transport percentage; MRT, mean residence time; NLC, nanostructured lipid carrier; NP, nanoparticle; PBCAs, polybutyl cyanoacrylates; PCL, polycaprolactone; PLA, poly(d,l-lactide); PLGA, poly(d,l-lactide-*co*-glycolide); SLN, solid lipid nanoparticles.

6.2   NANOPARTICULATE SYSTEMS

NPs are colloidal particles with dimensions less than 1000 nm and can be used for better drug delivery. It can be prepared either by encapsulating the drug within a vesicle and/or by dispersing the drug molecules within a matrix (Saha et al., 2010).

For the past decade or more, nanoparticulate drug-delivery systems have been studied in detail for spatial and temporal delivery, especially in tumor and brain targeting. NPs have shown great promise as an efficacious drug delivery as has been found in both pharmaceutical and clinical research. As a drug carrier, NPs have significant advantages over traditional drug-delivery systems, such as better systemic stability, entrapment efficiency, bioavailability, and long circulation time in vivo and targeted distribution with longer half-life. Thus, these systems have found increased application owing to its added advantages in brain delivery of the therapeutic agents. Majorly investigated nanoparticulate systems for aforementioned applications are polymeric NPs and SLNs, whose preparation methods, characterization, and in-vivo fate are discussed in detail in following sections.

6.2.1 POLYMERIC NANOPARTICLES

Polymeric NPs, in the size range of in and around 200 nm are generally composed of polymeric core matrix within which drugs or other biological molecules can be incorporated. Over the years, number of polymers has been investigated for preparation of the ideal polymeric NP with minimum deleterious effects on our body. These polymers can be divided chiefly on the basis of their source/origin, namely on the basis of natural, semisynthetic, and synthetic origin. But over the years, based on different clinical and pharmacokinetic studies, few of them gained a wider acceptance over others owing to its inherent characteristic advantages as well as limited deleterious effects. Among these wide variety of biodegradable/biocompatible polymers, PLGA, poly(D,L-lactide) (PLA), polybutyl cyanoacrylates (PBCAs), polycaprolactone (PCL), etc. are extensively reported as the constituent of the core polymer matrix used for preparation of NPs for the delivery of drugs to the CNS. They provide numerous advantages, which include increased stability by protecting the drug from in vivo degradation and releasing a therapeutic load at a controlled rate but enabling it to attain the optimal therapeutic range and at the targeted/preferential site of action at the same instant (Costantino and Boraschi, 2012).

Numerous investigations related to delivery of drug across BBB by incorporation in polymeric nanoparticulate drug-delivery systems have been reported. In a study conducted by our group, etoposide-loaded PLGA and PCL NPs were prepared and biodistribution and pharmacokinetics of radiolabeled etoposide and etoposide-loaded PCL and PLGA NPs were studied. Etoposide and etoposide-loaded NPs labeled with Tc-99m were intravenously administered and their respective pharmacokinetic and biodistribution parameters were determined. The results showed a higher residence of etoposide-containing NPs compared to etoposide and demonstrated the advantage of PCL and PLGA NPs as drug carrier for etoposide in enhancing the bioavailability with targeted distribution owing to higher brain permeability and with the potential of reducing the toxicity related to nonspecific etoposide distribution. In another study in a rat model, imatinib mesylate-loaded PLGA NPs were prepared for biodistribution and pharmacokinetic studies. The results showed that nanoparticulate formulations increased the extent of drug permeation to brain with nearly 100% increase in mean residence time and threefold increase in area under the curve as compared to pure drug (Snehalatha et al., 2008).

There are numerous evidences for the existence of a direct nose-to-brain delivery route whereby nanoparticulate systems on administration to the nasal cavity is transported via the olfactory epithelium and/or via the trigeminal nerves directly to the CNS. This approach was studied by Haque et al. whereby venlafaxine (VLF)-loaded chitosan NPs were delivered via intranasal (i.n.) route to enhance the uptake of VLF in brain. The higher %drug transport efficiency (for NPs: 508.59; for solution: 268.75) and %direct transport percentage (for NPs: 80.34; for solution: 62.76) of VLF chitosan NP over i.n. solution formulation suggested its efficiency in treatment of depression (Haque et al., 2012).

Coating of NPs with surfactants is another useful approach to induce increased brain uptake as mentioned in the earlier section. Numerous studies conducted in this area have reportedly found that coating with surfactant systems resulted in enhanced brain concentration of drug/dye as compared with uncoated systems (Ambruosi et al., 2006; Goldberg et al., 2009).

6.2.1.1 GENERAL METHODS OF POLYMERIC NANOPARTICLES PREPARATION

Various methods have been reported for the preparation of polymeric NP and can be classified into two main categories based on the criteria, whether preparation involves polymerization reaction or directly obtained from preformed polymer (Reis et al., 2006).

6.2.1.1.1 Nanoparticles Obtained by Polymerization of Monomer

6.2.1.1.1.1 Emulsion Polymerization

Emulsion polymerization is one of the low time-consuming methods for NP preparation (Mock et al., 2006). Continuous phase used in this method can be either organic or aqueous. The monomer is dispersed into an emulsion or inverse microemulsion or into a nonsolvent. Surfactants were also used during the preparation to prevent the aggregation during polymerization process. This method is less preferred for NP preparation recently because of the requirement of large amounts of organic solvents, which are highly toxic to the body. In addition, there is a requirement of surfactants and initiators, which have to be eliminated from the final NP. Furthermore, the polymers formed from these kinds of monomers are nonbiodegradable in nature, which also resulted in low acceptability of this preparation method. As a further improvement of this method, scientists have used polymethyl methacrylate, polyethyl cyano acrylate, and PBCA polymers to prepare NP using less toxic organic solvents, such as cyclohexane (International Conference on Harmonization (ICH) class 2) and *n*-pentane (International Conference on Harmonization (ICH) class 3) (Reis et al., 2006).

6.2.1.1.1.2   Interfacial Polymerization

In this method, monomer and drug are dissolved in a mixture, containing an oil and alcohol, which then extruded through a small needle into an aqueous solution under continuous stirring with or without organic solvents, such as acetone or ethyl alcohol. Spontaneous formation of NP occurs by the polymerization of monomer with the help of initiating ions in the aqueous phase. The concentration of the resultant colloidal suspension can be performed by evaporation under reduced pressure. Cyanoacrylate monomers are mainly used in this method and examples of drugs incorporated during NP preparation by this method include insulin, indomethacin, calcitonin, etc. (Krauel et al., 2005).

6.2.1.1.1.3   Interfacial Polycondensation

Interfacial polycondensation is another method for preparing polymeric NP using monomers, where lipophilic monomer, such as phtaloyldichloride, and hydrophilic monomer, such as diethylenetriamine, are used, with or without surfactants. In recent methods, polyether urethane was used for NP preparation by this method (Bouchemal et al., 2004).

6.2.1.1.2   Nanoparticles Obtained from Preformed Polymers

6.2.1.1.2.1   Emulsification/Solvent Evaporation

In this method, as a first step, polymer solution containing drug to be incorporated is emulsified into an aqueous phase and as a second step, the emulsion formed is evaporated thereby precipitating into nanospheres. High-energy homogenization techniques are utilized to disperse drug-containing polymeric organic solution into aqueous phase or nonsolvent organic phase such as ethyl acetate or chloroform. The evaporation of the solvent can be performed by increasing the temperature under vacuum or by continuous stirring. Polymers such as PLGA, polycaprolactone, polyhydroxybutyrate, ethyl cellulose, etc. have been used in this method for NP preparation and drugs such as testosterone, cyclosporin-A, indomethacin, albumin, etc. were incorporated into these systems (Jaiswal et al., 2004).

6.2.1.1.2.2   Solvent Displacement and Interfacial Deposition

Both these methods are based on spontaneous emulsification of an internal organic phase containing polymer into an external water phase. Solvent displacement method produces either nanospheres or nanocapsules; interfacial deposition produces only nanocapsules (Ribeiro et al., 2008).

In solvent displacement method, preformed polymer in organic solvent is precipitated and the solvent is diffused into aqueous phase in either the presence or absence of surfactants. The polymer deposited on the interface between the aqueous and organic phase is precipitated instantaneously due to the speedy diffusion of organic solvent. To achieve this, a completely miscible solvent is used and it should be nonsolvent to the polymer for maximum result. In this technique, if a small quantity of nontoxic oil is added in the organic phase, nanocapsules also can be prepared, where high drug-loading efficiency has been reported for hydrophobic drugs. Only water-miscible organic solvents such as acetone or dichloromethane are used in this method, and as per the diffusion rate, spontaneous emulsification occurs. This method is useful for the encapsulation of hydrophobic drugs and less suitable for hydrophilic drugs. Mechanism of NP must be similar to the diffusion process found in spontaneous emulsification. Various polymers such as PCL, PLA, PLGA, etc. are widely used for preparation of NP by this method. Drug such as cyclosporin was used for entrapping in NP by this method with very high entrapment efficiency (97%) (Reis et al., 2006). Interracial deposition, which is majority an emulsification/solidification technique, produces nanocapsules. In this method, an extra-component of oily nature, which is immiscible with mixture, but highly miscible with the solvent of polymer, is incorporated in the preparation. Nanocapsules were obtained by the deposition of polymer on the interface of fine oil droplets and aqueous phase.

6.2.1.1.2.3   Emulsification/Solvent Diffusion

In this technique, the polymer is solubilized in a partially water-soluble solvent like propylene carbonate, which is saturated with aqueous phase for thermodynamic equilibration of both liquids. As a next step, this saturated system is emulsified in an aqueous phase with stabilizer, thereby diffusing the solvent to the external phase and producing nanocapsules or nanospheres based on oil/polymer ratio. The solvent is then removed by evaporation or filtration. This technique exhibited high encapsulation efficiencies and good batch-to-batch uniformity with narrow size distribution. In case of incorporation of hydrophilic drugs, there is a chance of oozing out of drug into the external aqueous phase. A large number of polymers have been used for NP preparation by this method and these include PLA, PLGA, gelatin, etc. Drugs such as doxorubicin, coumarin, cyclosporin, etc. have been incorporated by this method (Murakami et al., 1999).

6.2.1.1.2.4   Salting Out with Synthetic Polymers

In this method of NP preparation, salting out technique is used to separate water-miscible solvent from aqueous solution. In a way, it can be considered as a modified procedure of emulsification/solvent diffusion technique. In this technique, polymer and drug are made to dissolve in an organic solvent such as acetone. This solution is further emulsified into an aqueous system containing electrolytes such as magnesium chloride, magnesium acetate, calcium chloride, etc. Sometimes, nonelectrolytes like sucrose also were used as salting out agent. Stabilizers such as polyvinyl pyrolidone were also used in the preparation as stabilizing agent. The oil/water emulsion is further diluted with sufficient aqueous phase to ease the diffusion of acetone into the aqueous external phase resulting in the formation of nanospheres. The solvent and salting out agent are removed further by techniques such as cross-flow filtration. Nanospheres with high encapsulation efficiencies are prepared by this technique using polymers such as PLA, ethyl cellulose, etc. (Kumar et al., 2004).

6.2.1.1.2.5   Desolvation of Macromolecules

This is another technique of NP preparation where desolvation is done by charge and pH changes or adding desolvating agents such as alcohol or concentrated solutions of inorganic salts. No heat process is involved during the preparation and therefore it is suitable for thermo-liable drugs. Macromolecular materials used for this method include gelatin, casein, serum albumin (human and bovine), etc. (Coester et al., 2000).

6.2.1.1.2.6 Supercritical Fluid Technology

This is a comparatively recent method of NP preparation where a combination of drug and polymer are made to solubilize in a supercritical or compressed fluid (supercritical CO_2) and subsequently passed through a nozzle where the solution is expanded. The supercritical fluid gets evaporated in this process and eventually solute particles get precipitated in nanosize. This technique produces particles free of solvent residues because of the complete evaporation of supercritical fluid during the expansion process. Numerous drugs have been incorporated by this technique including protein drug such as insulin and obtained highly promising results. Polymers such as PEG/PLA are used for NP preparation using this method. However, the capital investment required for initial setup is very high. In addition, the solubility of polar substances in supercritical fluid is very less extent only; therefore, this method is comparatively less suitable for hydrophilic drugs (Byrappa et al., 2008; Kim et al., 2008).

6.2.2 SOLID LIPID NANOPARTICLES

SLNs as colloidal drug-delivery system came into the foray first in the year 1991 and quickly attained recognition as a potential drug carrier system owing to several desirable properties for instance, least biotoxicity, increased drug stability, better drug loading, controlled drug release, targeted delivery, ease of commercialization owing to simple preparation, and associated sterilization technique among others. They are composed of biocompatible/biodegradable lipid matrix that is solid at body temperature and exhibit size in a range between that of 100 and 400 nm. SLNs offer several advantages owing to the use of solid lipid which limits the flow of drug in the nanoparticulate system owing to its solid conformation and thereby limits the phenomenon of drug leaching, previously observed in conventional lipid formulations. General ingredients used in the preparation of SLN are solid lipid(s), emulsifier(s), and water. The term "lipid" has a broader sense here and includes fatty acids (e.g., stearic acid), triglycerides (e.g., tristearin), steroids (e.g., cholesterol), waxes (e.g., cetyl palmitate), and partial glycerides (e.g., Imwitor).

SLNs have been widely used for the delivery of drugs, previously impermeable, to the brain owing to the advantages previously mentioned along with its augmented ability to cross BBB. For instance, in studies conducted by Martins et al. (2012), camptothecin-loaded SLNs were formulated using cetyl palmitate, Dynasan 114, and Witepsol E85 by hot high-pressure homogenization (HPH) technique. A higher affinity of the SLN to the porcine brain capillary endothelial cells over macrophages was observed. In vivo studies carried out on rat models using fluorescently labeled SLN showed higher accumulation in the brain post i.v. administration.

A more advanced version of SLN called nanostructured lipid carriers (NLCs), with increased drug loading are also becoming popular recently for brain targeting, chiefly due to reduced surface desorption characteristic in comparison to traditional SLNs owing to the higher retentive ability of the highly disordered core structure provided by NLCs. NLCs are composed of a solid lipid and a certain amount of liquid lipid (oil), maintaining the solid state at both room and body temperature (Müller et al., 2007; Tsai et al., 2011).

6.2.2.1 PREPARATION TECHNIQUES OF SOLID LIPID NANOPARTICLES

SLN are prepared from lipid, emulsifier, and water/solvent by various production techniques (Joseph and Saha, 2013; Mehnert and Mäder, 2001; MuÈller et al., 2000).

6.2.2.1.1 High-Pressure Homogenization

HPH technique is one of the most powerful and reliable methods of SLN preparation (Jenning et al., 2002). In this method, liquid is pushed through a very narrow gap of very small size with high pressure up to 2000 bar, thereby accelerating the fluid to a very high velocity of more than 1000 km/h. Because of very high shear stress and cavitation forces, particles disrupt down to a range of submicron size. There are two basic approaches generally used in HPH technique, namely hot homogenization and cold homogenization.

6.2.2.1.1.1 Hot Homogenization Technique

In this technique, the drug is initially solubilized in lipid melt. A pre-emulsion is prepared by dispersing this lipid melt into a surfactant aqueous solution, which is kept under stirring and heated, to a temperature above the melting point of the lipid. The resultant pre-emulsion is then subjected to HPH using a high-shear mixing device for many cycles (2–5) applying pressure in a range up to 1500 bar. The prepared nanosized emulsion is then allowed to cool down to room temperature or sometimes even to lower temperature (Jenning et al., 2002). In this cooling down process, the nanodroplets are solidified and produce an aqueous dispersion of SLN. Care should be taken while selecting the number of cycles in the homogenization process, since large number of cycles may result in contamination of SLN with homogenizer metal and may increase the particle size due to particle coalescence, which happens due to high-surface free energy of particles. More number of cycles may also increase the production cost of SLN prepared by this technique. This method is mainly suitable for SLN preparation of hydrophobic drugs and temperature-resistant drugs (Al-Haj et al., 2008; Jenning et al., 2002).

6.2.2.1.1.2   Cold Homogenization Technique

To overcome the problems associated with hot homogenization technique, such as temperature-induced degradation of active ingredient, distribution of active ingredient into aqueous phase during homogenization process, complexity of crystallization step, etc., cold homogenization technique has been introduced (Mehnert and Mäder, 2001). In this method, initially the active ingredient is made to dissolve/disperse in lipid melt, similar to that of hot homogenization process. Subsequently, the lipid melt with the active ingredient in it is rapidly cooled down with the help of liquid nitrogen/dry ice. This rapid cooling result in the homogeneous distribution of active ingredient within the lipid matrix. The solidified lipid mixture is then subjected to milling process by using a ball/mortar mill to obtain particles in a size range of 50–100 μm. These microparticles are then dispersed in chilled surfactant solution and subjected to HPH at or below room temperature to produce SLN (Mehnert and Mäder, 2001; MuÈller et al., 2000). This technique generally produces particles with higher size and broader distribution as compared to hot homogenization technique. Even though this technique was emerged as an alternative for hot homogenization to avoid high temperature, in initial steps, there is an involvement of heat, therefore makes this method unfavorable for very temperature sensitive drugs, which are unstable at a temperature of melting point of lipid. However, as compared to hot homogenization technique, this technique minimizes the exposure of active ingredient to a very high temperature.

6.2.2.1.2 Solvent Emulsification-Evaporation Technique

In this method, the solid lipid is solubilized in water immiscible organic solvent such as chloroform, cyclohexane, ethyl acetate, etc. and the active ingredient is either dissolved or dispersed in it (Mehnert and Mäder, 2012). This organic phase obtained is added into an aqueous system containing surfactant and emulsified by stirring/homogenization. The organic solvent present is subsequently removed by evaporation under reduced pressure or by mechanical stirring. SLNs are obtained by the precipitation of lipid phase in the aqueous phase upon removal of organic solvent. This technique is suitable for hydrophobic drugs; however, hydrophilic drugs also can be incorporated by this method by preparing water/oil/water emulsion and solubilizing hydrophilic drug in internal aqueous phase. This method is highly suitable for thermo-liable drugs since there is no involvement of heat during the preparation. However, the toxicity from residual organic solvent present in the final formulation must be carefully addressed.

6.2.2.1.3 Solvent Emulsification Diffusion Technique

The lipid is initially dissolved in a partially aqueous miscible solvent, such as isobutyric acid, benzyl alcohol, tetrahydro-furan, etc. and consequently emulsified in an aqueous solution containing surfactant by mechanical stirring. This emulsion is further diluted into an excess aqueous phase at a controlled temperature, which results in the diffusion of solvent into external aqueous phase, thereby precipitating the SLN. Various techniques such as distillation or ultrafiltration are used for the removal of solvent. The lipid concentration, nature, surfactant, rate of stirring, processing temperature, etc. are critical for optimizing the formation of SLN (Trotta et al., 2003).

6.2.2.1.4 Solvent Displacement/Nanoprecipitation Technique

This method is used widely in the preparation of polymeric NP, and later on, this technique has been modified and used for SLN preparation (Fonseca et al., 2002). In this technique, solid lipid and active ingredient are solubilized in water miscible organic solvents such as acetone, methanol, ethanol, isopropanol, etc., and subsequently injected/dispersed into an aqueous dispersion containing surfactant with stirring, thereby producing suspension of SLN (Noriega-Peláez et al., 2011). The solvent present is removed by different techniques such as distillation, ultracentrifugation, evaporation under reduced pressure, lyophilization, etc. This method is suitable for hydrophobic drugs, which are incorporated with high loading efficiencies. In addition, the heat involved is very less during the process, thereby suitable for temperature sensitive active ingredients (Mehnert and Mäder, 2012).

6.2.2.1.5   Microemulsion-Based Technique

This method is based on the dilution of microemulsions, which is a thermodynamically stable isotropic mixture made of oil/lipid, emulsifier, and water. In this technique, both lipid and aqueous phase with emulsifier are made to mix at a temperature above melting point of lipid, in suitable ratios under stirring (MuÈller et al., 2000). The hot microemulsion obtained is diluted into chilled aqueous phase, which is kept at constant stirring. Large quantity of aqueous phase is used for dilution, which sometimes may go up to 50 times of microemulsion. By dilution with cold aqueous phase, lipid droplets are solidified and produce SLN. One of the major disadvantage of this technique is it's proneness to phase transitions with minor changes in composition or thermodynamic variables. Further, large quantities of solvents have to be removed from final formulation, which is a tedious process.

6.2.2.1.6   Supercritical Fluid Method

In this method, drug and solid lipid are made to solubilize in a supercritical or compressed fluid such as supercritical CO_2 and subsequently passed through a nozzle where the solution is expanded. The supercritical fluid is evaporated in this process and eventually solute particles are precipitated in nanosize. This technique produces particles free of solvent residues because of the complete evaporation of supercritical fluid during the expansion process (Chen et al., 2006; Chattopadhyay et al., 2007).

6.2.3   SECONDARY PRODUCTION STEPS FOR PREPARATION OF POLYMERIC/SOLID LIPID NANOPARTICLES

6.2.3.1   LYOPHILIZATION

Lyophilization or freeze-drying is one of the most significant secondary production steps in NP preparation, which enhances the physical and chemical stability of the formulation obtained (Zimmermann et al., 2000). Lyophilization aids to increase the long-term stability for a NP preparation with hydrolyzable drugs. When the nanoparticulate systems are lyophilized into solid preparation, it prevents Ostwald ripening as well as hydrolytic degradation, thereby increasing the stability. After lyophilization, NP can be more easily incorporated into different dosage forms such as tablets, capsules, parental dispersions, pellets, etc. In lyophilization, the NP dispersion is freezed which is then subjected to evaporation under vacuum. Cryoprotectant such as sorbitol, mannitol, glucose, trehalose, etc. are added to the dispersion to prevent or minimize particle aggregation and to redisperse the lyophilizates in a more efficient way. These cryoprotectants aid in NP stability by minimizing osmotic activity of water. They act as placeholders and prevent the contact between discrete NP. They also interact with polar head groups of surfactants present and act as a pseudo-hydration shell (Abdelwahed et al., 2006).

6.2.3.2   SPRAY DRYING

Spray drying is a less frequently used method for transforming aqueous NP dispersion into a dry product. Spray dryers make use of hot gases and atomizers/nozzles to disperse effectively the NP dispersion and therefore sometimes result in aggregation and partial melting of NP. In case of SLN, melting can be minimized by incorporating ethanol in dispersion medium (Freitas and Müller, 1998).

6.2.3.3   STERILIZATION

Sterilization is another secondary production step which is highly desirable for NP meant for parental administration. Aseptic production of NP, filtration, gamma irradiation, heating, etc. are the general methods used for sterilizing nanoparticulate systems (Cavalli et al., 1997). In the case of heat resistant drugs and NP material, autoclaving is a good method of choice. It was also found that sterilizing of nanoparticulate systems by this method could slightly increase the particle size. Sterilization by filtration requires high pressure and should not cause any change in the nanoparticulate system with respect to stability and drug-release characteristics. In the case of gamma irradiation, free radicals are obtained and may undergo secondary reactions leading to chemical modifications. High molecular mobility and presence of oxygen enhances degradation reactions induced by gamma radiations (Konan et al., 2002).

6.2.4 CHARACTERIZATION OF NANOPARTICLES

To find out whether the obtained NPs meet the required criteria and objectives, several characterization methods are used as shown in Table 6.2. Most widely applied methods for characterization include particle size (techniques—dynamic light scattering, static light scattering, acoustic methods, etc.), zeta potential, electron microscopy (scanning, transmission, etc.), atomic force microscopy, differential scanning calorimetry, X-ray diffraction, entrapment efficiency, drug content, in vitro drug release (including study of release kinetics and mechanisms), etc. A few of those common analysis techniques used for particular parameters/characteristics of the NPs are mentioned in Table 6.2. The list of analytical techniques used for characterization of formulated NPs is nonexhaustive and not limited only to those mentioned in the table.

TABLE 6.2 List of Analytical Techniques for Characterization Studies.

Parameter	Principle/mechanism	Technique/instrument
Particle size	Dynamic light scattering	Zeta sizer
	Electron microscopy	SEM, TEM
	Laser diffraction	Photon correlation spectroscopy
	Acoustic methods/ultrasonic waves	AAS
	Sedimentation	ADS
	Diffusion property of particles	FFF coupled with MALS
Surface charge	Zeta potential	Zeta sizer
	Surface charge	AFM
Morphology	Electron microscopy	SEM, TEM
	Piezoelectric scanner for detection of surface charge	AFM
	Quantum tunneling	STM
Crystallographic orientation	X-ray	XRD
	Raman light scattering	Raman spectroscopy
Elemental composition	Magnetic resonance	NMR spectroscopy
	BSE, energy dispersive X-ray	XPS, SEM

TABLE 6.2 *(Continued)*

Parameter	Principle/mechanism	Technique/instrument
Entrapment efficiency, drug content, drug release studies	Absorbance/transmittance	Ultraviolet spectroscopy
	Chromatography technique	high performance liquid chromatography
Compatibility studies	Characteristic thermal peaks	DSC
	Functional peaks detection	Infrared radiation spectroscopy

AAS, acoustic attenuation spectroscopy; ADS, analytical disk centrifugation; AFM, atomic force microscopy; BSE, back-scattered electrons; DSC, differential scanning calorimetry; FFF, field flow fractionation; MALS, multi-angle light scattering detector; NMR, nuclear magnetic resonance; SEM, scanning electron microscopy; STM, surface tunneling microscopy; TEM, transmission electron microscopy; XPS, X-ray photoelectron spectroscopy; XRD, X-ray diffraction spectroscopy.

6.3 IN VIVO FATE AND TOXICITY ASPECTS OF NANOPARTICLES

Nanoparticulate systems administered through oral or transdermal route exhibit negligible/very few problems in biological systems. In addition, NP administered via intramuscular or subcutaneous route do not cause much problems if appropriate surfactants are used. The particle size is not that significant for these routes as compared to administration by parental route. Particle-size distribution is the major issue associated with intravenous injection of nanoparticulate systems. There is a potential danger of capillary blockage because of bigger particles or aggregation of particles, which may even lead to death due to embolism. The diameter of fine capillaries present in human is about 9 μm in diameter. Therefore, the NP must be in submicron size, if designed for parenteral administration (Wong et al., 2008).

The interaction of NP with phagocytizing cells has been studied extensively by many groups. The macrophages of mononuclear phagocyte system immediately remove/clear the NP present in the blood circulation, considering them as foreign bodies. This occurs more for NP with more than 200 nm and of hydrophobic nature. To overcome this, PEGylation has been done to obtain stealth particles with hydrophilic surface, and therefore exhibited long half-life in blood circulation. Various other polymers or surfactants such as tweens, PVA, polyacrylamides, etc. have also been used to overcome phagocytosis, thereby increasing circulation time and better targeting efficiency (Sheng et al., 2009; Wong et al., 2008).

The in vivo fate of administered NP is dependent on diverse factors including administration route, interaction with biological surroundings such as adsorption of biological material, enzymatic process, etc. When nanoparticulate systems were administered intravenously, numerous pharmacokinetics, and biodistribution studies suggested the prolonged presence of drug in blood circulation with NP than pure drug solution. NP also exhibited higher drug concentrations in lung, spleen, and brain, while pure drug solution distributed more into liver and kidneys (Kreuter, 2001; Yang et al., 1999). However, diverse reports are available in the literature regarding the pharmacokinetics and biodistribution profiles of administered NP depending on the drug, polymer or solid lipid, surfactants, particle size, charge, route of administration, and many other factors (Li and Huang, 2008; Owens and Peppas, 2005).

6.4   FUTURE PROSPECT

Recently, advances made in respect to nanotechnology have been translated in the field of drug delivery. One of the most innovative technologies implemented has to be the use of "Theranostics." The idea of imaging and treatment has changed the perception of medical fraternity and has the potential to transform the way we treat different ailments, especially malignant tumors or HIV, where the drug molecule administered is highly potent and pose serious threat to other normal living cells. Theranostic NPs for treatment of malignant tumors are being researched widely but only very recently, its application on brain targeted delivery is being considered (Fleuren et al., 2014). Theranostic NPs not only aid our motive of targeted drug delivery to sites of action but also help in our decision-making. For example, in case of treating brain tumors by the application of chemotherapeutic agent containing theranostic NPs, it would not only help in treatment through targeted on site delivery but also help us in mapping the disease progression and thereby make decisions regarding the dosage or treatment regimen to be implemented. Thereby, we can see how "theranostic nanoparticles" can provide an elegant solution to personalized treatment of patients in future by coupling diagnosis and therapeutics. Other than this, the development of NPs with multifunctionality owing to surface attached ligands or antibodies have also taken the world by storm and endeavors to find the best possible functionality agent is still continuing with promises of being able to grasp the evereluding solution.

6.5   CONCLUSION

As authors of this chapter, our chief objective was to articulate our thoughts and knowledge in the best possible way and put forward the same to give the reader a comprehensive idea about the various aspects behind development of brain-targeted nanoparticulate delivery systems. At the end of this chapter, we hope that the reader be able to comprehend the multifaceted approach required for effective delivery of potent drug molecules to the CNS, especially the brain, for efficient treatment of neuronal disorders and ailments. As of now, the scientific fraternity have not been completely successful in treating neuronal disorders, particularly owing to the several hurdles as has been mentioned at the beginning of this chapter. Simply opting to formulate NPs with good entrapment efficiency or drug loading is not the solution. A system with good entrapment and loading of potent drug molecules along with good permeability, targeted delivery, and sustained release pattern at concentrations whereby therapeutic levels be maintained is our primary goal, something which is still eluding us and making our endeavors in this regard inconsequential. Our enhanced understanding of neuronal disorders and its progression have helped us in finding solutions which are effective at molecular levels and in ex vitro systems. But the true success will only be possible to achieve if this success can be translated in vivo in patients suffering from the ailments arising due to progression of neuronal damages in the brain or CNS. And, thereby lies the prospect of noteworthy contribution that a formulation/ drug delivery scientist can make in the medical community for the greater good of the society. We hope, the complex nature of brain targeted delivery with its complexities, hurdles, and possible solutions could be answered to the best possible extent, although the studies and information mentioned here are not limited as formulation science is a dynamic and nonexhaustive domain, and probably at the time this chapter be published, we sincerely hope that the scientific fraternity might have made further notable progressions with regards to brain targeted delivery and open up a new dimension in medical and pharmaceutical sector.

KEYWORDS

- **nanoparticles**
- **brain targeting**
- **blood–brain barrier**
- **targeted drug-delivery systems**
- **polymeric**
- **solid lipid nanoparticles**

REFERENCES

Abdelwahed, W.; Degobert, G.; Stainmesse, S.; Fessi, H. Freeze–Drying of Nanoparticles: Formulation, Process and Storage Considerations. *Adv. Drug Deliv. Rev.* **2006,** *58* (15), 1688–1713.

Abdelbary, G. A.; Tadros, M. I. Brain Targeting of Olanzapine via Intranasal Delivery of Core–Shell Difunctional Block Copolymer Mixed Nanomicellar Carriers: In Vitro Characterization, Ex Vivo Estimation of Nasal Toxicity and In Vivo Biodistribution Studies. *Int. J. Pharm.* **2013,** *452* (1–2), 300–310.

Al-Haj, N. A.; Abdullah, R.; Ibrahim, S.; Bustamam, A. Tamoxifen Drug Loading Solid Lipid Nanoparticles Prepared by Hot High Pressure Homogenization Techniques. *Am. J. Pharmacol. Toxicol.* **2008,** *3* (3), 219–224.

Ambruosi, A.; Khalansky, A. S.; Yamamoto, H.; Gelperina, S. E.; Begley, D. J.; Kreuter, J. Biodistribution of Polysorbate 80-Coated Doxorubicin-Loaded (14C)-Poly(Butyl Cyanoacrylate) Nanoparticles after Intravenous Administration to Glioblastoma-Bearing Rats. *J. Drug Target.* **2006,** *14* (2), 97–105.

Batrakova, E. V.; Miller, D. W.; Li, S.; Alakhov, V. Y.; Kabanov, A. V.; Elmquist, W. F. Pluronic P85 Enhances the Delivery of Digoxin to the Brain: In Vitro and In Vivo Studies. *J. Pharmacol. Exp. Ther.* **2001,** *296,* 551–557.

Bellavance, M.-A.; Poirier, M.-B.; Fortin, D. Uptake and Intracellular Release Kinetics of Liposome Formulations in Glioma Cells. *Int. J. Pharm.* **2010,** *395* (1–2), 251–259.

Bhanushali, R. S.; Gatne, M.; Gaikwad, R. V.; Bajaj, N.; Morde, M. A. Nanoemulsion Based Intranasal Delivery of Antimigraine Drugs for Nose to Brain Targeting. *Indian J. Pharm. Sci.* **2009,** *71* (6), 707–709.

Borlongan, C. V.; Emerich, D. F. Facilitation of Drug Entry into the CNS via Transient Permeation of Blood Brain Barrier: Laboratory and Preliminary Clinical Evidence from Bradykinin Receptor Agonist, Cereport. *Brain Res. Bull.* **2003,** *60* (3), 297–306.

Bouchemal, K.; Briançon, S.; Perrier, E.; Fessi, H.; Bonnet, I.; Zydowicz, N. Synthesis and Characterization of Polyurethane and Poly(Ether Urethane) Nanocapsules Using a New Technique of Interfacial Polycondensation Combined to Spontaneous Emulsification. *Int. J. Pharm.* **2004,** *269* (1), 89–100.

Byrappa, K.; Ohara, S.; Adschiri, T. Nanoparticles Synthesis Using Supercritical Fluid Technology—Towards Biomedical Applications. *Adv. Drug Deliv. Rev.* **2008,** *60* (3), 299–327.

Cavalli, R.; Caputo, O.; Carlotti, M. E.; Trotta, M.; Scarnecchia, C.; Gasco, M. R. Sterilization and Freeze-Drying of Drug-Free and Drug-Loaded Solid Lipid Nanoparticles. *Int. J. Pharm.* **1997,** *148* (1), 47–54.

Chattopadhyay, P.; Shekunov, B. Y.; Yim, D.; Cipolla, D.; Boyd, B.; Farr, S. Production of Solid Lipid Nanoparticle Suspensions Using Supercritical Fluid Extraction of Emulsions (SFEE) for Pulmonary Delivery Using the AERx System. *Adv. Drug Deliv. Rev.* **2007,** *59* (6), 444–453.

Chen, Y.; Jin, R.; Zhou, Y.; Zeng, J.; Zhang, H.; Feng, Q. Preparation of Solid Lipid Nanoparticles Loaded with Xionggui Powder-Supercritical Carbon Dioxide Fluid Extraction and Their Evaluation In Vitro Release. *Zhongguo Zhong yao za zhi = Zhongguo zhongyao zazhi = Chin. J. Chin. Mater. Med.* **2006,** *31* (5), 376–379.

Coester, C.; Langer, K.; Von Briesen, H.; Kreuter, J. Gelatin Nanoparticles by Two Step Desolvation a New Preparation Method, Surface Modifications and Cell Uptake. *J. Microencapsul.* **2000,** *17* (2), 187–193.

Costantino, L.; Boraschi, D. Is There a Clinical Future for Polymeric Nanoparticles as Brain-Targeting Drug Delivery Agents? *Drug Discov. Today* **2012,** *17* (7), 367–378.

Elkharraz, K.; Faisant, N.; Guse, C.; Siepmann, F.; Arica-Yegin, B.; Oger, J. M.; Gust, R.; Goepferich, A.; Benoit, J. P.; Siepmann, J. Paclitaxel-Loaded Microparticles and Implants for the Treatment of Brain Cancer: Preparation and Physicochemical Characterization. *Int. J. Pharm.* **2006,** *314* (2), 127–136.

Fleuren, E. D. G.; Versleijen-Jonkers, Y. M. H.; Heskamp, S.; van Herpen, C. M. L.; Oyen, W. J. G.; van der Graaf, W. T. A.; Boerman, O. C. Theranostic Applications of Antibodies in Oncology. *Mol. Oncol.* **2014,** *8,* 799–812.

Fonseca, C.; Simoes, S.; Gaspar, R. Paclitaxel-Loaded PLGA Nanoparticles: Preparation, Physicochemical Characterization and In Vitro Anti-Tumoral Activity. *J. Control. Release* **2002,** *83* (2), 273–286.

Freitas, C.; Müller, R. H. Effect of Light and Temperature on Zeta Potential and Physical Stability in Solid Lipid Nanoparticle (SLN™) Dispersions. *Int. J. Pharm.* **1998,** *168* (2), 221–229.

Gabathuler, R. Approaches to Transport Therapeutic Drugs across the Blood–Brain Barrier to Treat Brain Diseases. *Neurobiol. Dis.* **2010,** *37* (1), 48–57.

Gajbhiye, V.; Jain, N. K. The Treatment of Glioblastoma Xenografts by Surfactant Conjugated Dendritic Nanoconjugates. *Biomaterials* **2011,** *32* (26), 6213–6225.

Gelperina, S.; Maksimenko, O; Khalansky, A.; Vanchugova, L.; Shipulo, E.; Abbasova, K.; Berdiev, R.; Wohlfart, S.; Chepurnova, N.; Kreuter, J. Drug Delivery to the Brain Using Surfactant-Coated Poly(Lactide-*co*-Glycolide) Nanoparticles: Influence of the Formulation Parameters. *Eur. J. Pharm. Biopharm.* **2010,** *74* (2), 157–163.

Goldberg, J. F.; Perlis, R. H.; Bowden, C. L.; Thase, M. E.; Miklowitz, D. J.; Marangell, L. B.; Calabrese, J. R.; Nierenberg, A. A.; Sachs, G. S. Manic Symptoms During Depressive Episodes in 1,380 Patients with Bipolar Disorder: Findings from the STEP-BD. *Am. J. Psychiatry* **2009,** *166* (2), 173–181.

Haque, S.; Md, S.; Fazil, M.; Kumar, M.; Sahni, J. K.; Ali, J.; Baboota, S. Venlafaxine Loaded Chitosan NPs for Brain Targeting: Pharmacokinetic and Pharmacodynamic Evaluation. *Carbohydr. Polym.* **2012,** *89* (1), 72–79.

Jaiswal, J.; Gupta, S. K.; Kreuter, J. Preparation of Biodegradable Cyclosporine Nanoparticles by High-Pressure Emulsification-Solvent Evaporation Process. *J. Control. Release* **2004,** *96* (1), 169–178.

Jenning, V.; Lippacher, A.; Gohla, S. Medium Scale Production of Solid Lipid Nanoparticles (SLN) by High Pressure Homogenization. *J. Microencapsul.* **2002,** *19* (1), 1–10.

Joseph, E.; Saha, R. N. Advances in Brain Targeted Drug Delivery: Nanoparticulate Systems. *J. PharmaSciTech* **2013,** *3,* 1–8.

Joseph, E.; Reddi, S.; Rinwa, V.; Balwani, G.; Saha, R. Design and In Vivo Evaluation of Solid Lipid Nanoparticulate Systems of Olanzapine for Acute Phase Schizophrenia Treatment: Investigations on Antipsychotic Potential and Adverse Effects. *Eur. J. Pharm. Sci.* **2017,** *104,* 315–325.

Joseph, E.; Saha, R. N. Investigations on Pharmacokinetics and Biodistribution of Polymeric and Solid Lipid Nanoparticulate Systems of Atypical Antipsychotic Drug: Effect of Material Used and Surface Modification. *Drug Dev. Ind. Pharm.* **2017,** *43* (4), 678–686.

Kim, M.-S.; Jin, S.-J.; Kim, J.-S.; Park, H. J.; Song, H.-S.; Neubert, R. H.; Hwang, S.-J. Preparation, Characterization and In Vivo Evaluation of Amorphous Atorvastatin Calcium Nanoparticles Using Supercritical Antisolvent (SAS) Process. *Eur. J. Pharm. Sci.* **2008,** *69* (2), 454–465.

Konan, Y. N.; Gurny, R.; Allémann, E. Preparation and Characterization of Sterile and Freeze-Dried Sub-200 nm Nanoparticles. *Int. J. Pharm.* **2002,** *233* (1), 239–252.

Krauel, K.; Davies, N.; Hook, S.; Rades, T. Using Different Structure Types of Microemulsions for the Preparation of Poly(Alkylcyanoacrylate) Nanoparticles by Interfacial Polymerization. *J. Control. Release* **2005,** *106* (1), 76–87.

Kreuter, J. Nanoparticulate Systems for Brain Delivery of Drugs. *Adv. Drug Deliv. Rev.* **2001,** *47* (1), 65–81.

Kreuter, J. Influence of the Surface Properties on Nanoparticle-Mediated Transport of Drugs to the Brain. *J. Nanosci. Nanotechnol.* **2004,** *4* (5), 484–488.

Kreuter, J. Nanoparticulate Systems for Brain Delivery of Drugs. *Adv. Drug Deliv. Rev.* **2012,** *64* (Suppl.), 213–222.

Kumar, M. R.; Bakowsky, U.; Lehr, C. Preparation and Characterization of Cationic PLGA Nanospheres as DNA Carriers. *Biomaterials* **2004,** *25* (10), 1771–1777.

Kumar, M.; Misra, A.; Babbar, A. K.; Mishra, A. K.; Mishra, P.; Pathak, K. Intranasal Nano-emulsion Based Brain Targeting Drug Delivery System of Risperidone. *Int. J. Pharm.* **2008,** *358* (1–2), 285–291.

Li, S.-D.; Huang, L. Pharmacokinetics and Biodistribution of Nanoparticles. *Mol. Pharm.* **2008,** *5* (4), 496–504.

Manjunath, K.; Venkateswarlu, V. Pharmacokinetics, Tissue Distribution and Bioavailability of Clozapine Solid Lipid Nanoparticles after Intravenous and Intraduodenal Administration. *J. Control. Release* **2005,** *107* (2), 215–228.

Martins, S.; Tho, I.; Reimold, I.; Fricker, G.; Souto, E.; Ferreira, D.; Brandl, M. Brain Delivery of Camptothecin by Means of Solid Lipid Nanoparticles: Formulation Design, In Vitro and In Vivo Studies. *Int. J. Pharm.* **2012,** *439* (1–2), 49–62.

Mehnert, W.; Mäder, K. Solid Lipid Nanoparticles: Production, Characterization and Applications. *Adv. Drug Deliv. Rev.* **2001,** *47* (2–3), 165–196.

Mehnert, W.; Mäder, K. Solid Lipid Nanoparticles: Production, Characterization and Applications. *Adv. Drug Deliv. Rev.* **2012,** *64* (Suppl.), 83–101.

Mock, E. B.; De Bruyn, H.; Hawkett, B. S.; Gilbert, R. G.; Zukoski, C. F. Synthesis of Anisotropic Nanoparticles by Seeded Emulsion Polymerization. *Langmuir* **2006,** *22* (9), 4037–4043.

MuÈller, R. H.; MaÈder, K.; Gohla, S. Solid Lipid Nanoparticles (SLN) for Controlled Drug Delivery—A Review of the State of the Art. *Eur. J. Pharm. Sci.* **2000,** *50* (1), 161–177.

Müller, R. H.; Petersen, R. D.; Hommoss, A.; Pardeike, J. Nanostructured Lipid Carriers (NLC) in Cosmetic Dermal Products. *Adv. Drug Deliv. Rev.* **2007,** *59* (6), 522–530.

Murakami, H.; Kobayashi, M.; Takeuchi, H.; Kawashima, Y. Preparation of Poly (dl-Lactide-*co*-Glycolide) Nanoparticles by Modified Spontaneous Emulsification Solvent Diffusion Method. *Int. J. Pharm.* **1999,** *187* (2), 143–152.

Noriega-Peláez, E. K.; Mendoza-Muñoz, N.; Ganem-Quintanar, A.; Quintanar-Guerrero, D. Optimization of the Emulsification and Solvent Displacement Method for the Preparation of Solid Lipid Nanoparticles. *Drug Dev. Ind. Pharm.* **2011,** *37* (2), 160–166.

Owens, D. E.; Peppas, N. A. Opsonization, Biodistribution, and Pharmacokinetics of Polymeric Nanoparticles. *Int. J. Pharm.* **2006,** *307* (1), 93–102.

Pavan, B.; Dalpiaz, A.; Ciliberti, N.; Biondi, C.; Manfredini, S.; Vertuani, S. Progress in Drug Delivery to the Central Nervous System by the Prodrug Approach. *Molecules* **2008,** *13* (5), 1035–1065.

Ramos-Cabrer, P.; Agulla, J.; Argibay, B.; Pérez-Mato, M.; Castillo, J. Serial MRI Study of the Enhanced Therapeutic Effects of Liposome-Encapsulated Citicoline in Cerebral Ischemia. *Int. J. Pharm.* **2011,** *405* (1–2), 228–233.

Rautio, J.; Laine, K.; Gynther, M.; Savolainen, J. Prodrug Approaches for CNS Delivery. *AAPS J.* **2008,** *10* (1), 92–102.

Reis, C. P.; Neufeld, R. J.; Ribeiro, A. J.; Veiga, F. Nanoencapsulation. I. Methods for Preparation of Drug-Loaded Polymeric Nanoparticles. *Nanomedicine* **2006,** *2* (1), 8–21.

Ribeiro, H. S.; Chu, B.-S.; Ichikawa, S.; Nakajima, M. Preparation of Nanodispersions Containing β-Carotene by Solvent Displacement Method. *Food Hydrocolloids* **2008,** *22* (1), 12–17.

Saha, R. N.; Vasanthakumar, S.; Bende, G.; Snehalatha, M. Nanoparticulate Drug Delivery Systems for Cancer Chemotherapy. *Mol. Membr. Biol.* **2010,** *27* (7), 215–231.

Sheng, Y.; Liu, C.; Yuan, Y.; Tao, X.; Yang, F.; Shan, X.; Zhou, H.; Xu, F. Long-Circulating Polymeric Nanoparticles Bearing a Combinatorial Coating of PEG and Water-Soluble Chitosan. *Biomaterials* **2009,** *30* (12), 2340–2348.

Snehalatha, M.; Venugopal, K.; Saha, R. N.; Babbar, A. K.; Sharma, R. K. Etoposide Loaded PLGA and PCL Nanoparticles. II: Biodistribution and Pharmacokinetics after Radiolabeling with Tc-99m. *Drug Deliv.* **2008,** *15* (5), 277–287.

Teow, H. M.; Zhou, Z.; Najlah, M.; Yusof, S. R.; Abbott, N. J.; D'Emanuele, A. Delivery of Paclitaxel across Cellular Barriers Using a Dendrimer-Based Nanocarrier. *Int. J. Pharm.* **2013,** *441* (1–2), 701–711.

Trotta, M.; Debernardi, F.; Caputo, O. Preparation of Solid Lipid Nanoparticles by a Solvent Emulsification–Diffusion Technique. *Int. J. Pharm.* **2003,** *257* (1), 153–160.

Tsai, Y.-M.; Chien, C.-F.; Lin, L.-C.; Tsai, T.-H. Curcumin and Its Nano-Formulation: The Kinetics of Tissue Distribution and Blood–Brain Barrier Penetration. *Int. J. Pharm.* **2011,** *416* (1), 331–338.

Tsai, M.-J.; Wu, P.-C.; Huang, Y.-B.; Chang, J.-S.; Lin, C.-L.; Tsai, Y.-H.; Fang, J.-Y. Baicalein Loaded in Tocol Nanostructured Lipid Carriers (Tocol NLCs) for Enhanced Stability and Brain Targeting. *Int. J. Pharm.* **2012,** *423* (2), 461–470.

Vyas, T. K.; Shahiwala, A.; Amiji, M. M. Improved Oral Bioavailability and Brain Transport of Saquinavir upon Administration in Novel Nanoemulsion Formulations. *Int. J. Pharm.* **2008,** *347* (1–2), 93–101.

Vykhodtseva, N.; McDannold, N.; Hynynen, K. Progress and Problems in the Application of Focused Ultrasound for Blood–Brain Barrier Disruption. *Ultrasonics* **2008,** *48* (4), 279–296.

Wang, Z. H.; Wang, Z. Y.; Sun, C. S.; Wang, C. Y.; Jiang, T. Y.; Wang, S. L. Trimethylated Chitosan-Conjugated PLGA Nanoparticles for the Delivery of Drugs to the Brain. *Biomaterials* **2010,** 31(5), 908–915.

Weiss, N.; Miller, F.; Cazaubon, S.; Couraud, P.-O. The Blood-Brain Barrier in Brain Homeostasis and Neurological Diseases. *Biochim. Biophys. Acta—Biomembr.* **2009,** *1788* (4), 842–857.

Wong, J.; Brugger, A.; Khare, A.; Chaubal, M.; Papadopoulos, P.; Rabinow, B.; Kipp, J.; Ning, J. Suspensions for Intravenous (IV) Injection: A Review of Development, Preclinical and Clinical Aspects. *Adv. Drug Deliv. Rev.* **2008,** *60* (8), 939–954.

Yang, S. C.; Lu, L. F.; Cai, Y.; Zhu, J. B.; Liang, B. W.; Yang, C. Z. Body Distribution in Mice of Intravenously Injected Camptothecin Solid Lipid Nanoparticles and Targeting Effect on Brain. *J. Control. Release* **1999,** *59* (3), 299–307.

Yoshida, J.; Mizuno, M.; Fujii, M.; Kajita, Y.; Nakahara, N.; Hatano, M.; Saito, R.; Nobayashi, M.; Wakabayashi, T. Human Gene Therapy for Malignant Gliomas (Glioblastoma Multiforme and Anaplastic Astrocytoma) by In Vivo Transduction with Human Interferon Beta Gene Using Cationic Liposomes. *Hum. Gene Ther.* **2004,** *15*, 77–86.

Zimmermann, E.; Müller, R.; Mäder, K. Influence of Different Parameters on Reconstitution of Lyophilized SLN. *Int. J. Pharm.* **2000,** *196* (2), 211–213.

CHAPTER 7

NANOCAPSULES IN BRAIN TARGETING

SOUMYA NAIR and JAYANTHI ABRAHAM*

Microbial Biotechnology Laboratory, School of Biosciences and Technology, VIT, Vellore, Tamil Nadu, India

Corresponding author. E-mail: jayanthi.abraham@gmail.com

ABSTRACT

Nanocapsules are nanoscale shells ranging from 10 to 1000 nm in a vesicular organization. The outer shell is made of biodegradable and non-toxic polymers whereas, the inner core carries the drug. The advantages of encapsulation and the nanoscale size of the nanocapsule its protection from the adverse biological environment, its controlled release, and accurate targeting in drug delivery. It is designed such that it has an enhanced penetrability to cross the blood–brain barrier (BBB), which protects the central nervous system (CNS) from the potential hazardous neurotoxic substance present in the connective tissue and penetrate the brain tissue to deliver the drug present within the capsule to the target cell. Nanocapsules decrease the peripheral toxicity by monitoring the drug release and by improving the photochemical constancy. The current chapter focuses on the BBB, its physiology, and the preparative methods of nanocapsules along with the various strategies of drug delivery via brain targeting using its modifications. Furthermore, the chapter evaluates the types of nanocapsules used for drug delivery in brain targeting and the mechanism involved in the process. Because of its promising applications, nanocapsules are emerging as the ultimate drug carrier system for brain targeting. They are promising as drug carriers to deliver therapeutic drugs across the BBB for the treatment of brain diseases and as an image augmentation agent.

7.1 INTRODUCTION

In the last three decades, an array of drug carrier systems has been thoroughly studied with the purpose of targeting and regulating the drug release. Researchers have also evaluated and refined the efficacy and selectivity of the targeted drug formulations (Barratt, 2000; Couvreur et al., 2002; Schaffazick et al., 2003). The controlled release of the therapeutic drug molecule from its carrier system is designed in such a way that it provides an appropriate response at the target site of action and for a prolonged period of time, thereby remodeling and improving the medication (Vauthier et al., 2003). These drug carrier systems can be administered into the body through different routes, such as the intravenous, ophthalmic, oral, intraperitoneal, intramuscular, and skin (Barratt, 2000).

Drug targeting or smart drug delivery is defined as a selective and an efficient release of the drug in the body at a specific anatomical site, organ, tissues, or cells, where the drug response and effect is required. Drug carrier systems are also used to increase the therapeutic index by reducing the congenital absorption and the toxic side effects of the applied drug, which may occur as a result of the drug acting at ubiquitous sites (Kreuter, 1994; Yokoyama and Okano, 1996; Kumaresh et al., 2001; Sílvia et al., 2007).

Nanomedicine is an emerging new field of research, which is created by the integration of nanotechnology and medical research. It has become an important and a promising gateway for developing and manufacturing new and potent targeted therapies and medications, with a particular impact on neuroscience and oncology. The major reason behind the gaining popularity is because, with the development of new targeted therapeutic strategies, the fundamental limitations of classical pharmacotherapy could be very well improved. In the last few years, an array of colloidal systems such as nanospheres and nanocapsules has been studied and reported as a promising and a potential carrier for the different therapeutic drug delivery in order to treat different diseases (Paola et al., 2012).

Nanopharmacology, formed by the integration of nanotechnology and nanomedicine, is a new division of pharmacology, progressively developing as a future step toward treating and preventing any disease or ailments, using molecular tools and knowledge about human body. Nanopharmacology mostly deals with the interactions between the therapeutic drugs with proteins, DNA, and RNA or the interactions between conventional drugs and the biological systems at nanoscale level. In the current scenario, the main attraction in the field of nanopharmacology at present is the synthesis of polymeric micro-sized therapeutic drug carrier system, which is self-assembled in nature (bottom-up assembly). These carrier systems have been exploited for their many applications in the field of health care. They are used for their controlled and targeted drug delivery to the target organ, reducing the risks of hazardous health effects caused because of drug toxicity. The drug concentration can also be modulated in order to increase its bioavailability at the therapeutic sites (Vishnu et al., 2013).

7.2 SYNTHESIS OF NANOCAPSULES

Nanocapsule synthesis can be achieved by a variety of approaches. One of the most primitive practices for synthesizing nanocapsules was first established in Rohm and Haas research laboratories. Their model involved a carboxylated core with one or more outer polymer shells. The ionization of core with an appropriate base under optimum temperature and pressure made it to expand due to the osmotic swelling in order to produce hollow particles with water and polyelectrolyte in the central core. Apart from these, there are other approaches which have been patented and are more complex in terms of synthesis and the overall chemistry behind it.

At present, there are six canonical methods of nanocapsule synthesis listed below:

1. Nanoprecipitation (solvent displacement method)
2. Solvent evaporation
3. Emulsion coacervation
4. Emulsion diffusion
5. Double emulsification
6. Polymer-coating method
7. Layer by layer nanocapsule synthesis

The above-mentioned processes depend on various characteristics such as the physiochemical properties of the polymer, inner core material, particle size and the nature of the drug encapsulated in the nanocapsule core. Apart from these classical methods, there are other approaches to nanocapsule synthesis, which are the modified versions of the existing methodologies, which will be discussed later.

7.2.1 NANOPRECIPITATION

It is another term for solvent displacement method or interfacial deposition. Nanocapsules are synthesized by forming a colloidal suspension flanked by two distinct phases—solvent and non-solvent phases. The solvent phase (organic phase) comprises either a solution in a particular solvent or a mixture of different solvents altogether, such as methanol, hexane, dichloromethane, dioxane, acetone, etc. These solvents are capable of forming a film. The non-solvent phase (aqueous phase) consists of a non-solvent solution or a mixture of non-solvents, which is also capable of forming a film (Fessi et al., 1988). They are often supplemented with one or more natural or artificial surfactants. The solvent phase is injected into the non-solvent phase. The resulting solution is agitated to produce a colloidal suspension. The process is repeated until the nanocapsules are synthesized from the colloidal suspension. The shape and size of the resulting nanocapsule is totally dependent on the rate of solvent injection and the colloidal suspension agitation (Sugimoto, 1987; Quintanar et al., 1998a; Chorny et al., 2002; Legrand et al., 2007; Lince et al., 2008). Different drugs used as an active ingredient, solvents, and non-solvent phases used in the nanoprecipitation method has been tabulated in Table 7.1. Different drugs for brain targeting, synthesized using the nanoprecipitation method is listed out in Table 7.2.

TABLE 7.1 Different Drugs Used as Active Ingredient, Solvents, and Non-solvent Phases Used in Nanoprecipitation Method.

Active ingredient	Polymer	Solvent	Non-solvent	Reference
Indomethacin	Pcl/pla	Acetone	Water	Pohlmann et al., 2002
Atovaquone	Pla	Acetone	Water	Dalenc, on et al., 1997
Fluconazole labeled with 99mtechnetium	Pla/pla–peg	Methanol	Water	Nogueira De Assis et al., 2008
Usnic acid	Plga 50/50a	Acetone	Phosphate buffer solution (pH 7.4)	Pereira et al., 2006
Melatonin	Eudragit s100	Acetone	Water	Schaffazick et al., 2008

TABLE 7.2 Different Drugs Synthesized Using the Nanoprecipitation (Solvent Displacement) Method.

Drug	Therapeutic activity	Reference
Olanzapine	Antipsychotic	Seju et al., 2011
Rivastigmine tartrate	Treating Alzheimer disease	Joshi et al., 2010
Loperamide	Opioid	Tosi et al., 2007
Risperidone	Antipsychotic	Muthu et al., 2009
Temozolomide	Chemotherapeutic agents for glioblastoma multiforme	Chooi and Ing, 2016
Doxorubicin	Antitumor agent	Gulyaev et al., 1999
		Steiniger et al., 2004
Tubocurarine	Antitumor agent	Alyautdin et al.,1998
Hexapeptide dalargin	Antitumor agent	Alyautdin et al., 1995
		Kreuter et al., 1995
Paclitaxel	Antitumor agent	Sanjeev and Padmanabha, 2016

The mean size of the nanocapsule synthesized using nanoprecipitation is usually 200–250 nm. The drug concentration in the core is usually maintained between 10–25 mg. Although a wide-range of raw materials can be used to prepare nanocapsule through the process of nanoprecipitation method, only a few methodologies are feasible to work upon, in laboratory scale (Devissaguet et al., 1991) or industrial scale. The synthetic and natural polymers can be used for the encapsulation (Nogueira de Assis et al., 2008). Synthetic polymers are usually preferred over the natural ones because of purity, durability, and reproducibility when compared to the natural polymers (Khoee and Yaghoobian, 2008). Figure 7.1 shows the flow chart of the preparative method of nanoprecipitation.

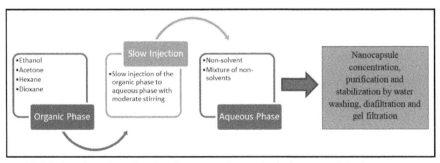

FIGURE 7.1 Flow chart showing the preparative method of nanoprecipitation.

7.2.2   SOLVENT EVAPORATION

The solvent evaporation method for drug encapsulation is one of the primitive techniques that has been extensively used for the preparation of nanocapsules for the controlled release of therapeutic drugs to the target cell. The major steps involved in the preparation method are the following: (1) dissolving the desired polymer and the active compound in the organic solvent, (2) emulsifying the organic phase (dispersed phase) in an aqueous phase (continuous phase), (3) extraction of the solvent, followed by its evaporation, and (4) recovery and drying of nanocapsules to eliminate any remaining solvent post evaporation. Examples of drugs encapsulated using this technique are the following: narcotic antagonists (naltrexone and cyclazocine), anticancer agents (Cisplatin and 5-fluorouracil), hydrophobic steroids (progesterone), and local anesthetics (lidocaine).

There are different modified versions of solvent evaporation technique which mainly depends on the hydrophilicity or the hydrophobicity of the therapeutic drug. Such a modified method can give an efficient drug encapsulation of the desired drug. Different methods to encapsulate hydrophilic drugs are discussed in the next section in brief. Different techniques used to encapsulate hydrophilic drugs in the nanocapsule have been tabulated in Table 7.3. Different drugs for brain targeting synthesized using the solvent evaporation method are listed out in Table 7.4.

TABLE 7.3   Different Techniques Used to Encapsulate Hydrophilic Drugs in the Nanocapsule.

Technique	Definition	Name of the drug encapsulated using the procedure	References
W/o/w double emulsion	The aqueous solution of hydrophilic drug is emulsified with organic phase (w/o emulsion), this emulsion is then dispersed into a second aqueous solution forming a second emulsion	Theophylline Proteins Vaccines Insulin	Herrmann and Bodmeier, 1998 Chattopadhyay et al., 2006 Furtado et al., 2006 Luan et al., 2006 Syamasri gupta, 2011 Navneet et al., 2016

TABLE 7.3 *(Continued)*

Technique	Definition	Name of the drug encapsulated using the procedure	References
O/w co-solvent method	When the drug is not soluble in the main organic solvent, a second solvent called co-solvent is necessary to dissolve the drug	Protein Paclitaxel Indomethacin Ketoprofen	Herrmann and Bodmeier, 1998 Luan et al., 2006 Navneet et al., 2016 P. Chattopadhyay et al., 2006 Syamasri gupta, 2011 Furtado et al., 2006
O/w dispersion method	The drug is dispersed in form of solid powder in the solution of polymer and organic solvent	Peptide	Herrmann and Bodmeier, 1998, Reithmeier et al., 2001
O/o non-aqueous solvent evaporation method	The aqueous phase is replaced by oil	Peptide	Herrmann and Bodmeier, 1998

TABLE 7.4 Different Drugs Synthesized Using the Solvent Evaporation Method.

Drug	Therapeutic activity	Reference
Rivastigmine	Treating Alzheimer disease	Jose et al., 2014
Risperidone	Antipsychotic	Sonal et al., 2011
Temozolomide	Chemotherapeutic agents for glioblastoma multiforme	Chooi and Ing, 2016
Dexamethazone	Antitumor	Zharapova et al., 2012

7.2.3 EMULSION COACERVATION METHOD

This process involves the solvent (organic) phase emulsification with a non-solvent (aqueous) phase. This is performed by stirring or ultrasound. After emulsifying, coacervation is carried out using electrolytes, any dehydration agent, or by varying the temperature (Krause and Rohdewald, 1985; Lutter et al., 2008). This step is followed by cross-linking in order to obtain thin solvated nanocapsule shells (Gander et al., 2002). The use of electrolyte dehydration of varying the optimum temperature and pH is taken into consideration to reduce the polymer solvation capacity. There are other factors apart from these that can be used to carry out coacervation process (Lutter et al., 2008). Figure 7.2 depicts the preparative method of emulsion coacervation. Different drugs used as active ingredient, solvents, and non-solvent phases used in emulsion coacervation method has been tabulated in Table 7.5.

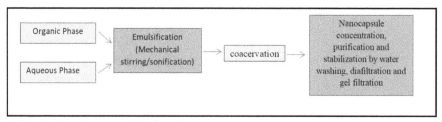

FIGURE 7.2   Flow chart showing the preparative method of emulsion coacervation.

TABLE 7.5   Different Drugs Used as Active Ingredient, Solvents, and Non-solvent Phases Used in Emulsion Coacervation Method.

Active ingredient	Polymer	Solvent	Non-solvent	Reference
Turmeric oil	Sodium alginate Sodium alginate-chitosan	Turmeric oil, ethanol or acetone	Sodium alginate polysorbate 80	Lertsuttiwong et al., 2008a
			Chitosan acetic acid water	Lertsuttiwong et al., 2008b
Triamcinolone acetonide	Swine skin gelatin type ii	Chloroform	Desolvation agents: sodium sulfate and isopropanol g	Krause and rohdewald, 1985
Hydrogen tetrachloroaureate	Poly (1,4 butadiene) (pb)-block-poly-styrene (ps)-block-poly(ethylene oxide) (peo) triblock terpolymer	W/o micro-emulsion of the pseudo-ternary system water/sds/ xylene– pentanol		Lutter et al., 2008

7.2.4 EMULSION DIFFUSION METHOD

This method of nanocapsule synthesis consists of three different working phases: solvent (organic phase), non-solvent (aqueous phase), and dilution phase (Quintanar et al., 1998b, 2005). The lipophilic and hydrophilic drug can be encapsulated using this method. The process consists of addition of solvent phase to the non-solvent phase with high agitation resulting in the formation of an emulsion. Water is added to the resulting emulsion making the solvent to diffuse completely in the solution or the external phase. The nanocapsule synthesized using this method depends on the rate of agitation used for the formation of the emulsion, the chemical composition of the solvent phase, the outer shell properties, and its final concentration followed by the size of the final emulsion (Guinebretière, 2001; Moinard-Chécot et al., 2008). The mean size of the nanocapsule produced is usually around 425–450 nm. This size is usually smaller than the emulsion droplet size. There are various modifications to the current emulsion diffusion methodology, which has been practiced by several researchers for better results (Ma et al., 2001; Perez et al., 2001; Hassou, 2007; Moinard-Chécot et al., 2008). Different drugs used as an active ingredient, solvents, and dilution phases used in the emulsion diffusion method have been tabulated in Table 7.6. Figure 7.3 depicts the flow chart involved in the preparative method of emulsion diffusion.

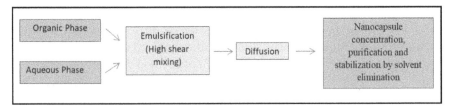

FIGURE 7.3 Flow chart showing the preparative method of emulsion diffusion.

TABLE 7.6 Different Drugs Used as Active Ingredient, Solvents, and Dilution Phases Used in Emulsion Diffusion Method.

Active ingredient	Polymer	Solvent 1	Solvent 2	Dilution phase	Reference
Indomethacine	Pcl/pla	Ethyl acetate	Water	Water	Limayem et al., 2004
Chlorambucil Clofibrate Vitamin e	Pla/ Eudragit	Ethyl acetate, propylene carbonate or benzyl alcohol	Water	Water	Quintanar et al., 1998b

TABLE 7.6   *(Continued)*

Active ingredient	Polymer	Solvent 1	Solvent 2	Dilution phase	Reference
Eugenol	Pcl	Ethyl acetate	Water	Water	Choi et al., 2009
Hinokitiol	Pcl	Ethyl acetate	Water	Water	Joo et al., 2008
4-nitroanisole	Pla	Dichloro-methane acetone	Water	Polyvinylal-cohol aqueous solution	Romero-cano and vincent, 2002
Sudan iii	Pcl/pla	Ethyl acetate	Water	Water	Chécot et al., 2008

*Pcl, poly (lactic acid); Pla, polycaprolactone.

7.2.5   DOUBLE EMULSIFICATION METHOD

This is a complex system, also termed as emulsions of emulsions (Garti, 1997; Grigoriev and Miller, 2009). It can be classified as either water/oil/water or oil/water/oil type. This presents many possibilities of controlled release of the drug from the droplet in the core. It is a more stable way of synthesizing nanocapsules. It is a two-step process of emulsification where surfactants of two different nature is are used: hydrophilic and hydrophobic (Khoee and Yaghoobian, 2008). Surfactants play an important role in emulsions. It behaves as a barrier for the release of drug toward the internal surface and can also behave as a steric stabilizer toward the external surface. The nanocapsule synthesized as a result of this method depends on water/oil emulsion concentration and the stabilizing agent used for the process. The mean particle size of the resultant nanocapsule is on an average between 150–200 nm. For nanocapsule synthesis by double emulsification, the primary emulsion formed because of ultrasound is stabilized by the addition of a stabilising agent to it. At the end of the dispersion process, the solvents are removed by either extraction or evaporation causing the nanocapsules to be hardened in the non-solvent (aqueous) medium (Bilati et al., 2005). The preparative method of double emulsion has been depicted in Figure 7.4. The different drugs used as an active ingredient, solvents and dilution phases used in double emulsion method has been tabulated in Table 7.7. The different drugs for brain targeting synthesized using the double emulsification method has been listed in Table 7.8.

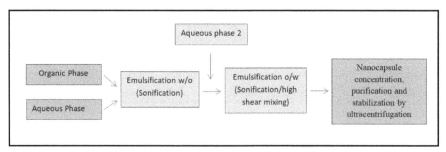

FIGURE 7.4   Flow chart showing the preparative method of double emulsion.

TABLE 7.7   Different Drugs Used as Active Ingredient, Solvents, and Dilution Phases Used in Double Emulsion Method.

Active ingredient	Polymer	Solvent 1	Solvent 2	Dilution phase	Reference
Plasmid DNA	Pla–Peg	Water	Methylene chloride	Water Ethanol	Perez et al., 2001
Insulin	Pla–peg–pla copolymers	Hydro-chloric acid	Acetone	Polysorbate 20, Dextrin Water	Ma et al., 2001
Ciprofloxacin.hci	Plga	Active ingredient	PVA		Jeong et al., 2008
Bovine serum albumin	Pla/pcl–peo block copo-lymer sorbitan monooleate methylene chloride	Protein Water	Polysorbate 80 glycerin:water (1:1)	Water	Lu et al., 1999
Tetanus toxoid	Pla/plga Ethyl acetate Methylene chloride	Protein water protein–sodium oleate water	PVA Water	PVA aqueous solution	Bilati et al. 2005a

*Pla, poly(lactic acid); Peg, poly(ethylene oxide); Pcl, poly(epsilon-caprolactone); Peo, poly(ethylene glycol); Plga, poly (D,L-lactic-co-glycolicacid).

TABLE 7.8   Different Drugs Synthesized Using the Double Emulsification Method.

Drug	Therapeutic activity	Reference
Triptorelin	Antitumor	Nicoli et al., 2001
Doxorubicin	Antitumor antibiotic	Jiang et al., 2011
Epirubicin	Anthracycline antibiotic	Zhou et al., 2006

7.2.6   POLYMER-COATING METHOD

Various methodologies have been formulated to deposit a thin layer of polymer coat on the surface of the nanocapsule. The formation of the film is brought about by directly absorbing the desired polymer onto the nanoparticle and incubating it with polymer dispersion and constantly stirring (Calvo et al., 1997). This method has also been modified by many researchers to get better outcome. One such modification is the preparation of a nanoemulsion template, which is coated by a polymer on the water/oil interface. To this solution, the polymeric materials are added which later gets precipitated against the nanoemulsion by a process known as solvent evaporation. Another such modification has been carried out where nanocapsules are fabricated with the help of sonification of the water/oil emulsion. The resultant is coated with the desired polymer and dichloromethane solution (Anton et al., 2008). Example of a drug synthesized using the polymer coating method is Amphotericin b. It is used for treating neurodegenerative diseases (Saad et al., 2012). Table 7.9 depicts the different drugs used as active ingredient, organic and inorganic phases in polymer coating method. Figure 7.5 shows the preparative method of polymer coating nanocapsule.

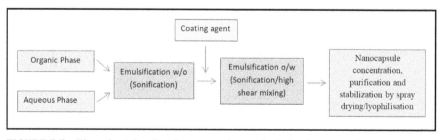

FIGURE 7.5   Flow chart showing the preparative method of polymer coating.

TABLE 7.9 Different Drugs Used as Active Ingredient, Organic and Inorganic Phases Used in Polymer Coating Method.

Active ingredient	Organic phase	Aqueous phase	Reference
Nanoemulsion salmon calcitonin	Active ingredient capric/ caprylic triglycerides Ethanol Soybean lecithin acetone	Poloxamer 188-water	Prego et al., 2006
Indomethacin	Pcl Capric/caprylic triglycerides Lecithin Acetone	Poloxamer 188-water	Calvo et al., 1997
Tetanus toxoid	Pla/lecithin/ethyl acetate or plgaa/lecithin/ethyl acetate		Vila et al., 2002

Pla, poly(lactic acid); Pcl, poly(epsilon-caprolactone); Plgaa, polylactic-co-glycolic acid.

7.2.7 LAYER BY LAYER NANOCAPSULE SYNTHESIS

This is carried out as a final step of all the classical methods of nanocapsule synthesis (double emulsification and nanoprecipitation). As the name suggests, in this step, the desired polymer is added layer by layer to the external non-solvent (aqueous) medium and an extra layer to form simultaneously. This is accompanied by the precipitation of the negatively charged polymer and solvent diffusion (Calvo et al., 1997; Radtchenko et al., 2002a; Vila et al., 2002). This kind of an assembly process was first developed by Sukhorukov et al. (1998). This method is used for the preparation of the colloidal particles called polyelectrolyte capsules. These capsules have well-defined chemical and structural properties. The adsorption of oppositely charged particles can be performed on the colloidal particle surface, whereas, the inner core is loaded with the therapeutic drug of interest (Antipov et al., 2002; Ai and Gao, 2004; Fan et al., 2002; Radtchenko et al., 2002b; Krol et al., 2004; Cui et al., 2009). Figure 7.6 depicts the preparative method of the layer by layer nanocapsules.

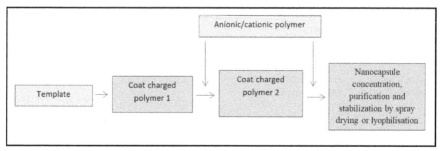

FIGURE 7.6   Flow chart showing the preparative method layer-by-layer nanocapsules.

7.3   MODIFICATIONS OF THE EXISTING METHODS OF SYNTHESISING NANOCAPSULES

McDonald et al. (2000) modified emulsion polymerization technique by using a water-soluble alcohol and a non-solvent hydrocarbon. As a result, the polymer could affect the formation of microvoids, hollow monodispersed particles, and its morphology. Hydrocarbon encapsulation can take place only if low molecular weight polymer is made at the beginning of the process.

Another modified method is the synthesis of alkyl-cyanoacrylate nanocapsules. These nanocapsules have thinner walls and its structure has high reproducibility. The capsule means size depends on the concentrations of the monomer components and the oil. Florence et al. (1979) and Wohlgemuth et al. (2000) were the first to describe the nanocapsules synthesis using polyalkylcyanoacrylates, by interfacial polymerization. Apart from these, there are other techniques used for the preparation of polyalkylcyanoacrylatenanocapsules (Al Khouri Fallouh et al., 1992).

Berg et al. (1989) had testified the synthesis of microcapsules by methyl methacrylate polymerization reaction in the presence of water/oil emulsion. The oil phase comprises an alkane and the water/oil emulsion was stabilized by a number of emulsifiers (Okubo et al., 1999). Oil and water-soluble initiators were used for the process. The monomer was introduced into the emulsion through the water phase. This method can be used to study the morphological characteristics of polymeric nanoparticles ranging from 1–100 μm in size.

Okubo et al. (1999) evaluated the penetration and/or release behaviors of various solvents into and/or from the interior of micrometer-sized monodisperse cross-linked polystyrene/polydivinylbenzene composite particles. The hollow particles were produced by the seeded polymerization utilizing the dynamic swelling method. Itou et al. (1999) prepared the crosslinked hollow polymer particles in the sub-micrometer size by means of a seeded emulsion polymerization. The morphology of the particles depends on the composition of divinylbenzene and methyl methacrylate.

7.4   BLOOD–BRAIN BARRIER

The blood–brain barrier (BBB) is a semipermeable active diffusion barrier lined by the endothelial cells, astrocytes and pericytes and is well connected via tight junctions, where increased electrical resistance is present, thereby obstructing the influx of most of the compounds from blood and other connective tissue to the brain (Butte, 1990; Schlosshauer, 1993). Adherent junction proteins function in cellular adherence. Astrocytes help in transportation of molecules over the endothelial cells (Roney et al., 2005; Bernacki et al., 2008; Birst 2011; Haque et al., 2012). P-glycoproteins present on the apical endothelial cell layer transport molecules from the brain to the circulatory system. Foot projections from astrocytes form a complex system by enveloping the vessels and function in the stimulation and sustenance of the hindrance properties. Whereas, the axonal projections from the neurons to the arterial smooth muscle comprises peptides, neurotransmitters, etc., and control the cerebral blood. The tight endothelial junctions are just about 100× tighter than the narrow endothelial junctions. It permits the inflow of water molecules, gases, and few lipid soluble molecules via passive diffusion (Pardridge, 2007; Gregoriadis, 2008). Certain molecules which are important for the proper functioning of CNS, for example, glucose and certain amino acids cross the BBB through selective transport mechanism. A few regions in the brain do not have a BBB. An effort to deliver the necessary therapeutic drug to the target brain cell/tissue across the BBB requires it to go against the normal functioning of the brain. Regardless of the development in the field of medical sciences and nanomedicine alongside the rapid advancement in our knowledge of the various BBB components, its molecular structure, its receptor-ligand expression, most of the diseases associated with the brain remain untreated because of the inability of many therapeutic drugs to cross the BBB. This challenge can be overcome by gaining insights of the BBB, its physiology, its response under the influence of physical and chemical stimulus, presence of the surface receptors and drug carrier technologies. Effective methodologies or techniques should be carefully considered and evaluated with regards to its influence on the overall function of BBB.

7.5   THE BBB PHYSIOLOGY AND BIOLOGY

The brain is well secured and actively regulated to preserve the microenvironment of the central nervous system (CNS). The two most important opening to pass through the brain parenchyma are the cerebral blood and the cerebro-spinal fluid (CSF). In humans, the brain consists of more than 100 billion capillaries on whole, providing a total surface area of 20 m² (Yan et al., 2012). Entry of any substances into the brain is stringently regulated by the BBB and the blood CSF barrier. The surface of brain CSF barrier is filled with CSF and not blood (Rip et al., 2009). This along with the expression of certain enzymes like peptidases and cytosolic enzymes leads to constant flushing of the injected molecule back to the blood, thereby protecting the brain microenvironment (Nabel et al., 1993; Budai et al., 2001; Pathan et al., 2009). The BBB occurs along all capillaries and consists of tight junctions around the capillaries that do not exist in normal circulation (De Vries et al., 1997). Endothelial cells curb the diffusion of minuscule objects (like bacteria) and large or hydrophilic particles, whereas, it allows the diffusion of hydrophobic molecules (some gases, water, lipid solvent substances, and hormones) into the CSF. It behaves actively as an to the brain from various bacterial diseases. Because of the presence of the CSF, brain contaminations are very rare. Nevertheless, because certain antibodies are too big in size to cross the BBB, certain diseases which do take place become severe and complicated to treat. Viruses have the potential to effectively cross the BBB and assemble themselves to the circulating immune cells (Shehata et al., 2008; Gupta et al., 2011; Ishii et al., 2013; Rim et al., 2013). Transport of such particles is guarded because of the reduced rate of trans-cytotic vesicles and pinocytotic action. Other reasons which can be stated are the increased rate of metabolic action of transporters and cytosolic compounds along with the active efflux pump of p-glycoprotein (Kaur et al., 2008). The BBB is therefore well thought-out as the most important brain barrier in preventing the inflow of any toxic molecules to the brain (Kumaresh et al., 2001; Gubha et al., 2004; Mori et al., 2014).

The exceptional biological features of the BBB that allows it to function as the physical, transport, metabolic, immunological, or enzymatic barrier are listed below as the following:

1)   Selective permeability to the molecules which are effluxing from the blood through the endothelial cells of the brain.
2)   Lack of apertures, the presence of fewer pinocytes, a larger number of mitochondria present in the endothelial cells (Stewart, 2000).

3) Presence of tight junctions between the adjacent endothelial cells of the brain parenchyma, formed by transmembrane adhesion molecules like occludin, and claudins, along with the cytoplasmic accessory proteins like zonulaoccludens 1/2, cingulin, AF-6, and 7H6 (Persidsky et al., 2006).

4) Presence and expression of carrier-mediated transport system such as GLUT1 (glucose carrier), LAT1 (amino acid carrier), efflux transporters (p-glycoprotein), MRP (multidrug resistance-associated proteins), insulin and lipoprotein receptors (Abbott et al., 2006).

5) Synergistic functions and regulation of BBB by tight junctions encompassing pericytes, perivascular macrophages, perivascular end feet, astrocytes and neurons (Dohgu et al., 2005).

6) The BBB is made out of an association between the endothelial cells of the brain along with the macrophages and the mast cells. It also has a stringent limit and selective permeability for the passage of immune cells (lymphocytes) (Williams et al., 2001).

7.6   NANOCAPSULES FOR BRAIN DRUG DELIVERY

Nanocapsules are the emerging class of drug carrier systems that can be easily designed to deliver therapeutic drugs to treat several parts of the body, along with the brain. Neural disorders are one of the major reasons for the increase in the rate of mortality. It comprises an average of 12% deaths worldwide (Zambaux et al., 1998). Amongst such disorders, Alzheimer's and various classes of dementia are considered to be the cause of 2.84% of the aggregate deaths, whereas, the cerebral vascular disease comprises around 8% of the total deaths in the metropolitan nations in 2005 (Chen et al., 2004). Nanocapsules in brain targeting have become an area of research, attracting increasing attention because of its use in drug transport across the BBB. This has been possible only because of the swift increase in our understanding about the brain receptors and the fast advancement in the field of polymer chemistry and nanobiotechnology. Nanocapsules are unique nanocarriers because of their discrete size, the material used and easily reproducible structures. They can be made to behave like macromolecules in certain environments. They can carry more concentration of the therapeutic drug in greater load and are capable of monitoring drug release. Nanocapsules can be made to carry an array of drugs on their surface. The surface properties can be modified according to the need. Such a gradient should support and improve the passive diffusion of the therapeutic drug (Bhandare et al., 2014; Dennis et al., 2015; Khetawat et al., 2015; Lloyd-Hughes et al., 2015; Matilda et al., 2015; Nikalje et al., 2015; Rakesh et al., 2015; Singh et al., 2015; Upadhyay et al., 2015). Such properties make nanocapsules an attractive alternate for transporting the drug across the BBB to treat various diseases and to reduce the mortality rate caused due to neurological diseases worldwide (Patel et al., 2015; Maroof et al., 2016).

The use of nanocapsules can improve the drug delivery to the target brain cell in several ways. A number of studies have been reported about the use of an array of nanocapsules (most of which are lipid soluble) in the delivery of a wide range of therapeutic or diagnostic drugs for the improvement of *in vitro* and *in vivo* BBB permeability and thereby, drug accumulation in the brain. Many of the drugs that are delivered are substrates for ABC transporters but they have weak permeability to cross BBB. Examples of such drugs are doxorubicin, rhodamine, fluorescein, vinblastine, methotrexate, etc. (Krishna and Mayer, 2000). But when the above-mentioned drugs are encapsulated in nanocapsules as the active ingredient, they are able to accomplish the desired therapeutic values for the treatment of the different CNS ailments. It has been observed that such encapsulated drug can actively work in lesser dose and with reduced length of the therapy when compared to the normal drug when administered in their true form. This approach reduces the adverse side effects of the drug (Ryan et al., 2008). Because only the carrier needs to be engineered without any modification of the drug, drug pharmacology can time and again be circumvented. In addition, nanocapsules can be engineered to achieve the required pharmacokinetic and biodistribution profiles for the optimum treatment of the CNS.

7.7 IDEAL PROPERTIES OF NANOCAPSULES AS THE NANOCARRIER SYSTEM

Ideal properties of nanocarriers for the drug delivery across the BBB are listed below (Lockman et al., 2002; Jean-Christophe Olivier, 2005; Béduneau et al., 2007):

1. Safe, biodegradable and biocompatible.
2. Particle size less than 100 nm (except if the drug transport is via monocytes or macrophages).
3. Stable in blood (no aggregation and dissociation).
4. Extended blood circulation time and no opsonization.
5. Non-immunogenic/non-thrombogenic/non-inflammatory.

6. BBB-targeted moiety (receptor or adsorptive mediated transport mechanism across the brain parenchyma, or uptake by monocytes or macrophages).
7. Well maintained parent drug stability.
8. Comparatively cheap in terms of manufacturing procedures.
9. Tuneable drug release profiles.
10. Applicable to carry small molecules, proteins, peptides, or nucleic acids.

Despite a large variety of nanocarriers developed so far, it is noteworthy that only amphiphilic molecule-formed liposomes and polymeric nanoparticles have been extensively exploited for brain drug delivery. Several such systems are now in clinical trials for anticancer drug delivery. Non-amphiphilic colloidal drug carrier systems such as dendrimers and microemulsions are still at relatively early development stage.

7.8 TYPES OF NANOCAPSULES USED IN BRAIN TARGETING

Varieties of nanocarrier systems have been studied to improve the knowledge of therapeutic drug delivery to the brain in order to treat neurological disorders. These nanocarriers fall into a few wide-ranging categories: polymer or dendrimer based (e.g., acrylicpolymers, polyesters of lactide units, and pluronic block copolymers), micelle-based and lipid-based (Wong et al., 2010).

7.8.1 POLYMER-BASED NANOCAPULES

For the past few decades, polymeric nanocapsules have been gaining attention in the field of brain targeting. They are microsized transporters, ranging from 1 to 1000 nm. This nanocarrier system is made of either natural or synthetic polymers, within which the desired drug can be encapsulated either in solid or in liquid state (mixture or solvents), or chemically connected to the surface or directly adsorbed (Morris et al., 2012). Most commonly used polymers are poly(lactide-co-glycolides), polylactides, polyglycolides, polycyanoacrylates, polyanhydrides, and polycaprolactone. Chitosan can also be used for the preparation of nanocapsules. Most commonly used method for the preparation of polymer-based nanocapsule is the polymerization of the monomer units into the dispersed phase of the emulsion

(Vijaya et al., 2011). Chen et al. (2004) has specified that polymeric nano-capsules are suitable carrier systems for brain targeting. Major properties of the polymeric nanocapsule are the following (Guo et al., 2015):

- Improved stability.
- Efficiently manufactured via number of strategies.
- Higher dose of therapeutic agent can be delivered to the brain across the BBB.

7.8.2   DENDRIMER-BASED NANOCAPSULE

Dendrimers consist of repeating units of branched monomeric structures. These are highly branched nanocarrier structures. When engineered with precision, they can form spherical structures of diameter ranging from 1–10 nm. They appear symmetric at the center but when expanded it takes a shape of a 3D structure in water. These structures may have void space within or functional groups attached to the surface for drug molecules to be conjugated or encapsulated. Dendrimeric nanocapsules have been reported to have higher BBB permeability when compared to its free drug solution. Some of the noted properties of the dendrimers are the following (Laine et al., 2012; Ostafin et al., 2012; Wang et al., 2013; Aghajanloo et al., 2014; Ahmed and Gao, 2014; Oliveira, 2014; Santos and Oliveira, 2014)

- They can be exploited for their spreading nature.
- They can bear more than one dose of a particular therapeutic drug to the brain.
- Stacking limit is high compared to other nanocarrier systems.
- Reduced toxicity.

Although the drug kinetics is unpredictable and its long-term viability profile is less known when compared to the polymeric nanocapsules (Dutta and Jain, 2007).

7.8.3   MICELLE-BASED NANOCAPSULE

A micelle is a structure made from 50 to 100 amphiphilic water dispersible molecules, which includes surfactants, block copolymers, etc., with their size ranging from 5 to 20 nm in diameter. In an aqueous solution, these amphiphilic molecules aggregate and position themselves in such a way that their hydrophilic heads are exposed to the outside and their hydrophobic part is toward the inner core region. Pluronic micelles have high effect in drug transport across BBB when studied *in vitro* and *in vivo*. Important properties of micelle, for which it is treated as a nanocarrier system are as follows (Batrakova et al., 2003; Menaa F, 2015):

- Small size,
- Good drug molecule solubilising,
- Drug permeability enhancement,
- Improved stability by cross linking shell and core chain,
- They can be made responsive to the external stimulus like pH, light, temperature, chemical gradient, ultrasound, etc.

7.8.4 LIPOSOMES

Liposomes are made of phospholipid bilayers (El-Said et al., 2013; Gowda et al., 2013). They can also be prepared from cholesterol and natural nontoxic phospholipid. They have been reported to be used for the delivery of antiretroviral drugs like stavudine and zidovudine, which was designed to treat HIV-associated neurological illnesses. Liposomes coupled with brain targeting peptides are used to deliver amphotericin B across the BBB model made of rat brain endothelial cells (Zhang et al., 2003). Some important characteristics of liposomes are as follows (Fresta et al., 1994; Anda et al., 1995):

- Long term usage,
- Low toxicity,
- Biocompatibility and biodegradability,
- Deliver lipophilic and hydrophilic compounds most efficient nanocarrier system,
- Used for the treatment of brain tumors and cerebral ischemia,
- Used to deliver antiretroviral drug and transport of opioid peptides.

Despite of being the best nanocarrier system for brain targeting, liposomes do have some limitations—speedy degradation of phospholipids, instability, cannot be stored for an extended period, quick elimination from the system etc. Some of these problems have been resolved by modifying the surface of these liposomes by using hydrophilic polymer like polyethylene glycol and conjugated phospholipids. Because of these modifications, liposomes display prolonged blood circulation, increased *in vivo* residential time and improved drug half-lives (Woodle, 1995).

7.8.5   CATIONIC LIPOSOMES

Cationic liposomes are a special type of nanocarrier system which contains positively charged lipids. They were first produced and exploited as transfection vehicles to transport genetic material into the cell. This was strategically prevented from the lysosomal degradation. In one case study, cationic liposome uses bolaphiles, containing hydrophilic groups enveloping a hydrophobic chain. Bolaphiles can cross the BBB, and allow controlled discharge of the drug to the desired target (Skaat et al., 2013)

7.8.6   SOLID LIPID NANOPARTICLES

These nanocapsules are made of stable hydrophobic lipid center (fatty acids, triglycerides, or waxes), in which the therapeutic drug is dispersed. The small size of such nanoparticles allows it to cross the BBB (Bhattarai et al., 2013; Toffoli et al., 2013). The mean size ranges from 40–1000 nm (Lucks and Muller, 1996). They are prepared through high-pressure homogenization or micro-emulsification.

7.9   MECHANISM OF NANOCAPSULE-MEDIATED BRAIN TARGETING

Figure 7.7 represents the various transport mechanism of the nanocapsule-mediated drug delivery to the brain across the BBB (Wong et al., 2010; Salunkhe et al., 2015)

1) Increase the drug gradient locally at the BBB by the process of passive targeting.
2) Allow the encapsulated drug-trafficking by general or receptor-mediated endocytosis (2a, 2b).
3) Block the nanocapsule efflux transporters at the periphery of the BBB.

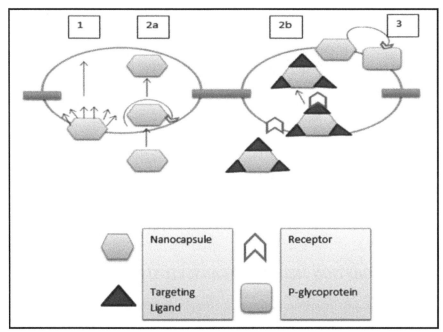

FIGURE 7.7 (See color insert.) Schematic diagram of different mechanisms employed in the brain targeting using nanocapsules.

7.9.1 LIPID-MEDIATED ENDOCYTOSIS

Liposomes are transported across the BBB through the process of lipid mediated free facilitated diffusion. It is also called lipid mediated endocytosis (Amiji et al., 2013). Lipoprotein receptors are present which binds to the substrate (lipoproteins) and form complexes which facilitate the transport of liposome nanocapsules across the BBB. Example: Apolipoprotein E (apoE) binds to the nanocapsules, thereby forming a complex which binds to the low-density lipoprotein receptor present on the BBB. This facilitates the transport of lipids and cholesterol across BBB.

7.9.2 RECEPTOR-MEDIATED ENDOCYTOSIS AND TRANSCYTOSIS

Polymer-based nanocapsules are transported across BBB by the endothelial cells present on the brain capillary, via the process of receptor-mediated endocytosis. This is followed by transcytosis, where the nanocapsules cross the tight junctions of the endothelial cells to enter the brain parenchyma.

Surface modification of the nanocapsules has been reported to increase the therapeutic drug uptake by the target brain cell. The surface coating can be performed with surfactants like poloxamer 188/polysorbate 80. Receptor-mediated endocytosis depends on the receptors that are present on the luminal surface of the BBB. Ligands coated on the surface of the nanocapsules binds to its specific receptor, causing a conformational change, followed by the start of transcytosis process. In this process, as soon as the encapsulated drug is internalized, the plasma membrane is pinched off to form a vesicle (Kreuter et al., 2013; Masserini, 2013). Other receptors which take part in receptor mediated endocytosis are transferrin receptor, SR-BI (scavenger receptor class B type I) and insulin receptor. Endocytosis can take place as long as the receptor (present on the luminal surface of BBB) is bound to its specific ligand (present on the nanocapsule surface).

7.9.3   ADSORPTION-MEDIATED TRANSCYTOSIS

In this process, the transportation is facilitated by the electrostatic interaction between the nanocapsules and the BBB. Cationic nanocapsules and cationic liposomes are used for this specific method because of the presence of the positive charge which facilitates the binding to the endothelial cells. TAT peptides can be used to improve the drug transportation into the brain as they function as cell penetrating peptides on the surface of the cationic nanocapsules and liposomes.

7.9.4   APPLICATION OF A MAGNETIC FIELD GRADIENT

Nanocapsules can be transported across the BBB by the application of the magnetic field. These capsules can be made to enter the brain or removed by simply controlling the direction of the magnetic field gradient. For this mechanism to take place, the nanocapsule must possess a non-zero magnetic moment and should have a diameter of less than 50 nm (Nair et al., 2013; Guduru et al., 2015).

7.10   DRAWBACK OF NANOCAPSULES AS THE BRAIN TARGETING AGENT

Despite having numerous advantages as the ultimate drug delivery system, it has some disadvantages few are listed below:

1) Physical handling of the nanocapsules is difficult, when present in liquids and in dry form.
2) Because of the smaller size and larger surface area, nanocapsules may tend to form aggregates and ultimately causing its accumulation in the body, toxic effects of which is not clearly understood till date.

7.11 FUTURE PROSPECTS OF BRAIN TARGETING BY NANOCAPSULES

The BBB inhibits the passage of many drugs which creates a substantial threat for the targeted therapy. It is one of the limiting factors for the development of new therapeutic drugs for the treatment of the brain diseases. Presently, the nanocapsules are used as a therapeutic vehicle, providing a significant strategy to pass through the BBB and serve as a targeted drug delivery. The major advantage of using nanocapsules for targeted treatment of brain is that it carries the properties of the drug molecule. This helps in the reduction of the peripheral toxicity caused due to leaching. Nanocapsules are reported to be used in the treatment of brain tumors, Alzheimer disease and other neurodegenerative diseases. Brain targeting by nanocapsules rely totally on the interaction between the specific receptors on the molecule and the BBB. This could be a potential area of nanopharmacology research. Existing nanocapsule technology requires adequate research before it can be used for clinical application.

7.12 CONCLUSION

From the above discussion, it is presumed that nanocapsules are used to transport therapeutic drugs across the BBB in order to treat various neurological diseases. At present, there are many medications which are used to treat the ailments. Although the available clinical data is limited for the complete treatment of the disorder, evidence suggests that nanocapsules have the potential to be used as medication from a clinical point of view due to features such as minimized reactivity, low dose of the drug, greater half-life and improved permeability of the drug across BBB. Because the field of nanomedicine is in the developing stage, the long-term health effects of these nanocapsules are unknown. More study on the drug pharmacokinetics, toxicity and its role in therapeutics must be well ascertained before clinical trials.

In the current article, we have provided the different techniques which can be used in brain targeting and to improve the drug delivery across BBB in CNS. Several types of nanocapsules have been used for drug delivery like polymer based nanocapsule, liposome, micelle dendrimer, etc., all of these mentioned above have been reported in the treatment of diseases such as Parkinson's diseases, Alzheimer's, Epilepsy, etc. Modifications of the existing methods have also been discussed in this review for a better understanding of the nanocapsule, its preparative methods and its possible applications in the field of brain drug targeting. There are many studies which show the use of nanocapsule as a nanocarrier system and as a platform to deliver the drugs appropriately to the target cell in the brain. Drugs like tacrine, quinolone, rivastigmine, etc., have been studied and used as the active ingredient in the nanocapsule for the treatment of Alzheimer's disease. Post injecting the drugs, the results showed remarkable progress in the effects of the drugs when compared to its non-nanocarrier system. Thereby, suggesting that nanocapsules can be used as a promising tool and a solution to treat drugs by crossing the BBB. One factor which needs to be given special attention is the accumulation of the drugs in the body and its health effects on a long term basis. This area of concern must be further analysed properly in order to improve them.

KEYWORDS

- **brain targeting**
- **blood–brain barrier**
- **drug delivery**
- **encapsulation**
- **nanocapsule**
- **image augmentation**

REFERENCES

Acharya, S. R.; Reddy, P. R. V.; Brain-targeted Delivery of Paclitaxel Using Endogenous Ligand. *AJPS* **2016,** *11* (3), 427–438.

Abbott, N. J., Ronnback, L.; Hansson, E.; Astrocyte-Endothelial Interactions at the Blood–Brain Barrier. *Nat. Rev. Neurosci.* **2006,** *7*, 41–53.

Aghajanloo, M.; Rashidi, A. M. ; Moosavian M. A.; Synthesis of Zinc- Organic Frameworks Nano Adsorbent and Their Application for Methane Adsorption. *J. Chem. Eng. Process Technol.* **2014,** *5* (5), 1000203.

Ahmed, S. S.; Gao, G.; Examination of the Blood Brain Barrier Integrity in a Mouse Model of the Neurodegenerative Canavan's Disease. *J Neurol Disord.* **2014,** *2* (6), 10.4172/2329-6895.1000i105.

Ai, H.; Gao J.; Size-Controlled Polyelectrolyte Nanocapsules Via Layer-By-Layer Self-Assembly. *J. Mater. Sci.* **2004,** *39*, 1429–1432.

Al Khouri Fallouh, N.; Roblot-Treupel, L.; Fessi, H.; Development of a New Process for the Manufacture of Polyisobutylcyanoacrylate Nanocapsules. *Int. J. Pharm.* **1986,** *28*, 125–132.

Alyautdin, R. N.; Petrov, V. E.; Langer, K.; Berthold, A.; Kharkevich, D. A.; Kreuter, J.; Delivery of Loperamide Across the Blood-Brain Barrier with Polysorbate 80-Coated Poly-butylcyanoacrylate Nanoparticles. *Pharm. Res.* **1997,** *14*, 325–328.

Alyautdin, R. N.; Tezikov, E. B.; Ramge, P., Kharkevich, D. A.; Begley, D. J., Kreuter, J.; Significant Entry of Tubocurarine into the Brain of Rats by Adsorption to Polysorbate Study. *J. Microencapsul.* **1998,** *15*, 67–74.

Anda, T.; Yamashita, H.; Fujita, H.; Tokunaga, Y.; Shibata, S.; Basic Experiment of BBB Permeability of Cisplatin Encapsulated in Liposome By Means of In Vitro BBB. *DrugDe-livSyst* **1995,** *10*, 425–430.

Antipov, A. A.; Sukhorukov, G. B.; Leporatti, S., Radtchenko, I. L.; Donath, E.; Mohwald, H.; Polyelectrolyte Multilayer Capsule Permeability Control. *Colloid Surf.* **2002,** *198–200*, 535–541.

Anton, N.; Benoit, J. P.; Saulnier, P.; Design and Production of Nanoparticles Formulated from Nanoemulsion Templates—A Review. *J. Control. Release* **2008,** *128*, 185–199.

Azevedo, A. F.; Galhardas, J.; Cunha, A.; Cruz, P.; Goncalves, L. M. D.; Almeida, A J.; Microencapsulation of *Streptococcus equi* Antigensin Biodegradable Microspheres and Preliminary Immunisation Studies. *Eur. J. Pharm. Biopharm.* **2006,** *64*,131–137.

Barratt, G. M.;Therapeutic Applications of Colloidal Drug Carriers. *PSTT* **2000,** *3*, 163–71.

Batrakova, E. V.; Li, S., Alakhov, V. Y.; Miller, D. W.; Kabanov, A. V.; Optimal Structure Requirements for Pluronic Block Copolymers in Modifying P-Glycoprotein Drug Efflux Transporter Activity in Bovine Brain Micro Vessel Endothelial Cells. *J. Pharmacol. Exp Ther.* **2003,** *304*, 845–854.

Beduneau, A.; Saulnier, P.; Anton, N.; Hindré, F.; Passirani, C.; Rajerison, H.; Noiret, N.; Benoit, J. P.; Pegylated Nanocapsules Produced by an Organic Solvent Free Method: Eval-uation of Their Stealth Properties. *Pharm. Res.,* **2006,** *23* (9), 2190–2199.

Berg J.; Sundberg, D; Kronberg, B. J.; Microencapsulation. *Polym. Mater. Sci. Eng.* **1986,** *3*, 327.

Bernacki, J.; Dobrowolska, A.; Nierwińska, K., Małecki, A.; Physiology and Pharmacolog-ical Role of the Blood-Brain Barrier. *Pharmacol Rep.* **2008,** *60* (5), 600–622.

Bhandare, N.; Narayana, A.; Applications of Nanotechnology in Cancer: A Literature Review of Imaging and Treatment. *J. Nucl. Med. Radiat. Ther.,* **2014,** *5*, 195.

Bhattarai, S. R., Bhattarai, N.; Biodegradable and Bioabsorbable Inorganic Particles in Cancer Nanotechnology. *J. Nanomed. Nanotechnol.* **2013,** *4*, 170.

Bilati, U.; Allémann, E.; Doelker, E.; Nanoprecipitation Versus Emulsion-Based Techniques for the Encapsulation of Proteins into Biodegradable Nanoparticles and Process-Related Stability Issues. *AAPS Pharmscitech.* **2005,** *6*, E594–E604.

Birst, R.; Brain Drug Delivery System: A Comprehensive Review on Recent Experimental and Clinical Finding. *IJPSR*, **2011,** *2* (4), 792–806.

Budai, M.; Szogyi, M.; Liposomes as Drug Carrier Systems: Preparation, Classification and Therapeutical Advantages of Liposomes. *Acta Pharm. Hung.* **2001,** *71* (1),114–118.

Butte, A. M.; Jones, H. C.; Abbot, N. J.; Electrical Resistance Across the Blood-Brain Barrier in Anaesthetized Rats: A Developmental Study. *J Physiol.* **1990,** *429,* 47–62.

Calvo, P.; Vila-Jato, J. L.; Alonso, M. J.; Evaluation of Cationic Polymer-Coated Nanocapsules as Ocular Drug Carriers. *Int. J. Pharm.* **1997,** *153,* 41–50.

Chattopadhyay, P.; Shekunov, B. Y.; Seitzinger, J.; Huff, R.; Particles from Supercritical Fluid Extraction of Emulsion. US Patent. 6,998,051, **2006.**

Chécot, F.; Lecommandoux, S.; Klok, H. A.; Gnanou, Y.; From Supramolecular Polymersomes to Stimuli-Responsive Nano-Capsules Based on Poly(Diene-Bpeptide) Diblock Copolymers. *Eur. Phys. J.* **2003,** E *10,* 25–35.

Chen, Y; Dalwadi, G.; Benson, H. A. E.; Drug Delivery Across the Blood-Brain Barrier. *Curr. Drug Deliv.* **2004,** 361–376.

Choi, M. J.; Soottitantawat, A.; Nuchuchua, O.; Gi, M. S.; Ruktanonchai, U.; Physical and Light Oxidative Properties of Eugenol Encapsulated by Molecular Inclusion and Emulsion–Diffusion Method. *Food Res. Int.,* **2009,** *42,* 148–156.

Chorny, M.; Fishbein, I.; Danenberg, H. D.; Golomb, G.; Lipophilic Drug Loaded Nanospheres Prepared by Nanoprecipitation: Effect of Formulation Variables on Size; Drug Recovery and Release Kinetics. *J. Control. Release* **2002,** *83,* 389–400.

Couvreur, P.; Barratt, G.; Fattal, E.; Legrand, P.; Vauthier, C.; Nanocapsule Technology: A Review. *Crit. Rev. Ther Drug Carrier Syst.* **2002,** *19,* 99–134.

Cui, J., Fan, D.; Hao, J.; Magnetic (Mo72Fe30)-Embedded Hybrid Nanocapsules. *J. Colloid Interface Sci.* **2009,** *330,* 488–492.

Dalenc, O. F.; Amjaud, Y.; Lafforgue, C.; Derouin, F.; Fessi, H.; Atovaquone and Rifabutine-Loaded Nanocapsules: Formulation Studies. *Int. J. Pharm.* **1997,** *153,* 127–130.

De Vries, H. E.; Blom-Roosemalen; M. C., Van Oosten, M.; De Boer, A. G.; Van Berkel, T. J.; Breimer, D. D.; Kuiper, J.; The Influence of Cytokines on the Integrity of the Blood-Brain Barrier In Vitro. *J. Neuroimmunol.* **1996,** *64* (1), 37–43.

Dennis, E. et al.; Utilizing Nanotechnology to Combat Malaria. *J. Infect. Dis. Ther.* **2015,** *3,* 229.

Devissaguet, J. P.; Fessi, H.; Puisieux, F.; Process for the Preparation of Dispersible Colloidal Systems of a Substance in the Form of Nanocapsules. US Patent 5049322, **1991.**

Diwan, M.; Park, T. G.; Pegylation Enhances Protein Stability During Encapsulation in PLGA Microspheres. *J. Contr. Rel.* **2001,** *73,* 233–244.

Dohgu, S. et al.; Brain Pericytes Contribute to the Induction and Up-Regulation of Blood–Brain Barrier Functions Through Transforming Growth Factor-Beta Production, *Brain Res.,* **2005,** *1038,* 208–215.

Dutta, T.; Jain, N. K.; Targeting Potential and Anti-HIV Activity of Lamivudine Loaded Mannosylated Poly (Propyleneimine) Dendrimer. *Biochim. Biophys. Acta* **2007,** *1770,* 681–686.

El-Said, N.; Kassem, A. T.; Aly, H. F.; Nanoemulsion for Nanotechnology Size-Controlled Synthesis of pd (ii) Nanoparticles Via Nanoemulsion Liquid Membrane. *J. Membrane Sci. Technol.* **2013,** *3,* 125.

Fan, J., Bozzola, J. J., Gao, Y.; Encapsulation of Uranyl Acetate Molecules Using Hollow Polymer Templates. *J. Colloid Interf. Sci.* **2002,** *254,* 108–112.

Feng, L. et al.; Pharmaceutical and Immunological Evaluation of a Single-Dose Hepatitis B Vaccine Using PLGA Microspheres. *J. Contr. Rel.* **2006,** *112,* 35–42.

Fessi, H.; Puisieux, F.; Devissaguet, J. P.; Procédé de Préparation de Systèmes Colloïdaux Dispersibles d'une Substance Sous Forme de Nanocapsules. *European Patent 274961 A1,* **1988**.

Florence A. T.; Whateley T. L.; Wood D. A.; Potentially Biodegradable Microcapsules with Poly (Alkyl 2-Cyanoacrylate) Membranes. *J. Pharm. Pharmacol.* **1979,** *31,* 422–424.

Fresta, M.; Puglisi G.; Giacomo, C. D.; Russo, A.; Liposomes as In Vivo Carriers for Citicoline: Effects on Rat Cerebral Post-Ischaemic Reperfusion. *J Pharm Pharmacol.* **1994,** *46,* 974–981.

Furtado, S.; Abramson, D.; Simhkay, L.; Wobbekind, D.; Mathiowitz, E.; Subcutaneous Delivery of Insulin Loaded Poly(Fumaric-Co-Sebacic Anhydride) Microspheres to Type 1 Diabetic Rats. *Eur. J. Pharm. Biopharm.,* **2006,** *63,* 229–236.

Gander, B.; Blanco-Príeto, M. J.; Thomasin, C.; Wandrey, Ch.; Hunkeler, D.; Coacervation/ Phase Separation. In: Swarbrick, J., Boylan, J. C.; Eds., *Encyclopedia of Pharmaceutical Technology.* Marcel Dekker, New York, 2002; pp 481–496.

Garti N.; Double Emulsions—Scope, Limitations and New Achievements. *Colloid Surf.,* **1997,** 233–246.

Gowda R. et al.; Use of Nanotechnology to Develop Multi-Drug Inhibitors for Cancer Therapy. *J. Nanomed. Nanotechnol.* **2013,** *4,* 184.

Grigoriev, D., Miller, R.; Mono- and multilayer covered drops as carriers. *Curr. Opin. Colloid Interf. Sci.* **2009,** *14,* 48–59.

Gubha, S., Mandal, B.; Dispersion Polymerization of Acrylamide. *J. Colloid Interface Sci.* **2004,** *271,* 55–59.

Guduru R. et al; Magnetoelectric "Spin" on Stimulating the Brain. *Nanomedicine* **2015,** *10* (13), 2051–2061.

Guinebretière, S.; Nanocapsules Par Emulsion–Diffusion de Solvant: Obtention, Caracterisation et Mecanisme de Formation. *Ph.D. Thesis.* Université Claude Bernard-Lyon 1, **2001**.

Gulyaev, A. E.; Gelperina, S. E.; Skidan, I. N.; Antropov, A. S.; Kivman, G. Y., Kreuter, J.; Significant Transport of Doxorubicin into the Brain with Polysorbate 80-Coated Nanoparticles. *Pharm. Res.* **1999,** *16,* 1564–1569.

Guo, P.; Studies and Application of Nanomotor for Single Pore Sensing, Single Fluorescence Imaging, and RNA Nanotechnology. *Biochem. Anal. Biochem.* **2015,** *4,* i105.

Gupta, M.; Sharma, V.; Targeted Drug Delivery System: A Review. *Res. J. Chem. Sci.* **2011,** *1* (2).

Guterres, S. S. et al.; Polymeric Nanoparticles, Nanospheres and Nanocapsules, for Cutaneous Applications. *Drug Target Insights* **2007,** *2,* 147–157.

Hassou, M.; Modelisation and Simulation de la Formation des Nanocapsules Polymeriques Par la Methode d'Emulsion–Diffusion. Ph.D. Thesis. Université Claude Bernard-Lyon 1, Francia. **2007**.

Haque, S., Md, S., Alam, M. I. et al.; Nanostructure Based Drug Delivery Systems for Brain Targeting. *Drug Dev. Ind. Pharm.* **2012,** *38* (4), 387–411.

Herrmann, J.; Bodmeier, R.; Biodegradable Somatostatin Acetate Containing Microspheres Prepared by Various Aqueous and Non-Aqueous Solvent Evaporation Methods. *Eur. J. Pharm. Biopharm.* **1998,** *45,* 75–82.

Ishii, T., Asai, T., Oyama, D. et al.; Treatment of Cerebral Ischemia-Reperfusion Injury with PEGylated Liposomes Encapsulating FK506. *FASEB J.* **2013,** *27* (4),1362–70.

Itou, N.; Masukawa, T.; Ozaki, I.; Hattori, M.; Kasai, K.; Cross-Linked Hollow Polymer Particles by Emulsion Polymerization. *Colloids Surf.* **1999**, *153*, 311.

Jeong, Y. I. et al.; Ciprofloxacin-Encapsulated Poly(Dl-Lactide-Co-GlyColide) Nanoparticles and Its Antibacterial Activity. *Int. J. Pharm.* **2008**, *352*, 317–323.

Jiang, H. H.; Kim, T. H.; Lee, S.; Chen, X.; Youn, Y. S.; Lee, K. C.; PEGylated TNF Related Apoptosis Inducing Ligand (TRAIL) for Effective Tumor Combination Therapy. *Biomaterials* **2011**, *32* (33), 8529–8537.

Jose Prakash, D.; Ravichandiran, V.; Arunachalam, G.; Targeting the Brain with Rivastigmine Loaded PEG-PLGA NanoParticles for Alzheimer's Disease. *Int. J. Pharm. Pharm. Sci.* **2014**, *4* (4), 82–87.

Joshi, S. A.; Chavhan, S. S.; Sawant, K. K.; Rivastigmine-Loaded PLGA and PBCA Nanoparticles: Preparation, Optimization, Characterization, In Vitro and Pharmacodynamic Studies. *Eur. J. Pharm. Biopharm.* 2010, *76* (2), 189–199.

Joo, H. H.; Lee, H. Y.; Guan, Y. S.; Kim, J. C.; Colloidal Stability and In Vitro Permeation Study of Poly(e-Caprolactone) Nanocapsules Containing Hinokitiol. *J. Ind. Eng. Chem.* **2008**, *14*, 608–613.

Kaur, I. P; Bhandari, R.; Bhandari, S. et al.; Potential of Solid Lipid Nanoparticles in Brain Targeting. *J. Control Release* **2008**, *127* (2), 97–109.

Khetawat, S., Lodha, S.; Nanotechnology (Nanohydroxyapatite Crystals): Recent Advancement in Treatment of Dentinal Hypersensitivity. *J, Interdiscipl. Med. Dent. Sci.* **2015**, *3*, 181.

Khoee, S.; Yaghoobian, M.; An Investigation into the Role of Surfactants in Controlling Particle Size of Polymeric Nanocapsules Containing Penicillin-G in Double Emulsion. *Eur. J. Med. Chem.* **2008**, *44* (6), 2392–2399.

Krause H. J., Rohdewald, P.; Preparation of Gelatin Nanocapsules and Their Pharmaceutical Characterization. *Pharm. Res.* **1985**, *5*, 239–243.

Jörg, K.; Drug Delivery to the Central Nervous System by Polymeric Nanoparticles: What do We Know? *Adv. Drug Deliv. Rev.* **2014**, *7*, 2–14.

Kreuter J.; Nanoparticles. In: Kreuter J, editor. Colloidal drug delivery systems. Marcel Dekker: New York; **1994**; pp 219–342.

Kreuter, J.; Nanoparticulate Systems for Brain Delivery of Drugs. *Adv. Drug Deliv. Rev.* **2001**, *47*, 65–81.

Krishna, R.; Mayer, L. D; Multidrug Resistance (MDR) in Cancer: Mechanisms, Reversal Using Modulators of MDR and the Role of MDR Modulators in Influencing the Pharmacokinetics of Anticancer Drugs. *Eur. J. Pharm. Sci.* **2000**, *11*, 265–283.

Krol, S. et al.; Nanocapsules—A Novel Tool for Medicine and Science. In: Frontiers of Multifunctional Integrated Nanosystems; Buzaneva, E., Scharff, P.; Eds., Academic Publishers: The Netherlands, 2004; pp 439–446.

Laine, R.; Unlap, M. T.; IPX-750, A Dopamine Gluconamine that Binds d1/d5 Receptors and Has Anti-Parkinsonian Effects in Three Animal Models, Is Transported Across the Blood Brain Barrier. *J. Biotechnol. Biomater.* **2012**, *2*, 142.

Lee, c. Y.; Ooi, I. H.; Preparation of Temozolomide-Loaded Nanoparticles for Glioblastoma Multiforme Targeting—Ideal Versus Reality. *Pharmaceuticals* **2016**, *9*, 54.

Legrand, P.; Lesieur, S.; Bochot, A.; Gref, R.; Raatjes, W.; Barratt, G.; Vauthier, C.; Influence of Polymer Behaviour in Organic Solution on the Production of Polylactide Nanoparticles by Nanoprecipitation. *Int. J. Pharm.* **2007**, *344*, 33–43.

Lertsutthiwong, P.; Noomun, K.; Jongaroonngamsang, N.; Rojsitthisak, P., Nimmannit, U.; Preparation of Alginate Nanocapsules Containing Turmeric Oil. *Carbohydr. Polym.* **2008a,** *74*, 209–214.

Lertsutthiwong, P.; Rojsitthisak, P., Nimmannit, U.; Preparation of Turmeric Oil-Loaded Chitosan-Alginate Biopolymeric Nanocapsules. *Mater. Sci. Eng. C.* **2008b,** *29* (3), 856–860.

Li, W.; Zhang, F.; Zhao, M.; et al.; Effects of Intracellular Process on the Therapeutic Activation of Nanomedicine. *Pharm. Anal. Acta* **2015,** *6*, 368.

Lince, F.; Marchisio, D. L.; Barresi, A. A.; Strategies to Control the Particle Size Distribution of Poly-E-Caprolactone Nanoparticles for Pharmaceutical Applications. *J. Colloid Interface Sci.* **2008,** *322*, 505–515.

Limayem, I.; Charcosset, C.; Fessi, H.; Purification of Nanoparticle Suspensions by A Concentration/Diafiltration Process. *Sep. Purif. Technol.* **2004,** *38*, 1–9.

Little, S. R.; Lynn, D. M.; Puram, S. V.; Langer, R.; Formulation and Characterization of Poly (Beta Amino Ester) Microparticles for Genetic Vaccine Delivery. *J. Contr. Rel.* **2005,** *107*, 449–462.

Lloyd-Hughes, H.; Shiatis, A. E.; Pabari, A. et al.; Current and Future Nanotechnology Applications in the Management of Melanoma: A Review. *J. Nanomed. Nanotechnol.* **2015,** *6*, 334.

Lockman, P. R; Mumper, R. J.; Khan, M. A; Allen, D. D.; Nanoparticle Technology for Drug Delivery across the Blood-Brain Barrier. *Drug Dev. Ind. Pharm.* **2002,** *28*, 1–13.

Lu, Z.; Bei, J.; Wang, S.; A Method for the Preparation of Polymeric Nanocapsules without Stabilizer. *J. Control. Release* **1999,** *61*, 107–112.

Luan, X.; Skupin, M.; Siepmann, J.; Bodmeier, R.; Key Parameters Affecting the Initial Release (Burst) and Encapsulation Efficiency of Peptide-Containing Poly(Lactide-Co-Glycolide) Microparticles. *Int. J. Pharm.* **2006,** *324*, 168–175.

Lucks, J. S.; Müller, R. H.; Medication Vehicles Made of Solid Lipid Particles (Solid Lipid Nanospheres SLN). German Patent 1996, EP0000605497.

Lutter, S.; Koetz, J.; Tiersch, B.; Boschetti de Fierro, A.; Abetz, V.; Formation of Gold Nanoparticles in Triblock Terpolymer-Modified Inverse Microemulsions. *Colloid Surf.* **2008,** *329*, 160–176.

Ma, J., Feng, P.; Ye, C.; Wang, Y.; Fan, Y.; An Improved Interfacial Coacervation Technique to Fabricate Biodegradable Nanocapsules of an Aqueous Peptide Solution from Polylactide and its Block Copolymers with Poly(Ethylene Glycol). *Colloid Polym. Sci.* **2001,** *279*, 387–392.

Massimo, M.; Nanoparticles for Brain Delivery. *ISRN Biochem.* **2013,** *18*.

Maroof, K.; Zafar, F.; Ali, H. et al.; Scope of Nanotechnology in Drug Delivery. *J. Bioequiv. Availab.* **2016,** *8,* 1–5.

Matilda, A.; Oskari, E.; Topias, S. et al.; A Review on Ophthalmology Using Nanotechnology. *J. Nanomed. Nanotechnol.* **2015,** *6*, 272.

McDonald, C. J.; Bouck, K. J.; Chaput, A. B.; Stevens, C. J.; Emulsion Polymerization of Voided Particles by Encapsulation of a Nonsolvent. *Macromolecules* **2000,** *33*, 1593.

Meinel, L.; Illi, O. E.; Zapf, J.; Malfanti, M.; Merkle, H. P.; Gander, B.; Stabilizing Insulin Next-Like Growth Factor-I in Poly(d,l-Lactide-Co-Glycolide) Microspheres. *J. Control. Release* **2001,** *70*, 193–202.

Menaa, B.; The Importance of Nanotechnology in Biomedical Sciences. *J. Biotechnol. Biomater.* **2011,** *1*,105e

Mori, N. M.; Sheth, N. R.; Mendapara, V. P. et al.; SLS Brain Targeting Drug Delivery for CNS: A Novel Approach. *Int. Res. J. Pharm.* **2014,** *5* (9), 658–662.

Morris, M. C; Fluorescent Biosensors – Promises for Personalized Medicine. *J. Biosens. Bioelectron.* **2012,** *3*,e111.

Müller, R. H.; Mäder, K.; Lippacher, A.; Jenning, V.; Solid-Liquid (Semi-Solid) Liquid Particles and Method of Producing Highly Concentrated Lipid Particle Dispersions. **2000,** *199* (45,203.2).

Muthu, M. S.; Rawat, M. K.; Mishra, A.; Singh, S.; PLGA Nanoparticle Formulations of Risperidone: Preparation and Neuropharmacological Evaluation. *Nanomed. Nanotechnol. Biol. Med.,* **2009,** *5* (3), 323–333.

Nabel, G. J.; Nabel, E. C.; Yang, Z. Y. et al; Direct Gene Transfer with DNA-Liposome Complexes in Melanoma: Expression, Biologic Activity, and Lack of Toxicity in Humans. *Proc. Natl. Acad. Sci. USA* **1993,** *90* (23), 11307–11311.

Nair, M.; Guduru, R.; Liang, P.; Hong, J.; Sagar, V.; Khizroev, S.; Externally-Controlled On-Demand Release of Anti-HIV Drug AZTTP Using Magneto-Electric Nanoparticles as Carriers. *Nat. Commun.* **2013,** *4*, 1707.

Nikalje, A. P.; Nanotechnology and its Applications in Medicine. *Med. Chem.* **2015,** *5*, 81–89.

Nicoli, S. et al.; Design of Triptorelin Loaded Nanospheres for Transdermal Iontophoretic Administration, *Int. J. Pharm.* **2001,** *214* (1–2), 31–35.

Nogueira de Assis, D.; Furtado, V. C.; Carneiro, J. M.; Spangler, M.; Nascimento, V.; Release Profiles Andmorphological Characterization by Atomic Forcemicroscopic and Photon Correlation Spectroscopy of 99mtechnetium-Fluconazole Nanocapsules. *Int. J. Pharm.* **2008,** *349*, 152–160. Okubo, M.; Minami, H.; Ynamoto, Y.; Penetration/Release Behaviors of Various Solvents into/from Micron-Sized Monodispersed Hollow Polymer Particles. *Colloids Surf.,* **1999,** *153* (1–3), 405–411.

Olivier, J.-C.; Drug Transport to Brain with Targeted Nanoparticles, *NeuroRX*, **2005,** *2* (1), 108–119.

Ostafin, A. E.; Batenjany, M. M.; Nanomedicine Making *Headway across* **the** Blood Brain Barrier. *J. Nanomed. Nanotechnol.* **2012,** *3*, e123.

Patel, S.; Nanda, R.; Sahoo, S.; Nanotechnology in Healthcare: Applications and Challenges. *Med. chem.* 2015, *5,* **528–533***.*

Pathan, S. A. et al.; CNS Drug Delivery Systems: Novel Approaches. *Recent Pat. Drug Deliv. Formul.* **2009,** *3*, 71–89.

Pardridge, W. M.; Blood–Brain Barrier Delivery. *Drug Discov. Today* **2007,** *12* (1–2), 54–61

Pereira, N. et al.; Nanoencapsulation of Usnic Acid: an Attempt to Improve Antitumor Activity and Reduce Hepatoxicity. *Eur. J. Pharm. Biopharm.* **2006,** *64*, 154–160.

Perez, C. et al.; Poly(Lactic Acid)-Poly(Ethylene Glycol) Nanoparticles as New Carriers for the Delivery of Plasmid DNA. *J. Control. Release* **2001,** *75*, 211–224.

Persidsky, Y.; Ramirez, S. H.; Haorah, J.; Kanmogne, G. D.; Blood–Brain Barrier: Structural Components and Function Under Physiologic and Pathologic Conditions. *J. Neuroimmune Pharmacol.* **2006,** *1*, 223–236.

Plasari, E.; Grisoni, P. H.; Villermaux, J.; Influence of Process Parameters on the Precipitation of Organic Nanoparticles by Drowning-Out. *Chem. Eng. Res. Des.* **1997,** *75*, 237–244.

Pohlmann, A. R.; Weiss, V.; Mertins, O.; Pesce da Silveria, N.; Guterres, S. S.; Spray-Dried Indomethacin-Loaded Polyester Nanocapsules and Nanospheres: Development, Stability Evaluation and Nanostructure Models. *Eur. J. Pharm. Sci.* **2002,** *16*, 305–312.

Prego, C.; Fabre, M.; Torres, D.; Alonso, M. J.; Efficacy and Mechanism of Action of Chitosan Nanocapsules for Oral Peptide Delivery. *Pharm. Res*. **2006,** *23*, 549–556.

Quintanar, D.; Allémann, E.; Fessi, H.; Doelker, E.; Preparation Techniques and Mechanisms of Formation of Biodegradable Nanoparticles from Preformed Polymers. *Drug Dev. Ind. Pharm*. **1998a,** *24*, 1113–1128.

Quintanar, D.; Allémann, E.; Doelker, E.; Fessi, H.; Preparation and Characterization of Nanocapsules from Preformed Polymers by a New Process Based on Emulsification–Diffusion Technique. *Pharm. Res*. **1998b,** *15*, 1056–1062.

Quintanar, D.; Fessi, H.; Doelker, E.; Alleman, E.; Method for Preparing Vesicular Nanocapsules. US Patent 6884438, 2005.

Radtchenko, I. L.; Sukhorukov, G. B.; Möhwald, H.; Incorporation of Macromolecules into Polyelectrolyte Micro- and Nanocapsules Via Surface Controlled Precipitation on Colloidal Particles. *Colloid Surf.* **2002a,** A *202*, 127–133.

Radtchenko, I. L.; Sukhorukov, G. B.; Möhwald, H.; A Novel Method for Encapsulation of Poorly Water-Soluble Drugs: Precipitation in Polyelectrolyte Multilayer Shells. *Int. J. Pharm*. **2002b,** *242*, 219–223.

Rakesh, M.; Divya, T. N.; Vishal, T. et al.; Applications of Nanotechnology. *J. Nanomed. Biotherap. Discov*. **2015,** *5*, 131.

Reithmeier, H.; Herrmann, J.; Göpferich, A.; Lipid Microparticles as a Parenteral Controlled Release Device for Peptide. *J. Contr. Rel*. **2001,** *73*, 339–350.

Rim, H. S.; Kwangmeyung, K.; Nano-Enabled Delivery Systems Across the Blood-Brain Barrier. *Arch. Pharm. Res*. **2013,** *37* (1), 24–30.

Rip, J.; Schenk, G. J.; de Boer, A. G.; Differential Receptor-Mediated Drug Targeting to the Diseased Brain. *Expert Opin. Drug Deliv*. **2009,** *6*, 227–237.

Romero-Cano, M. S.; Vincent, B.; Controlled Release of 4-Nitroanisole from Poly(Lactic Acid) Nanoparticles. *J. Control. Release* **2002,** *82*, 127–135.

Roney, C.; Kulkarni, P., Arora, V. et al.; Targeted Nanoparticles for Drug Delivery through the Blood-Brain Barrier for Alzheimer's Disease. *J. Controlled Release* **2005,** *108* (2–3), 193–214.

Gregoriadis, G.; Liposome Research in Drug Delivery: the Early Days. *J. Drug Target*. **2008,** *16* (7–8), 520–524.

Ryan, S. M.; Mantovani, G.; Wang, X.; Haddleton, D. M.; Brayden, D. J.; Advances in PEGylation of Important Biotech Molecules: Delivery Aspects. *Exp. Opin. Drug Deliv*. **2008,** *5*, 371–383.

Saad, M. Z. H.; Jahan, R.; Bagul, U.; Nanopharmaceuticals: a New Perspective of Drug Delivery System. *Asian J. Biomed. Pharm. Sci*. **2012,** *2* (14).

Salunkhe, S. S. et al.; Development of Lipid-Based Nanoparticulate Drug Delivery Systems and Drug Carrier Complexes for Delivery to Brain, *JAPS*, **2015,** *5* (5), 110–129.

Sánchez-Moreno, P. et al.; Characterization of Different Functionalized Lipidic Nanocapsules as Potential Drug Carriers. *Int. J. Mol. Sci*. **2012,** *13*, 2405–2424.

Santos-Oliveira, R. Pharmaceutical Equivalence and Bioequivalence of Radiopharmaceuticals: Thinking the Possibility of Generic Radiopharmaceuticals and Preparing for New Technology as Nanotechnology Drugs. *J. Bioequiv. Availab*. **2014,** *6*, 23–23.

Schlosshauer, B.; The Blood-Brain Barrier: Morphology, Molecules, and Neurothelin. *Bioassays* **1993,** *15* (5), 341–346.

Schaffazick, S. R.; Guterres, S. S.; Freitas, L. L.; Pohlmann, A. R.; Aracterizacaoe Estabilidade Fisico-Quimica de Sistemas Polimericos Nanopaticulados Para Administracao de Farmacos. *Química Nova* **2003**, *26*, 726–737.

Schaffazick, S. R.; Siqueira, I. R.; Badejo, A. S.; Jornada, D. S.; Pohlmann, A. R.; Netto, C. A.; Guterres, S. S.; Incorporation in Polymeric Nanocapsules Improves the Antioxidant Effect of Melatonin Against Lipid Peroxidation in Mice Brain and Liver. *Eur. J. Pharm. Biopharm.* **2008,** *69*, 64–71.

Seju, U.; Kumar, A.; Sawant, K. K.; Development and Evaluation Of Olanzapine-Loaded PLGA Nanoparticles for Nose-To-Brain Delivery: In Vitro and In Vivo Studies. *Acta Biomater.* **2011,** *7* (12), 4169–4176.

Shehata, T.; Ogawara, K.; Higaki, K. et al.; Prolongation of Residence Time of Liposome by Surface-Modification with Mixture of Hydrophilic Polymers. *Int. J. Pharm.* **2008,** *359*, 272–279.

Soppimath, K. S.; Aminabhavi, T. M.; Kulkarni, K. R.; Rudzinski, W. E.; Biodegradable Polymeric Nanoparticles As Drug Delivery Devices. *J. Control. Release* **2001,** *70*, 1–20.

Sharma, N.; Effect of Process and Formulation Variables on the Preparation of Parenteral Paclitaxel-Loaded Biodegradable Polymeric Nanoparticles: A Co-Surfactant Study. *AJPS* **2016,** *11* (3), 404–416.

Singh, M. et al.; Controlled Release of Recombinant Insulin-Like Growth Factor from a Novel Formulation of Polylactide-Co-Glycolide Microparticles. *J. Control Release* **2001,** *70*, 21–28.

Singh, R. K.; Bansode, F. W.; Sharma, S. et al.; Development of a Nanotechnology-Based Biomedicine RISUG-M as a Female Contraceptive in India. *J. Nanomed. Nanotechnol.* **2015,** *6*, 297.

Skaat, H., Margel, S.; Newly Designed Magnetic and Non-Magnetic Nanoparticles for Potential Diagnostics and Therapy of Alzheimer's Disease. *J. Biotechnol. Biomater.* **2013,** *3*, 156.

Patel, S.; Chavhan, S.; Soni, H.; Babbar, A. K.; Mathur, R.; Mishra, A. K.; Sawant, K.; Brain Targeting of Risperidone-Loaded Solid Lipid Nanoparticles By Intranasal Route. *J. Drug Target.* **2011,** *19* (6), 468–474.

Steiniger, S. C.; Kreuter, J.; Khalansky, A. S.; Skidan, I. N.; Bobruskin, A. I.; Smirnova, Z. S.; et al.; Chemotherapy of Glioblastoma in Rats Using Doxorubicin-Loaded Nanoparticles. *Int. J. Cancer* **2004,** *109*, 759–767.

Stewart, P. A.; Endothelial Vesicles in the Blood–Brain Barrier: Are They Related to Permeability? *Cell. Mol. Neurobiol.* **2000,** *20*,149–163.

Sugimoto, T.; Preparation of Monodispersed Colloidal Particles. *Adv. Colloid Interf. Sci.* **1987,** *28*, 65–108.

Sukhorukov, G. B. et al.; Layer-By-Layer Self-Assembly of Polyelectrolytes on Colloidal Particles. *Colloid Surf.* **1998,** *137*, 253–266.

Gupta, S.; Biocompatible Microemulsion Systems for Drug Encapsulation and Delivery. *Curr. Sci.* **2011,** *101* (2).

Toffoli, G.; Rizzolio, F.; Role of Nanotechnology in Cancer Diagnostics. *J. Carcinogene Mutagene* **2013,** *4*, 135.

Tosi, G.; Costantino, L.; Rivasi, F.; Ruozi, B.; Leo, E.; Vergoni, A. V. et al; Targeting the Central Nervous System: In Vivo Experiments with Peptide-Derivatized Nanoparticles Loaded with Loperamide and Rhodamine-123. *J. Control Release* **2007,** *122* (1), 1–9.

Upadhyay, S.; Ganguly, K.; Palmberg, L.; Wonders of Nanotechnology in the Treatment for Chronic Lung Diseases. *J. Nanomed. Nanotechnol.* **2015,** *6,* 337.

Vauthier, C.; Dubernet, C.; Fattal, E.; Pinto-Alphandary, H.; Couvreur, P.; Poly (Alkylcyano-acrylates) as Biodegradable Materials for Biomedical Applications. *Adv. Drug Deliv. Rev.* **2003,** *55,* 519–548.

Vila, A.; Sánchez, A.; Tobío, M.; Calvo, P.; Alonso, M. J; Design of Biodegradable Particles for Protein Delivery. *J. Control. Release* **2002,** *78,* 15–24.

Vijaya Shanti, B.;, Mrudula, T. et al.; Novel Applications of Nanotechnology in Life Sciences. *J. Bioanal. Biomed.* **2011,** *S11,* 1.

Prabhakara, V.; Bibib, T.; Aslamc, M.; Nanocapsules - A Self-Assembly Approach Towards Future Medicine, *IJCET,* **2013,** *3* (3).

Wang, W.; Chen, G.; Chen, Y.; Nanotechnology as a Platform for Thermal Therapy of Prostate Cancer. *J. Mol. Biomark. Diagn.* **2013,** *4,* e117.

Williams, K.; Alvarez, X.; Lackner, A. A.; Central Nervous System Perivascular Cells Are Immunoregulatory Cells that Connect the CNS with the Peripheral Immune System. *Glia* **2001,** *36,*156–164.

Wohlgemuth, M.; MacHtle, W.; Mayer, C.; Improved Preparation and Physical Studies of Polybutylcyanoacrylate Nanocapsules. *J. Microencapsulation* **2000,** *17* (4), 437–448

Woodle, M. C.; Sterically Stabilized Liposome Therapeutics. *Adv. Drug Deliv. Rev.* **1995,** *16,* 249–265.

Wong, H. L.; Chattopadhyay, N.; Wu, X. U.; Bendayan, R.; Nanotechnology Applications for Improved Delivery of Antiretroviral Drugs to the Brain. *Adv. Drug Deliv. Rev.* **2010,** *62,* 503–517.

Chen, Y.; Modern Methods for Delivery of Drugs across the Blood–Brain Barrier. *Adv. Drug Deliv. Rev.* **2012,** *64* (7), 640–665.

Yokoyama, M., Okano, T.; Targetable Drug Carriers: Present Status and a Future Perspective, *Adv. Drug Deliv. Rev.* **1996,** *21,* 77–80.

Zambaux, M., Bonneaux, F., Gref, R.; Influence of Experimental Parameters on the Characteristics of Poly(Lactic Acid) Nanoparticles Prepared by Double Emulsion Method. *J. Control. Release* **1998,** *50,* 31–34.

Zharapova, L.; Synthesis of Nanoparticles and Nanocapsules for Controlled Release of the Antitumor Drug "Arglabin" and Antituberculosis Drugs. 2012; Technische Universiteit Eindhoven: Eindhoven, ISBN: 978-90-386-3129-5.

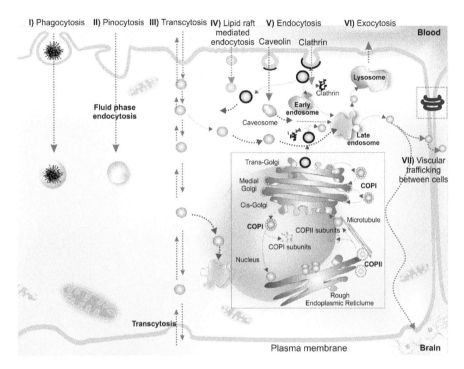

FIGURE 2.3 Schematic representation of vesicular trafficking at the BBB. Macromolecules are internalized through phagocytosis (I). Engulfed particles in the phagosomes may be subjected to fusion with lysosomes. The surrounding substances are often internalized by pinocytosis/fluid phase endocytosis (II). The main route for trafficking of large molecules is transcytosis (III). Lipid rafts are involved in signal transduction and internalization of some macromolecules like cholera toxin (IV). The membranous caveolae and CCPs are engaged in endocytosis process (V). Various biomolecules such as Tf, LDL may be colocalized with CCPs. The internalized macromolecules within vesicles may interact with endosomes and lysosomes and exocytose out of the cell (VI) or even associate with the neighboring cells (VII). BBB, blood–brain barrier; CCPs, clathrin-coated pits; Tf, transferrin; LDL, low-density lipoprotein. Image was adapted with permission from our previously published work (Barar et al., 2016). Note: Not drawn to scale.

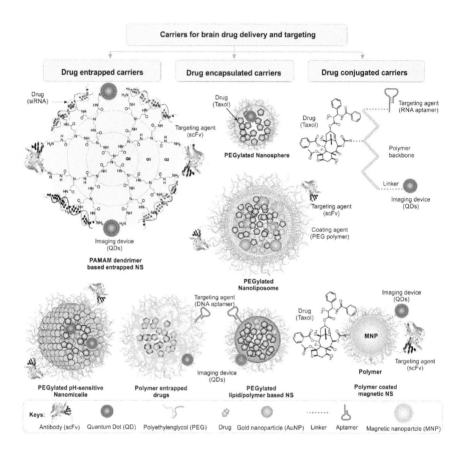

FIGURE 2.4 Schematic representation for various types of carriers for brain drug delivery and targeting. Image was adapted and modified with permission from our previously published work (Barar and Omidi, 2014).

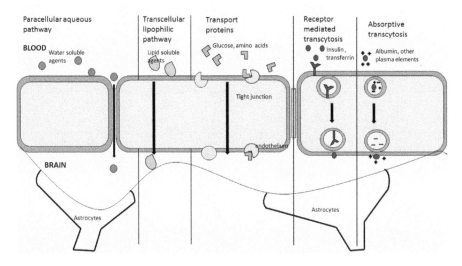

FIGURE 3.2   Interaction of various kinds of drugs with the blood–brain barrier.

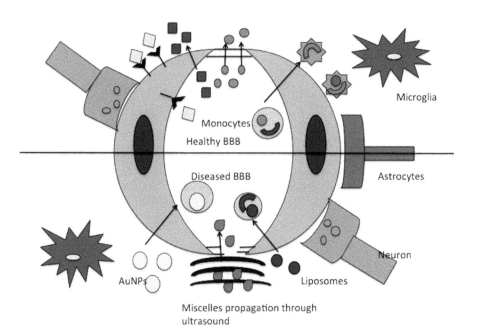

FIGURE 3.3   Different mechanisms for drug penetration by nanoparticles across the blood–brain barrier.

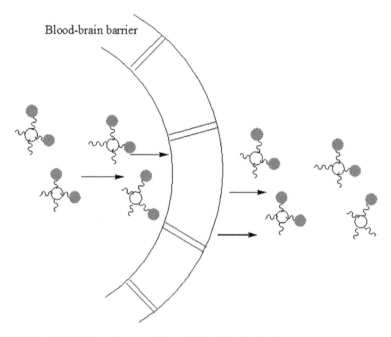

FIGURE 4.2   Mechanism of nanocarriers via blood–brain barrier.

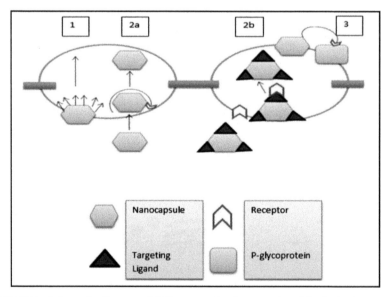

FIGURE 7.7   Schematic diagram of different mechanisms employed in the brain targeting using nanocapsules.

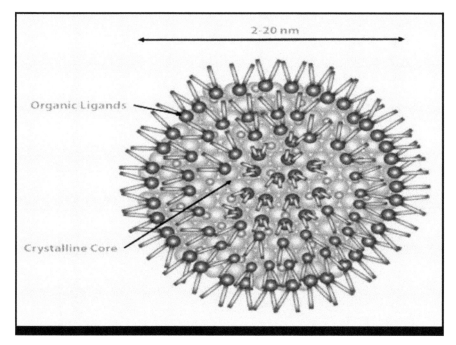

FIGURE 9.1   Quantum dots dimensions.

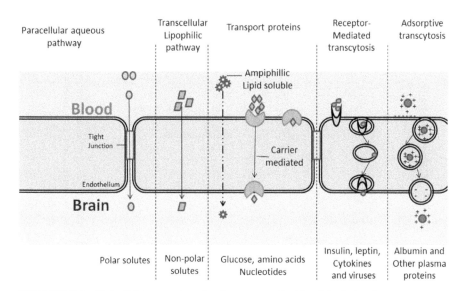

FIGURE 10.1   Potential transport mechanisms across the blood–brain barrier.

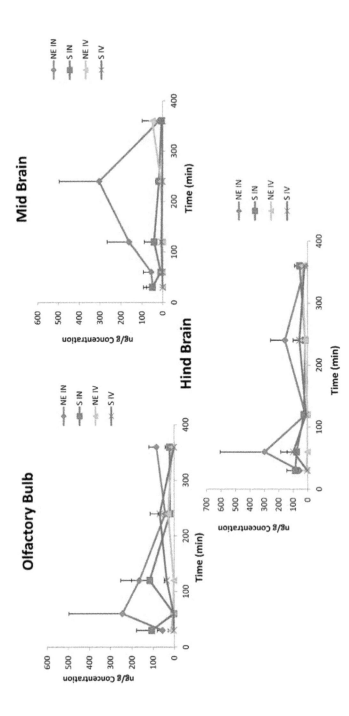

FIGURE 14.1 Mean (±SD) concentration–time curves of cyclosporine-A (CsA) in different regions of rat brain after intranasal (IN) or intravenous (IV) administration of CsA-nanoemulsion (CsA-NE) or CsA-solution (CsA-S) at a dose of 5 mg/kg.

Source: Yadav et al., 2015.

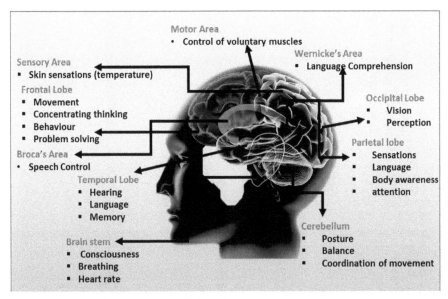

SCHEME 14.1 Arius components of the brain.

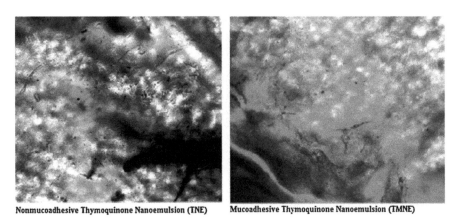

Nonmucoadhesive Thymoquinone Nanoemulsion (TNE)   Mucoadhesive Thymoquinone Nanoemulsion (TMNE)

FIGURE 14.5 CLSM images of thymoquinone nonmucoadhesive nanoemulsion (TNE) and thymoquinone mucoadhesive nanoemulsion (TMNE).
Source: Ahmad et al., 2016.

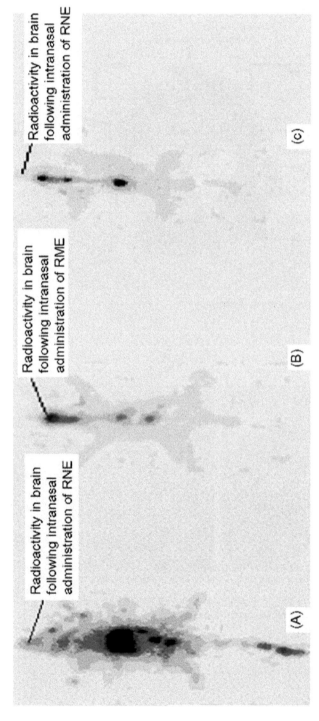

FIGURE 14.6 Gamma scintigraphy images of rat (A/P view) showing the presence of radioactivity (A) RNE (i.v.), (B) RMNE (i.n.), and (C) RNE (i.n.).

Source: Kumar et al., 2008.

CHAPTER 8

SOLID LIPID NANOPARTICLES FOR BRAIN TARGETING

M. M. DE ARAUJO, L. B. TOFANI, I. L. SUZUKI, P. D. MARCATO, and M. V. L. B. BENTLEY

Department of Pharmaceutical Sciences, School of Pharmaceutical Sciences of Ribeirão Preto, University of São Paulo, Ribeirão Preto, SP, CEP 14040-903, Brazil

Corresponding author. E-mail: margaretemoreno@gmail.com

ABSTRACT

The central nervous system, one of the most delicate microenvironments of the body, is protected by the blood–brain barrier (BBB) regulating its homeostasis. However, the BBB also significantly precludes the delivery of drugs to the brain, thus, preventing the therapy of a number of neurological disorders. As a consequence, several strategies are currently being sought after to enhance the delivery of drugs across the BBB. The present chapter discusses the potential use of solid lipid nanoparticles (SLN) for brain drug targeting purposes. The structures, preparation techniques, and physicochemical characterization of SLN are systematically elucidated. The potential advantages of the use of solid lipid nanoparticles over others nanosystems are accounted on the bases of a lower cytotoxicity, higher drug loading capacity and best production scalability. Solid lipid nanoparticles physicochemical characteristics are also particularly regarded in order to address the critical issues related to the development of suitable brain targeting formulations. Supplemented with small size which prolongs the circulation time in blood, feasible scale up for large scale production and absence of burst effect makes them interesting candidates for study. Finally, future technological approaches are described. The strong efforts to allow the translation from preclinical to concrete clinical applications are worth the economic investments.

8.1 INTRODUCTION

The blood–brain barrier (BBB) is a dynamic barrier, composed of a capillary system of endothelial cells that protects the brain against invading organisms and harmful substances while supplying the brain with the required nutrients for proper function (Kaur et al., 2008). It is also important to prevent drug transport into the brain via blood circulation. BBB limits the transport into the brain through physical (tight junctions) and metabolic barriers (presence of enzymes), unlike peripheral capillaries that allow the exchange of substances between cells (Begley, 2004). Thus, the BBB is determinant in the drugs permeation in the brain (Kaur et al., 2008).

The cerebral capillaries of BBB are created by the process of angiogenesis, during this process, endothelial cells (ECs) traverse the extracellular matrix and degrade the basement membrane to create a microvascular network (Nag et al., 2003; Reese and Karnovsky, 1967). These confer the characteristic physiological properties upon the BBB, distinguishing the cerebral EC from the periphery EC. The brain ECs have fewer endocytotic vessels than peripheral ECs, which limits the transcellular flux at the BBB (Roney et al., 2005). The characteristics of these brain EC structures help to define the selectiveness at the BBB; besides the endocytotic vessels, cerebral ECs have more mitochondria (Nag, 2003) than peripheral ECs, which drives the increased metabolic workload necessary to maintain ionic gradients across the BBB (Roney et al., 2005). The electron-dense layers of the basement membrane fuse EC and astrocytes, which involve more than 99% of the basal capillary membrane and also play an important role in the BBB induction of high paracellular electrical resistance (Johanson, 1980).

The BBB restricts solute entry into the brain via the transcellular route due to an increased electrical resistance between the ECs at the tight junctions (TJ) (Roney et al., 2005). Indispensable proteins, including claudin and the junctional adhesion molecule (JAM), compose the TJ. The claudins form the seal of the TJ by homotypically binding to each other on adjacent EC cells (Wolburg et al., 2003). The JAM regulates leukocyte transmigration at the BBB (Martin-Padura et al., 1998). Between the ECs, there is an expression of TJ, which is one of the most critical characteristics, because of their consequences on the function of the BBB (Reichel et al., 2003). These features provide almost complete restriction of the paracellular pathway, control over the CNS penetration, differential expression of transporters, receptors and enzymes at the luminal or abluminal cell surface, allowing the BBB to act as a dynamic interface between the periphery of the body (blood) and the central compartment (brain) (Roney et al., 2005).

To overcome the BBB, the molecule should present some physicochemical characteristics, such as high lipophilicity and a molecular weight <500 Da (Pardridge, 2003). Others factors that influence the drug transport into the brain are drug affinity for efflux proteins, cellular enzymatic stability, drug charge, and affinity for receptors or carriers. Small lipophilic molecules most easily pass through from the capillaries. Molecules with charge (negative or positive) or large or hydrophilic require gated channels, ATP, proteins, and/or receptors to facilitate their passage through the BBB Transport mechanisms at the BBB can be manipulated for cerebral drug targeting (Roney et al., 2005).

Due to efficient BBB, the treatment of brain diseases is a challenge (Juillerat-Jeanneret, 2008). The use of nanoparticles, out of the several different approaches that exist, has been regarded as great potential for drug delivery into the central nervous system (CNS).

Solid lipid nanoparticles (SLNs) constitute an attractive colloidal drug carrier system that improves the drug delivery into the BBB and is a promising drug targeting system for the treatment of CNS disorders (Wang et al., 2002). Besides, lipids are rightly being considered as safe and biodegradable materials for drug delivery (Blasi et al., 2007). SLNs are spherical solid lipid particles in the nanometer range, which are dispersed in water or in an aqueous surfactant solution (Chen et al., 2004, Westesen et al., 1993).

There are many approaches for SLN preparation. The most common methods consist in high-pressure homogenization (HPH) at elevated or low temperatures, high shear homogenization or ultrasonication, and phase inversion temperature (PIT) method (Mader and Mehnert, 2001; Muller et al., 2000). These methods will be covered in this chapter.

SLN is an innovative way to administer molecules into the brain because it can overcome the solubility, permeability, and toxicity problems associated with the drug. This is an advantage over the conventional invasive methods of drug delivery to the brain. Besides, these systems have high physical stability. The production of SLNs, on a large scale, can be performed in HPH , which is an economical and simple method. Moreover, the possibility of incorporating lipophilic and hydrophilic molecules and the possibility of several administration routes make this type of delivery system more promising. SLNs open a new channel for an effective delivery of several kinds of drug including analgesics, antitubercular, anticancer, antiaging, antianxiety, neuroleptics, antibiotics, and antiviral agents into the brain (Kaur et al., 2008).

8.2 SOLID LIPID NANOPARTICLES

SLNs were introduced in 1991 by Muller et al., being used as alternative carrier systems to traditional colloidal carriers, such as liposomes, polymeric micro- and nanoparticles, and emulsions (Schwarz et al., 1994). The lipids used in the manufacturing of SLNs are solid at room and body temperature (Muller 2000).

Many methods have been developed for the SLNs preparation using biocompatible and safe lipids to use in medicine. The essential excipients of SLNs are solid lipid as matrix material (0.1–30% w/w), surfactants (0.5–5% w/w), and water. Besides, cosurfactants can also be used. The term lipid is used in a broader sense, including triglycerides (e.g., tri-stearin), partial glycerides (e.g., imwitor), fatty acids (e.g., stearic acid), steroids (e.g., cholesterol), and waxes (e.g., cetyl palmitate) (Mehnert and Mader 2012; Yadav et al., 2013). Many classes of emulsifiers (relative to charge and molecular weight) have been used to stabilize the lipid dispersion. It has been found that the combination of emulsifiers might prevent particle agglomeration more efficiently (Yadav et al., 2013).

SLNs combine the advantages of the traditional systems, including high biocompatibility, high bioavailability, controlled release, physical stability, protection from degradation of incorporated drugs, great tolerability, and no problems with multiple routes of administration such as oral, intravenous (i.v.), pulmonary, and transdermal administration. Furthermore, SLN avoids some traditional nanometric systems major disadvantages, such as it is possible to produce SLN without organic solvent (contrary to existing polymeric nanoparticles), and present reduced adverse effects and a high stability in vitro and in vivo. The latter characteristic made them an alternative not only to polymeric nanoparticles but also to previous lipid-based formulations (e.g., liposomes) (Parhi and Suresh 2012; Rostami et al., 2014).

However, an important drawback appeared to compromise the future applicability of the formulation due to the crystallinity of solid lipids: the low drug loading capacity and drug expulsion during storage. This problem occurs when the low ordered lipid α-modification of the particle matrix transforms to the highly ordered ß-modification during storage. The ß-modification is characterized by a perfect crystal lattice with few imperfections and hence little room for drug accommodation; thus the incorporated drug may be expelled from the lipid matrix. Further investigations on the formulation helped on the improvement of the SLNs. The incorporation of a liquid lipid to the solid matrix of the nanoparticle was found to increase the number of imperfections in the core solid matrix, facilitating the incorporation of the high amount of drug, while preserving the physical stability of the nanocarriers. This new unstructured lipidic matrix, which do not form a highly ordered crystalline arrangement, is known as nanostructured lipid carriers (NLCs) (Weber et al., 2014; Iqbal et al., 2012).

Thus, NLCs are the second-generation lipid nanocarriers and possess a solid lipid matrix at room and body temperature that consists of a blend of a solid lipid and oil, preferably in a ratio of 70:30 up to a ratio of 99.9:0.1, whereas the surfactant content ranges 1.5–5% (w/v) (Beloqui et al., 2016).

NLCs can strongly immobilize drugs and prevent the particles coalescence as compared to emulsions. Furthermore, the liquid oil droplets in the solid matrix increase the drug loading capacity as compared to SLNs. Recently, NLCs have been intensively studied as delivery carriers for hydrophilic and hydrophobic drugs. The SLN and NLCs were developed with a perspective to meet industrial needs in terms of qualification and validation, scale up, simple technology, and low cost (Naseri et al., 2015; Beloqui et al., 2016; Iqbal et al., 2012; Rostami et al., 2014).

8.2.1 PREPARATION OF SLNs

SLN or NLC are generally obtained by different methodologies, including techniques based on solvents and nonsolvents. The use of the solvents to originate SLNs consists in the dissolution of the solid lipid in an organic solvent followed by their evaporation. In this chapter, we will focus on nonsolvents techniques based on high- and low-energy methods to produce the SLNs. HPH and shear homogenization to disperse the lipid in the aqueous phase being associated with low production cost and low citotoxicity compared to the methods that use organic solvents (Elbahwy et al., 2017). Typically, the type of lipid and the technique chosen for preparation can influence the properties of the SLNs such as encapsulation efficiency, crystallinity and release pattern, size, and particle charge (Teeranachaideekul et al., 2017).

8.2.1.1 HIGH SHEAR HOMOGENIZATION OR ULTRASONICATION

High shear homogenization and ultrasonication are dispersing techniques that are used to produce SLNs (Speiser, 1990; Domb, 1993). SLN can be obtained by the dispersion of a melted lipid in a warm aqueous phase (containing surfactants) by high-sheer homogenization or by ultrasonication or by high sheer homogenization followed by ultrasonication (Shah et al., 2015). Both methods are considered easy to handle. However, sometimes the dispersion quality is compromised by the presence of some microparticles and by metal contamination from the tip of equipment (Mader and Mehnert, 2001).

Both techniques produce SLN by melt emulsification (Mukherjee et al., 2009). In this case, SLN can be obtained due to the use of an appropriate emulsifying agent, which allows the formation of the nanoemulsion under the high shear (Carlotti et al., 2007). These methods involve an aqueous surfactant solution at the same heating temperature as a solid lipid (approximately 5–10°C above its melting point), then, the aqueous solution is dispersed at this melt lipid in under high-speed stirring to form an emulsion. Afterward, the ultrasonication or high-shear homogenization is used to reduce the size of this emulsion into nanoparticles (Shah et al., 2015).

The ultrasonication technology is based on cavitation process where microbubbles are produced, and when they collapse, the emulsion globules are broken down leading to size particle reduction. The method employs the same procedure as hot HPH, except using ultrasonication device in place of a homogenizer (Mader and Mehnert, 2001).

Initially, high-shear homogenization technique was used for the production of solid lipid nanodispersions; however, dispersion quality is generally compromised by the presence of microparticles.

The ultrasound process or high-shear homogenization is fast and highly reproducible if the operating parameters are optimized (Chaturvedi and Kumar, 2012). There some parameters involved in the preparation of SLN that can be improved: the concentration and the type of lipid and surfactant, lipid/surfactant ratio and sonication time (for ultrasonication method) or speed or agitation time, for high shear homogenization, to reduce the size of nanoparticles, and the polydispersity index (PdI) (Golmohammadzadeh et al., 2012). Ultrasound probes are very easy to clean, and sample losses are negligible; however, the dispersion quality is frequently compromised by the presence of microparticles that can lead physical instabilities by particle growth upon storage. Furthermore, metal contamination should be considered if these methods are used (Venkateswarlu and Manjunath, 2004).

8.2.1.2 HIGH-PRESSURE HOMOGENIZATION

HPH was the first technique used for SLN production, representing the main method to stabilize them (Battaglia et al., 2011). In this technique, particles dispersion passes through a narrow cavity under high-pressure (100–200 bar), being accelerated in a short distance with high speed (around 100 km/h) and colliding a barrier. All these processes together lead to an efficient reduction of particle size (Marcato, 2009).

The HPH involves multistep, and it is performed using hot or cold homogenization. The drug is dissolved, solubilized, or dispersed in a melting lipid combined with an aqueous solution containing a surfactant (Patil et al., 2014). However, the high pressure generates a high temperature that can cause thermodynamically stress in the final product leading the formation of microparticles or dispersion with high PdI. Thus, an optimization in the process parameters is necessary. Furthermore, due to high temperature in the equipment, this technique is not recommended to thermosensitive drugs (Battaglia et al., 2011).

In this method, the drug is dispersed in the melted lipid and after that it dispersed in hot surfactant solution under high-speed stirring. Both phases should be in the same temperature. The preemulsion formed is submitted to HPH using, in most of cases, two or three homogenization cycles at 500 bar or more (Radtke and Wissing, 2002). The primary product formed after the HPH process is a nanoemulsion due to liquid character of the lipid. Solid particles are formed after the cooling of the sample at room temperature or in a temperature below. Besides that, the sample can remain in its state supercooled for several months due to high able in prolong the lipid crystallization by the presence of emulsifiers and small size (Mehnert and Mäder, 2012).

Cold homogenization is characterized by cooling the mixture of drug and melted lipid until its solidification. Afterward, the solidified lipid-drug is dispersed in cold surfactant solution by stirring, forming a microsuspension. At the end, the microsuspension is added in an HPH, where they are breaking down to SLNs (Radtke and Wissing, 2002). This process avoids or reduces the melting of the lipid, minimizing the loss of hydrophilic drugs to the aqueous phase. Furthermore, this method is better for encapsulation of thermosensitive drugs in SLN than hot homogenization (Muller, 2000).

The HPH has several advantageous compared to others methodologies to produce SLNs including easy scale up, short production time, no organic solvents, or metal contamination. Due to these approaches, this method is widely used in several areas into the pharmaceutical industry, such as the production of nanoemulsions for parenteral nutrition (Pardeike et al., 2009).

8.2.1.3   PIT METHOD

The PIT method was first introduced by Shinoda and Saito in 1969 being used for the preparation of nanoemulsions, and it was recently adapted for the SLN preparation. PIT is defined as the "temperature or temperature range at which the hydrophilic and lipophilic properties of a nonionic surfactant just balance" (Friberg et al., 1976). It is based on changes in the properties of nonionic surfactants (e.g., molecular geometry, packing, and oil–water partitioning) when the temperature is changed. These types of surfactant become lipophilic with increasing temperature because of the dehydration of polyoxyethylene chains due to the breakdown of hydrogen bonds with water molecules (Huynh et al., 2009).

An oil phase, composed by solid lipids and nonionic surfactants, and an aqueous phase containing surfactant is separately heated above the PIT. The aqueous phase is added dropwise to the oil phase, at constant temperature and under agitation, forming an opaque W/O emulsion. The opaque appearance of the W/O emulsion can be attributed to strong light scattering due to large droplets of the emulsion. At low temperatures, the surfactant monolayer has a large, positive, spontaneous curvature forming O/W emulsions, characterized by high conductivity values. By increasing the temperature, the spontaneous curvature becomes negative, leading to W/O emulsion formation as a rapid decrease in conductivity. Then, this emulsion is cooled to room temperature under slow and continuous stirring. At the PIT, the turbid mixture becomes translucent, which is indicative of the bicontinuous microemulsion formation. Below the PIT is formed an O/W nanoemulsion, which turns in SLN below the lipid melting point (Gao and McClements, 2016; Sarpietro et al., 2014; Huynh et al., 2009).

Thus, in this method, surfactants are used, which can lead to a phase inversion from O/W emulsion to a W/O nanoemulsion at high temperature above the PIT, and to the formation of an O/W nanoemulsion at lowered temperature below the PIT (Gao and McClements, 2016). This is an easy and scalable method interesting to produce SLN.

Table 8.1 demonstrates different methods and lipids matrices used to prepare SLNs as a delivery system to several drugs.

TABLE 8.1 Different Lipid Matrices and Methods Used to Prepare SLNs with Different Drugs.

Lipid matrix	Surfactant	Preparation method	Drug	Reference
Cetyl palmitate	Polysorbate 60 and 80	HPH	Camptothecin	(Martins et al., 2013) camptothecin-loaded SLN with mean size below 200 nm, low polydispersity index (<0.25
Cetyl palmitate	Ceteth-20, Isoceteth-20, and Oleth-20	PIT	Idebenone	(Sarpietro et al., 2014)
Glyceril monoestearate/ Compritol 888 ATO® Precirol AT 5®/Stearic acid/Palmitic acid	Tween 80®	Solvent emulsification-diffusion	Haloperidol	(Yasir and Sara, 2014)
Resveratrol/ Compritol 888 ATO®	Tween 80®/ Polyvinyl alcohol (PVA)	High shear homogeneization/ Sonication	Resveratrol	(Jose et al., 2014) because of its low half-life (<0.25 h
Witepsol E 85®	Polyvinyl alcohol (PVA)	Double-emulsion	RVG-9R/ BACE1 siRNA	(Rassu et al., 2017)
Glyceryl monoestearate (GMS)	Lecithin and Pluronic F127	Solvent diffusion	Rizatriptan	(Singh et al., 2015)
Behenic acid/ Tripalmitin/ Cacao butter	Cholesteryl Hemisuccinate/Taurocholate/ Tween 80®	High shear homogeneization	TX-BCNU-SLN	(Kuo & Cheng, 2016)
Compritol 888 ATO®/Cetyl palmitate/Softisian 142®	Tween 80®	HPH	-	(Blasi et al., 2011)

TABLE 8.1 *(Continued)*

Lipid matrix	Surfactant	Preparation method	Drug	Reference
Stearic acid	Epikuron 200/tauro-cholate sodium salt	High shear homogeneization	Doxorubicin	(Fundaró et al., 2000)
Glycerol Mono Stearate	Poly Sorbate-80/ Epikuron 200	Emulsification-solvent diffusion	Piperine	(Yusuf et al., 2013)
Octanoic acid	Pluronic F-68/ Pluronic F-127	Emulsification-solvent diffusion	3´,5´-dioc-tanoyl-5-fluoro-2´-deoxyuridine	(Wang at al., 2002)
Cetyl palmitate	polysorbate 80	High shear homogenization followed by sonication	Lucifer yellow	(Neves et al., 2015)
Tripalmitin/ Gelucire 48/9 and Gelucire 62/5	Hydrogenated soybean phosphatidylcholine 80%/poloxamer 188	High shear homogenization followed by sonication	Baicalein	(Tsai et al., 2012)
Precirol ATO5/ Glyceryl Monostearate/Glyeryl Tripalmitate	Tween 80/ Pluronic F68/ Pluronic F127	High shear homogenization followed by sonication	Vinpocetine	(Morsi et al., 2013)

Table 8.2 summarizes some of the main characterization techniques and specific parameters detected by each one used for SLNs.

TABLE 8.2   Techniques for Assessment of the Physicochemical Characteristics of SLNs.

Characterization technique	Parameters analyzed
X-ray diffraction	Crystallite size, crystallinity degree and orientation, crystalline phase, and chemical composition
Small-angle X-ray scattering (SAX)	Particle size and shape, spatial distribution of particles in a medium, particle interactions, inter-atomic distances
Dynamic light scattering (DLS) or Photon correlation spectroscopy (PCS)	Particle size and distribution
Zeta potential	Surface charge
Atomic Force Microscopy (AFM)	Particle size, size distribution, structure/ shape, stability
Differential scanning calorimetry (DSC)	Measurement of crystallinity, lipid modification and assessment of alternative colloidal structures

8.3   SLNs AS DRUG DELIVERY SYSTEM TO THE CNS

8.3.1   SLNs: DRUG DELIVERY SYSTEM OF CHOICE TO BRAINS TARGET

Delivery of drugs to the brain is a challenge due to limitation mediated by the BBB. Considering the success of the nanoparticles to pass through the BBB and aspects of their stability and toxicity, a suitable option for drug delivery to the brain are SLNs (Kaur et al., 2008). The advantages of the use SLNs to brain target delivery can be demonstrated below.

8.3.1.1   PARTICLE SIZE

The size of the particles has a crucial role in their clearance by the sinusoidal spleen of human and rats. SLNs in range of 100–200 nm are not taken up readily by cells of reticuloendothelial system (RES), passing to the liver and spleen filtration (Kaur et al., 2008). Jose et al. (2014) observed that SLNs smaller than 250 nm presented a longer time in the blood; consequently, more amount of the drug could be taken up to brain. SLNs with size from 38 to 6 nm were observed within in the neurons of the rats and mice 30 min after i.v. administration, while SLNs above 250 nm cannot be transported to the brain (Kreuter, 2014). Martins et al. (2013) observed that camptothecin-SLNs with homogeneous size below 200 nm were taken up by EC with high accumulation in the brain. Sarpietro et al. (2014) demonstrated the success of the SLNs prepared by PIT method, using cetyl-palmitate as lipid, with small size (<100 nm) to brain delivery of the idebenone.

8.3.1.2   PHYSICAL–CHEMICAL AND BIOLOGICAL STABILITY OF ENCAPSULATED DRUGS IN SLN

The SLNs present the advantages of the protection of incorporated drugs from biological and chemical degradation (Singh et al., 2015). Many drugs can across the BBB, but show low in vivo efficacy, since these compounds can be eliminated rapidly of the plasma. The small interfering RNA (siRNA) molecule presents lack of tissue target and low stability in serum due to its negative charge and hydrophilic nature (Sardo et al., 2015). RVG-9R cell penetration peptide and BACE1 siRNA complex were encapsulated in SLNs coated with chitosan (CS) to the treatment of Alzheimer´s disease. The results showed that the use of SLNs as delivery system allowed the transport of the siRNA molecule across the olfactory epithelium, with an increase of the penetration across the mucosal surface of the respiratory epithelium to the trigeminal nerve endings (Rassu et al., 2017).

8.3.1.3   PROFILE THE DRUG´S PHARMACOKINETIC AND BIODISTRIBUTION

SLNs improve the drug´s pharmacokinetic and biodistribution, increasing its half-life in the bloodstream and, consequently, increase the accumulation of drugs in tissues (Gastaldi et al., 2014). Campt-loaded SLNs could alter the profile drug's biodistribution in vivo, increasing its retention time in the brain and its antitumor effect against glioma (Martins et al., 2013). A lipophilic prodrug 5-fluoro-2'deoxyuridine (FUdr) was incorporated into SLNs, and its concentration in the brain was higher at the same point when compared with FUdr solution (Wang et al., 2002). Yang et al. (1999) demonstrated the influence in the biodistribution profile of the antitumor molecule isolated from *Campotheca acuminata* (Campothecin) loaded in SLNs. The authors observed that Campothecin-SLNs presented a prolonged time of residence in several organs (more than four times) when in solution (Yang et al., 1999). Piperedine-SLNs prepared by emulsification-solvent diffusion were investigated in artificial Alzheimer model, and the results demonstrated a reduction in the oxidative stress and in the cholinergic degradation after 7, 14, and 28 days when compared with any control used (Yusuf et al., 2013).

8.3.1.4   INCREASE PERMEATION OF DRUGS THROUGH THE BBB

About 98% of new drugs development to central nervous system (CNS) falling as potential drug since they do not across the BBB in therapeutic concentrations (Blasi et al., 2011). SLNs have attracted many attentions in the last 10 years as a drug delivery system to brain target since they have to able to pass the BBB (Jose et al., 2014). Due to their small size, the SLNs can themselves diffuse via transcellular or paracellular pathways though the mucosa and release the drug inside the cells target (Rassu et al., 2017). SLNs loaded with riluzole-drug used in the treatment of aminotropic lateral by injection subcutaneous presented better ability in delivery to the rat brain when compared with the free drug aqueous dispersion (Fabiola et al., 2013).

8.3.2   STRATEGIES TO INCREASE THE HALF-LIFE AND BRAIN RETENTION OF SLNs

The body distribution of SLNs is strongly dependent on their surface characteristics like size, surface hydrophobicity, surface mobility, etc. These carriers can access the blood compartment easily (because of their small size) but the detection of these particles by the immune cells is a major limitation for their use (Fang et al., 2013). Uptake of nanoparticles by RES could result in therapeutic failure due to insufficient pharmacological drug concentration in the target tissue (e.g., brain). To overcome these limitations, various researchers have tried to increase the plasma half-life of SLNs by the following strategies (Kaur et al., 2008; Fang et al., 2013).

8.3.2.1   SURFACE COATING WITH HYDROPHILIC POLYMERS/ SURFACTANTS

The high rates of RES mediated detection and clearance of colloidal carriers by the liver, reduce the half-life of the drug. The interaction of the colloidal carriers with opsonins and thus with the membranes of macrophages (opso-nization) is believed to be the major criteria for clearance of these systems from the bloodstream. Hence, to prevent this clearance and to increase their bioavailability at the target site, the opsonization should be avoided or decreases. This RES recognition can be prevented by coating the surface particles with a hydrophilic and flexible polymer and/or surfactant (Kaur et al., 2008; Fang et al., 2013; Gastaldi et al., 2014.

Coating with polyethylene glycol (PEG), a polymer of hydrophilic nature showed promising results. It has been suggested that the PEG's with a molecular weight between 2000 and 5000 are necessary to suppress plasma protein adsorption. Furthermore, it has been observed that the thicker the particles coating, the slower the clearance, and hence a better protec-tion against liver uptake (Kaur et al., 2008). Fundarò et al. (2000) prepared nonstealth and stealth SLN, containing doxorubicin as an ion-pair complex. Stearic acid-PEG 2000 was used as stealth agent. These SLN showed signifi-cantly higher drug concentration in the rat brain (nearly 10 µg/g) compared to nonstealth SLN (2 µg/g) and doxorubicin solution, after i.v. administra-tion. The pegylated surface combined with the lipophilicity of the SLN may explain the presence of doxorubicin in the brain tissues (Fundarò et al., 2000).

In this connection, other hydrophilic molecules have been tried are Pluronic F-68®, Brij 78®, Brij 68®, or Tween-80® or hyaluronic acid. The particles surface modification with these polymers causes a steric hindrance effect, decreasing adsorption of opsonin in the SLN, as well as slowing removal of the particles by the RES.

In a study aimed at enhancing brain-specific targeting distribution, employing surface-modified SLN, Yusuf et al. (2013) prepared piperine SLN by an emulsification–solvent diffusion technique, with Tween-80® coating. Piperine is a natural alkaloid having a potent antioxidant effect, with poten-tial applications in Alzheimer's disease, since it readily crosses the BBB. Due to intense first-pass metabolism, the administration of piperine for brain delivery is not straightforward. Piperine encapsulates in SLN-Tween 80 was successfully targeted to the brain and was found to be effective at a low dose (2 mg/kg body weight). This result shows the successf of an effective delivery across the BBB with a generous payload and good delivery capa-bilities (Yusuf et al., 2013).

Wang et al. (2002) employed SLN modified with Pluronic F-68®, into which was incorporated 5-Fluoro-20-deoxyuridine (FUdR), a derivative of 5-fluorouracil with a significant cytotoxic activity that crosses the BBB to a moderate extent. However, this drug is rapidly metabolized after administration, particularly by the liver. The study demonstrated that SLN incorporation enhanced penetration and transportation of the FUdR into the brain. The study authors put forth two explanations: surface modification of the SLN with Pluronic F-68® could cause a steric hindrance effect that would decrease the adsorption of opsonin in the SLN surface, reducing RES uptake and prolonging retention time in plasma. Second, increased retention of the SLN in the brain capillaries combined with adsorption onto the capillary walls might create a higher concentration gradient, which would enhance transport across the EC layer, and consequently delivery the drug to the brain. The SLN might also be endocytosed by the EC, followed by drug release within these cells and delivery to the brain (Wang et al., 2002).

8.3.2.2   USE OF LIGANDS

Ligands or homing devices that specifically bind to surface epitopes or receptors on the target sites can be coupled to the surface of the long-circulating carriers. Modified SLN with an active targeting mechanism may have great potential, and represent a new challenge in SLN formulation (Kaur et al., 2008).

Neves et al. (2015) developed a new system to enter the brain by functionalizing SLN with apolipoprotein E, aiming to enhance their binding to low-density lipoprotein receptors on the BBB EC. No toxicity in the human cerebral microvascular EC (hCMEC/D3), a human BBB model, was observed up to 1.5 mg/mL of SLN over 4 h of incubation. The brain permeability was evaluated in transwell devices with hCMEC/D3 monolayers, and a 1.5-fold increment in barrier transit was verified for functionalized nanoparticles when compared with nonfunctionalized ones. The results suggested that these novel apolipoprotein E-functionalized nanoparticles resulted in dynamic stable systems able to be used for an improved and specialized brain delivery of drugs through the BBB (Neves et al., 2015).

Tsai et al. (2012) investigated the brain targeting ability of baicalein-loaded NLCs, containing vitamin E, and poloxamer 188, via i.v. route. The authors concluded that vitamin E helped increasing baicalein's stability in vivo pharmacokinetics studies. Moreover, baicalein-loaded NLCs exhibited a 7.5- and 4.7-fold higher baicalein accumulation in the brain compared to baicalein in solution in the cerebral cortex and brain system, respectively (Tsai et al., 2012).

8.3.2.3 PARTICLE SIZE

As mentioned in the previous topic, the size and the deformability of particles play a critical role in their clearance by the sinusoidal spleen. Particles must be either small or deformable enough to avoid the splenic filtration process in the walls of venous sinuses. Therefore, the size of a nanoparticle should not exceed 200 nm ideally.

Vinpocetine (VIN) is a derivative of vincamine alkaloid used for chronic cerebral vascular ischemia. However, it suffers from low bioavailability and short half-life. To overcome this issue, Morsi et al. (2013) formulated SLNs with VIN to be used as a brain-targeted sustained drug-delivery system. They observed a correlation between the particle size (z-average) and the cumulative released percent after 96 h. They observed that increasing surfactant concentration significantly decreased the particle size and significantly increased the cumulative released percent (Morsi at al., 2013).

8.3.3 PRECLINICAL STUDIES

There are already preclinical studies of SLN formulations to brain-target in the literature. This type of evaluation is extremely important for new drug development, especially for the emergence of therapies not yet explored that can effectively reach the market. Zara et al. (2002) made SLN nonstealth and stealth with doxorubicin for i.v. administration. As the stealthing agent, they used PEG 2000 at various concentrations. The SLNs and stealth SLN containing increasing amounts of stealthing agent, allowed doxorubicin-loaded nanoparticles to be transported through the BBB. They observed an increase of doxorubicin concentration in the brain on increasing the stealthing agent (PEG 2000). The amount of drug present in the rabbit brain ranged from 27.5 ng/g for nonstealth SLN to 242.0 ng/g for stealth SLN with 0.45% of PEG after 30 min of administration. After 6 h, doxorubicin was only detected in the group treated with stealth SLNs loaded with 0.45% of PEG (Zara et al., 2002).

Wang et al. (2002) incorporated 3',5'-dioctanoyl-5-fluoro-2'-deoxyuridine into SLNs. The drug solution and SLNs with the drug were administered intravenously, then it was determined that the AUC values achieved with SLNs were twice as high as that obtained by injecting plain drug solution (Wang et al., 2002). An i.v. injection of 1.3 mg/kg of camptothecin (an anticancer drug) into mice resulted in a prolonged drug resistance time in the body when loaded in SLNs compared with the drug solution. Furthermore, an increase of five folds in plasma AUC and of 10 folds in brain AUC was observed (Yang et al., 1999).

Reddy and Venkateshwarlu (2004) found a relationship between the charge on the SLN and the brain drug levels. They studied i.v. administration of etoposide loaded in SLN prepared with tripalmitin and etoposide solution to evaluate the brain levels. The etoposide loaded in SLN with positive charge achieved highest brain concentration (0.07% of injected dose per gram of organ/tissue) when compared to etoposide loaded in SLN with negative charge (0.02% of injected dose per gram of organ/tissue) and etoposide solution (0.01% of injected dose per gram of organ/tissue) (Reddy and Venkateshwarlu, 2004).

Manjunath and Venkateshwarlu (2006) produced SLNs of nitrendipine (lipophilic drug) for improving its bioavailability upon i.v. administration. Different types of triglycerides, such as tripalmitin, trimyristin and tristearin, emulsifiers (soy lecithin, poloxamer 188), and charge modifiers (dicetyl phosphate; DCP and stearylamine, SA), were used to prepare SLN. Nitrendipine loaded in SLNs was more extent by the brain and maintained higher drug levels for 6 h as compared to 3 h with the suspension (Manjunath and Venkateshwarlu, 2006).

Preclinical studies in 2008 showed that the blocking of VEGF-mediated neoangiogenesis could promote tumor infiltration (possibly by overexpression of proinvasive molecules or by cooption of existing cerebral blood vessels) and recruitment of circulating EC into the neoplasm (Norden et al., 2008). These data suggest that anti-VEGF signaling pathway inhibition could optimally work only if combined to other cytotoxic chemotherapeutics, to non-VEGF-mediated antiangiogenetic factors, or to radiotherapy.

A preclinical study showed that monoclonal antibodies directed against the extracellular portion of EGFR-VIII and PDGFR could be effective in the treatment of gliomas and could be used to target SLN with different drugs to glioma cells (Rich and Bigner, 2004). Similarly, antibodies directed to different target involved in the glioma VEGF signaling pathway (i.e., VEGFR) could be used to target SLN to glioma EC, possibly, interfering with the angiogenic process (Rich and Bigner, 2004). However, ligand–receptor and antibody–antigen recognition could interact and activate systemic and local host biological reactions, interfering immunological response. Because of this, an MAb directed to sites of the targeted molecule not involved in the endogenous ligand recognition was developed (Beduneau et al., 2007). MAb-conjugated liposomes, known as immunoliposomes, proved effective as brain drug delivery systems. Zhang et al. developed immunoliposomes carrying a plasmid DNA encoding the EGF receptor antisense mRNA, conjugated with two MAb directed to the mouse, Tf receptor (to pass through the BBB), and to human insulin receptor (to intratumor cell delivery). This study showed that these immunoliposomes are effective through i.v. administration in mice with U87 brain tumors (Zhang et al., 2002). A similar result was obtained in the same brain glioma model using immunoliposomes carrying a short hairpin RNA targeting EGFR mRNA (Boado, 2005; Zhang et al., 2004). Various colloidal carriers (including pegylated nanoparticles and NLC) conjugated with MAb antirat Tf (OX26) are under study and show promising results as brain drug delivery systems (Beduneau et al., 2007; Pardridge, 2007).

Blasi et al. (2013) prepared SLN with polysorbate® 80 (P80) intended to brain drug delivery. The in vivo toxicity was evaluated in rats. The results confirmed that SLN-P80 was effectively taken up by the BBB by endocytosis/transcytosis, after i.v. administration (Blasi et al., 2013).

Devkar et al. (2014) produce ondansetron (OND) hydrochloride loaded mucoadhesive NLCs for efficient delivery to brain through nasal route (i.n.). It was studied that the profile of OND-loaded NLC in brain and plasma after i.n. and i.v. administration. The results suggested that OND could be transported to the CNS after intranasal delivery, increasing drug concentration in brain and its bioavailability after nasal administration, being a promising approach for brain targeting via i.n.

Gartziandia et al. (2015) designed and optimized a suitable CS-coated NLC (CS-NLC) formulation to delivery of drugs to the brain after intranasal administration. The nasal mucosa toxicity studies after being exposed to CS-NLCs, by i.n. administration, for 15 consecutive days showed no histopathological lesions that indicated toxicity. CS-NLCs were loaded with the near infrared dye, DiR, and a high fluorescence by the nanoparticles was detected in mice brain after 0.5, 4.5, 6.5, and 23.5 h post-administration of CS-NLS. This CS-NLC formulation showed effective delivery to the brain after a single dose by i.n. administration, using less invasive administration routes (Gartziandia et al., 2015).

Esposito et al. (2015) produced NLC containing cannabinoid drugs. To this, rimonabant (RMN) was employed as model cannabinoid antagonist due to its physicochemical characteristics. RMN concentration in the brain and in the plasma 6 h after i.n. administration was performed. The RMN concentration in brain of animals treated with RMN-NLC was 5.8 $\mu g/g$, while the animals that received free RMN was 5.1 $\mu g/g$. The ratio of RMN concentration between brain and plasma ([RMNbrain]/[RMNplasma]) was calculated.

In the case of NLC administration, ([RMNbrain]/[RMNplasma]) was 17.11, while in the case of the drug administered by the reference solution, the ratio was 11.74 (Esposito et al., 2015).

Meng et al. (2016) developed a novel curcumin-loaded protein-free LDL resembling nanocarrier (PS80-NLC) to deliver Cur beyond the BBB. The NLC and PS80-NLC were intravenously administered to rats to investigate the brain-targeting capability of nanocarriers. DiR-loaded NLC and PS80-NLC were intravenously administered to rats to investigate the brain-targeting capability of nanocarriers. The PS80-NLC group showed higher fluorescence intensity (nearly 2.35 times greater) in the brain than the NLC group, suggesting that PS80-NLC accumulated more than NLC in the brain. The results demonstrated that PS80-NLC could cross the BBB and penetrate the brain more efficiently than NLC, being an assuring choice for the brain targeted delivery of drug. PS80-NLC and NLC were distributed in the liver and spleen, which agrees with the literature (Martins et al., 2013).

Esposito et al. (2017) studied the potential of SLN and NLC as nano-formulations to administer to the CNS poorly water soluble drugs, the URB597. As a strategy to alter SLN biodistribution and enhancing blood circulation time, the SLN surface was modified with P80. The tests were conducted in rats to study the efficacy of SLN-P80-URB597 to affect brain function and behavior. The results exploited a noninvasive intranasal (i.n.) administration route as an alternative to intraperitoneal administration. The data demonstrated that SLN-P80-URB597 increased social play behavior in rats compared to SLN-P80. The results suggest that the i.n. route could be proposed as an alternative administration route to exploit the therapeutic potential of anandamide hydrolysis inhibitors in social dysfunctions, such as autism (Esposito et al., 2017).

8.4 PERSPECTIVES OF THE DELIVERY SLNs TO BRAIN TARGET

Due to poor specificity of drugs to the treatment of the brain diseases in terms of the biodistribution and pharmacology, improvements in the development and researches to new drugs delivery systems are need to decrease the side effects and improve the survival of patients (Jose et al., 2014). The low number of new drugs reaching market demonstrates the complexity of CNS drug development. Thus, optimal strategies not exploited to development CNS drug delivery need to be considered, for increasing the number of therapeutics drugs that effectively reach in the market (Blasi et al., 2007).

The use of the nanoparticles to brain target has demonstrated potential delivery to a great variety of drugs, such as analgesics, anti-Alzheimer's disease, protease inhibitors, anticancer drugs, and several macromolecules (Kreuter 2014). The potential of lipid nanoparticles as drug delivery of various agents therapeutics has been successfully demonstrated, since that this carrier presents several advantages over others systems (Rostami et al., 2014). Many companies have encapsulated drugs within SLNs to enhance their penetration in the CNS (Gastaldi et al., 2014). Complete patent of products based on SLN had been acquired by Sky Pharma AG (Muttens, Switzerland). Vectorpharma (Triste, Italy) is developing SLNs products by microemulsion technique (Muller, 2000).

However, to that nanoparticles can be used as brain-target delivery systems several parameters should be considerate in the future: (1) the drug delivery system that would provide the brain-target should be biodegradable with effective biological safety; (2) the method to obtain the delivery system should produce nanoparticles more homogeneous and predictable; (3) factors that influence *in vivo* behavior of nanoparticles should be well evaluated and elucidated; and (4) considerations to improve the targeting efficiency should be made in the brain target system before their clinical application (Gao and McClements, 2016).

In the near future, neurologists and patients will benefit from nanotechnology-based drug delivery systems that will improve therapeutic response with reduced costs (Fonseca-Santos, 2015). In relation to SLNs as drug delivery systems, the focus will be the development of the self-actuated therapy considering the industrial needs such as scale-up, simple technology, qualification and validation, low cost, regulatory concerns (e.g., GRAS), etc. (Jaiswal and Gupta, 2013).

8.5 CONCLUDING REMARKS

SLNs can be a desirable and innovative drug system delivery drugs into the brain due to their ability overcoming the solubility, toxicity, and permeability of the drugs to the brain. SLNs shows promising as a potential therapeutic agent in the treatment of many diseases located in the brain tissue. However, the structure and dynamics of SNLs in different levels in vitro and in vivo, as well their stability, toxicity, and interaction with biological tissue should be continuously elucidated. In association, further research works should be conducted to explored brain target, so that suitable carriers can be developed soon.

KEYWORDS

- **solid lipids nanoparticles**
- **brain target**
- **preparation of SLN**
- **SLN for brain target**
- **nanotechnology**
- **characterization of SLNs**

REFERENCES

Battaglia, L.; et al. Techniques for the Preparation of Solid Lipid Nano and Microparticles. In *World's Largest Science, Technology & Medicine Open Access Book Publisher*: 2011. DOI: 10.5772/58405.

Beduneau, A.; Saulnier, P.; Benoit, J. P. Active Targeting of Brain Tumors Using Nanocarriers. *Biomaterials* **2007**, *28*, 4947–4967.

Begley, D. J. Delivery of Therapeutic Agents to the Central Nervous System: the Problems and the Possibilities. *Pharmacol. Ther.* **2004**, *104* (1), 29–45.

Beloqui, A.; et al. Nanostructured Lipid Carriers: Promising Drug Delivery Systems for Future Clinics. *Nanomedicine* **2016**, *12* (1), 143–161.

Blasi, P.; et al. Solid Lipid Nanoparticles for Targeted Brain Drug Delivery. *Adv. Drug Deliv. Rev.* **2007**, *59* (6), 454–477.

Blasi, P.; et al. Lipid Nanoparticles for Brain Targeting I. Formulation Optimization. *Int. J. Pharm.* **2011**, *419* (1–2), 287–295.

Blasi, P.; et al. Lipid Nanoparticles for Brain Targeting III. Long-Term Stability and In Vivo Toxicity. *Int. J. Pharm.* **2013**, *454* (1), 316–323.

Boado, R. J. RNA Interference and Nonviral Targeted Gene Therapy of Experimental Brain Cancer. *NeuroRx.* **2005**, *2* (1), 139–150.

Carlotti, M. E.; et al. Photostability and Stability over Time of Retinyl Palmitate in an O/W Emulsion and in SLN Introduced in the Emulsion. *J. Disper. Sci. Technol.* **2005**, *26* (2), 125–138.

Chaturvedi, S. P.; Kumar, V. Production Techniques of Lipid Nanoparticles: A Review. *RJBCS* **2012**, *3* (3), 525–541.

Chen, Y.; Dalwadi, G.; Benson, H. A. E. Drug Delivery Across the Blood-Brain Barrier. *Curr. Drug Deliv.* **2004**, *1* (4), 361–76.

Devkar, T. B.; Tekade, A. R.; Khandelwal, K. R. Surface Engineered Nanostructured Lipid Carriers for Efficient Nose to Brain Delivery of Ondansetron HCL Using Delonix Regia Gum as a Natural Mucoadhesive Polymer. *Colloids Surf B Biointerfaces.* **2014**, *122*, 143–150.

Domb, A. J. Lipospheres for Controlled Delivery of Substance. **1993**. Microencapsulation. Methods and Industrial Applications, 1st ed., Lipospheres for Controlled Delivery of Substances. U.S. Patents 5,188,837; 5,227,165; 5,221,535; 5,340,588. 2.

Elbahwy, I. A.; et al. Enhancing Bioavailability and Controlling the Release of Glibenclamide from Optimized Solid Lipid Nanoparticles. *J. Drug Deliv. Sci. Tecnhol.* **2017**, *38*, 78–89.

Esposito, E.; et al. Cannabinoid Antagonist in Nanostructured Lipid Carriers (NLCs): Design, Characterization and In Vivo Study. *Mater Sci. Eng. C Mater. Biol. Appl.* **2015**, *48*, 328–336.

Esposito, E.; et al. Lipid Nanoparticles for Administration of Poorly Water Soluble Neuroactive Drugs. *Biomed. Microdevices.* **2017**, *19* (3), 44.

Bondì, M. L.; Craparo, E. F.; Giammona, G.; Drago F. Brain-Targeted Solid Lipid Nanoparticles Containing Riluzole: Preparation, Characterization and Biodistribution Research Article. *Nanomedicine* **2013**, *5* (1), 25–32.

Fang, C.-L.; Al-Suwayeh, S. A.; Fang, J.-Y. Nanostructured Lipid Carriers (NLCs) for Drug Delivery and Targeting. *Recent Pat. Nanotechnol.* **2013**, *7* (1), 41–55.

Friberg, S.; Irena, L.; Gunilla, G. Microemulsions Containing Nonionic Surfactants-The Importance of the Pit Value. *J. Colloid Interface Sci.* **1976**, *56* (1), 19–32.

Fonseca-Santos, B.; Daflon Gremião, M. P.; Chorilli, M. Nanotechnology-Based Drug Delivery Systems for the Treatment of Alzheimer's Disease. *Int. J. Nanomed.* **2015**, *10*, 4981–5003.

Fundarò, A.; et al. Non-Stealth and Stealth Solid Lipid Nanoparticles (SLN) Carrying Doxorubicin: Pharmacokinetics and Tissue Distribution after I.v. Administration to Rats. *Pharmacol. Res.* **2000**, *42* (4), 337–343.

Gao, S.; David, J. Mc. Formation and Stability of Solid Lipid Nanoparticles Fabricated Using Phase Inversion Temperature Method. *Colloids Surf A* **2016**, *499*, 79–87.

Gartziandia, O.; et al. Chitosan Coated Nanostructured Lipid Carriers for Brain Delivery of Proteins by Intranasal Administration. *Colloids Surf B Biointerfaces* **2015**, *134*, 304–313.

Gastaldi, L.; et al. Solid Lipid Nanoparticles as Vehicles of Drugs to the Brain: Current State of the Art. *Eur. J. Pharm. Biopharm.* **2014**, *87* (3), 433–444.

Golmohammadzadeh, S.; Mokhtari, M.; Jaafari, M. R. Preparation, Characterization and Evaluation of Moisturizing and UV Protecting Effects of Topical Solid Lipid Nanoparticles. *Braz. J. Pharm. Sci.* **2012**, *48* (4), 683–690.

Huynh, N. T.; et al. Lipid Nanocapsules: A New Platform for Nanomedicine. *Int. J. Pharm.* **2009**, *379* (2), 201–209.

Iqbal, Md A.; et al. Nanostructured Lipid Carriers System: Recent Advances in Drug Delivery. *J. Drug Targ.* **2012**, *20* (10), 813–830.

Jaswal, S.; Gupta, G. D. Recent Advances in Solid Lipid Nanoparticles and Challenges. *Indo Am. J. Pharm. Res.* **2013**, *3* (12), 1601–1610.

Jose, S.; et al. In Vivo Pharmacokinetics and Biodistribution of Resveratrol-Loaded Solid Lipid Nanoparticles for Brain Delivery. *Int. J. Pharm.* **2014**, *474* (1–2), 6–13.

Johanson, C. Permeability and Vascularity of the Developing Brain: Cerebellum vs. Cerebral Cortex. *Brain Res.* **1980**, *190*, 3–16.

Juillerat-Jeanneret, L. The Targeted Delivery of Cancer Drugs Across the Blood-Brain Barrier: Chemical Modifications of Drugs or Drug-Nanoparticles? *Drug Discov. Today.* **2008**, *13* (23–24), 1099–1106.

Kaur, I. P.; et al. Potential of Solid Lipid Nanoparticles in Brain Targeting. *J. Control Release.* **2008**, *127* (2), 97–109.

Kreuter, J. Drug Delivery to the Central Nervous System By Polymeric Nanoparticles: What Do We Know? *Adv. Drug Deliv. Rev.* **2014**, *7*, 2–14.

Kuo, Y.-; Cheng, S.-J. Brain Targeted Delivery of Carmustine Using Solid Lipid Nanoparticles Modified With Tamoxifen and Lactoferrin for Antitumor Proliferation. *Int. J. Pharm.* **2016,** *499* (1–2), 10–19.

Mäder, K.; Mehnert, W. Solid Lipid Nanoparticles: Production, Characterization and Applications. *Adv. Drug Deliv. Rev.* **2001,** *47* (2–3), 165–96.

Manjunath. K; Venkateswarlu ,V. Pharmacokinetics, Tissue Distribution and Bioavailability of Nitrendipine Solid Lipid Nanoparticles After Intravenous and Intraduodenal Administration. *J. Drug Target.* **2006,** *14,* 632–645.

Marcato, P. D. Preparação, Caracterização e Aplicações em Fármacos e Cosméticos de Nanopartículas Lipídicas Sólidas. *REF* VI. **2009,** *2,* 1–37.

Martin-Padura, I.; Lostaglio, S.; Schneemann, M. Junctional Adhesion Molecule, a Novel Member of the Immunoglobulin Superfamily that Distributes at Intercellular Junctions and Modulates Monocyte Transmigration. *J. Cell Biol.* **1998,** *142,* 117–127

Martins, S. M.; et al. Brain Targeting Effect of Camptothecin-Loaded Solid Lipid Nanoparticles in Rat After Intravenous Administration. *Eur. J. Pharm. Biopharm.* **2013,** *85* (3 PART A), 488–502.

Mehnert, W; Karsten, M. Solid Lipid Nanoparticles: Production, Characterization and Applications. *Adv. Drug. Deliv. Rev.* **2012,** *64,* 83–101.

Meng, F.; et al. Design and Evaluation of Lipoprotein Resembling Curcumin-Encapsulated Protein-Free Nanostructured Lipid Carrier for Brain Targeting. *Int J Pharm.* **2016,** *506* (1–2), 46–56.

Morsi, N. M; Dalia, M. G; Hany, A. B. Brain Targeted Solid Lipid Nanoparticles for Brain Ischemia: Preparation and in Vitro Characterization. *Pharm. Dev. Technol.* **2013,** *18* (3), 736–744.

Mukherjee, S.; Ray, S; Thakur, R. S. Solid Lipid Nanoparticles: A Modern Formulation Approach in Drug Delivery System. *Indian. J. Pharm. Sci.* **2009,** *71* (4), 349–358.

Müller, R. Solid Lipid Nanoparticles (SLN) for Controlled Drug Delivery: A Review of the State of the Art. *Eur. J. Pharm. Biopharm.* **2000,** *50* (1), 161–177.

Nag, S. *Morphology and Molecular Properties of Cellular Components of Normal Cerebral Vessels.* Humana Press, Totowa, NJ, 2003.

Naseri, N.; Hadi V.; Parvin, Z.-M. Solid Lipid Nanoparticles and Nanostructured Lipid Carriers: Structure Preparation and Application. *Adv. Pharm. Bull.* **2015,** *5* (3), 305–313.

Neves, A. R.; et al. Solid Lipid Nanoparticles as a Vehicle for Brain-Targeted Drug Delivery: Two New Strategies of Functionalization with Apolipoprotein E. *Nanotechnology* **2015,** *26* (49), 495103.

Norden, A. D.; et al. Bevacizumab for Recurrent Malignant Gliomas: Efficacy, Toxicity, and Patterns of Recurrence. *Neurology* **2008,** *70,* 779–787.

Pardeike, J.; Hommoss, A.; Müller, R. H. Lipid Nanoparticles (SLN, NLC) in Cosmetic and Pharmaceutical Dermal Products *Int. J. Pharm.* **2009,** *366,* 170–184

Pardridge, W. M. Blood-Brain Barrier Drug Targeting: the Future of Brain Drug Development. *Mol. Interv.* **2003,** *3* (2), 90–105.

Pardridge, W. M. Drug Targeting to the Brain. *Pharm Res.* **2007,** *24* (9) 1733–1744.

Parhi, R.; Padilama, S. Preparation and Characterization of Solid Lipid Nanoparticles—A Review. *Curr. Drug Discov. Technol.* **2012,** *9* (1), 2–16.

Patil, H.; et al. Continuous Manufacturing of Solid Lipid Nanoparticles by Hot Melt Extrusion. *Int J Pharm.* **2014.** *471* (1–2), 153–156.

Radtke, M.; Wissing, S. A. Solid Lipid Nanoparticles (SLN) and Nanostructured Lipid Carriers (NLC) in Cosmetic and Dermatological Preparations. *Adv. Drug Deliv. Rev.* **2002,** *1*, 131–155.

Rassu, G.; et al. Nose-to-Brain Delivery of BACE1 siRNA Loaded in Solid Lipid Nanoparticles for Alzheimer's Therapy. *Colloids Surf B Biointerfaces* **2017,** *152*, 296–301.

Reddy, J. S.; Venkateshwarlu, V. Novel Delivery Systems for Drug Targeting to the Brain. *Drugs of Future.* **2004,** *29*, 63–83.

Reichel, A.; Begley, D.; Abbott, N. *Methods in Molecular Medicine: The Blood–Brain Barrier: Biology and Research Protocols. Humana Press*, Totowa, NJ, 2003; pp 307–325.

Reese, T.; Karnovsky, M. Fine Structural Localization of a Blood–Brain Barrier to Exogenous Peroxidase. *J. Cell Biol.* **1967,** *34*, 207–217.

Rich, J. N.; Tejas Bigner, D. D. Development of Novel Targeted Therapies in the Treatment of Malignant Glioma. *Nat. Rev. Drug Discov.* **2004,** *3* (5), 430–446.

Roney, C.; et al. Targeted Nanoparticles for Drug Delivery Through the Blood-Brain Barrier for Alzheimer's Disease. *J. Control Release.* **2005,** *108* (2–3), 193–214.

Rostami, E.; et al. Drug Targeting Using Solid Lipid Nanoparticles. *Chem. Phys. Lipids* **2014,** *181*, 56–61.

Sardo, C.; et al. Development of a Simple, Biocompatible and Cost-Effective Inulin-Diethylenetriamine Based siRNA Delivery System. *Eur J Pharm Sci.* **2015,** *75*, 60–71

Sarpietro, M. G.; et al. Idebenone Loaded Solid Lipid Nanoparticles: Calorimetric Studies on Surfactant and Drug Loading Effects. *Int. J. Pharm.* **2014,** *471* (1–2), 69–74.

Schwarz, C.; et al. Solid Lipid Nanoparticles (SLN) for Controlled Drug Delivery. I. Production, Characterization and Sterilization. *J. Control Release* **1994,** *30* (1), 83–96.

Shah, R.; Eldridge, D.; Palombo, E.; Harding, I. Lipid Nanoparticles: Production, Characterization and Stability. *Springer Briefs in Pharmaceutical Science & Drug Development*; 2015. DOI:10.1007/978-3-319-10711-0_1.

Shinoda, K; Hiroshi, S. The Stability of O/W Type Emulsions as Functions of Temperature and the HLB of Emulsifiers: The Emulsification by PIT-Method. *J. Colloid Interface Sci.* **1969,** *30* (2), 258–263.

Singh, A.; et al. Preparation and Characterization of Rizatriptan Benzoate Loaded Solid Lipid Nanoparticles for Brain Targeting. *Mater Today* **2015,** 2 (9), 4521–4543

Speiser, P. Lipid nanopellets als Tra¨gersystem fu¨r Arzneimit- I. Preparation and Particle Size Determination. Pharm. Phartel zur peroralen Anwendung. European Patent EP 0167825. **1990**

Teeranachaideekul, V.; et al. Influence of State and Crystallinity of Lipid Matrix on Physicochemical Properties and Permeation of Capsaicin-Loaded Lipid Nanoparticles for Topical Delivery. *J. Drug Deliv. Sci. Tecnhol.* **2017,** *39*, 300–307.

Tsai, M. J.; et al. Baicalein Loaded in Tocol Nanostructured Lipid Carriers (Tocol NLCs) for Enhanced Stability and Brain Targeting. *Int. J. Pharm.* **2012,** *423* (2), 461–470.

Venkateswarlu, V. ; Manjunath, K. "Preparation, Characterization and In Vitro Release Kinetics of Clozapine Solid Lipid Nanoparticles. *J. Control Release* **2004,** *95* (3), 627–38.

Wang, J. X.; Xun, S.; Zhi, R. Z. Enhanced Brain Targeting by Synthesis of 3',5'-Dioctanoyl-5-Fluoro-2'-Deoxyuridine and Incorporation into Solid Lipid Nanoparticles. *Eur. J. Pharm. Biopharm.* **2002,** *54* (3), 285–290.

Weber, S.; Zimmer, A.; Pardeike, J. Solid Lipid Nanoparticles (SLN) and Nanostructured Lipid Carriers (NLC) for Pulmonary Application: A Review of the State of the Art. *Eur. J. Pharm. Biopharm.* **2014,** *86* (1), 7–22.

Westesen, K.; Siekmann, B.; Koch, M. H. J. Characterization of Submicron-Sized Drug Carrier Systems Based on Solid Lipids by Synchrotron Radiation X-Ray Diffraction. *Trends Colloid Interfac Sci VII* **1993,** 356–356.

Wolburg, H.; et al. Localization of Claudin-3 in Tight Junctions of the Blood–Brain Barrier is Selectively Lost During Experimental Autoimmune Encephalomyelitis and Human Glioblastoma Multiforme. *Acta Neuropathol.* **2003,** *105,* 586–592.

Yadav, N.; et al. Solid Lipid Nanoparticles—A Review. *Int. J. App. Pharm.* **2013,** *5* (2), 8–18.

Yang, S. C.; et al. Body Distribution in Mice of Intravenously Injected Camptothecin Solid Lipid Nanoparticles and Targeting Effect on Brain. *J. Control Release* **1999,** *59* (3), 299–307.

Yasir, M.; Sara, U. V. S. Solid Lipid Nanoparticles for Nose to Brain Delivery of Haloperidol: In Vitro Drug Release and Pharmacokinetics Evaluation. *Acta Pharm. Sin. B.* **2014,** *4* (6), 454–463.

Yusuf, M.; et al. Preparation, Characterization, in Vivo and Biochemical Evaluation of Brain Targeted Piperine Solid Lipid Nanoparticles in an Experimentally Induced Alzheimer's Disease Model. *J. Drug Target.* **2013,** *21* (3), 300–311.

Zara, G. P.; Cavalli, R.; Bargoni, A. Intravenous Administration to Rabbits of Non-1308stealth and Stealth Doxorubicin-Loaded Solid Lipid Nanoparticles at Increasing Concentrations of Stealth Agent: Pharmacokinetics and Distribution of Doxorubicin in Brain and Other Tissues. *J. Drug Target.* **2002,** *10,* 327–335.

Zhang, Y.; et al. Suppression of EGFR Expression by Antisense or Small Interference RNA Inhibits U251 Glioma Cell Growth In Vitro and In Vivo. *Cancer Gene Ther.* **2004,** *13,* 530–538.

Zhang, Y.; Jeong Lee, H.; Boado, R. J.; Pardridge, W. M. Receptor-Mediated Delivery of an Antisense Gene to Human Brain Cancer Cells. *J. Gene Med.* **2002,** *4* (2), 183–94.

CHAPTER 9

AN INTRODUCTORY NOTE ON QUANTUM DOTS FOR BRAIN TARGETING

KIRTI RANI[*]

Amity Institute of Biotechnology, Amity University, Sector 125, Noida 201303, Uttar Pradesh, India

[*]*E-mail: krsharma@amity.edu; Kirtisharma2k@rediffmail.com*

ABSTRACT

Quantum dots (QDs) are nanoparticles or nanocrystals of a semiconducting material with diameters in the range of 2–10 nm and fluorescent semiconductor QDs have attracted lot of scientific attention over the last decade for their use in biomedical applications, whereas their potential toxicity and instability in biological environment has puzzled clinical researchers. Their surface modification and functionalization have been subjected for significant progress over last several years to make them versatile probes for biomedical applications, especially in biological imaging. This practice is facilitated with biological molecules with broad absorption spectra, bright emission, and physical and photophysical robustness along with current state of the art of the synthesis, modification, and bioconjugation of prepared QDs. These surface-modified QDs could be further used for brain-targeted imaging for getting precise localization of tumor tissues within normal brain parenchyma to achieve accurate diagnostic biopsy and complete surgical resection for improved surgical management of brain tumors as potent optical imaging and optical spectroscopy tools.

9.1 INTRODUCTION

Quantum dots (QDs) are nanosized particles or nanocrystals of a semiconducting material with diameters in the range of 2–10 nm, which were first discovered in 1980. QDs are photostable nanomaterials that have fluorescence lifetimes in the nanosecond range. Especially the semiconductor nanocrystals, called QDs, are a newly emerging nanomaterial, which has attracted many interests and has excitons confined in all three spatial dimensions (Fig. 9.1). QDs have properties that are between those of bulk semiconductors and discrete molecules. After excitation of the QDs, the emitted energy by QDs can be adjusted by controlling the composition and particle size of QDs due to the well-known effect of quantum-size confinement effect.

Although toxicity of QDs offers potentially invaluable societal benefits, such as in vivo biomedical imaging and detection, they may also pose risks to human health and the environment under certain conditions via at least three different pathways by which QDs can interfere with organism function and lead to metabolic disability or death. First is their composition, and upon their catalytic degradation processes inside the organism, toxic ions could be released from QDs and poison the cells and tissues or organisms in which QDs are posed to be administrated for molecular and cell imaging.

Compared with the bulk material, QD nanoparticles are more likely to have partial decomposition and release of ions due to their high surface-to-volume ratio. Another possible negative effect of QDs results from their small size, regardless of the composition of material. Particles can stick onto the surface of cell membranes or can be ingested and retained inside cells, causing the impairing and deleterious effects.

Researchers have demonstrated that the QD photoluminescence (PL) is effectively suppressed by an electric field that is commonly seen in the brain's neuronal membranes. This study revealed that QD PL is capable of monitoring a firing neuron's action potential profile with millisecond time resolution. These bright, small, and crystalline semiconductors have several beneficial photophysical features. As a result, they are now being considered for the potential imaging of neuronal action, and QDs are easily interfaced with experimental brain systems that tend to have a low cytotoxicity. They can also be easily localized either on or within the cells and plasma membranes. QDs have two-photon action cross-sections, which are several orders of magnitude larger than the fluorescent proteins or organic dyes currently used for biological imaging (Rowland et al., 2016).

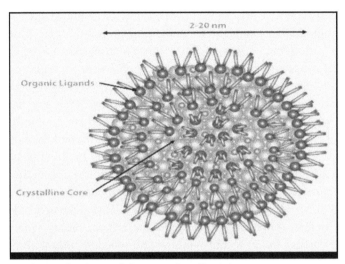

FIGURE 9.1 **(See color insert.)** Quantum dots dimensions.

QDs are novel class of inorganic fluorophore, which are recognized these days for their exceptional photophysical properties and exploited to be applied to various recent existing and emerging biomedical technologies. QDs have been proposed for the task of imaging action potentials as more effective nanoscale semiconductors with unique photophysical properties, which is found to observe the size-dependent PL. These QD nanoscale semiconductors have attracted great interest due to their unique structural, optical, and electronic properties, which arise due to their large surface-to-volume ratio and quantum-confinement effect. QDs, zero-dimensional semiconductor nanomaterials, are confined to a size of 2–8 nm particles that are smaller than the exciton. "Bohr radius" and various modifications were opted for the synthesis of very unique semiconductor QDs when doped with different dopants to be used as more advanced, safe, and potent nanophotocatalysts. The recent introduction of QDs into biological imaging was facilitated by their ability to functionalize their surfaces with wide variety of compatible biological molecules having unique broad absorption spectra, bright emission, and photophysical robustness. Photogenerated electrons and holes were experimentally observed for their quantum-confinement effect, which was observed in the "blue shift" in optical absorption spectra and enhancement of excitonic peak with reduction in particle size. The particle size below which the blue shift occurs is often termed as "critical size," which is the Bohr radius of excitons in semiconductors (Abbott, 2013). QDs, rods, and tetrapods have diameters that fall below the material's Bohr-exciton radius. When a photon hits such a semiconductor, some of their electrons are excited into higher energy states. When they return to their ground state, a photon of a frequency characteristic of that material is emitted. Quantum confinement is a phenomenon, which is the widening of the band-gap energy of the semiconductor material when its size has been shrunken to nanoscale.

The band gap of QD material is the energy required to create an electron and a hole with zero kinetic energy at a distance that their Coulombic attraction could be ignored. A bound electron–hole pair, termed exciton, would be generated if one carrier approaches the other. This exciton behaves like a hydrogen atom, except that a hole, which is not a proton, forms the nucleus. These QD particles then show a quantum-confinement effect due to localization of charge carriers hereafter called QDs or quantum particles and size of a semiconductor nanoparticle falls below the critical radius. Their respective charge carriers begin to behave quantum mechanically and the charge confinement leads to a series of discrete electronic states, and therefore, free-standing nanoparticles are preferred to understand the basic quantum-size effects in these materials (Jamieson, 2007).

In addition, QDs are designed for environmental monitoring, sensing, and testing as they have efficient semiconductor nanophotocatalystic properties. But they would need to have requirements similar to those that can be used in biological applications, such as biocompatible fluorescent labels for diagnosis and imaging of cells, tissues, and organs under physiological conditions. Previously, QDs have been shown to be useful and durable nanophotocatalysts for various problems of environmental interest such as water and air purification. Due to higher surface area-to-volume ratio than their bulk counterparts, the recombination of the electron–hole pair within the QDs drastically reduces as particle size, and therefore, QDs are expected to have higher photocatalytic activity than its bulk.

Moreover, QDs also have more attention because their confinement by the excited electrons and holes leads to optical and electronic properties different from those in bulk semiconductors. For QDs, the transportation length of electron/hole from crystal interface to the surface is short, which helps to accelerate the migration rate of electron/hole to the QD's surface to participate in the various chemical and biological reaction processes due to exhibiting increasing accessible surface of photocatalysts, as nanoparticles lead to increasing photocatalytic activity of QDs. Very tiny nanosized QDs can be easily incorporated into a cellular membrane that have low cytotoxicity and possess an intrinsic and specific sensitivity to surrounding electric fields (Agarwal et al., 2015; Spira and Hai, 2013; Tsytsarev et al., 2014).

When QDs were doped with various photostable or magnetostable metal ions, the energy from absorbed photons can be efficiently transferred to the impurity, quickly localizing the excitation and suppressing undesirable reactions on the nanocrystal surface. For example, incorporation of ZnS with other transition metals such as manganese, nickel, and copper can have a beneficial effect on the photoreactivity of photocatalysts, and doped ZnS semiconductor materials have a wide range of applications in electroluminescence devices, phosphors, light-emitting displays, and optical sensors. As mentioned, the number and the lifetime of free carriers (electrons/holes) are particle size- and dopant-dependent. Doping of ZnS with other transition metal ions offers a way to trap charge carriers and extend the lifetime of one or both of the charge carriers. The absorption and emission spectra have been tailored by choice of maintaining the nanocrystal size and shape by surface modification and PL efficiency that are determined by using surface traps.

These have been tuned via appropriate selection of the nanocrystal-capping ligands, and using various capping ligand-exchange process, the surface of the CdSe QDs can be modified by replacing the longer chain ligands of conventional trioctyl phosphine oxide (TOPO) or oleic acid (OA) with shorter chain ligand of butyl amine. This imparts colloidal stability and water solubility to CdSe QDs for its potential applications in biosensors and biological imaging. So, the reported study is well versed to be coined that crystallite sizes, oxidation potential of CdSe QDs, and stereochemical compatibility of ligands (TOPO, OA, and butyl amine) greatly influence the photophysics and photochemistry of CdSe QDs (Sharma et al., 2010). On the other hand, surface defects in QDs are only due to adopting surface modification or surface functionalization processes in which various capping agents or ligands are used to maintain their very tiny nanosized particles that act as temporary "traps" for the electron, hole, or excitons, quenching radiative recombination, and reducing the quantum yields. Surface passivation of QDs can confine the carrier inside the core and change the size and optical properties of QDs. Therefore, capping or passivation of the surface is crucial for development of photostable QDs. In the capping agent effect on the physical dimensions of QDs, it is possible to suggest that the capping agents can cover the particles, and therefore, the particles do not coalesce to form bigger particles, even after an extensive period. The application of very tiny nanosized QDs has been proposed for imaging of neuronal action potentials as well as it may therefore be possible to overcome several key limitations associated with current brain-imaging techniques to achieve improved resolution-imaging map to analyze brain function (Empedocles and Bawendi, 1997; Rowland et al., 2015).

From the last decades, semiconductor and fluorescent QDs are found to be considered very important optical-imaging tools with their surface modifications for surgical management of brain tumors, because successful surgical management of brain tumors requires the precise localization of tumor cells and cancerous tissues within normal brain parenchyma to achieve accurate diagnostic biopsy and complete surgical resection. The intravenous injection of QDs has been accompanied by reticuloendothelial system and macrophage sequestration for getting improved brain-imaging map for tracking the optically labeling of the tumors. Hence, macrophage-mediated delivery of QDs to brain tumors might be chosen as a novel biomedical technique to label tumors preoperatively through which tumors may be detected with optical imaging and chosen optical spectroscopy tools that become helpful to provide the surgeon with real-time optical feedback during the resection and biopsy of brain tumors (Popescu and Toms, 2006). Bioconjugated QDs are also excellent fluorescent probes and nanovectors that especially designed to transverse across the blood–brain barrier (BBB) and visualize drug delivery inside the brain to treat many neurodegenerative disorders and brain tumors. The theranostic QDs also offer a strategy to significantly improve the effective dosages of drugs to transverse across the BBB and to target inside the brain. Previously, targeted semiconductor QDs were investigated with the combination of fluorescence microscopy, as optical tools for discriminating glioblastoma cells from normal brain tissue. And, it was concluded that QDs can be coupled to biomolecules and, thus, target specific molecules on cells by using living cultured cells. Epidermal growth factor receptor (EGFR, Her1) was chosen as the specific target on glioma cells to observe upregulated activity of Her1 expression activity in all gliomas and glioblastoma multiformes as well as mapped the low-grade oligodendroglial tumors. The intraoperative diagnosis of brain tumors and timely evaluation of their respective biomarkers can be better diagnostic and therapeutic tools that might be helpful for adopting improved therapeutic and clinical measures to prepare effective guidelines for arranging rapid adjunctive clinical studies. So, this proposed study was evaluated for feasibility and specificity of using of streptavidin-coated conjugated anti-EGFR antibodies QD-labeled antibodies QD for rapid visualization of EGFR expression in human brain tumor cells and human glioma tumor cell lines (frozen tissue sections of glioblastoma multiforme and of oligodendroglioma) in which elevated levels of EGFR expression (SKMG-3, U87) was reported. These findings demonstrated that QD-labeled antibodies can provide an instant and precise nanotechnology-based biomedical technique for characterizing the presence or absence of a specific predictive biomarker (Wang et al., 2007).

Polymeric materials are not found to employed compatible matrices for QDs surface modifications for getting improved optical applications, but also provide mechanical and chemical stability to QDs as well as preventing nanocrystal agglomeration that offered them good processability into technologically relevant nanostructures. Using polymers as the modifier of QDs, surface modifications are especially required to permit monodispersion of materials and further surface functionalization is necessary to facilitate specific biological and chemical interactions in reaction or targeting medium. Various innovative practices are still under considerations to develop a variety of polymer-modified QD materials, using nanofabrication methods, as well as resulting in their excellent optical properties and applications (Schipper et al., 2009). Surface modification of QDs with a combination of polyethylene glycol (PEG) and other polymers has also been investigated. In the previous study, the use of an ABC triblock copolymer in addition to PEG for QD coating aims to minimize the aggregation and fluorescence loss of QDs when they are stored in physiological buffer or injected into live animals. Various experiments in cells and animal models confirmed the stability and brightness of these surface-modified fluorescent QDs. New type of quantoplexes incorporating near-infrared (NIR)-emitting CdTe QDs, polyethylenimine (PEI), and plasmid DNA were also assembled. After their intravenous injection, the quantoplexes accumulated rapidly in the lung, liver, and spleen, and the fluorescence signal could be detected for at least a week. Tracking of these quantoplexes immediately after intravenous injection was performed and rapid redistribution was observed from the lungs to the liver, which was dependent on the PEI topology and the quantoplex formulation. In addition, a similar quantoplex has been also assembled to couple with PEI where the PEI was replaced by PEG–PEI conjugate, which exhibited passive tumor accumulation in nude mice bearing subcutaneous tumors. Using a solid dispersion technique, hydrophobic 10,12-pentacosadiynoic acid (PCDA) was incorporated to assemble QD-loaded micelles and both PEG–PCDA (polyethylene glycol–polydiacetylene conjugates) and PCDA–herceptin conjugates participated in the micelle assembly which was subjected to intramicellar cross-linking between PCDAs upon ultraviolet (UV) irradiation. Noninvasive fluorescent imaging showed that with herceptin as the targeting ligand for human EGFR 2 (Her2), these QD-loaded micelles exhibited high antitumor activity and selective toxicity that led to a marked reduction in tumor volume (Gao et al., 2008). Surface coating of QDs with amphiphilic polymers is employed to achieve multivalent polymerization of QDs surface, synthesis of end-functionalized polymers on the QDs surface, and preparation of the dendrimer–QD materials. Surface coating of QDs with amphiphilic polymers would improve water solubility and chemical as well as biological functionality of the prepared very tiny nanosized QDs. Multidentate polymeric ligands have the ability to retain the luminescence quantum yields and to simultaneously provide the necessary and efficient colloidal stability with chemical and biological functionality. Attachment of end-functionalized polymers to the QD surfaces leads to many new applications for such advanced materials, and in addition, dendrimer-encapsulated approaches have been used for the controlled synthesis of advanced and more potent QDs. Surface modification of QDs can improve their respective aqueous solubility, especially via the organometallic route, as well as protect them from any degradation and fluorescence-quenching phenomenon. In addition, their surface modification with PEG can also help to minimize the nonspecific uptake in normal organs and provide functional groups for further choosing any bioconjugation process (Schipper et al., 2009).

Delivery of imaging agents to the brain is highly important for the diagnosis and treatment of various central nervous system (CNS) disorders to elucidate their pathophysiology. Optically active QDs can provide a novel nanoprobe with unique physical, chemical, and optical properties as a more promising tool for molecular and cellular imaging apart from their poor stability and low BBB permeability that severely limits their ability to enter into and act on their target sites in the CNS following parenteral administration. So, recently, QD-based imaging platform was developed for brain imaging by incorporating QDs into the core of PEG–poly(lactic acid) nanoparticles which was surface functionalized with wheat germ agglutinin and further delivered into the brain via nasal application. The resulting surface-modified QDs were found to have high payload capacity, improved aqueous behavior, and showed safe brain targeting and imaging properties (Gao et al., 2008).

9.2   CONCLUSION

The chapter provided precise knowledge to describe the use of QDs for brain imaging as effective optical tools, making them more advanced multifunctional nanoprobes by using various surface modifications and bioconjugation processes. These semiconductor QDs have generated extensive interest for biomedical and clinical applications due to high brightness, long-term stability, simultaneous detection of multiple signals, tunable emission spectra. So, many considerable clinical and innovative practices must be adopted to functionalize the QDs as more potent nanoprobes for improving their applicability in brain imaging to treat many neurodegenerative and CNS disorders, brain tumors as well as achieving brain surgical management strategies.

ACKNOWLEDGMENT

The author would like to thank Amity University, Noida, Uttar Pradesh, India.

KEYWORDS

- **quantum dots**
- **nanoparticles**
- **brain imaging**
- **optical imaging**
- **semiconducting material**
- **optical spectroscopy**

REFERENCES

Abbott, A. Solving the Brain. *Nature* **2013,** *499,* 272–274.

Agarwal, R.; Domowicz, M. S.; Schwartz, N. B.; Henry, J.; Medintz, I.; Delehanty, J. B.; Stewart, M. H. Delivery and Tracking of Quantum Dot Peptide Bioconjugates in an Intact Developing Avian Brain. *ACS Chem. Neurosci.* **2015,** 6, 494–504.

Empedocles, S. A.; Bawendi, M. G. Quantum-Confined Stark Effect in Single CdSe Nanocrystalline Quantum Dots. *Science* **1997,** 278, 2114–2117.

Gao, X.; Chen, J.; Chen, J.; Wu, B.; Chen, H.; Jiang, X. Quantum Dots Bearing Lectin-Functionalized Nanoparticles as a Platform for In Vivo Brain Imaging. *Bioconj. Chem.* **2008,** 19 (11), 2189–2195. **DOI**: 10.1021/bc8002698.

Jamieson, T.; Bakhshi, R.; Petrova, D.; Pocock, R.; Imani, M.; Seifalian, A. M. Biological Applications of Quantum Dots. Biomaterials **2007,** 28, 4717.

Popescu, M. A.; Toms, S. A. In Vivo Optical Imaging Using Quantum Dots for the Management of Brain Tumors. Expert Rev. Mol. Diagn. **2006,** 6 (6), 879–890.

Rowland, C. E.; Susumu, K.; Stewart, M. H.; Oh, E.; Makinen, A. J.; O'Shaughnessy, T. J.; Kushto, G. Imaging Cellular Membrane Potential through Ionization of Quantum Dots. *Proc. SPIE* **2016,** 9722, 97220S.

Rowland, C. E.; Susumu, K.; Stewart, M. H.; Oh, E.; Makinen, A. J.; O'Shaughnessy, T. J.; Kushto, G. Electric Field Modulation of Semiconductor Quantum Dot Photoluminescence: Insights into the Design of Robust Voltage-Sensitive Cellular Imaging Probes. *Nano Lett.* **2015,** 15, 6848–6854.

Schipper, M. L.; Iyer, G.; Koh, A. L.; Cheng, Z.; Ebenstein, Y, Aharoni, A.; Keren, S.; Laurent, A.; Li, B.; Rao, J.; Chen, X.; Banin, U.; Wu, A. M.; Sinclair, R.; Weiss, S. Particle Size, Surface Coating, and PEGylation Influence the Biodistribution of Quantum Dots in Living Mice. Small **2009,** 5, 126–134.

Sharma, S. N.; Kumara, U.; Singh, V. N.; Mehta, B. R.; Kakkar, R. Surface Modification of CdSe Quantum Dots for Biosensing Applications: Role of Ligands. Thin Solid Films **2010,** 519 (3), 1202–1212.

Spira, M. E.; Hai, A. Multi-Electrode Array Technologies for Neuroscience and Cardiology. *Nat. Nanotechnol.* **2013,** 8, 83–94.

Tsytsarev, V.; Liao, L. D.; Kong, K. V.; Liu, Y. H.; Erzurumlu, R. S.; Olivo, M.; Thakor, N. V. Recent Progress in Voltage-Sensitive Dye Imaging for Neuroscience. J. *Nanosci. Nanotechnol.* **2014,** 14, 4733–4744.

Wang, J.; Yong, W. H.; Sun, Y.; Vernier, P. T.; Koeffler, H. P.; Gundersen, M. A.; Marcu, L. J. Receptor-Targeted Quantum Dots: Fluorescent Probes for Brain Tumor Diagnosis. *Biomed. Opt.* **2007,** 12 (4), 044021.

CHAPTER 10

GRAPHENE FOR BRAIN TARGETING

S. J. OWONUBI[1], B. A. ADERIBIGBE[2,*], V. O. FASIKU[1], E. MUKWEVHO[1], and E. R. SADIKU[3]

[1]Department of Biological Sciences, North-West University, Private Bag X2046, Mmabatho 2735, South Africa

[2]Department of Chemistry, University of Fort Hare, Alice Campus, Eastern Cape, South Africa

[3]Department of Chemical, Metallurgical and Materials Engineering, Tshwane University of Technology, CSIR Campus, Building 14D, Private Bag X025, Lynwood Ridge 0040, Pretoria, South Africa

*Corresponding author. E-mail: blessingaderibigbe@gmail.com

ABSTRACT

Delivery of drugs to a target locale has become the major objective of drug delivery systems (DDSs); this is because in addition to other benefits, it reduces toxicity to the barest minimum. The brain is an essential organ and thus is very well protected. To deliver drugs to the brain, fondly called brain targeting, therapeutic formulations need to possess the capacity to circumvent the barriers, which protect the brain from possible invaders or unwanted molecules. Thus, a clear understanding of the barriers protecting targeted delivery to the brain is of utmost significance for successful targeting to the brain.

In this chapter, we introduce DDSs and targeted drug delivery, mapping out targets drug formulations aiming for in the body. Then, we focus on brain targeting and the strategies, which are employed to deliver drugs to the brain, indicating properties which nanocarriers possess that allow the blood–brain barrier to be transversed. And finally, with the immense research successes of graphene and its derivatives, it is no surprise that its capacity to circumvent the protective barrier has been identified by researchers and highlighted by relevant references made.

10.1   INTRODUCTION

The crucial objective for any drug-delivery-based study involves providing assistance to patients through the designing of therapeutically beneficial formulations; this is because the means by which a drug formulation is conveyed to the patient can have a substantial influence on the drugs' efficacy. The techniques involved to deliver drugs have over time been accepted by researchers around the world as the missing puzzle to effectively treat any patient, because it is the connecting factor between the formulation of the drug and the treatment of the patient (Owonubi et al., 2017a). While the administration of some formulations can result in an optimum effective concentration range favoring maximum benefit to the patient, alternate concentrations could be toxic or yield no medical benefit to the patient (Reddy and Swarnalatha, 2010). Over decades, as a result of the much needed improvement to the efficacy of therapeutic formulations to treat severe diseases, a multidisciplinary methodology for the delivery of the therapeutic formulation has over time been developed.

10.2   DRUG DELIVERY SYSTEM

The term "drug delivery systems" (DDSs) has evolved and become devices or formulations of drugs which allow for the conveyance of clearly defined substance(s) into the body, providing improvement to its effectiveness, reduced toxicity, with an apt impact on the quantity of the substance released and how frequent the release occurs (Owonubi et al., 2017a). Over the years, utilization of controlled means to deliver drugs has advanced significantly with researchers, resulting in the design of numerous therapeutic formulations which lead to improved patient convenience and compliance (Hoffman, 2008). More recently, researchers are able to deliver therapeutic formulations at desired rate for lengthy periods of time, which could range from mere days (Aderibigbe et al., 2015a) to much longer (Aderibigbe et al., 2015b), sometime perceived as sustained delivery systems. Although generally, delivery via transdermal and oral approaches is known to occur within 24 h substantially improving the delivered therapeutics' effectiveness yet reducing the side-effects, delivery systems which employ implants tend to be capable of delivering therapeutic formulations for months, and in some other cases lasting for years (Huebsch and Mooney, 2009; Owonubi et al., 2017b; Saini et al., 2015; Teo et al., 2016). During the last two decades, pharmaceuticals and researchers worldwide focused on the delivery systems called targeted DDS. Drug targeting, as some researchers call it, evolved as a result of a few challenges encountered by researchers using the conventional drug dosage delivery forms (Muller and Keck, 2004); these challenges are as follows:

- Solubility of the therapeutic formulations
- Low absorption of the drug formulations
- Existence of high-bound membranes making diffusion very difficult
- Biological instability of therapeutic formulations
- Short half-life of the therapeutic formulation
- Large volume of distribution of the therapeutic formulation
- Low specificity of therapeutic formulation
- Low therapeutic index of formulation
- Potential side-effects.

10.3 TARGETED DRUG DELIVERY

This unique form of drug delivery is one where the therapeutic formulation is deliberately targeted or delivered purposely to a site of activity without affecting neighboring organs, tissues, or cells. Some researchers describe the targeted drug delivery (TDD) as a smart DDS, and this technique changed the perception of researchers, for it greatly improves the effectiveness of a therapeutic formulation and reduces the side-effects as a result of the concentration of the therapeutics in the specific required location, thus reducing the presence of the therapeutic formulation in other body tissues or organs where it is not required, leading to less toxicity. Usually, certain characteristics are accepted as ideal for a TDDS to perform optimally (Chen and Liu, 2012), and these ideal characteristics are as follows:

- Biochemically nontoxic
- Biocompatibility (nonimmunogenic)
- Physiochemically stable in vivo and in vitro
- Capacity to restrict the distribution of the drug only to the targeted cells
- Possess constant capillary circulation
- Possess drug release rates which are controllable and predictable
- Drug release rates should not have an effect on the drug action
- Amount of drug released should be therapeutic, no excesses
- Minimal drug leakage during the drug transition
- The system must be biodegradable without challenges
- The design of the TDDs should be easy, reproducible, and cost-effective

Despite the advances made with TDD, more research need to be done to achieve breakthrough clinical trials. This is a result of the existence of certain hindrances to the advancement of drug targeting, and one of the most prominent factors is the existence of mucus lining in the body (Kumar Khanna, 2012). The coatings (mucus layer) protect certain sensitive tissues in the body but also hinders the access of these tissues to beneficial therapeutic formulations. These protective coatings famous for confining and finally eliminating foreign organisms tend to limit the conveyance of drugs formulated to the various sections of the body, as well as the digestive tract, female reproductive system, eyes, and lungs. Another hindrance to mention is the blood–brain barrier (BBB), a semipermeable capillary membrane consisting of a lone sheet of endothelial cells lining the inner capillary surfaces of the brain, permitting only defined molecules through. As such, molecules which are large and lowly fat soluble are restricted from entry into the brain, and molecules which possess high electric charge are slowed down. But those molecules which are soluble in lipids easily transverse the BBB and get into the brain. This BBB is known to be penetrable by glucose, oxygen, and carbon dioxide but will resist hydrogen ions from passing through (Kumar Khanna, 2012). As the skull shields the brain externally, so the BBB protects the brain internally. Figure 10.1 depicts the basics about the BBB, trying to show different means by which molecules could be permitted to go through the barrier.

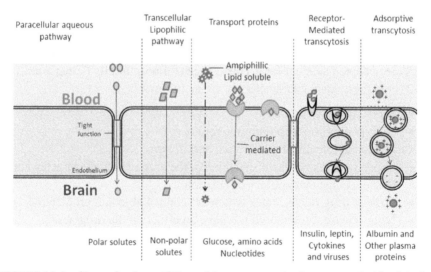

FIGURE 10.1 **(See color insert.)** Potential transport mechanisms across the blood–brain barrier.

Two major classifications of drug targeting exist, in the form of active and passive.

10.3.1   PASSIVE TARGETING

Targeting in the passive form, sometimes called "physical targeting," is grounded on the design of a drug–carrier complex which evades elimination from the body via excretion, phagocytosis, metabolism, and opsonization in order that the therapeutic formulation resides flowing in the blood stream, thereby allowing its spread to the defined target locale by stimuli such as pH, temperature, shape, or molecular size. Examples of passive drug targeting are direct drug injection and catheterization. Passive targeting exploits the possible alteration to the physiology of the tissues which are diseased in altered pathological situations.

10.3.2   ACTIVE TARGETING

This form of drug targeting relates to specific antibodies, antibody fragments, and peptides that are coupled (a form of surface modification) with possible DDS or drug candidates and assist them to behave like "homing devices" for attachment to receptor-type molecules existing at the known target location. Active targeting offers a wide array of prospects, having the benefit of the ability to modify the surfaces of the potential carrier systems. Vasir et al. (2005) indicated that to target by the use of magnetic field or some form of ultrasonic energy still classifies such targeting as an active targeting. Quite a number of drug conjugate employing active targeting use a three-tier arrangement made up of a target, a polymer, or lipid acting in place of the carrier and an active chemotherapeutic formulation (Kumar Khanna, 2012). For potential targets which are specific to certain tumors, the antigens on the cell surfaces will have to show positive expression completely and consistently in the tumor cells, but negative expression in the cells without tumor.

The active targeting approach further classified into three categories.

10.3.2.1   FIRST-ORDER TARGETING

This form of targeting tends to restrict the spreading of potential drug carrier systems to the capillary bed of a prearranged target locale, tissue, or organ, for example, sectional targeting in joints, plural cavity, lymphatics, cerebral ventricles, eyes, and peritoneal cavity.

10.3.2.2   SECOND-ORDER TARGETING

This form of targeting denotes the transport of drugs to defined cells, such as tumor cells, epidermal Langerhans cells, osteoclast cells (in bone), or dendritic cells (in lymphoid tissues), for example, selective drug delivery to microglial cells in central nervous system (CNS).

10.3.2.3   THIRD-ORDER TARGETING

This category of targeting relates specifically to delivery of drugs to intracellular sites of target cells, for example, entrance of drug complexes using ligand-based receptors and endocytosis.

10.4   DRUG TARGETS IN THE BODY

In the body, the main points of targets of drugs are the receptors on cell membranes, the cell membranes with lipids constituents, and the antigens or proteins on cell surfaces.

10.4.1   RECEPTOR ON CELL MEMBRANES

These receptors tend to allow for defined interaction between the drug carriers and the cells, thus giving room for their absorption through receptor-mediated endocytosis (Vasir et al., 2005). In the case of cancer-related treatments, receptors called "folate receptors" are known to be overexpressed in the cancer cells with epithelial origin. These folate receptors aid in cancer treatments by aiding in the delivery of drugs to specific tumor sites, for example, ovary cancers, breast cancers, brain, cancer tumors, and lung malignancies. Furthermore, another means to TDD is via peptide receptors. These receptors are highly expressed in quite a number of tumor cells; thus, a potential to conjugate drug carrier to analogs of peptides could result in a potential to enable specific targeting on interaction with the peptide receptors.

10.4.2   CELL MEMBRANES WITH LIPID COMPOSITION

Lipids, the permeability of their membranes, and fluid nature or their artificial phospholipid analogs are influenced by the interactions with cellular membranes; these lead to the mechanisms of signal transduction affected and it induces apoptotic cell death.

10.4.3   CELL SURFACES WITH PROTEINS OR ANTIGENS

In comparison to normal cells, diseased cells tend to either over- or under-express new proteins or they find themselves expressing entirely different proteins altogether. This phenomenon is what monoclonal antibodies takes advantage for designing drug formulations which target these proteins. For example, a growth factor, erbB2, is known to be overexpressed on the tumor cell surfaces in certain breast cancers (Iqbal and Iqbal, 2014); this allows for its use for treatment together with a doxorubicin (DOX) drug liposomal formulation attached to anti-erbB2 antibody (Belmonte et al., 2015; Rocic, 2015). This form of immunotherapy is attractive as it is easy to access, expresses low within in the normal cells, and is distributed evenly within the tumor.

10.5   BRAIN TARGETING

Vaccine delivery via the mucosal membranes has over the years been of interest to researchers; this is due to the fact that the surfaces of these membranes double as a key entry location for quite a number of pathogens. The nasal delivery is one of the most attractive mucosal sites for immunization and this is solely because the nasal epithelium is known for its relatively low enzymatic activity, high permeability, and its capacity to host significant number of cells with normal immune response. Furthermore, the nasal mucosal surface route is advantageous because it has the capacity to offer basic cost-effective procedures for immunization, with marked improvement to the compliance of patients. More recently, polymer-based nanocarriers have been employed as they provide alternate but feasible means to deliver antigenic molecules to the nasal route (Ali et al., 2010; Bernocchi et al., 2016; Csaba et al., 2009; Tobío et al., 1998; Vila et al., 2004). Apart from using nanocarriers to improve protection by encapsulating the antigen and assisting in transport facilitation, drug delivery employing nanoparticles has the capacity to provide more effective and efficient immune cell antigen recognition capacity. These additions are important for optimum transportation and delivery of the antigens leading to the ideal immune response. This reveals that vaccination via the mucosal nasal route can be achieved and optimized by designing ideal vaccine nanocarriers.

The brain is an organ which is the center of the CNS. Being the safe haven of the CNS is dynamically regulated and well protected. Although there exists a number of means by which molecules in circulation progress into the brain, the two main means are via circulation of blood within the brain and the movement of the cerebrospinal fluid (CSF). As highlighted beforehand, this protective BBB guards the brain by limiting entrance to certain organisms, only allowing defined substances. In relation to drug delivery and targeting, it is the most arduous barrier to circumvent, mainly hindering transportation of drugs through blood circulation into the brain. Although researchers over time have gathered better insight into the behavior of the components of the BBB components molecular structure, receptor expressions at the barrier, progresses made by researchers, and innovations in techniques based on nanotechnology, most of the disease relating to the CNS or the brain still linger untreated by effective therapies to date. This is solely as a result of potential candidate drugs being not able to transverse the BBB, other specialized CNS barriers or the blood cerebrospinal fluid barrier (BCSFB), the to reach the particular regions of brain (Chen and Liu, 2012; Neuwelt et al., 2008). The total understanding of the physiology of the BBB, response to external stimulus, permeability under diverse conditions, and behavior of various receptors at the BBB needs to be achieved for a solution to circumvent the barrier. The BBB and the BCSFB strictly control entrance of molecule(s) through parenteral means into the brain. The barrier is hence generally referred to as the essential barrier that prevents molecule(s) from getting into the brain through widespread outlets of networks of blood capillary and thus an understanding of its unique biological characteristics contributing to its low selective permeability is paramount to evading it.

The most common noninvasive method employed as a solution to deliver drug(s) to the brain of any particular patient is to transform the drug molecule to become more fat soluble, a process known as lipidization. Usually, the water-soluble drugs limit transport through the BBB, but on lipidization of these drug parts, a solution to the BBB issue by the use employing traditional chemistry means is achieved. Treatment of diseases associated to the CNS is mainly perplexing; this is a result of the default preventive measures setup by the BBB. Delivering drug molecule(s) to the brain parenchyma is mostly prevented by range of biochemical, physiological, and metabolic hindrances which altogether makeup the BCSFB, BBB, and possibly the blood–tumor barrier. Only more recently with the developments in drug-delivery techniques have sufferers of multiple forms of brain diseases witnessed reasonable successes to prove that the difficult barriers protecting entrance into the brain might eventually have been overcome. With all these factors to consider, generally it has been indicated that for a DDS to be ideal for brain targeting, it should possess two main characteristics (Gao, 2016):

- Penetrate effectively through or bypass stealthily the BBB.
- Target-specific diseased locale, with little or no interference with normal cells.

Recent innovations in designing polymer-based DDSs have improved the conveyance of potential target drugs into the parenchyma of the brain. Significant improvement can only result with research efforts focused at the design of more therapeutics with less toxic drug molecules and a more forceful quest of more effective mechanism of delivering these drugs to brain targets (Gokhale et al., 1994; Kshirsagar et al., 2005). Various techniques to aid to deliver drug(s) to the CNS and possibly transverse the barrier of the brain in the process are shown in Figure 10.2.

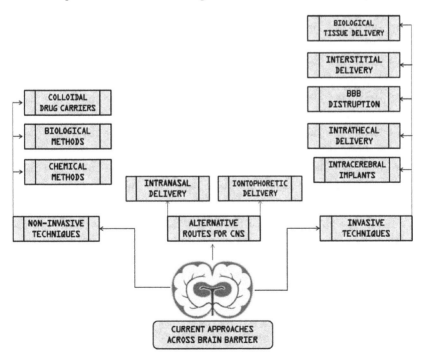

FIGURE 10.2 Current approaches to delivery to central nervous system.

10.5.1   APPROACHES TO DELIVER DRUG(S) TO THE BRAIN

To successfully transverse the BBB for potential TDD to the brain, many drugs have required formulation to have sufficient physical and chemical characteristics to enable passage. Characteristics such as positive charge, high lipid solubility, and low molecular size are crucial for its feasibility (Banerjee et al., 1996), but other attempts to transverse or bypass the BBB over time have been attempted with some level of success.

Temporal disruption of the BBB could be achieved in the presence of gas bubbles, by a pulsed ultrasound technique (Etame et al., 2012; Hynynen et al., 2005, 2006; Kinoshita, 2006), or in the midst of circulating microbubbles by focused ultrasound as achieved by Alkins et al. (2013) and Fan et al. (2013). A different method to temporarily disrupt the BBB is by the injection of mannitol solution into the neck via the arteries (Upadhyay, 2014), causing an increased sugar uptake by the brain capillaries and eventually leading to shrinkage of endothelial cells by loss of water causing them to opens tight junction, which are the physical barriers of the BBB. This effect typically takes about 20–30 min, in the course of which drugs, which normally do not have the capacity to go across the BBB, get across freely. Other researchers have rather utilized intraventricular/intrathecal delivery to bypass the BBB, by implanting a connector to the ventricles subcutaneously inside the brain via the scalp by an outlet catheter. This technique was employed by researchers of the Hotchkiss Brain Institute insulin delivery into the brain (Derakhshan and Toth, 2013). By intranasal delivery, some researchers have as well been able to bypass the BBB (Illum, 2004). Although not all drug forms have been successful with this technique (Yoffey, 1958), cases have occurred wherein nasal drug formulations have been conveyed to locations which are high and deep in the nasal cavity; some other cases have had the formulations reaching the olfactory mucosa; thus, transportation of the drug(s) to the brain or in other cases to the CSF through the neurons of the olfactory receptor might happen (Sharma et al., 1996).

Although the aim is to be able to transverse or bypass the BBB, there are significant aspects to consider which relate to BBB's integrity throughout treatment phase. Three key approaches are used to inspect the BBB, and they are (Mendonça et al., 2015) as follows:

- Total anatomical observation centered on the bordering infusion of Evans blue dye.
- Assessment by the transmission electron microscopy at the subcellular level using the tracer lanthanum nitrate found outside the cell.

This will aid to find out if the tracer transversed and got through the BBB to the brain parenchyma.

- The use of western blot to find out potential matrix–cell or cell–cell disorder in adhesion junctional proteins and endothelial tight proteins in brain parenchyma.

These approaches are necessary because quite a number of stimuli or agents could upset the BBB's integrity and possible transportation of drug through it.

Other researchers have employed the use of drug carriers—colloidal or nanodrug carriers, tailored to convey drug(s) into the brain parenchyma as they possess attractive properties, which allow for transportation of drug(s) across the BBB (Chen and Liu, 2012). The properties of potential nanocarriers which enable them transverse the BBB are:

- Nontoxic nature
- Biodegradable and biocompatible
- Stability in blood (no potential to dissociate or aggregate)
- Possess particle sizes less than 100 nm (except if it is transported by monocytes or macrophages)
- Prolonged circulation time in the blood
- Capacity to be stealth, not resulting in an immune response
- Possess a targetable BBB moiety (potential receptors or capacity to be taken up by monocytes or macrophages)
- Maintenance of inbred drug stability
- Programable drug discharge profile
- Capacity to encapsulate smaller proteins, molecules, nucleic acids, or peptides.

Of the numerous carriers, it is significant to mention that only molecules formed from chemical species possessing polar heads and hydrophobic tails (amphiphilic molecules—polymeric nanoparticles and liposomes) have been comprehensively researched upon target the brain (Chen and Liu, 2012; Garcia-Garcia et al., 2005). In recent times, a researcher developed a unique biodegradable polymer hydrogel consisting of poly(lactic-*co*-glycolic acid) and chitin to aid to deliver therapeutics to tumors of the brain (Krebs, 2014). This gel was detailed in the patent allowing for the release of anti-vascular endothelial growth factor (VEGF) to the periphery of the resected tumor site in a localized manner over a sustained period having steady release rate. In another patent by Bae et al. (2007), these researchers focused on using a pH-sensitive polymer to release a drug in the acidic microenvironment of endosomes and solid tumors. Weiss et al. (2010) researched upon ligands that can be used to target such as folate and mixed micelles were attached to them to enhance delivery into brain cells. Yet, another filed patent focused on development of small, less aggregable polymer nanoparticles encapsulated with drugs for penetration of the brain (Zhou et al., 2015). Another researcher filed a patent, where they utilized polymethacrylic acid grafted starch nanoparticles containing polysorbate moieties to target the polymer to brain tissues (Wu and Shalviri, 2013).

But with the recent advent of researchers employing graphene oxide (GO) as a nanocarriers for the TDD (Liu et al., 2013a, 2013b; Wang et al., 2013c), some level of success has been observed.

10.6   GRAPHENE FOR BRAIN TARGETING

Graphene has been the recent catch for about a decade as a widely researched upon material, having two dimensions with sp^2-hybridized carbon structure. This structure endows graphene with excellent mechanical properties, electrical conductivity and electron transport capabilities, very high aspect ratio, large surface area, pliability and impermeability, and low coefficient of thermal expansion (Balandin et al., 2008; Bolotin et al., 2008; Bunch et al., 2008; Geim, 2009; Huang et al., 2011; Lee et al., 2008; Novoselov et al., 2005; Stankovich et al., 2006; Zhu et al., 2010). It has been reported that the theoretic exact surface area of graphene reaches 2630 m^2/g, even greater when compared to theoretic surface area of single-walled carbon nanotube (1300 m^2/g) (Stankovich et al., 2006).

GO is regarded mostly as a forerunner for the production of graphene by thermal or chemical reduction. It is synthesized through chemical curing of graphite via oxidation and dispersal in water/appropriate organic solvents (Park and Ruoff, 2009; Stankovich et al., 2007). GO synthesized from oxidation of graphite in acidic circumstances is rather more utilized because it is more advantageous than using pure graphene (Jiang et al., 2016; Loh et al., 2010). GO allows for chemical functionalization as a result of its hydrophilic functional groups; also, it possesses a much broader range of physical properties when compared to pure graphene; this is as a result of having diverse structural states (Dreyer et al., 2010). The pi-structure of graphene enables the capacity to interface with drugs which are water loving and these feature makes graphene materials ideal drug carriers, especially for hydrophobic drugs (Goenka et al., 2014; Kuila et al., 2012; Liu et al., 2013b; Rana et al., 2011). Numerous means by which GO can be functionalized using click chemistry has provided the capacity of its application for various purposes as well as to deliver drugs to targeted locations. Other derivatives of graphene, nanographene, reduced-GO, and nano-GO have also been employed for brain targeting.

In recent times, quite a lot of researchers have utilized graphene or its derivatives for targeting brain related locations or cells. In 2015, Li and colleagues at the Sun Yat-Sen University in Guangzhou successfully utilized functional nano-GO particles for targeting brain glioma U251 cells (Li et al., 2015), which are the most malignant tumors accounting for over 50% of intracranial tumors (Deorah et al., 2006). Li and his colleagues prepared functionalized nano-GO particles and incubated them with U251 glioma cells, which are from human brain, pleomorphic, and adherent. They observed under near-infrared exposure, the targeted fluorescence imaging and photothermy of the incubated cells, and also their cytotoxicity was estimated by Cell Counting Kit-8. Using anti-transferrin receptor (CD71)–fluorescein (FITC) as a control with a fluorescence microscope, a significant killing efficacy was visible after 808 nm near-infrared exposure using flow cytometry. It was reported that both the nano-GO–transferrin–FITC group, together with the CD71–FITC group, showed fluorescence, which was greenish-yellow, whereas the control group not having the target molecule nano-GO–FITC was negative. With results showing that the nano-GO–transferrin–FITC incubated after 48 h with U251 cells at 0.1, 1, 3, and 5 mg/ml, the absorbance was 0.747 ± 0.031, 0.732 ± 0.043, 0.698 ± 0.051, and 0.682 ± 0.039, while the absorbance of control group is 0.759 ± 0.052. This showed no significant differences between the nano-GO–FITC groups and the control group. Also reported was that the nano-GO–transferrin–FITC had significantly greater death index and apoptosis when compared to the nano-GO–FITC group and the blank control group ($P < 0.05$). The findings showed these nano-GO–transferrin–FITC particles had an observable effect of target imaging and photothermal therapy on glioma U25 cells.

The synthesis of porphyrin-functionalized graphene oxide (PGO) and confirmation of its success in the removal of substantial brain cancer cells targeted in vitro was reported in the *New Journal of Chemistry* (Su et al., 2015). Using noninvasive phototermal therapy with near-infrared laser irradiation, these researchers synthesized a biocompatible PGO with high absorbance of 808 nm as a brain cancer therapy model. It was reported that PGO is more stable than reduced-GO in aqueous solution by two times. Also it was highlighted that the photothermal conversion of PGO increased by 33% and 89% in comparison to reduced-GO and GO, respectively, under the 808 nm irradiation, resulting in destruction of a large number of brain cells in vitro. The researchers came to a conclusion that the synthesized PGO model with active functional groups aids to specifically target brain cancer cells without affecting the surrounding cells.

In a further study, a collaboration between researchers of the Texas Tech University and Texas A&M, Su et al. (2016) focused on targeted therapy to treat brain cancer by synthesizing "porphyrin-immobilized nanographene oxide" (PNG) conjugated to a peptide. The researchers synthesized PNG and bioconjugated a peptide to it to potentially improve its capacity for photothermal brain cancer therapy. It was reported in the research that the control PNG, which was spread into an agar-based synthetic tissue prototype, showed a phototermal conversion efficiency of 19.93%, having PNG concentration of only 0.5 wt%, with 0.6°C/s heating rate at the onset of the irradiation, whereas with GO at 0.5 wt%, the resultant phototermal conversion efficiency was 12.20% and 0.3°C/s heating rate. This leads to the grafting of the PNG to a peptite, to specifically target brain tumor cells and the research reported that the photothermal therapy effects of the PNG–peptide totally eliminated the tumor cells in vivo using rat models.

As reported in the *Journal of Nanobiotechnology*, Mendonça et al. (2015) in Brazil showed how the utilization of reduced-GO induced the transient passage through the BBB allowing potential targeting to the brain, utilizing the matrix-assisted laser desorption/ionization mass spectrometry imaging (MALDI–MSI), a recent method for identifying the progressive dispersal of nanomaterials through the brain. This study reported that systemically injected reduced-GO was traced mainly to the hippocampus and thalamus of rats and its entrance caused a reduction in the paracellular tightening of the BBB. This decrease was evident by Evans blue dye infusion, junctional protein expression levels, and subcellular transmission electron microscopy. The researchers also studied the utilization of MALDI for detection of temporal spreading of potential nanomaterials all over the brain. They reported that reduced-GO was successfully identified and examined over time in the brain and confirmed that MALDI could be an important tool to treat disorders to the brain, which are normally not responsive as a result of the BBB impermeability.

Researchers revealed that treating with GO is adequate to prevent tumor development in glioblastoma cells involved in brain cancers (Fiorillo et al., 2015). They used two grades of GO (0.2–2-μm flake-sized GO called s-GO and 2–20 μm flake-sized GO called b-GO) to access the effects of GO on the proliferation of cancer stem cell (using MCF7 cells) by tumor-sphere assay. The s-GO showed inhibition of tumor-sphere formation, which was dose-dependent in the range of 1.25–25 µg/mL having an IC-50 of 12.5 µg/mL. But, although the b-GO was dose dependent as well, they occurred in the ranges of 6.25–100 µg/mL and having a similar IC-50 as s-GO. Since this showed that regardless of the GO flake size, there was similar potency, they went further to use b-GO to evaluate efficacy against multiple cancer types: ovarian, lung, prostrate, pancreatic, and most importantly brain cancers. Using 25 and 20 µg/mL doses to test efficacy, the researchers confirmed that GO is sufficient in preventing tumor-sphere formation across six independent cancer lines, suggesting that GO targets phenotypic properties unique to cancer cell lines.

The design of a photoresponsive protein–graphene–protein (PGP) shell for the transportation of anticancer cargo to tumor location and later unload these target payloads was reported by Asian scientists (Hu et al., 2014). These researchers designed a photothermal capsule, using reduced-GO nanosheets and lactoferrin by double emulsion method. This capsule was then used to successfully transport DOX payload of 9.43 µmol/g and subsequently triggered the burst release using near-infrared irradiation photoactuation. The PGP capsules were biocompatible having great cancer-targeting efficacy and they successfully eradicated tumors in 10 days after lone 5 min near-infrared irradiation with no distal harm confirming their potential for photoresponsive-targeted chemotherapy for tumor treatment.

The development of a lactoferrin-modified multifunctional glioma-targeted DDS was reported in the *Materials Science and Engineering C* journal. These researchers employed GO-containing biotargeting ligand and superparamagnetic iron oxide nanoparticles on the surface of the GO for ability to target by magnetism. The targeted delivery technique demonstrated a release reliant on the pH condition and expected superparamagnetic behavior, with potential therapeutic application to the treatment of gliomas (Song et al., 2017). Song and his colleagues loaded superparamagnetic iron oxide nanoparticles on the surface of GO through chemical precipitation. They then used the ligand lactoferrin for the construction of the targeted delivery system, as it has the capacity to target glioma cells and cells of the BBB. Using VSM, TEM, and DLS, the distribution size of 200–1000 nm was confirmed for the designed lactoferrin-modified multifunctional glioma-targeted DDS and when this delivery system was compared against C6 glioma cells for cytotoxicity and intracellular delivery efficiency with free DOX, loaded with the GO–iron oxide, the DOX loaded on this newly designed lactoferrin-conjugated GO–iron oxide nanocomposite delivery system showed superior performance.

Researchers have successfully fashioned transferrin conjugated DOX-encapsulated PEGylated nanoscaled graphene oxide (Tf-PEG-GO-DOX) to target glioma cells and deliver DOX, an anticancer drug (Liu et al., 2013a). By evaluation of tumor volume after treatment, the effect on the glioma cells of rat models was compared. The comparison between the saline, free DOX, PEG–GO–DOX, and Tf–PEG–GO–DOX-loaded nanocarriers for treatment revealed volumes of 326.6, 303.9, 259.7, and 180.8 mm^3, respectively, at day 14 of postintracranial implantation, with Tf–PEG–GO–DOX showing the strongest preventive influence on the tumor development. These researchers also reported the survival rates of the different treatment groups after treatment, with mean survival times physiological saline at 17 days, DOX treatment at 19 days, PEG–GO–DOX at 21 days, and Tf–PEG–GO–DOX at 25 days relating to approximately 41.7% lifespan extension for the Tf–PEG–GO–DOX treatment group, when compared to the 31.6% and 19% for PEG–GO–DOX for DOX treatments, respectively.

The brain tumor targeting ability of IL-13 by modification and conjugation to polymer nanoparticles and mesoporous silica-coated graphene nanosheets was demonstrated (Wang et al., 2013a). After successfully designing the multifunctional DDS and adsorption of DOX with high loading efficiency, the researchers confirmed high absorption in the near-infrared irradiation and superior photothermal efficiency, even when compared to carbon nanotubes. They also reported substantial targeting of glioma cells and excellent sustained release properties, pH-response, and heat-stimuli response.

Many more researchers have been involved in the use of graphene of some form or the other for brain targeting for treatment with levels of success (Lu et al., 2012; Liu et al., 2013a; Robinson et al., 2011; Wang et al., 2013b; Yang et al., 2013a, 2013b).

10.7   CONCLUSION AND PERSPECTIVES

In addition to the growing use of graphene, researchers have found interesting the utilization of derivatives of graphene in the past decade with numerous publications and patents alike. From its use in TDD to its use for many other applications, its success has been very evident, with its potential being appreciated more by each new publication and patent. But it is noteworthy to mention that there are quite some discrepancies with the results obtained by some scientists in the context of its toxicity to cells. Most scientists have reported very good biocompatibility with no significant cellular damage on treatment, but some researchers have reported cell toxicity, enhancing apoptosis, and necrosis (Sasidharan et al., 2011; Vallabani et al., 2011). It has been identified that this disparity in research findings is a result of the dimensions, differences in size, method of functionalization, and possibly purification technique used for the graphene-based materials. An additional important aspect to mention is its method of elimination from the living system. As the method of elimination, degradation or excretion is one other factor that is not fully understood by researchers.

Improved knowledge about the distinct properties of graphene and the distinct behavior of its derivatives in varied biological conditions is necessary for not just to improved research but to better understand its application. So, better understanding of the functionalization of the derivatives and conjugation modalities will lead to improved knowledge about its tunable biocompatibility and related general pharmacokinetics.

ACKNOWLEDGMENTS

The financial assistance of the Medical Research Council (MRC) South Africa and National Research Foundation, South Africa toward this research is hereby acknowledged.

KEYWORDS

- **drug delivery**
- **graphene**
- **brain targeting**
- **tumors**
- **cancer**
- **blood–brain barrier**

REFERENCES

Aderibigbe, B.; Varaprasad, K.; Sadiku, E.; Ray, S.; Mbianda, X.; Fotsing, M.; Owonubi, S.; Agwuncha, S. Kinetic Release Studies of Nitrogen-Containing Bisphosphonate from Gum Acacia Crosslinked Hydrogels. *Int. J. Biol. Macromol.* **2015a,** *73,* 115–123.

Aderibigbe, B. A.; Owonubi, S. J.; Jayaramudu, J.; Sadiku, E. R.; Ray, S. S. Targeted Drug Delivery Potential of Hydrogel Biocomposites Containing Partially and Thermally Reduced Graphene Oxide and Natural Polymers Prepared via Green Process. *Colloid Polym. Sci.* **2015b,** *293* (2), 409–420.

Ali, J.; Ali, M.; Baboota, S.; Sahani, J. K.; Ramassamy, C.; Dao, L.; Bhavna. Potential of Nanoparticulate Drug Delivery Systems by Intranasal Administration. *Curr. Pharm. Des.* **2010,** *16* (14), 1644–1653.

Alkins, R.; Burgess, A.; Ganguly, M.; Francia, G.; Kerbel, R.; Wels, W. S.; Hynynen, K. Focused Ultrasound Delivers Targeted Immune Cells to Metastatic Brain Tumors. *Cancer Res.* **2013,** *73* (6), 1892–1899.

Bae, Y. H.; Na, K.; Lee, E. S. pH-Sensitive Polymeric Micelles for Drug Delivery. *Google Patents,* 2007.

Balandin, A. A.; Ghosh, S.; Bao, W.; Calizo, I.; Teweldebrhan, D.; Miao, F.; Lau, C. N. Superior Thermal Conductivity of Single-Layer Graphene. *Nano Lett.* **2008,** *8* (3), 902–907.

Banerjee, G.; Nandi, G.; Mahato, S. B.; Pakrashi, A.; Basu, M. K. Drug Delivery System: Targeting of Pentamidines to Specific Sites Using Sugar Grafted Liposomes. *J. Antimicrob. Chemother.* **1996,** *38* (1), 145–150.

Belmonte, F.; Das, S.; Sysa-Shah, P.; Sivakumaran, V.; Stanley, B.; Guo, X.; Paolocci, N.; Aon, M. A.; Nagane, M.; Kuppusamy, P.; Steenbergen, C.; Gabrielson, K. ErbB2 Overexpression Upregulates Antioxidant Enzymes, Reduces Basal Levels of Reactive Oxygen Species, and Protects against Doxorubicin Cardiotoxicity. *Am. J. Physiol. Heart Circ. Physiol.* **2015,** *309* (8), H1271–H1280.

Bernocchi, B.; Carpentier, R.; Lantier, I.; Ducournau, C.; Dimier-Poisson, I.; Betbeder, D. Mechanisms Allowing Protein Delivery in Nasal Mucosa Using NPL Nanoparticles. *J. Control. Release* **2016,** *232,* 42–50.

Bolotin, K. I.; Sikes, K. J.; Jiang, Z.; Klima, M.; Fudenberg, G.; Hone, J.; Kim, P.; Stormer, H. L. Ultrahigh Electron Mobility in Suspended Graphene. *Solid State Commun.* **2008,** *146* (9–10), 351–355.

Bunch, J. S.; Verbridge, S. S.; Alden, J. S.; van der Zande, A. M.; Parpia, J. M.; Craighead, H. G.; McEuen, P. L. Impermeable Atomic Membranes from Graphene Sheets. *Nano Lett.* **2008,** *8* (8), 2458–2462.

Chen, Y.; Liu, L. Modern Methods for Delivery of Drugs across the Blood-Brain Barrier. *Adv. Drug Deliv. Rev.* **2012,** *64* (7), 640–665.

Csaba, N.; Garcia-Fuentes, M.; Alonso, M. J. Nanoparticles for Nasal Vaccination. *Adv. Drug Deliv. Rev.* **2009,** *61* (2), 140–157.

Deorah, S.; Lynch, C. F.; Sibenaller, Z. A.; Ryken, T. C. Trends in Brain Cancer Incidence and Survival in the United States: Surveillance, Epidemiology, and End Results Program, 1973 to 2001. *Neurosurg. Focus* **2006,** *20* (4), E1.

Derakhshan, F.; Toth, C. Insulin and the Brain. *Curr. Diabetes Rev.* **2013,** *9* (2), 102–116.

Dreyer, D. R.; Park, S.; Bielawski, C. W.; Ruoff, R. S. The Chemistry of Graphene Oxide. *Chem. Soc. Rev.* **2010,** *39* (1), 228–240.

Etame, A. B.; Diaz, R. J.; Smith, C. A.; Mainprize, T. G.; Hynynen, K.; Rutka, J. T. Focused Ultrasound Disruption of the Blood-Brain Barrier: A New Frontier for Therapeutic Delivery in Molecular Neurooncology. *Neurosurg. Focus* **2012,** *32* (1), E3.

Fan, C. H.; Ting, C. Y.; Lin, H. J.; Wang, C. H.; Liu, H. L.; Yen, T. C.; Yeh, C. K. SPIO-Conjugated, Doxorubicin-Loaded Microbubbles for Concurrent MRI and Focused-Ultrasound Enhanced Brain-Tumor Drug Delivery. *Biomaterials* **2013,** *34* (14), 3706–3715.

Fiorillo, M.; Verre, A. F.; Iliut, M.; Peiris-Pages, M.; Ozsvari, B.; Gandara, R.; Cappello, A. R.; Sotgia, F.; Vijayaraghavan, A.; Lisanti, M. P. Graphene Oxide Selectively Targets Cancer Stem Cells, across Multiple Tumor Types: Implications for Non-Toxic Cancer Treatment, via "Differentiation-Based Nano-Therapy". *Oncotarget* **2015,** *6* (6), 3553–3562.

Gao, H. Progress and Perspectives on Targeting Nanoparticles for Brain Drug Delivery. *Acta Pharm. Sin. B* **2016,** *6* (4), 268–286.

Garcia-Garcia, E.; Andrieux, K.; Gil, S.; Couvreur, P. Colloidal Carriers and Blood-Brain Barrier (BBB) Translocation: A Way to Deliver Drugs to the Brain? *Int. J. Pharm.* **2005,** *298* (2), 274–292.

Geim, A. K. Graphene: Status and Prospects. *Science* **2009,** *324* (5934), 1530.

Goenka, S.; Sant, V.; Sant, S. Graphene-Based Nanomaterials for Drug Delivery and Tissue Engineering. *J. Control. Release* **2014,** *173,* 75–88.

Gokhale, P. C.; Kshirsagar, N. A.; Khan, M. U.; Pandya, S. K.; Meisheri, Y. V.; Thakur, C. P.; Choudhary, C. B. Successful Treatment of Resistant Visceral Leishmaniasis with Liposomal Amphotericin B. *Trans. R. Soc. Trop. Med. Hyg.* **1994,** *88* (2), 228.

Hoffman, A. S. The Origins and Evolution of "Controlled" Drug Delivery Systems. *J. Control. Release* **2008,** *132* (3), 153–163.

Hu, S.-H.; Fang, R.-H.; Chen, Y.-W.; Liao, B.-J.; Chen, I. W.; Chen, S.-Y. Photoresponsive Protein–Graphene–Protein Hybrid Capsules with Dual Targeted Heat-Triggered Drug Delivery Approach for Enhanced Tumor Therapy. *Adv. Funct. Mater.* **2014,** *24* (26), 4144–4155.

Huang, X.; Yin, Z.; Wu, S.; Qi, X.; He, Q.; Zhang, Q.; Yan, Q.; Boey, F.; Zhang, H. Graphene-Based Materials: Synthesis, Characterization, Properties, and Applications. *Small* **2011,** *7* (14), 1876–1902.

Huebsch, N.; Mooney, D. J. Inspiration and Application in the Evolution of Biomaterials. *Nature* **2009,** *462* (7272), 426–432.

Hynynen, K.; McDannold, N.; Sheikov, N. A.; Jolesz, F. A.; Vykhodtseva, N. Local and Reversible Blood-Brain Barrier Disruption by Noninvasive Focused Ultrasound at Frequencies Suitable for Trans-Skull Sonications. *NeuroImage* **2005,** *24* (1), 12–20.

Hynynen, K.; McDannold, N.; Vykhodtseva, N.; Raymond, S.; Weissleder, R.; Jolesz, F. A.; Sheikov, N. Focal Disruption of the Blood-Brain Barrier due to 260-kHz Ultrasound Bursts: A Method for Molecular Imaging and Targeted Drug Delivery. *J. Neurosurg.* **2006,** *105* (3), 445–54.

Illum, L. Is Nose-to-Brain Transport of Drugs in Man a Reality? *J. Pharm. Pharmacol.* **2004,** *56* (1), 3–17.

Iqbal, N.; Iqbal, N. Human Epidermal Growth Factor Receptor 2 (HER2) in Cancers: Overexpression and Therapeutic Implications. *Mol. Biol. Int.* **2014,** *2014,* 9.

Jiang, Y.; Biswas, P.; Fortner, J. D. A Review of Recent Developments in Graphene-Enabled Membranes for Water Treatment. *Environ. Sci.: Water Res. Technol.* **2016,** *2* (6), 915–922.

Kinoshita, M. Targeted Drug Delivery to the Brain Using Focused Ultrasound. *Top. Magn. Reson. Imaging* **2006,** *17* (3), 209–215.

Krebs, M. D. Biodegradable Polymers for Delivery of Therapeutic Agents. *Google Patents,* 2014.

Kshirsagar, N.; Pandya, S.; Kirodian, B.; Sanath, S. Liposomal Drug Delivery System from Laboratory to Clinic. *J. Postgrad. Med.* **2005,** *51* (5), 5.

Kuila, T.; Bose, S.; Mishra, A. K.; Khanra, P.; Kim, N. H.; Lee, J. H. Chemical Functionalization of Graphene and Its Applications. *Prog. Mater. Sci.* **2012,** *57* (7), 1061–1105.

Kumar Khanna, V. Targeted Delivery of Nanomedicines. *ISRN Pharmacol.* **2012,** *2012,* 9.

Lee, C.; Wei, X.; Kysar, J. W.; Hone, J. Measurement of the Elastic Properties and Intrinsic Strength of Monolayer Graphene. *Science* **2008,** *321* (5887), 385–388.

Li, Z. J.; Li, C.; Zheng, M. G.; Pan, J. D.; Zhang, L. M.; Deng, Y. F. Functionalized Nano-Graphene Oxide Particles for Targeted Fluorescence Imaging and Photothermy of Glioma U251 Cells. *Int. J. Clin. Exp. Med.* **2015,** *8* (2), 1844–1852.

Liu, G.; Shen, H.; Mao, J.; Zhang, L.; Jiang, Z.; Sun, T.; Lan, Q.; Zhang, Z. Transferrin Modified Graphene Oxide for Glioma-Targeted Drug Delivery: In Vitro and In Vivo Evaluations. *ACS Appl. Mater. Interfaces* **2013a,** *5* (15), 6909–6914.

Liu, J.; Cui, L.; Losic, D. Graphene and Graphene Oxide as New Nanocarriers for Drug Delivery Applications. *Acta Biomater.* **2013b,** *9* (12), 9243–9257.

Loh, K. P.; Bao, Q.; Eda, G.; Chhowalla, M. Graphene Oxide as a Chemically Tunable Platform for Optical Applications. *Nat. Chem.* **2010,** *2* (12), 1015–1024.

Lu, Y. J.; Yang, H. W.; Hung, S. C.; Huang, C. Y.; Li, S. M.; Ma, C. C.; Chen, P. Y.; Tsai, H. C.; Wei, K. C.; Chen, J. P. Improving Thermal Stability and Efficacy of BCNU in Treating Glioma Cells Using PAA-Functionalized Graphene Oxide. *Int. J. Nanomed.* **2012,** *7,* 1737–1747.

Mendonça, M. C. P.; Soares, E. S.; de Jesus, M. B.; Ceragioli, H. J.; Ferreira, M. S.; Catharino, R. R.; da Cruz-Höfling, M. A. Reduced Graphene Oxide Induces Transient Blood–Brain Barrier Opening: An In Vivo Study. *J. Nanobiotechnol.* **2015,** *13* (1), 78.

Muller, R. H.; Keck, C. M. Challenges and Solutions for the Delivery of Biotech Drugs—A Review of Drug Nanocrystal Technology and Lipid Nanoparticles. *J. Biotechnol.* **2004,** *113* (1–3), 151–170.

Neuwelt, E.; Abbott, N. J.; Abrey, L.; Banks, W. A.; Blakley, B.; Davis, T.; Engelhardt, B.; Grammas, P.; Nedergaard, M.; Nutt, J.; Pardridge, W.; Rosenberg, G. A.; Smith, Q.; Drewes, L. R. Strategies to Advance Translational Research into Brain Barriers. *Lancet Neurol.* **2008,** *7* (1), 84–96.

Novoselov, K. S.; Geim, A. K.; Morozov, S. V.; Jiang, D.; Katsnelson, M. I.; Grigorieva, I. V.; Dubonos, S. V.; Firsov, A. A. Two-Dimensional Gas of Massless Dirac Fermions in Graphene. *Nature* **2005,** *438* (7065), 197–200.

Owonubi, S.; Agwuncha, S.; Mukwevho, E.; Aderibigbe, B.; Sadiku, E.; Biotidara, O.; Varaprasad, K. Application of Hydrogel Biocomposites for Multiple Drug Delivery. *Handb. Compos. Renew. Mater., Polym. Compos.* **2017a,** *6,* 139.

Owonubi, S. J.; Agwuncha, S. C.; Fasiku, V. O.; Mukwevho, E.; Aderibigbe, B. A.; Sadiku, E. R.; Bezuidenhout, D. Biomedical Application. In *Polyolefin Fibres,* 2nd ed.; Ugbolue, S. C. O., Ed.; Birmingham, AL: Woodhead Publishing, 2017b; p 592.

Park, S.; Ruoff, R. S. Chemical Methods for the Production of Graphenes. *Nat. Nanotechnol.* **2009,** *4* (4), 217–224.

Rana, V. K.; Choi, M. C.; Kong, J. Y.; Kim, G. Y.; Kim, M. J.; Kim, S. H.; Mishra, S.; Singh, R. P.; Ha, C. S. Synthesis and Drug-Delivery Behavior of Chitosan-Functionalized Graphene Oxide Hybrid Nanosheets. *Macromol. Mater. Eng.* **2011,** *296* (2), 131–140.

Reddy, P. D.; Swarnalatha, D. Recent Advances in Novel Drug Delivery Systems. *Int. J. PharmTech Res.* **2010,** *2* (3), 2025–2027.

Robinson, J. T.; Tabakman, S. M.; Liang, Y.; Wang, H.; Sanchez Casalongue, H.; Vinh, D.; Dai, H. Ultrasmall Reduced Graphene Oxide with High Near-Infrared Absorbance for Photothermal Therapy. *J. Am. Chem. Soc.* **2011,** *133* (17), 6825–6831.

Rocic, P. Can ErbB2 Overexpression Protect against Doxorubicin Cardiotoxicity? *Am. J. Physiol. Heart Circ. Physiol.* **2015,** *309* (8), H1235–H1236.

Saini, M.; Singh, Y.; Arora, P.; Arora, V.; Jain, K. Implant Biomaterials: A Comprehensive Review. *World J. Clin. Cases* **2015,** *3* (1), 52–57.

Sasidharan, A.; Panchakarla, L. S.; Chandran, P.; Menon, D.; Nair, S.; Rao, C. N.; Koyakutty, M. Differential Nano-Bio Interactions and Toxicity Effects of Pristine versus Functionalized Graphene. *Nanoscale* **2011,** *3* (6), 2461–2464.

Sharma, D.; Chelvi, T. P.; Kaur, J.; Chakravorty, K.; De, T. K.; Maitra, A.; Ralhan, R. Novel Taxol Formulation: Polyvinylpyrrolidone Nanoparticle-Encapsulated Taxol for Drug Delivery in Cancer Therapy. *Oncol. Res.* **1996,** *8* (7–8), 281–286.

Song, M.-M.; Xu, H.-L.; Liang, J.-X.; Xiang, H.-H.; Liu, R.; Shen, Y.-X. Lactoferrin Modified Graphene Oxide Iron Oxide Nanocomposite for Glioma-Targeted Drug Delivery. *Mater. Sci. Eng. C* **2017,** *77,* 904–911.

Stankovich, S.; Dikin, D. A.; Dommett, G. H. B.; Kohlhaas, K. M.; Zimney, E. J.; Stach, E. A.; Piner, R. D.; Nguyen, S. T.; Ruoff, R. S. Graphene-Based Composite Materials. *Nature* **2006,** *442* (7100), 282–286.

Stankovich, S.; Dikin, D. A.; Piner, R. D.; Kohlhaas, K. A.; Kleinhammes, A.; Jia, Y.; Wu, Y.; Nguyen, S. T.; Ruoff, R. S. Synthesis of Graphene-Based Nanosheets via Chemical Reduction of Exfoliated Graphite Oxide. *Carbon* **2007,** *45* (7), 1558–1565.

Su, S.; Wang, J.; Vargas, E.; Wei, J.; Martínez-Zaguilán, R.; Sennoune, S. R.; Pantoya, M. L.; Wang, S.; Chaudhuri, J.; Qiu, J. Porphyrin Immobilized Nanographene Oxide for Enhanced and Targeted Photothermal Therapy of Brain Cancer. *ACS Biomater. Sci. Eng.* **2016,** *2* (8), 1357–1366.

Su, S.; Wang, J.; Wei, J.; Martínez-Zaguilán, R.; Qiu, J.; Wang, S. Efficient Photothermal Therapy of Brain Cancer through Porphyrin Functionalized Graphene Oxide. *New J. Chem.* **2015,** *39* (7), 5743–5749.

Teo, A. J. T.; Mishra, A.; Park, I.; Kim, Y.-J.; Park, W.-T.; Yoon, Y.-J. Polymeric Biomaterials for Medical Implants and Devices. *ACS Biomater. Sci. Eng.* **2016,** *2* (4), 454–472.

Tobío, M.; Gref, R.; Sánchez, A.; Langer, R.; Alonso, M. J. Stealth PLA-PEG Nanoparticles as Protein Carriers for Nasal Administration. *Pharm. Res.* **1998,** *15* (2), 270–275.

Upadhyay, R. K. Drug Delivery Systems, CNS Protection, and the Blood Brain Barrier. *BioMed Res. Int.* **2014,** *2014,* 869269.

Vallabani, N. V.; Mittal, S.; Shukla, R. K.; Pandey, A. K.; Dhakate, S. R.; Pasricha, R.; Dhawan, A. Toxicity of Graphene in Normal Human Lung Cells (BEAS-2B). *J. Biomed. Nanotechnol.* **2011,** *7* (1), 106–107.

Vasir, J. K.; Reddy, M. K.; Labhasetwar, V. D. Nanosystems in Drug Targeting: Opportunities and Challenges. *Curr. Nanosci.* **2005,** *1* (1), 47–64.

Vila, A.; Sánchez, A.; Janes, K.; Behrens, I.; Kissel, T.; Jato, J. L. V.; Alonso, M. A. J. Low Molecular Weight Chitosan Nanoparticles as New Carriers for Nasal Vaccine Delivery in Mice. *Eur. J. Pharm. Biopharm.* **2004,** *57* (1), 123–131.

Wang, Y.; Wang, K.; Zhao, J.; Liu, X.; Bu, J.; Yan, X.; Huang, R. Multifunctional Mesoporous Silica-Coated Graphene Nanosheet Used for Chemo-Photothermal Synergistic Targeted Therapy of Glioma. *J. Am. Chem. Soc.* **2013a,** *135* (12), 4799–4804.

Wang, Z.; Zhou, C.; Xia, J.; Via, B.; Xia, Y.; Zhang, F.; Li, Y.; Xia, L. Fabrication and Characterization of a Triple Functionalization of Graphene Oxide with Fe_3O_4, Folic Acid and Doxorubicin as Dual-Targeted Drug Nanocarrier. *Colloids Surf. B: Biointerfaces* **2013b,** *106,* 60–65.

Weiss, N.; Miller, F.; Cazaubon, S.; Couraud, P. O. Blood-Brain Barrier Part III: Therapeutic Approaches to Cross the Blood-Brain Barrier and Target the Brain. *Rev. Neurol. (Paris)* **2010,** *166* (3), 284–288.

Wu, X. Y.; Shalviri, A. Polymeric Nanoparticles Useful in Theranostics. *Google Patents*, 2013.

Yang, H. W.; Hua, M. Y.; Hwang, T. L.; Lin, K. J.; Huang, C. Y.; Tsai, R. Y.; Ma, C. C. M.; Hsu, P. H.; Wey, S. P.; Hsu, P. W.; Chen, P. Y.; Huang, Y. C.; Lu, Y. J.; Yen, T. C.; Feng, L. Y.; Lin, C. W.; Liu, L.; Wei, K. C. Non-Invasive Synergistic Treatment of Brain Tumors by Targeted Chemotherapeutic Delivery and Amplified Focused Ultrasound-Hyperthermia Using Magnetic Nanographene Oxide. *Adv. Mater.* **2013a,** *25* (26), 3605–3611.

Yang, H. W.; Lu, Y. J.; Lin, K. J.; Hsu, S. C.; Huang, C. Y.; She, S. H.; Liu, H. L.; Lin, C. W.; Xiao, M. C.; Wey, S. P.; Chen, P. Y.; Yen, T. C.; Wei, K. C.; Ma, C. C. EGRF Conjugated PEGylated Nanographene Oxide for Targeted Chemotherapy and Photothermal Therapy. *Biomaterials* **2013b,** *34* (29), 7204–7214.

Yoffey, J. M. Passage of Fluid and Other Substances through the Nasal Mucosa. *J. Laryngol. Otol.* **1958,** *72* (5), 377–84.

Zhou, J.; Patel, T. R.; Piepmeier, J. M.; Saltzman, W. M. Highly Penetrative Nanocarriers for Treatment of CNS Disease. *Google Patents*, 2015.

Zhu, Y.; Murali, S.; Cai, W.; Li, X.; Suk, J. W.; Potts, J. R.; Ruoff, R. S. Graphene and Graphene Oxide: Synthesis, Properties, and Applications. *Adv. Mater.* **2010,** *22* (35), 3906–3924.

SECTION III
VESICULAR CARRIERS

CHAPTER 11

A BRIEF ACCOUNT OF LIPOSOMES FOR BRAIN DELIVERY

D. B. BORIN[1] and R. C. V. SANTOS[2,*]

[1] Campus Tuxtepec - Biotecnology Department,
Universidad del Papaloapan, Franciscan University,
Santa Maria, Rio Grande do Sul, Brazil

[2] Microbiology and Parasitology Department, Federal University of Santa Maria, Santa Maria, Rio Grande do Sul, Brazil

*Corresponding author. E-mail: robertochrist@gmail.com

ABSTRACT

Liposomes are very studied structures and used in research, because they have a great ability to carry different compounds, thus being able to be adapted for the most diverse applications. In this chapter we discuss quickly and clearly the possible ways to produce and characterize liposomes in order to reach the central nervous system (CNS). Production and characterization methods are discussed, as well as modifications performed on the surface of liposomes to provide the vectorization to CNS. Although much research has been successful in this process, the task of achieving such a complex system is not easy and new strategies are emerging every day to carry the assets for this system, and new modifications in the structure of liposomes are among them.

11.1 INTRODUCTION

Liposomes are interesting nanostructures when we think of delivery of treatments to the central nervous system (CNS), since they present the possibility of drugs carrying hydrophilic, lipophilic, and hydrophobic substances. They also present vesicular structures, biocompatible, biodegradable, nontoxic, capable of directing, and releasing drugs in a controlled way, which are synthesized from lipids, such as phospholipids and sphingolipids. These systems are organized in a self-sustaining structure formed by phospholipid bilayers that can be produced with different sizes, surface charge, and easily bind to the surface different compounds capable of directing the delivery of drugs to the CNS (Garcia-Garcia et al., 2006; Gregoriadis, 1976; Noble et al., 2014; Vieira and Gamarra, 2016).

Liposomes can have different sizes and number of phospholipid bilayers, being obtained by several methods of production, which allow to obtain a certain "model" of carrier, each being indicated for different applications and better for certain type of drug to be carried (Alam et al., 2010; Lasic, 1993).

In general, the most used to reach the CNS are the liposomes with modifications in their surface like the addition of polyethylene glycol (PEG) to increase the live time of the carrier, making it furtive or like to use the term "Stealth," thus avoiding its clearence by the reticuloendothelial system (RES) cells. This greater time in the circulation increases the chances of reaching the brain (passive carrying), but more modifications and molecules are added to make this active or vectorized drive, making the liposomes more efficient in reaching the target tissue (Immordino et al., 2006; Lai et al., 2013; Uster et al., 1996).

Different molecules have been used in this active ligand-targeted drugs, such as albumin, glycides, transferrin, polysaccharides, peptides, antibodies, or aptamers. Using different mechanisms and strategies to reach the brain achieves success in different proportions (Lajoie and Shusta, 2015; Modi et al., 2009).

In spite of this, it is known that the task of reaching the CNS is not simple; currently, new strategies have been used, and yet in several studies, its effectiveness has been limited due to the great selectivity imposed by endothelial and CNS cells that form the blood–brain barrier (BBB) cells responsible for preventing the exchange of compounds between the blood and the CNS (Cramer et al., 2011; Dinda and Pattnaik, 2013; Pinzon-Daza et al., 2013; Vieira and Gamarra, 2016).

11.2   WHAT ARE LIPOSOMES?

Phospholipids when subjected to excess water (>95%) give rise to lamellar structures concentrically around an aqueous compartment, called liposomes, and are similar to cell membranes, so they were initially used for the study of these structures (Gregoriadis, 2007; Lasic and Papahadjopoulos, 1998). Liposomes are vesicular, biocompatible, biodegradable, nontoxic structures synthesized from lipids, such as phospholipids and sphingolipids; usually, sterols and antioxidants are added to improve the stability of the formulations (Virauri and Rhodes, 1995). These systems are organized in a self-sustaining structure formed by phospholipid bilayers (Fig. 11.1). They can be small unilamellar vesicle (SUV), large unilamellar vesicle (LUV), and multilamellar vesicle (MLV); the most used methods to obtain the liposomes are obtained by reverse-phase evaporation vesicles (REV), French press vesicles (FPV), and ethanol injection vesicles (EIV) (Gregoriadis, 2007; Lasic and Papahadjopoulos, 1998; Vemuri and Rhodes, 1995).

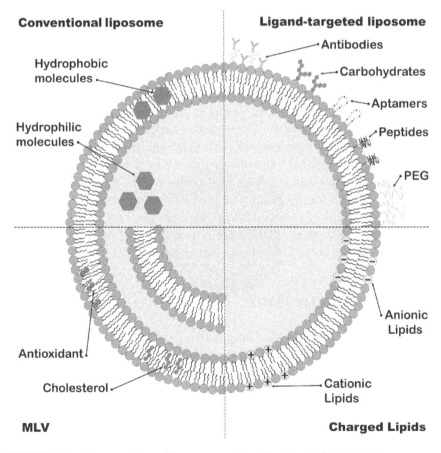

FIGURE 11.1   Representation of liposomes and targeting ligands for brain delivery.

Liposomes serve as carriers of drugs, biomolecules, or diagnostic agents. The stability of the liposomes can be affected by chemical, physical, and biological factors. Conventional liposomes are thermodynamically and chemically stable over a wide pH range of 3–9. However, their chemical instability manifests in the hydrolysis of the ester bonds and oxidation of unsaturated acyl bonds accelerates liposome breakdown; the greater stability is achieved at a pH around 6. The physical instability is mainly present in vesicle aggregation, changing the characteristics of the formulation as size and association rate (Gregoriadis, 2007; Saerten et al., 2005). Since the biological instability occurs due to the recognition of the liposomes by cells of the RES, conventional liposomes are rapidly captured. To avoid such capture, the surface can be modified with hydrophilic components such as PEG (Immordino et al., 2006). To improve chemical stability, inert atmospheres such as argon and nitrogen can be used as well as antioxidants and the lyophilization process. In order to generate physical stability, charged lipids are used in order to generate an electrostatic repulsion, avoiding vesicles aggregation, and to avoid the encapsulated drug extravasation, cholesterol and sphingomyelin are added to the formulation, reducing membrane permeability (Sapra and Allen, 2003; Sharma and Sharma, 1997).

Liposomes are characterized for size, Zeta potential, polydispersity index, and drug association rate. Interesting parameters evaluate the stability and application of the formulations. Although liposomes have been studied for some decades and there are already treatments in the market, and also because of technological and biological problems, the liposomes are still extensively studied for the development of stable formulations in the body aiming at the therapy of various diseases, such as cancer and neurodegenerative diseases (Batista et al., 2007; Cramer et al., 2011; Dinda and Pattnaik, 2013; Pinzon-Daza et al., 2013; Vieira and Gamarra, 2016).

11.3 HISTORY OF LIPOSOMES

The idea of using liposomes for drug delivery dates back to the late 1960s when it was thought of using these models of cell membranes as drug carriers. Over the next few decades, studies have gone through ups and downs with little expression in the academic setting. In the mid-1990s with the successful launch of a liposome-based drug (Doxil®) in the United States and Europe, research with these carriers returned with full force (Lasic and Papahadjopoulos, 1998).

Alec Bangham observed 50 years ago that phospholipids in aqueous solutions could form vesicles; since then, these carriers have been evolving in different areas for clinical applications (Torchilin, 2005).

One of the first investigations with liposomes for brain targeting in animals was with phosphatidylserine for the treatment of ischemic damage in 1979 by Bigon et al. (1979). From there, others studies have shown interest in using liposomes to target the brain, especially in the treatment of ischemia and glioma (Chapat et al., 1991; Firth et al., 1988; Imaizumi et al., 1990; Laham et al., 1988). After the launch of Doxil® in 1995, liposome research increased exponentially; it demonstrated improvements in anticancer therapy and reduction of drug toxicity of doxorubicin. It is currently being evaluated in clinical studies for the treatment of glioblastoma multiforme and pediatric brain tumors (Lasic, 1996; Vieira and Gamarra, 2016). This increase in liposome research has shown interest in the treatment of neurodegenerative diseases such as Parkinson (Kucherianu et al., 1997; Migliore et al., 2014; Yurasov et al., 1996) and Alzeimer (Bana et al., 2014; Mutlu et al., 2011; Rotman et al., 2015). Since then, the research has been growing up till now.

11.4 METHODS OF PREPARATION OF LIPOSOMES

Liposomes can be produced by different methods obtaining interesting characteristics for their application, as to their size, number of layers, and association rate. The basis for liposome formation is usually composed of phospholipids, cholesterol, and antioxidants, but for brain targeting, polymers and vectors such as PEG, proteins, and antibodies are usually added (Lasic and Papahadjopoulos, 1998).

Several methods have been described in the literature for the preparation of liposomes and are still being modified and adapted. Some relevant methods are REV, freeze-dried rehydration vesicles, EIV, hand-shaken MLVs, extrusion procedures, sonicated small, mostly unilamellar vesicles, and high-pressure homogenization vesicles. The protocols for all these techniques can be found in Lasch et al. (2003) and Gregoriadis (2007).

These methods produce liposomes with different sizes and number of lamellae. In addition, some methods cited the use of organic solvents that must be removed after the formation of the vesicles. This complicates the scheduling process (Vemuri and Rhodes, 1995). The size of the liposomes can be adjusted to the end of the process by extrusions, and it is also good to consider that the final sterilization method can alter the desired and obtained characteristics of the liposomes (Batista et al., 2007).

The main methods for producing liposomes are REV, EIV, and FPV. In the REV method, liposomes are prepared by dissolving the lipid components in organic solvent, such as chloroform and diethyl ether, in a round-bottomed flask, and then by adding the aqueous phase to give an organic phase volume ratio around 6:1. The flask is sealed with nitrogen and sonicated to produce an emulsion. Removal of organic solvent results in production of liposomes by REVs. In the EIV method, the lipids are dissolved in the solvent and then injected into heated aqueous solution, and shortly after, evaporation of the solvent is carried out. This method is used to obtain SUV and LUV liposomes. The conversion of MLV into SUV can be achieved after passage of the liposome suspension through a small orifice under pressure. This method is known as FPV, and for good homogeneity, the process is repeated sometimes. It can also be obtained by repeated extrusions in polycarbonate membrane filters. MLV is usually obtained by lipid film hydration where the phospholipids are dissolved in organic solvent, which upon removal causes the deposition of the lipid film on the walls of the flask. Thereafter, the aqueous solution is added and the liposomes form spontaneously with stirring. The preparation can be subjected to ultrasonic irradiation (Gregoriadis, 2007; Lasch et al., 2003; Vemuri and Rhodes, 1995). Table 11.1 shows some methods used to determine the characteristics of the liposomes obtained by different production methods.

TABLE 11.1   Characterization Methods.

Assays	Methodology
Basic assays	
Lamellarity determination	UV (TNBS), fluorescence (NBD), cryo-EM, NMR, SAXS
Size analysis	DLS, TEM, cryo-EM, AFM, SEC, HPLC-SEC, FFF, NMR
Zeta potential	Electrophoretic mobility
PDI or Đ	DLS, GPC, SEC
Lipid analysis	HPLC, TLC, GC, lipid phosphorus content, enzymatic assay
Lipid bilayer phase behavior	DSC, FTIR, NMR, Raman spectroscopy
pH	pH meter

TABLE 11.1   *(Continued)*

Assays	Methodology
Other assays	
Encapsulation efficiency	Minicolumn centrifugation, dialysis membrane, ultracentrifugation
In vitro drug release	Dialysis tubing
Residual organic solvente and heavy metals	GC, GC–MS, NMR

AFM, atomic force microscopy; *cryo-EM*, cryo-electron microscopy; *DLS*, dynamic light scattering; *FFF*, field-flow fractionation; *FTIR*, Fourier transform infrared spectroscopy; *GC*, gas chromatography; *GC–MS*, gas chromatography–mass spectrometry, *DSC*, differential scanning calorimetry; *GPC*, gel permeation chromatography; *HPLC*, high-performance liquid chromatography; *NBD*, *N*-(7-nitrobenz-2-oxa-1,3-diazol-4-yl); *NMR*, nuclear magnetic resonance; *PDI*, polydispersity index (or more recently dispersity without the poly, as per IUPAC recommendation), Ð, dispersity; *SAXS*, X-ray scattering; *SEC*, size-exclusion chromatography; *TEM*, transmission electron microscopy; *TLC*, thin-layer chromatography; *TNBS*, 2,4,6-trinitrobenzensulfonic acid.
Source: Gregoriadis (2007) and Laouini et al. (2012).

11.5   LIPOSOME SURFACE MODIFICATIONS

Different compounds are used to alter the surface of liposomes either to increase their stability, to stay longer in the blood stream, or to generate a specific vectorization. Conventional liposomes composed of phospholipids and cholesterol do not show good stability. When loaded lipids are added, this stability is increased, but in vivo they are more easily recognized by the RES, the greater its charge in the modulus (Soema et al., 2015).

The time in the circulation can be increased by modifying its surface by adding sulfatide, ganglioside monosialoganglioside galactose (GM1), mannose, or PEG, among other compounds, to avoid aggregation of the vesicles and decrease their recognition and removal by the RES (Modi et al., 2009). Surface modification by PEG usually occurs by the addition of synthetic phospholipids such as 1,2-distearoyl-*sn*-glycero-3-hosphoethanol-amine-*n*-[methoxy(PEG)-2000] (DSPE-PEG 2000). The addition of PEG in liposomes has technological and biological advantages, and one of the main properties is the reduction of liposome uptake by the RES, increasing the circulation time in the blood stream (Immordino et al., 2006; Wohlfart et al., 2012). The increased time of liposomes in the circulation occurs due to reduced interaction with plasma proteins and cell-surface proteins (Allen et al., 2002; Sercombe et al., 2015).

Different polymers have already been tested to generate PEG-like effects. Several polymers have been shown to be effective; among them, we can mention polyvinyl alcohol (Takeuchi et al., 2001), DSPE-poly(2-methyl-2-oxazoline) or poly(2-ethyl-2-oxazoline) (Woodle et al., 1994), poly(vinyl pyrrolidone) (PVP), poly(acryl amide) (PAA) (Torchilin et al., 1994, 2001), and phosphatidyl polyglycerols (Maruyama et al., 1994).

Liposomes coated with glycolipid GM1 and GM type III (a brain-tissue-derived monosialoganglioside) demonstrated a significant reduction in recognition by the RES as well as greater resistance to the gastrointestinal tract and also demonstrated good results in accumulation of the encapsulated drug in some brain regions such as cortex, basal ganglia, and mesencephalon (Mora et al., 2002; Taira et al., 2004).

Antibodies can be bound to the surface of liposomes by being called immunoliposomes, the terminal end of PEG containing the maleimide group is normally attached. Reduced thiol groups of Fab or scFv fragments were joined to the maleimide group of PEG-liposome, obtaining a stable thioether bond (Immordino et al., 2006; Lukyanov et al., 2004).

Another promising ligand is oligonucleotide aptamers, which are single-stranded DNA or RNA bound to the surface of the liposomes or in PEG and are capable of binding to different targets ranging from small molecules to cells. Aptamers will be important in the future of research, clinical diagnosis, and therapy. They may be an alternative for molecular recognition (McConnell et al., 2014).

Cationic lipids are also incorporated into liposomes to carry out transfection into gene therapy; the positive charge of the cationic liposomes interacts with the plasmidial DNA molecules with negative charges, condensing them and protecting against degradation in the systemic circulation, preserving the gene that will encode determined therapeutic protein. The most used cationic lipids are 1,2-dioleoyl-*sn*-glycero-3-phosphoethanolamine and 1,2-dioleoyl-3-trimethylammonium-propane (Mahato, 2005).

Liposomes can be coated by contrast agents for ultrasonography, magnetic resonance imaging, computed tomography imaging, and gamma-scintigraphy. DTPA–PLL–NGPE is a polychelating amphiphilic polymer capable of binding various metal ions to the surface of liposomes making them interesting as contrast agents (Torchilin, 2000).

The cationized bovine serum albumin bound to the surface of the liposomes succeeded in carrying liposomes through the BBB (Helm and Fricker, 2015).

Other possible applications for liposomes are in the induction of immune response known as virosomes, which are incorporated into the bilayer of liposomes' viral proteins such as hemagglutinin and neuraminidase, increasing the immune stimulus generated by the liposomes acting as adjuvants, and cationic liposomes have shown this same profile (Chen and Huang, 2005; Felnerova et al., 2004).

11.6   VECTORIZATION OF LIPOSOMES FOR THE CENTRAL NERVOUS SYSTEM

The CNS has a selective permeability, able to select only the entry of molecules of interest to the tissue. This is due to the cellular complex formed around the endothelium which form the BBB. This barrier is able to limit or prevent the arrival of some drugs to the CNS; some of them even when absorbed by endothelial cells are expelled by an efflux system, to which we can highlight the P-glycoprotein responsible for the ineffectiveness of several drugs (Alam et al., 2010; Wohlfart et al., 2012).

The encapsulation of a drug in a liposome can prolong the time of drug circulation, as well as reduce the systemic side-effects and increase the therapeutic effects in the CNS. For these reasons, liposomes are being used in several studies and clinical trials for the delivery of drugs to the CNS. Currently, liposomes are the most studied nanoparticles for brain delivery, being the most promising nanocarrier for clinical applications in brain pathologies (Rip, 2016; Vieira and Gamarra, 2016).

Among these studies, we can highlight some promising ones that are currently in clinical trials in different fields: pegylated doxorubicin liposomes (Doxil®) are in clinical trials for the treatment of glioblastoma multiforme and pediatric brain tumors (Ananda et al., 2011; Marina et al., 2002). Another nonconventional doxorubicin liposome (Myocet®) is in clinical trials for the treatment of glioblastoma multiforme (Di Legge et al., 2011). Another conventional liposome from daunorubicin (DaunoXome®) is in clinical trials for the treatment of pediatric brain tumors (Lippens, 1999). Conventional liposomes of amphotericin B (Abelcet® and Ambisome) are being tested for the treatment of cryptococcal meningitis (Loyse et al., 2013).

It can be seen that clinical trials are using only conventional and steroid liposomes, but other strategies to carry out active driving are being studied widely such as cation vectors, ligand-targeting, and triggered drug release among others.

Cationic liposomes demonstrate better interaction with negatively charged cell membranes, as well as those present in the capillary endothelium; with high levels of lecithins, the interaction with this membrane can lead to endocytosis (Joshi et al., 2015; Scherrmann et al., 2002). This feature was explored in a research where glial cell line-derived neurotrophic factor was carried out for the treatment of a model of Parkinson's disease where the neurotropic effect in the rat brain was induced (Migliore et al., 2014). Another application of cationic liposomes was able to generate neuroprotection in an animal model of cerebral focal transient ischemia by transporting a plasmid carrying the bcl-2-regulating apoptosis gene (Cao et al., 2002). Another study demonstrated the feasibility of doxirubicin-loaded cationic liposomes and ultrasound-activated quantum dots for the treatment of glioma (Lin et al., 2015).

Immunoliposomes have been extensively studied for the targeting of drugs to the CNS due to specificity for receptors of the cerebral endothelial cells; some widely used are the antibodies against the transferrin receptor called OX26 (for rats). They were used in a model of ischemia in rats, where they were effective in attenuating the damage caused to the brain, being efficient in transporting the vascular endothelial growth factor to the target tissue (Zhao et al., 2011). Immunoliposomes were also efficient in reducing hemorrhagic transformation after thrombolytic therapy with tissue plasminogen activator in cerebral ischemia (Asahi et al., 2003). Another study was successful against glioma, with immunoliposomes that were efficient in transporting paclitaxel, using arginine–glycine–aspartic acid (RGD) peptides that are specific targeting ligands of cancer cells, peptide, and transferrin (Qin et al., 2006). An anticancer therapy study used an immunoliposome against the epidermal growth factor receptor, being an insulin receptor expressed in several brain tumors (Mamot et al., 2005).

Other alternatives for drug release in the CNS are through stimuli such as ultrasound, alternating magnetic field, temperature, light pulses, or pH change. The focused ultrasound can be used to release drugs such as doxorubicin into the glioma generating better therapeutic efficacy (Treat et al., 2012). In a study with magnetoliposomes to release doxorubicin, an alternating magnetic field was used to release the drug in the region of interest (Guo et al., 2015). The thermosensitive liposomes can be used for the treatment of glioma; after localized brain warming, the liposomes release the drug at the target site, leading to tumor cells to death (Kakinuma et al., 1996). Photodynamic therapy can also be used to orientate the target cells more precisely; the use of liposomes with photofrin in the treatment of glioma has been studied (Jiang et al., 1998). Liposome pH-sensitive can be used against glioblastoma, because the inside of the tumor presents a more acidic pH; in contact with this medium, the liposomes release the drug (Pacheco-Torres et al., 2015).

Several studies have used liposomes to screen for tumors because of their targeting characteristics, which can incorporate contrast agents and be monitored in real time due to their pharmacokinetics (Krauze et al., 2005). Studies have used liposomes to locate gliomas in rats with high efficiency and optimal definition of tumor margins (Qin et al., 2015).

11.7 CONCLUDING REMARKS

The major problem with current treatments of brain pathologies is to provide the appropriate amount of drugs in the regions of interest in the brain; this problem occurs due to the BBB. More and more research is being done on how to transpose this barrier and liposomes are among the most studied, efficient, and promising methods. Currently, there are several options to reach the CNS with the use of liposomes. The best strategy should be chosen according to its applicability, costs, and efficiency for particular drug and pathology. In addition, it must be perfected so that we can see new and more efficient treatments emerge.

KEYWORDS

- **liposome**
- **immunoliposomes**
- **brain delivery**
- **targeting**
- **nanocarriers**
- **blood–brain barrier**

REFERENCES

Alam, M. I.; Beg, S.; Samad, A.; Baboota, S.; Kohli, K.; Ali J.; Ahuja, A.; Akbar, M. Strategy for Effective Brain Drug Delivery. *Eur. J. Pharm. Sci.* **2010,** *40,* 385–403.

Allen, C.; Dos, S. N.; Gallagher, R.; et al. Controlling the Physical Behavior and Biological Performance of Liposome Formulations through Use of Surface Grafted Poly(Ethylene Glycol). *Biosci. Rep.* 2002, *22*, 225–250.

Ananda, S.; Nowak, A. K.; Cher, L.; et al. Phase 2 Trial of Temozolomide and PEGylated Liposomal Doxorubicin in the Treatment of Patients with Glioblastoma Multiforme Following Concurrent Radiotherapy and Chemotherapy. *J. Clin. Neurosci.* 2011, *18*, 1444–1448.

Asahi, M.; Rammohan, R.; Sumii, T.; et al. Antiactin-Targeted Immunoliposomes Ameliorate Tissue Plasminogen Activator-Induced Hemorrhage after Focal Embolic Stroke. *J. Cereb. Blood Flow Metab.* 2003, *23*, 895–899.

Bana, L.; Minniti, S.; Salvati, E.; et al. Liposomes Bi-Functionalized with Phosphatidic Acid and an ApoE-Derived Peptide Affect Aβ Aggregation Features and Cross the Blood-Brain Barrier: Implications for Therapy of Alzheimer Disease. *Nanomedicine* 2014, *10*, 1583–1590.

Batista, C. M.; Carvalho, C. M. B.; Santos, N. S. Liposomes and Their Therapeutic: State of Art Applications. *Braz. J. Pharm. Sci.* 2007, *43*, 167–179.

Bigon, E.; Boarato, E.; Bruni, A.; Leon, A.; Toffano, G. Pharmacological Effects of Phosphatidylserine Liposomes: Regulation of Glycolysis and Energy Level in Brain. *Br. J. Pharmacol.* 1979, *66*, 167–174.

Cao, Y. J.; Shibata, T.; Rainov, N. Liposome-Mediated Transfer of the bcl-2 Gene Results in Neuroprotection after In Vivo Transient Focal Cerebral Ischemia in an Animal Model. *Gene Ther.* 2002, *9*, 415–419.

Chapat, S.; Frey, V.; Claperon, N.; et al. Efficiency of Liposomal ATP in Cerebral Ischemia: Bioavailability Features. *Brain Res. Bull.* 1991, *26*, 339–342.

Chen, W. C.; Huang, L. Non-Viral Vector as Vaccine Carrier. *Adv. Genet.* 2005, *54*, 315–37.

Craparo, E. F.; Bondi, M. L.; Pitarresi, G.; Cavallaro, G. Nanoparticulate Systems for Drug Delivery and Targeting to the Central Nervous System. *CNS Neurosci. Ther.* 2011, *17* (6), 670–677.

Dinda, S. C.; Pattnaik, G. Nanobiotechnology-Based Drug Delivery in Brain Targeting. *Curr. Pharm. Biotechnol.* 2013, *14* (15), 1264–1274.

Di Legge, A.; Trivellizzi, I. N.; Moruzzi, M. C.; Pesce, A.; Scambia, G.; Lorusso, D. Phase 2 Trial of Non-PEGylated Doxorubicin (Myocet) as Second-Line Treatment in Advanced or Recurrent Endometrial Cancer. *Int. J. Gynecol. Cancer* 2011, *21*, 1446–1451.

Felnerova, D.; Viret, J. F.; Gluck, R.; et al. Liposomes and Virosomes as Delivery Systems for Antigens, Nucleic Acids and Drugs. *Curr. Opin. Biotechnol.* 2004, *15*, 518–529.

Firth, G. B.; Firth, M.; McKeran, R. O.; et al. Application of Radioimmunoassay to Monitor Treatment of Human Cerebral Gliomas with Bleomycin Entrapped within Liposomes. *J. Clin. Pathol.* 1988, *41*, 38–43.

Garcia-Garcia, E.; Andrieux, K.; Gil, S.; Couvreur, P. Colloidal Carriers and Blood-Brain Barrier (BBB) Translocation: A Way to Deliver Drugs to the Brain? *Int. J. Pharm.* 2005, *298* (2), 274–292.

Gregoriadis, G. The Carrier Potential of Liposomes in Biology and Medicine. *New Engl. J. Med.* 1976, *295*, 704–765.

Gregoriadis, G. *Liposome Technology, Third Edition, Volume I. Liposome Preparation and Related Techniques*; London, UK: Lipoxen PLC, 2007.

Guo, H.; Chen, W.; Sun, X.; Liu, Y. N.; Li, J.; Wang, J. Theranostic Magnetoliposomes Coated by Carboxymethyl Dextran with Controlled Release by Low-Frequency Alternating Magnetic Field. *Carbohydr. Polym.* 2015, *118*, 209–217.

Helm, F.; Fricker, G. Liposomal Conjugates for Drug Delivery to the Central Nervous System. *Pharmaceutics* 2015, *7* (2), 27–42.

Imaizumi, S.; Woolworth, V.; Fishman, R. A.; Chan, P. H. Liposome-Entrapped Superoxide Dismutase Reduces Cerebral Infarction in Cerebral Ischemia in Rats. *Stroke* 1990, *21*, 1312–1317.

Immordino, M. L.; Dosio, F.; Cattel, L. Stealth Liposomes: Review of the Basic Science, Rationale, and Clinical Applications, Existing and Potential. *Int. J. Nanomed.* 2006, *1* (3), 297–315.

Jiang, F.; Lilge, L.; Grenier, J.; Li, Y.; Wilson, M. D.; Chopp, M. Photodynamic Therapy of U87 Human Glioma in Nude Rat Using Liposome-Delivered Photofrin. *Lasers Surg. Med.* 1998, *22*, 74–80.

Joshi, S.; Singh-Moon, R. P.; Ellis, J. A.; Chaudhuri, D. B.; Wang, M.; Reif, R.; Bruce J. N.; Bigio, I. J. Straubinger RM Cerebral hypoperfusion-Assisted Intra-Arterial Deposition of Liposomes in Normal and Glioma-Bearing Rats. *Neurosurgery* 2015, *76* (1), 92–100.

Kakinuma, K.; Tanaka, R.; Takahashi, H.; Watanabe, M.; Nakagawa, T.; Kuroki, M. Targeting Chemotherapy for Malignant Brain Tumor Using Thermosensitive Liposome and Localized Hyperthermia. *J. Neurosurg.* 1996, *84*, 180–184.

Krauze, M. T.; Mcknight, T. R.; Yamashita, Y.; et al. Real-Time Visualization and Characterization of Liposomal Delivery into the Monkey Brain by Magnetic Resonance Imaging. *Brain Res. Brain Res. Protoc.* 2005, *16*, 20–26.

Kucherianu, V. G.; Iurasov, V. V.; Kryzhanovskiĭ, G. N.; et al. The Effect of Liposomal Form of l-DOPA on the Development of Parkinsonian Syndrome in Mice. *Bull. Exp. Biol. Med.* 1997, *123*, 29–33.

Lajoie, J. M.; Shusta, E. V. Targeting Receptor-Mediated Transport for Delivery of Biologics across the Blood-Brain Barrier. *Annu. Pharmacol. Toxicol.* 2015, *55*, 613–631.

Lai, F.; Fadda, A. M.; Sinico, C. Liposomes for Brain Delivery. *Expert Opin. Drug Deliv.* 2013, *10* (7), 1003–1022.

Laham, A.; Claperon, N.; Durussel, J. J.; et al. Intracarotidal Administration of Liposomally Entrapped ATP: Improved Efficiency against Experimental Brain Ischemia. *Pharmacol. Res. Commun.* 1988, *20* (8), 699–705.

Laouini, A.; Jaafar-Maalej, C.; Limayem-Blouza, I.; et al. Preparation, Characterization and Applications of Liposomes: State of the Art. *J. Colloid Sci. Biotechnol.* 2012, *1* (2), 147–168.

Lasch, J.; Weissig, V.; Brandl, M. Preparation of Liposomes. In *Liposomes*, 2nd ed.; Torchilin, V. P., Weissig, V., Eds.; New York: Oxford University Press, 2003; Chap. 1.

Lasic, D. D. *Liposomes: From Physics to Applications*, 1st ed.; Amsterdam: Elsevier Science Publishers B.V., 1993; Chap. 3, p.63–90.

Lasic, D. D. Papahadjopoulos, D., Eds. *Medical Applications of Liposomes*; Amsterdam: Elsevier, 1998.

Lasic, D. D. Improved Liposome Stability and Drug Retention Significantly Increase the Anticancer Activity of Encapsulated Doxorubicin (Doxil), Enhancing the Effectiveness of Chemotherapy and Potentially Reducing Its Toxicity. *Nature* 1996, *380*, 561–562.

Lippens, R. J. Liposomal Daunorubicin (DaunoXome) in Children with Recurrent or Progressive Brain Tumors. *Pediatr. Hematol.* 1999, *16*, 131–139.

Loyse, A.; Thangaraj, H.; Easterbrook, P.; et al. Cryptococcal Meningitis: Improving Access to Essential Antifungal Medicines in Resource-Poor Countries. *Lancet Infect. Dis.* 2013, *13*, 629–637.

Mahato, R. I. Water Insoluble and Soluble Lipids for Gene Delivery. *Adv. Drug Deliv. Rev.* 2005, *57*, 699–712.

McConnell, E. M.; Holahan, M. R.; DeRosa, M. C. Aptamers as Promising Molecular Recognition Elements for Diagnostics and Therapeutics in the Central Nervous System. *Nucl. Acid Ther.* 2014, *24*, 388–404.

Mamot, C.; Drummond, D. C.; Noble, C. O.; Kallab, V.; Guo, Z.; Hong, K.; Kirpotin, D. B.; Park, J. W. Epidermal Growth Factor Receptor-Targeted Immunoliposomes Significantly Enhance the Efficacy of Multiple Anticancer Drugs In Vivo. *Cancer Res.* 2005, *65*, 11631–11638.

Marina, N. M.; Cochrane, D.; Harney, E.; et al. Dose Escalation and Pharmacokinetics of PEGylated Liposomal Doxorubicin (Doxil) in Children with Solid Tumors: A Pediatric Oncology Group Study. *Clin. Cancer Res.* 2002, *8*, 413–418.

Maruyama, K.; Okuizumi, S.; Ishida, O.; et al. Phosphatidyl Polyglycerols Prolong Liposome Circulation In Vivo. *Int. J. Pharm.* 1994, *111*, 103–107.

Migliore, M. M.; Ortiz, R.; Dye, S.; Campbell, R. B.; Amiji, M. M.; Waszczak, B. L.; Neurotrophic and Neuroprotective Efficacy of Intranasal GDNF in a Rat Model of Parkinson's Disease. *Neuroscience* 2014, *274*, 11–23.

Modi, G.; Pillay, V.; Choonara, Y. E.; Ndesendo, V. M. K; Du Toit, L. C.; Naidoo, D. Nanotechnological Applications for the Treatment of Neurodegenerative Disorders. *Prog. Neurobiol.* 2009, *88*, 272–285.

Mutlu, N. B.; Değim, Z.; Yılmaz, S.; Eşsiz, D.; Nacar, A. New Perspective for the Treatment of Alzheimer Diseases: Liposomal Rivastigmine Formulations. *Drug Dev. Ind. Pharm.* 2011, *37*, 775–789.

Noble, G. T.; Stefanick, J. F.; Ashley, J. D.; Kiziltepe, T.; Bilgicer, B. Ligand-Targeted Liposome Design: Challenges and Fundamental Considerations. *Trends Biotechnol.* 2014, *32*, 32–45.

Pacheco-Torres, J.; Mukherjee, N.; Walko, M.; et al. Image Guided Drug Release from pH-Sensitive Ion Channel-Functionalized Stealth Liposomes into an In Vivo Glioblastoma Model. *Nanomedicine* 2015, *11*, 1345–1354.

Pinzon-Daza, M. L.; Campia, I.; Kopecka, J.; et al. Nanoparticle- and Liposome-Carried Drugs: New Strategies for Active Targeting and Drug Delivery across Blood-Brain Barrier. *Curr Drug Metab.* 2013, *14* (6), 625–640.

Qin, L.; Wang, C. Z.; Fan, H. J.; et al. A Dual-Targeting Liposome Conjugated with Transferrin and Arginine–Glycine–Aspartic Acid Peptide for Glioma-Targeting Therapy. *Oncol. Lett.* 2014, *8*, 2000–2006.

Qiu, L. H.; Zhang, J. W.; Li, S. P.; et al. Molecular Imaging of Angiogenesis to Delineate the Tumor Margins in Glioma Rat Model with Endoglin-Targeted Paramagnetic Liposomes Using 3T MRI. *J. Magn. Reson. Imag.* 2015, *41*, 1056–1064.

Rip, J. -Liposome Technologies and Drug Delivery to the CNS. *Drug Discov. Today Technol.* 2016, *20*, 53–58.

Rotman, M.; Welling, M. M.; Bunschoten, A.; et al. Enhanced Glutathione PEGylated Liposomal Brain Delivery of an Anti-Amyloid Single Domain Antibody Fragment in a Mouse Model for Alzheimer's Disease. *J. Control. Release* 2015, *203*, 40–50.

Saetern, A. M.; Skar, M.; Braaten, A.; Brandl, M. Camptothecin Catalyzed Phospholipid Hydrolysis in Liposomes. *Int. J. Pharm.* 2005, *288*, 73–80.

Sapra, P.; Allen, T. M. Ligand-Targeted Liposomal Anticancer Drugs. *Prog. Lipid Res.* 2003, *42*, 439–462.

Sercombe, L.; Veerati, T.; Moheimani, F.; Wu, S. Y.; Sood, A. K.; Hua, S. Advances and Challenges of Liposome Assisted Drug Delivery. *Front. Pharmacol.* 2015, *6*, 286.

Sharma, A.; Sharma, U. S. Liposome in Drug Delivery: Progress and Limitations. *Int. J. Pharm.* 1997, *154*, 123–140.

Scherrmann, J. M. Drug Delivery to Brain via the Blood-Brain Barrier. *Vascul. Pharmacol.* 2002, *38* (6), 349–354.

Takeuchi, H.; Kojima, H.; Yamamoto, H.; et al. Evaluation of Circulation Profiles of Liposomes Coated with Hydrophilic Polymers Having Different Molecular Weights in Rats. *J. Control. Release* 2001, *75*, 83–91.

Torchilin, V. P.; Shtilman, M.; Trubetskoy, V.; et al. Amphiphilic Vinyl Polymers Selectively Prolong Liposome Circulation Time In Vivo. *Biochim. Biophys. Acta* 1994, *1195*, 181–184.

Torchilin, V. P.; Levchenko, T. S.; Whiteman, K. R.; et al. Amphiphilic Poly-*N*-Vinyl Pyrrolidones: Synthesis, Properties and Liposome Surface Modification. *Biomaterials* 2001, *22*, 3035–3044.

Torchilin, V. P. Recent Advances with Liposomes as Pharmaceutical Carrier. *Nat. Rev. Drug Discov.* 2005, *4*, 145–160.

Uster, P. S.; Allen, T. M.; Daniel, B. E.; Mendez, C. J.; Newman, M. S.; Zhu, G. Z. Insertion of Poly(Ethylene Glycol) Derivatized Phospholipid into Pre-Formed Liposomes Results in Prolonged In Vivo Circulation Time. *FEBS Lett.* 1996, *386*, 243–246.

Vemuri, S.; Rhodes, C. T. Preparation and Characterization of Liposomes as Therapeutic Delivery Systems: A Review. *Pharm. Acta Helv., Berne* 1995, *70*, 95–111.

Vieira, D. B.; Gamarra, L. F. Getting into the Brain: Liposome-Based Strategies for Effective Drug Delivery across the Blood Brain Barrier. *Int. J. Nanomed.* 2016, *11*, 5381–5414.

Wohlfart, S.; Gelperina, S.; Kreuter, J. Transport of Drugs across the Blood-Brain Barrier by Nanoparticles. [*J. Control. Release* 2012, *161*, 264–273.

Woodle, M.; Engbers, C.; Zalipsky, S. New Amphipatic Polymer–Lipid Conjugates Forming Long-Circulating Reticuloendothelial System-Evading Liposomes. *Bioconj. Chem.* 1994, *5*, 493–496.

Xiang, Y.; Wu, Q.; Liang, L.; et al. Chlorotoxin-Modified Stealth Liposomes Encapsulating Levodopa for the Targeting Delivery against the Parkinson's Disease in the MPTP-Induced Mice Model. *J. Drug Target.* 2012, *20*, 67–75.

Yurasov, V. V.; Podgornyi, G. N.; Kucheryanu, V. G.; et al. Effects of l-DOPA-Carrying Liposomes on Striatal Concentration of Dopamine and Its Metabolites and Phospholipid Metabolism in Experimental Parkinson's Syndrome. *Bull. Exp. Biol. Med.* 1996, *122*, 1180–1183.

Zhao, H.; Bao, X. J.; Wang, R. Z.; et al. Postacute Ischemia Vascular Endothelial Growth Factor Transfer by Transferrin-Targeted Liposomes Attenuates Ischemic Brain Injury after Experimental Stroke in Rats. *Hum. Gene Ther.* 2011, *22*, 207–215.

CHAPTER 12

LIPOSOMAL DRUG CARRIERS: PRINCIPLES, FORMULATION PERSPECTIVES, AND POTENTIALS FOR BRAIN DRUG DELIVERY

JULIANA PALMA ABRIATA*, MARCELA TAVARES LUIZ, GIOVANNI LOUREIRO RASPANTINI, PATRÍCIA MAZUREKI CAMPOS, and JULIANA MALDONADO MARCHETTI

School of Pharmaceutical Sciences of Ribeirao Preto, University of Sao Paulo, Sao Paulo, Brazil

Corresponding author. E-mail: julianafarma@gmail.com

ABSTRACT

The inherent anatomy and physiology of the brain poses obstacles in the way of drug delivery to it. The existence of lipidic blood–brain barrier confers the ability of selective permeation of lipophilic molecules across the brain biological membranes. The nanotechnology based drug carriers have demonstrated success as compared to their conventional counterparts. Among the diverse nanocarriers, liposomes have been explored extensively due to versatility and biocompatibility. The present chapter is aimed to provide an overview of liposomal drug carriers commencing with introduction and embracing manufacturing techniques, applications in variety of brain ailments.

12.1 INTRODUCTION

The treatment of brain disorders such as Alzheimer's disease, Parkinson's disease, multiple sclerosis, brain tumor, and others is currently a challenge because of the difficult diagnosis and drug delivery into the central nervous system (CNS) when compared with other organs in the human body. This impairment occurs due to the presence of blood–brain barrier (BBB) that is formed by brain capillary endothelial cells (BCECs). The BCECs, different from other endothelial cells present in the body, have a high concentration of tight junctions (TJ) and adhesion proteins, which promote a high resistance to the transport of substances from the bloodstream into the brain.

Although BBB is important to avoid the transport of unwanted substances and organisms into the brain and to maintain CNS homeostasis, its presence becomes a challenge for brain disorders treatment. One of the approaches to overcome the BBB limitation is the use of nanotechnology-based drug delivery systems. In this way, several strategies have been studied to overcome this barrier and thereby improve the targeted drug delivery (Blakeley, 2008; Hu et al., 2017). One of the ways to beat this barrier is by the use of modified surface nanocarriers with ligands that have high specificity and selectivity for receptors or transporters expressed at the BCEBs or at immune system cells for cell-mediated transport. The use of nanocarriers promotes large advantages when compared to administration of drugs by conventional formulations. Among these advantages, nanocarriers can prolong the drug circulation time in the blood, protect drugs from degradation inside the organism, enhance stability and bioavailability, and reduce side effects. Also, these systems are able to have their surface modified for the promotion of active targeting into target tissues (Comoglu et al., 2016). Therefore, a large number of researches used nanocarriers, such as polymeric nanoparticles, liposomes, carbon nanotubes, and solid lipid nanoparticles for the transport of drugs into the CNS to brain disease treatment.

Among these nanocarriers, liposomes are widely used as drug delivery systems for brain targeting because of their versatility and biocompatibility. Moreover, as liposomes have a vesicular structure similar to cell membranes, with an amphiphilic membrane with hydrophilic borders and an aqueous core, they can encapsulate both hydrophilic and lipophilic drugs. So, different liposomes preparation techniques are described in the literature and can influence the size, polydispersity index and encapsulation efficiency of hydrophilic and lipophilic drugs. Therefore, this chapter will address the main techniques used in the development of liposomes for drug delivery.

In addition to the advantages already mentioned, liposomes' surface can be easily modified with different ligands to promote active targeting of drugs into the brain, such as peptides, antibodies, proteins, glucose, and others. In this context, the present chapter will address the most widely used ligands for targeting liposomes into the brain. So, the present chapter will expose some strategies that have been developed to increase the transport of drugs through the blood to the brain. In addition, the chapter will expose some researches and clinical trials.

12.2 BIOLOGICAL BARRIERS TO BRAIN DRUG DELIVERY

Transport of molecules and organisms between blood and CNS is regulated and limited by two main barriers: the BBB and blood–cerebrospinal fluid barrier. Among these, BBB is the most efficient and regulated barrier present in the human body. The BBB existence was first described by Ehrlich in 1885, who saw that the intravenous injection of aniline dye has a quick bodily distribution, except in the brain. Later studies of the brain showed that BBB is formed by a group of highly specialized multicellular structures that protect the CNS against unwanted molecules and organisms and maintain its homeostasis (Sarkar et al., 2017).

The BBB is composed of BCECs, astrocytes, pericytes, and neuronal cells. Among these, BCECs are the most important BBB cell components because of the presence of TJ proteins (like Claudins and Occludin) and adherent junction proteins (like Junctional Adhesion Molecule-1) that are responsible for maintaining brain endothelial cells strongly linked, forming non-fenestrated endothelial cells (Grabrucker et al., 2016). TJs are responsible for the high transendothelial electrical resistance (TEER), which prevents paracellular transport of solutes. Besides that, the entrance and outcome of molecules between brain and blood are regulated by the expression of several receptors, ion channels, and influx/efflux transport proteins expressed on BCECs. Thus, BBB is essential to limit and regulate the entrance of immune system cells, xenobiotics, nutrients, and endogenous compounds (Wong et al., 2012).

Although the BBB is extremely important to maintain CNS homeostasis, the presence of this barrier troubles the effective diagnosis and treatment of neuronal disorders, such as Alzheimer's disease, Parkinson's disease, brain tumors, and others (Gao, 2016). Furthermore, the expression of efflux pumps, like P-glycoprotein (P-gp), multidrug-resistance-associated protein and breast cancer resistance protein (BCRP) can impair the treatment, leading to return of the drug to the bloodstream and reducing brain drug concentration. In this context, nanotechnology has been studied aiming to increase the drug delivery into the brain.

12.2.1 TRANSPORT ROUTES ACROSS THE BBB

Although the BBB promotes a restrict transport of solutes, some nutrients and endogenous compounds that are essential to great brain function can be transported to the brain using multiple mechanisms by passive diffusion or active transport. Some of these mechanisms have been studied such as strategies for drug delivery from blood to the brain, as shown in Figure 12.1. Thus, this chapter addresses the most important transport routes across the BBB that have been researched to promote a specific nanocarrier delivery for the treatment of brain disorders (Grabrucker et al., 2016).

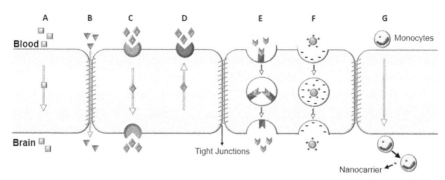

FIGURE 12.1 Transport routes across the blood–brain barrier. A: Transcellular transport, B: Paracellular transport, C: Carrier-mediated influx transport, D: Carrier-mediated efflux transport, E: Receptor-mediated transcytosis, F: Adsorptive-mediated transcytosis, and G: Cell-mediated transport.

The passive diffusion occurs due to a different concentration gradient across cell membranes, involving both paracellular (via TJ) and transcellular transport (travel through cells). However, paracellular and transcellular transport routes allow diffusion of very small solutes and hydrophilic or lipophilic properties, respectively. Aiming increased drug delivery to the brain, BBB disruption has evaluated to promote reversibly open TJ, allowing an increased drug paracellular transports (Sarkar et al., 2017). For this purpose, focused ultrasound (FUS) has been widely used for non-invasive opening of BBB. Lin and coworkers (2016) researched the influence of FUS induced with microbubbles on the transport of liposomes containing gene vectors for CNS therapy. Their results showed that it was an effective strategy to transport gene vector into the brain and may have a potential for neurodegenerative disease treatment (Lin et al., 2016). However, BBB disruption has some disadvantages, such as damage to the neuronal cells and entrance of unwanted molecules to the brain (Sarkar et al., 2017).

Thus, other approaches have been studied to increase drug delivery into the brain. In this way, the main routes used to transpose the BBB are carrier-mediated transport (CMT), receptor-mediated transcytosis (RMT), adsorptive-mediated transcytosis (AMT), and cell-mediated transport.

12.2.2 CARRIER-MEDIATED TRANSPORT

CMT or transporter-mediated transport is responsible for transporting solutes between the bloodstream and CNS using facilitated or active transport. There are two types of carriers present on BCECs cell membrane: carrier-mediated influx and carrier-mediated efflux. The carrier-mediated influx involves the uptake of essential endogenous substances, such as glucose (via glucose transporter-1, GLUT-1), amino acids (via large neutral amino acid transporter, LAT-1 and cationic amino acid transporter-1, CAT-1), monocarboxylic acids (via monocarboxylic acid transporter-1, MCT1) and nucleosides (via nucleoside transporter-2, CNT2) into the brain. Thus, CMT has been studied to promote specific targeting of nanocarriers into CNS (Gao 2016; Bhaskar et al., 2010). For this propose, using nanocarriers-functionalized with a specific ligand for CMT have been investigated. Li and coworkers (2016) evaluated BBB penetration of docetaxel-loaded functionalized glutamate-targeting liposomes (DOX-TGL) and docetaxel-loaded unmodified liposomes using an animal model. They showed a significant higher BBB penetration of DOX-TGL when compared with unmodified liposomes. These results demonstrated that LAT-1 is a promising transporter to promote drug delivery for glioma treatment (Li et al., 2016).

Carrier-mediated efflux belongs to the ATP-binding cassette family and it includes P-gp, multidrug resistance-mediated proteins, and BCRP. They are responsible for maintaining the homeostasis of the CNS through metabolites transport from the brain back to the bloodstream. However, active efflux transport removes out of CNS a large number of drugs and it reduces the treatment efficiency of CNS disorders (Meng, et al., 2017). In this context, nanotechnology has been a strategy to overcome efflux transport. Tang and coworkers (2014) developed liposomes for codelivery of doxorubicin and a P-gp inhibitor (verapamil). They showed that the system could overcome the P-gb because of synergistic effects: P-gp bypassing effect and P-gp inhibition effect (Tang et al., 2015).

12.2.3 RECEPTOR-MEDIATED TRANSCYTOSIS

RMT as well as CMT enables the transportation of large endogenous substances, such as nutrients, hormones, lipoproteins, and growth factors, from the bloodstream to the brain. The RMT route is based on the specific binding of ligands to cell surface that induces an endocytosis that triggers an endocytic vesicle formation. Then, exocytose process transports the ligand to the extravascular space in CNS. This transport occurs due to the expression of a large number of peptide-specific receptors on BCECs, such as transferrin receptor, insulin receptor, lactoferrin receptor, low-density lipoprotein receptor-related protein, and neonatal Fc receptor (Zhang et al., 2015; Glaser et al., 2017). RMT has become a strategy for targeting liposomes functionalized with a specific ligand to promote drug delivery into the brain. This kind of strategy was studied by Chen and coworkers (2016) when they functionalized liposomes with transferrin to promote α-Mangostin delivery into CNS. They showed that liposomes functionalized with transferrin improved the brain delivery of α-Mangostin and it is a potential carrier of α-Mangostin for treatment of Alzheimer's disease (Chen et al., 2016).

12.2.4 ADSORPTIVE-MEDIATED TRANSCYTOSIS

AMT involves a nonspecific electrostatic interaction between cationic molecules in the bloodstream with anionic BCECs surface (e.g., glycoprotein and heparin sulfate proteoglycans). Briefly, the adsorptive transcytosis of positively charged molecules is mediated by clathrin-dependent endocytosis. Thus, AMT has been investigated for nanocarriers' transportation (Comoglu et al., 2016; Bhaskar et al., 2010). Li and coworkers (2014) associated RMT (via GLUT-1 receptor) and AMT using dequalinium-lipid modified to prepare liposomes for paclitaxel delivery into the brain. They showed high efficacy of this nanocarrier to induce apoptosis in brain tumor cells both in vitro and in vivo (Li et al., 2014).

12.2.5 CELL-MEDIATED TRANSPORT

Cell-mediated transport is used to transport cells, such as macrophages, monocytes, and neutrophils, across BBB during inflammations processes in the brain. Generally, these inflammation processes are common to be observed in brain disorders, like Alzheimer's disease, Parkinson's disease, multiple sclerosis, and brain tumors. Because of the extensive capacity of these cells to transpose the BBB and endocytic nanocarriers, they have been used to transport nanocarriers from the bloodstream into the brain. So, these cells serve as "Trojan horses" to promote drug delivery into CNS (Patel and Patel 2017; Chen and Liu 2012). Qin and coworkers (2014) evaluated edavarone-loaded liposomes functionalized with RGD peptide (Arg-Gly-Asp) to improve depression treatment. Because of the affinity of RGD peptide to a receptor expressed on the leukocytes cell membrane, liposomes functionalized with RGD showed higher brain delivery because of leukocytes-mediated transport (Qin et al., 2014).

12.3   LIPOSOMES

Liposomes are small (ranging from approximately 20 nm to several micrometers), artificial, semipermeable, spherically shaped vesicles composed by phospholipid bilayers that were first described by Bangham et al. (1965). They are typically made of a variety of natural or synthetic phospholipids with a low intrinsic toxicity and cholesterol, which assemble into closed structures when hydrated by an aqueous solution (Akbarzadeh et al., 2013). The phospholipids structure and proportion at the mixture affects their morphological and physicochemical properties (such as size, surface charge, porosity, flexibility, stability, pharmacokinetics, and others). Cholesterol is also commonly used and plays an important role by raising flexibility by creating spaces between each phospholipid. These vesicles present an amphiphilic membrane with hydrophilic borders (because of the phosphate groups at the phospholipidic head) and a hydrophobic central layer, in which lipophilic drugs and other non-polar compounds may be present (Rip 2016; Huang et al., 2014) as schematized in Figure 12.2. They also present a hydrophilic inner core, usually hydrated with a polar solvent that may carry a polar molecule. Since their first report and subsequent studies, liposomes have been described as a versatile delivery system with several structural modulation capabilities, as their composition variety and synthesis processes enable a high engineering freedom.

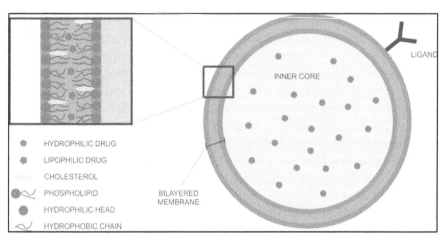

FIGURE 12.2   Liposome structure scheme showing the detailed bilayer general composition with the coordinated phospholipid assembly, drug entrapping possibilities (both hydrophilic and lipophilic), as well as functionalization capabilities pointed by the ligand at the surface.

Liposomes are generally classified on basis of their size and number of layers enclosing the aqueous solution in two big groups: the large multilamellar vesicles (LMVs) and unilamellar vesicles (ULVs), as shown in Figure 12.3. Multilamellar liposomes present an onion-like structure, with two or more (up to several) units of bilayers (Fig. 12.3) (Wagner and Vorauer-Uhl, 2011; Allen and Cullis, 2013). Unilamellar liposomes, on the other side,

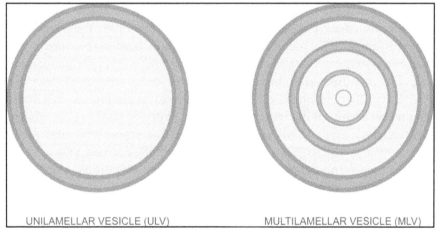

FIGURE 12.3   Liposomes structures scheme showing the unilamellar (ULVs) and multilamellar vesicles (MLVs), respectively.

present only one bilayer entrapping the aqueous solution and they are usually the goal for most drug delivery systems. The last group, unilamellar vesicles, is subcategorized according to its size in small unilamellar vesicles (SUVs) and large unilamellar vesicles (LUVs; Akbarzadeh et al., 2013). Most of the liposome preparation methods produce multilamellar vesicles, so in cases that unilamellar vesicles are the goal, subsequent processes are needed to reduce the number of bilayers and liposome sizes.

12.4   PREPARATION TECHNIQUES FOR LIPOSOMES

12.4.1   SYNTHESIS

This chapter will present liposomes synthesis methods that are used at the development of drug delivery systems, focusing on well-known processes and protocols. However, it is important to mention that methods and processes regarding the development of liposomes are in constant evolution; since the pioneering discovery of liposomes by Bangham, several methods have been reported (Huang et al., 2014; Allen, 1997). The main ones are lipid thin-film hydration (Bangham method), reverse phase evaporation and solvent injection.

Most of them produce multilamellar vesicles, so in order to achieve unilamellar vesicles processes like sonication, homogenization and high-pressure extrusion are performed afterward.

12.4.1.1   LIPID THIN-FILM HYDRATION

Lipid thin-film hydration or Bangham method was the first method described in literature for the synthesis of liposomes. It's an extremely simple method that is accomplished by three steps: solubilization of lipids and lipid-soluble substances into an organic solvent; drying the solvent under vacuum and consequent formation of a thin-film; and hydration of the film by an aqueous medium containing water-soluble substances. The temperature of the medium must be above the glass transition temperature (T_g) of the lipid with the higher T_g during the medium addition and hydration time. Hydration may be performed by vigorous shaking, mixing or stirring. The products of the hydration are multilamellar vesicles, which need further sizing processes—such as homogenization, (extrusion and sonication) to obtain unilamellar vesicles. However, one of the main limitations of this technique is the scale-up procedure. Although it is really practical and easy to reproduce the method in laboratory scale, scaling it up to industrial level is a challenge (Jaafar-Maalej et al., 2010). A general scheme of the method is illustrated in Figure 12.4.

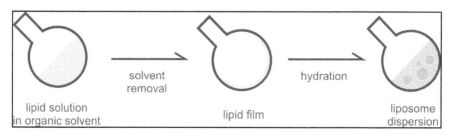

FIGURE 12.4   Large multilamellar vesicles (LMVs) produced by the lipid thin-film hydration method in three steps: lipid solubilization; solvent removal and thin-film formation; hydration and liposome formation.

12.4.1.2   REVERSE-PHASE EVAPORATION

The reverse-phase evaporation method was a great achievement in liposome development technologies, as it enabled the capability of a large aqueous medium entrapment—generating a higher aqueous-to-lipid ratio. This method is based on the formation of inverted micelles by the creation of a water-in-oil emulsion by brief sonication of a two-phase system containing the aqueous medium and lipids solubilized on organic solvents, such as chloroform, isopropyl alcohol or diethyl ether (Akbarzadeh et al., 2013). Afterward, the organic solvents are removed by a low-pressure environment and the liposomes are formed. That happens due to high lipid content at the aqueous media, which tends to create the vesicles by the interaction of the hydrophobic chains from the lipids with the inverted micelles. That method presents a moderate encapsulation efficiency (up to 65%). One of the main limitations is the contact of encapsulating material with organic solvents, which may be harmful to them depending on the material characteristics. Figure 12.5 shows the schematic procedure.

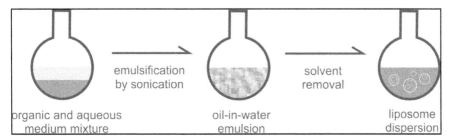

FIGURE 12.5 Large multilamellar (LMVs) produced by the reverse phase evaporation method. Briefly, two phases must form an emulsion water-in-oil emulsion (usually by sonication) and the solvent removal is carried further on.

12.4.1.3 SOLVENT INJECTION

Solvent injection methods were developed as an alternative to overcome the issues of heterogeneity of the liposomal dispersions that other methods guaranteed. The main relevance of the method relies on injecting an ethanolic or ether lipid solution in water or aqueous buffer, obtaining a narrow dispersion and without the necessity of subsequent steps (Sonar et al., 2008). The method consists on quickly injecting lipids through a low-gauge needle into the aqueous medium (Fig. 12.6). As the solvents mix, liposomes are formed immediately as small unilamellar vesicles. Still, a limitation of the technique is the removal of ethanol from the resultant dispersion, as it forms an azeotropic mixture with water—turning the separation processes extremely difficult to perform (Jaafar-Maalej et al., 2010).

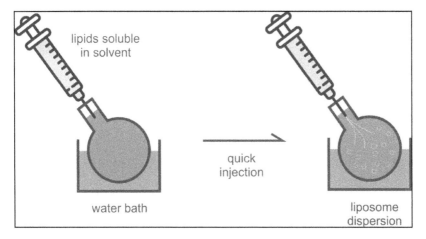

FIGURE 12.6 Small unilamellar vesicles (ULVs) formed by the solvent injection method in a simple step.

12.4.2 SIZING

As mentioned before, most of the times a sizing process is needed to reduce the liposome size and turn the dispersity as homogenous as possible. Several processes have been used at liposomal technologies, but three of them are the predominant: extrusion, sonication, and high-pressure homogenization.

12.4.2.1 EXTRUSION

The extrusion technique consists on forcing the large multilamellar vesicles to pass through a filter with a defined pore size, reducing them to a smaller particle size and on a unilamellar conformation (Mayer et al., 1986; Mui et al., 2003). This technique was first described by Olson and collaborators (Olson et al., 1979) and it is commonly used even today due to its easiness to perform, high reproducibility and good results. Still, there are some limitations to the technique and some cautions have to be taken in order to carry the method on good standards. Prior to the final extrusion (which determines the resultant particle size distribution), LMVs are generally passed through bigger pores in order to prevent membrane disrupting and granting homogeneity. Also, conducting the process in temperatures above the glass transition temperature is important, as it improves the membrane flexibility and eases the passing through the pores. Figure 12.7 schematically demonstrate how the process works. The main problem with extrusion is the upscaling from the laboratory to industrial batch sizes.

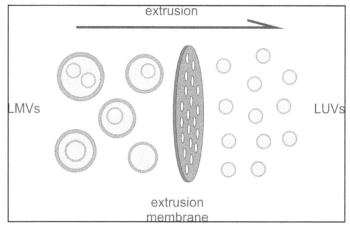

FIGURE 12.7 Large multilamellar vesicles passing through an extrusion membrane in order to achieve unilamellar vesicles.

12.4.2.2   SONICATION

The sonication method uses acoustic (also called sonic) energy to diminish particle size. It is generally applied to LMVs and widely used for reducing the sizing of liposomes and can be performed either by probe tip sonicators or bath sonicators. The acoustic energy generates a positive pressure that breaks the particles into smaller vesicles, with one or more bilayers (Fig. 12.8). The main advantage of the process is the time consumption, as it is way faster than the extrusion process (Lapinski et al., 2007). However, the reproducibility of the process is not a strong point, as it can be quite inconvenient even on a laboratory scale.

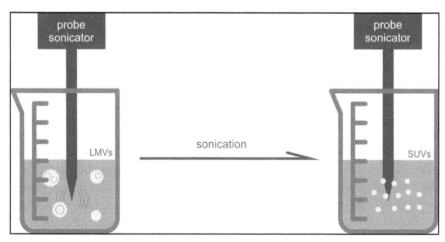

FIGURE 12.8   Large multilamellar vesicles being sonicated by a probe tip sonicator and cleaved by acoustic-generated pressure into small unilamellar vesicles.

12.4.2.3   HIGH-PRESSURE HOMOGENIZATION

One of the best tools for sizing liposomes in large scale, high-pressure homogenization plays an important role in the tools set for shrinking and narrowing the size distribution (Barnadas-Rodríguez and Sabés 2001). It works by shocking the particles with a high pressure, and factors as previous particle size, inlet pressure and a number of cycles may influence the final result. One of the main inconveniences related to the process is the cost: devices for high-pressure homogenization are usually big and expensive for some labs, needing a bigger investment.

12.5   LIPOSOMES-MEDIATED BRAIN DRUG DELIVERY

As described in the previous sections of this chapter, the transport of drugs across BBB is limited due to features of this barrier, in particular, the presence of high concentrations of tight junctions (TJ) and adhesion proteins between BCECs. Furthermore, drug concentration in the brain is limited by efflux transports that transport drugs from the brain back to the bloodstream. Therefore, the drug delivery to the brain is the biggest challenge in the treatment of CNS diseases. In this way, nanotechnology has gained research focus for drug delivery into the brain using the transport routes described in the previous section. For this purpose, nanocarriers such as liposomes can be modified with targeting moieties to bind specifically in receptor or transporter expressed at the BBB or at immune system cells for cell-mediated transport, thus enhanced CNS permeability. Then, some ligands that can be functionalized at liposomes surface will be shown in the following sections.

12.5.1   ANTIBODIES

In order to further improve the specificity of liposomes for receptors presents on BCECs and promote an increased drug delivery across BBB, some researchers have been investigating the targeting of liposomes modified with antibodies on its surface, called immunoliposomes. These liposomes can be functionalized using different techniques. Some of them involve the attachment of antibodies to lipid before liposomes preparation and another technique involves a couple of ligands to the surface of liposomes using thioether bonds. However, the technique of conjugation used can affect the cellular uptake efficiency because of the random organization of antibody on liposomes surface. Among these techniques, functionalization using thioether bonds, like reaction with maleimide group has been widely used to produce immunoliposomes (Paszko and Senge 2012; Gao et al., 2013). Qin and coworkers (2014) prepared immunolipossomes using the post-insertion method. In this technique, the antibody anti-transferrin was reacted with Traut's reagent to yield transferrin-SH. After this procedure, transferrin-SH was reacted with DSPE-PEG2000-Maleimide and it was incubated with liposomes pre-formed. The results obtained in this research showed that immunolipossomes prepared have a higher cellular uptake than liposomes non-functionalized in C6 cell line because of the efficient process of functionalization (Qin et al., 2014).

Among the antibodies used to target nanocarriers into the brain, the anti-transferrin antibody has been the most classical and widely used ligand to promote immunoliposomes targeting due to the presence of transferrin receptor (or CD71) on BCECs cell membrane surface, a receptor-mediated transcytosis. Therefore, liposomes functionalized with anti-transferrin receptor have been investigated to deliver drugs across the BBB and optimize treatments of many CNS disorders. Another antibody used to brain treatment is the anti-EGFR. This antibody is used to target nanocarriers into tumor cells because of the overexpression of epidermal growth factor receptor (EGFR) on the cellular membrane in many tumor cells, like glioma cells, and promote the uptake of nanocarriers. Thus, the next section will address some conducted researches using antibodies that have been developed in recent years for the treatment of brain disorders (Feng et al., 2009).

12.5.2   PEPTIDES

The first cell-penetrating peptide (CPP) discovery was Tat peptide, a basic domain of HIV-1 Tat protein (Green et al., 1989). Since then, a large number of peptides have been described in the literature for targeting drug delivery systems, including for brain targeting. The advantages of using peptides as a targeting agent are the safety, specificity, non-invasive mode of transport and high internalization efficiency. Thus, it id a promising strategy to promote liposomes delivery across the BBB for treatment of many brain disorders.

The most important mechanism of transport nanocarriers-modified peptide across BBB is through an endocytic pathway involving RMT, such as transferrin, integrin, and low-density lipoprotein receptors, CMT, and/ or AMT (Oller-Salvia et al., 2016). Besides that, some peptides like RGD peptide can promote cell-mediated transport across BBB. Qin and coworkers (2015) developed liposomes modified with cyclic RGD peptide that have a high affinity for integrin receptor present on monocytes cellular membrane. They showed that this nanocarrier can promote high internalization of trefoil factor 3 (TFF3) due the transport of liposomes by monocytes mediated transport into the brain. Thus, the cyclic RGD peptide and others that have been investigated in the literature are promising ligands to be attached to liposomes for increase drug delivery across BBB and improve the treatment of CNS disorders (Qin et al., 2015).

12.5.3 OTHER SURFACE MODIFICATIONS

The PEGylation strategy hampers the liposomes recognition, so they are not promptly captured by the macrophages of the reticuloendothelial system, what makes they remain in the blood circulation for long periods. This extended circulation time because of the polyethylene glycol (PEG) molecules on the surface of the liposome implement the passive gathering, named the enhanced permeability and retention (EPR) effect related to nanoparticle drug delivery systems (Koren et al., 2011). This has been used to overcome the limiting factors such as difficulty to cross the BBB and reach the brain for the treatment of brain pathologies. In addition to the EPR effect of nanoparticles for the accumulation at the target, other modes can be used to accomplish the treatments (Wei et al., 2016). In this way, the effectiveness of brain targeting can be dependent on the surface vectors. These strategies have been applied to overcome the barrier-limiting function and to decrease the local toxicity induced by the drug flow through the endothelial regions (Dinda and Pattnaik 2013). PEGylation modified the surface of the liposome to deliver dopamine at the brain tissue in a rat model of Parkinson's disease. It was obtained better uptake for dopamine encapsulated into liposomes and PEGylated liposomes with higher volume distribution into the brain. The membrane crossing through the BBB, the ability of PEGylated liposomes, increased the blood circulation time and promoted successful drug transportation to the brain (Kang et al., 2016). Long circulation time was also found for glutathione PEGylated liposomes. They investigated the pharmacokinetics and organ distribution of a fluorescent tracer after intraperitoneal and intravenous administrations reaching significant levels of the fluorescent tracer in the brain, which allowed prolonged dosing intervals and enhanced drug delivery in a safe treatment. Glutathione covering nanoparticles have been applied to several drug deliveries to the brain because glutathione exerts a central role to detoxify the cellular metabolism plus the antioxidant property, useful feature for Parkinson disease and cancer therapy (Rip et al., 2014; Koren et al., 2011).

The liposomes functionalization with polyethylene glycol (PEG) and polysaccharides molecules can improve the pharmacokinetics and biodistribution into the brain. These bindings constitute a protective layer on the liposomes, preventing the clearance and increasing the specificity. Furthermore, there are others types of ligands such as stimuli-responsive including magnetic field, temperature, ultrasound intensity, light, and electric pulses.

All of them based on active targeting and to enhance better brain drug delivery (Vieira and Gamarra 2016).

The increase of the capacity to transport across the BBB was achieved with glycosylated and PEGylated liposomes to deliver sertraline. The glycosylation attributed specificity by modifying the surface of the liposome with homing devices toward glucose transporters present in the brain, as glucose is an essential fuel for the brain metabolism and brain endothelial cells. Therefore, these different ligands may generate high levels of the drug into the brain because of the modification of transendothelial permeability (Harbi et al., 2016).

Coating with stearylamine and MPEG-DSPE molecules were applied on liposomes surface to enhance brain penetration through intranasal route. It was showed prolonged diffusion, controlled release of risperidone and higher brain exposure. Nasal instillation directly transported risperidone to the brain, via olfactory nerves and trigeminal pathway, circumventing the BBB (Narayan et al., 2016).

12.6   GENE THERAPY

Gene therapy is a new tool able to reach specific targets as genes or pathways associated with diseases, where conventional therapies with drugs do not reach. The gene therapy offers a great potential in treating several diseases, even for the brain. One of its weaknesses is that the genetic material usually gets degraded when it's not contained within a drug delivery system. Liposomes bearing siRNA for tumor therapy were applied on a nude mouse glioma model using ultrasound-sensitive nanobubbles to improve the efficacy delivery to glioma cells. They obtained high rates of siRNA transfection, causing enhanced gene silencing effect and elevated level of cancer cells apoptosis (Yin et al., 2013).

Dual receptor-specific peptides were anchored on liposomes surface to deliver siRNA with vascular endothelial growth factor (VEGF) sequence to inhibit angiogenesis of tumors. Gene transfection and silencing were evaluated in an orthotopic xenograft model of glioblastoma cells and the cell uptake in tumor showed superiority for the modified liposomes. The VEGF target promoted better anti-angiogenesis and apoptosis effect presenting a good platform for brain cancer therapy (Yang et al., 2014).

Another study using PEGylated liposomes functionalized with a 36-amino acid peptide via disulfide bridge and a small peptide OX26 were performed to deliver a plasmid DNA. Significant gene expressions of hTERTC27 in

C6-glioma-bearing rats was obtained, which was responsible for inducing telomere alteration and inhibiting tumor cell growth without causing side effects on normal cells. The developed delivery system was able to cross the BBB and target the glioma cells at the brain, what encourages a non-invasive targeting therapy via endovenous administration (Yue et al., 2014).

Since the brain is a multi-cell type organ, it is necessary the improvements in therapies to overcome the restricted management of brain diseases (Hayward et al., 2016). Gene therapy can propitiate long-term expression on genes of interest and if chemotherapy is combined, two different sites are reached in a synergic effect. Considering liposomes, the cationic ones are able to enter into the CNS because of the AMT and gene transfer by the positively charged lipids and more gene loaded in the complexes. Therefore, it could have gene transfection with low toxicity and immunogenicity in clinical applications (Sun et al., 2011).

12.7 APPLICATIONS IN BRAIN DISORDERS

As previously discussed, liposomes are considered versatile nanocarriers to encapsulate different types of drugs and they have been used to treat different diseases as brain disorders. Brain disorders' treatment is hampered by BBB, so several strategies have been studied to overcome these barriers and thereby improve the targeted drug delivery (Blakeley, 2008; Hu et al., 2017). This section will address some brain disorders applications with liposomes targeted with antibodies, peptides, and others surface modification approached in Section 12.5.

12.7.1 BRAIN TUMOR

According to National Brain Tumor Society, glioblastoma multiforme (GBM), a type of glioma, is the most common and malignant tumors of brain tumors in adults, in which 2 to 3 per 100,000 people in the United States and Europe are diagnosed per year, representing 12% to 15% of all intracranial tumors and 50% to 60% of astrocytic tumors (National Brain Tumor Society 2016).

Due to impaired therapeutic efficacy, some approaches have been developed to overcome the low brain drug accumulation and low selectivity (Chen et al., 2016). One promisor approach is transferrin. Song and coworkers (2017) developed a liposome modified with transferrin-containing vincristine and tetrandrine to reach the brain tumor and they showed good results in treating brain glioma. It demonstrated a transportation enhancement across the BBB via receptor-mediated endocytosis, increasing the cellular uptake and inhibiting multidrug resistance by suppressing the overexpressed P-gp proteins. Besides, the block of cancer invasion and VM channels formation was reached by downregulating the expression of PI3K and MMP2. This work showed important targeting liposomes for brain glioma (Song et al., 2017).

In the same way, Liu and coworkers (2017) developed liposome loaded transferrin peptide sterically stabilized to improve gliomas treatment. Such activity is possible due to the molecular recognition of transferrin receptors (TF-Rs) at the BBB and glioma cells, resulting on internalization and lysosomal escape of cell-penetrating peptide. Furthermore, this system was covered with PEG to protect it from opsonin binding in the bloodstream and also promote conjugation to ligands on the liposomes providing molecular recognition of receptor on surface cells. They showed the efficacy of liposome penetration into the BBB, long-term blood circulation in vivo, and specific tumor targeting, as well as high cell penetration and intracellular lysosomal escape capability (Liu et al., 2017).

Jiao and coworkers (2017) developed a liposomal system composed by CGT and Pep-1 peptide for the effective treatment of malignant brain glioma. The system increased the anticancer efficacy in glioma due to their peptide conjugation. The targeted system cellular uptake was increased from 47.5% to 89.8% compared to the non-targeted system. Besides it, the cell cancer toxicity and apoptosis were higher (Jiao et al., 2017).

nRGD peptide is largely used to transport drugs across the BBB as approached before in Section 5.2. Chen and coworkers (2017) developed a nRGD-peptide liposome containing lycobetaine and the conjugation formulation showing higher targeting to the tumor site. The inhibitory effect of VM channels was observed due to the association of iRGD with both drugs in the formulation. Besides, this portion of iRGD targeted the formulation to integrin receptors, increasing cellular internalization, and cellular growth inhibition by the drug's actions (Chen et al., 2017).

Another approach used is the antibodies conjugation as an epidermal growth factor (EGF) which receptors (EGFR) are overexpressed in glioma cells and non-expressed in normal brain cells. Feng and coworkers (2009) developed a liposomal formulation composed of nickel lipid, and anti-EGFR antibodies containing sodium borocaptate. They showed the immunoliposomes was targeted to tumor due to recognition by the anti-EGFR in cell tumor surface and it is an effective system to release boron into glioma cells in boron neutron capture therapy (Feng et al., 2009).

In addition to research, there are many reviewed clinical trials for drugs encapsulated in liposome for a brain tumor (Table 12.1). Depocyte® is a sustained-release formulation of the chemotherapeutic agent cytarabine, available in the market for patients with neoplastic meningitis from primary brain tumors (Gaviani et al., 2013; Glantz et al., 1999). Myocet® is another conventional liposome containing doxorubicin used for a brain tumor (Gaillard et al., 2012). Both conventional formulations have been found in trial together with others drugs. Among them is one that combines intrathecal liposomal cytarabine (Depocyte®) and temozolomide. It is currently in clinical trial for glioblastoma multiforme and it is in phase 1 and 2 (ClinicalTrials.gov Identifier NCT01044966). Another clinical trial with Depocyte® is on phase 1 and it is administrated together or in sequential of whole brain radiotherapy. This treatment can be used for a solid tumor neoplastic meningitis in patients with or without brain metastasis (ClinicalTrials.gov Identifier NCT00854867). Another one is on phase 2 and combines liposomal cytarabine and high-dose of methotrexate for patients with breast cancer and that disseminated to CNS. The association is important to stop the tumor cells division and kill the cells (ClinicalTrials.gov Identifier NCT00992602). These studies are important due to the lack of significant treatment options for patients with glioblastoma multiforme or metastasis from another type of tumor.

TABLE 12.1   Clinical Trials with Liposomal Formulations for Brain Tumor.

Drug (liposomal formulation)	Phase	Condition	ClinicalTrials. gov Identifier
Liposomal encapsulated **Ara-C** (**Depocyite®**)	1/2	Glioblastoma multiforme Glioma Astrocytoma Brain tumor	NCT01044966
Vincristine sulfate **liposome** (**Marqibo®**)	1/2	Sarcoma Neuroblastoma Wilms tumor Leukemia Lymphoma Brain tumors	NCT01222780

TABLE 12.1   *(Continued)*

Drug (liposomal formulation)	Phase	Condition	ClinicalTrials. gov Identifier
Doxorubicin HCl Liposome	1	Child-hood soft tissue sarcoma Childhood liver cancer Bone cancer Brain tumor Kidney tumor	NCT00019630
Intrathecal Liposomal Cytara-bine (DepoCyte®)	1	Solid tumor neoplastic meningitis brain metastases	NCT00854867
Glutathione pegylated lipo-somal **doxorubicin hydrochloride**	1/2	Brain metastases Lung cancer Breast cancer Melanoma Malignant glioma	NCT01386580
Liposomal Doxorubicin	2	Breast cancer (brain cancer metastases)	NCT00465673
Rhenium Liposome	1/2	Glioblastoma Astrocytoma	NCT01906385
Liposomal Cytarabine	2	Central nervous system metastases Leptomeningeal metastases Recurrent breast cancer Stage IV breast cancer Tumors metastatic to brain	NCT00992602
Liposomal Doxorubicin	2	Meningeal carcinomatosis	NCT01818713
Pegylated Liposomal Doxoru-bicine	1/2	Glioblastoma	NCT00944801
Cytarabine Liposome Injection	4	Meningeal neoplasms	NCT00029523
Liposomal Cytarabine	Termi-nated	Neoplastic meningitis	NCT00964743

12.7.2   NEUROLOGICAL DISEASES

Among neurological diseases, Alzheimer's and Parkinson's diseases are the most commons neurodegenerative diseases due to the high global prevalence (Kang et al., 2016). Alzheimer's disease is a neurodegenerative disorder and it does not have the exact cause and depends on age, genetic and family historic, but apolipoprotein E1 and 4 (ApoE 1-e4) genes are related with it. This disease affects the CNS work mainly the memory impairment, progressive cognitive neuronal dysfunction, thinking and behavioral disturb and others problems in patient routine. According to World Alzheimer report 2016, about 47 million people, all over the world suffering from dementia which is estimated to increase by 131 million till 2050 (Agrawal et al., 2017; Prince et al., 2016). The difficult to across and reach therapeutically effective concentrations in the CNS is the same as found in brain tumors. So many approaches have been studied to overcome these difficulties.

As Alzheimer's disease is related with accumulation of β-amyloid (Ab) plaques and neurofibrillary tangles in specific regions of the brain, several strategies have been developed to target it (Re et al., 2010; Ordóñez-Gutiérrez et al., 2017). Re and coworkers (2011) developed liposomes containing phosphatidic acid or cardiolipin to target amyloid-β (Aβ) that is highly produced in this disease. The system was functionalized with peptides derived from apolipoprotein E (ApoE). They observed, in vitro assays, a high uptake in brain capillary cells, up to 60%, due to the targeted liposomes with ApoE-derived peptide (Re et al., 2010).

Ordonez-Gutierrez and coworkers (2017) developed and characterized an immunoliposome PEGylated with a high capacity for capturing antibody (Ab) in the periphery. They observed the action of multivalent immunoliposomes to reducing circulating and brain levels of Ab1-40 and Ab1-42 by in vivo studies. Furthermore, the immunoliposome PEGylated mediated reduction in amyloidosis was correlated with lower levels of glial fibrillary acidic protein and reactive glia. Besides that, this system was able to be on bloodstream for a prolonged because of a PEGylation performed (Ordóñez-Gutiérrez et al., 2017).

Liposomes functionalized with transferrin have been extensively studied to release drugs into the brain. Chen and coworkers (2016) developed a liposome containing α-Mangostin targeted with transferrin and evaluated it by in vitro and in vivo studies for Alzheimer's disease. As result, they showed the system was able to cross the BBB due to their transferrin receptors in brain endothelial capillary cells, accumulating α-Mangostin in the brain. α-Mangostin is a potential drug used in Alzheimer's disease but it has low penetration in the brain due to the BBB (Chen et al., 2016).

Parkinson's disease is another neurodegenerative disturb with high global prevalence ranging from 100 to 200 per 100,000 people and the annual incidence could be 15 per 100,000 (Tysnes and Storstein 2017). The selective cell death of dopaminergic neurons in the substantia nigra pars compacta with reduction of brain dopamine concentration could be the causes of this disease. In consequence of this reduction, movement disorders such as akinesia/bradykinesia, rigidity, and tremor are observed in these patients (Kang et al., 2016).

Dopamine was encapsulated into PEGylated liposomes and this system was functionalized with OX26 Mab, antitransferrin receptor monoclonal antibody. This system was able to increase the brain cell uptake about 8-fold compared with dopamine alone and about 3-fold compared with the immunoliposome PEGylated containing dopamine. In pharmacokinetics study, the volume of distribution of immunoliposome PEGylated containing dopamine in the brain by the perfusion method was 4-fold higher than that system without PEG, indicating the effective conjugation of OX26 MAb to the transferrin receptor of brain capillary endothelium in brain tissue (Kang et al., 2016).

Another approach used is a peptide conjugation in liposomes containing levodopa to promote the target in brain tumor. This study proved the high cellular uptake with chlorotoxin-modified stealth liposome containing levodopa and by in vivo study dopamine and its metabolites were found in the substantia nigra and striata. In the methyl-phenyl-tetrahydropyridine-induced Parkinson's disease mice model, the behavioral disorders were reduced by the liposomal formulation and it is potential targeting delivery system to Parkinson's disease (Xiang et al., 2012).

12.8   FUTURE PROSPECTS

Considering the challenges to treat brain diseases much has been made to the better understanding of main impairments to cross BBB and rational design of drug delivery systems to accomplish the development of new formulations, specifically, liposomes. These prospects should be intensified in the next years because of the increase of global life expectancy and by the need to treat the elderly, who may develop disorders or tumors of the brain. Therefore, expressive advances and tools will be implemented to treat these progressive cases, to decrease the symptoms and health complications of the patients and to improve the patient's quality of life.

12.9   CONCLUSION

The drug delivery to the brain is the most challenge in the CNS diseases treatment due to the BBB. In this way, nanotechnology has gained research focus for drug delivery into the brain using the transport routes. Among the nanocarriers, liposomes are well-established systems and their preparation technique has been increasingly improved by modified with targeting moieties to bind specifically in receptor or transporter expressed at the BBB or at immune system cells for cell-mediated transport. Overall, studies showed better efficacy with great potential for clinical application, and some clinical trials are ongoing, in phase 1, 2, 4, or terminated.

KEYWORDS

- **blood–brain barrier**
- **brain disorder**
- **nanotechnology**
- **nanocarrier**
- **active targeting**
- **liposome synthesis**

REFERENCES

Agrawal, M.; Ajazuddin; Tripathi, D. K.; Saraf, S.; Saraf, S.; Antimisiaris, S. G.; Mourtas, S.; Hammarlund-Udenaes, M.; Alexander, A. Recent Advancements In Liposomes Targeting Strategies to Cross Blood-Brain Barrier (BBB) for the Treatment of Alzheimer's Disease. *J. Control. Release* **2017,** *260*, 61–77.

Akbarzadeh, A.; Rezaei-Sadabady, R.; Davaran, S.; Joo, S. W.; Zarghami, N.; Hanifehpour, Y.; Samiei, M.; Kouhi, M.; Nejati-Koshki, K. Liposome: Classification, Preparation, and Applications. *Nanoscale Res. Lett.* **2013,** *8* (1), 102.

Allen, T. M. Liposomes: Opportunities in Drug Delivery. *Drugs* **1997,** *54* (4), 14–24.

Allen, T. M.; Cullis, P. R. Liposomal Drug Delivery Systems: from Concept to Clinical Applications. *Adv. Drug Deliv. Rev.* **2013,** *65* (1), 36–48.

Bangham, A. D.; Standish, M. M.; Watkins, J. C. Diffusion of Univalent Ions Across the Lamellae of Swollen Phospholipids. *J. Mol. Biol.* **1965,** *13* (1), 238–252.

Barnadas-Rodríguez, R.; Sabés, M. Factors Involved in the Production of Liposomes with a High-Pressure Homogenizer. *Int. J. Pharm.* **2001,** *213* (1–2), 175–186.

Bhaskar, S.; Tian, F.; Stoeger, T.; Kreyling, W.; de la Fuente, J. M.; Grazú, V.; Borm, P.; Estrada, G.; Ntziachristos, V.; Razansky, D. Multifunctional Nanocarriers for Diagnostics,

Drug Delivery and Targeted Treatment Across Blood-Brain Barrier: Perspectives on Tracking and Neuroimaging. *Part. Fibre Toxicol.* **2010,** *7*, 3

Blakeley, J. Drug Delivery to Brain Tumors. *Curr. Neurol. Neurosci. Rep.* **2008,** *8* (3), 235–241

Chen, T.; Song, X.; Gong, T.; Fu, Y.; Yang, L.; Zhang, Z.; Gong, T. nRGD Modified Lycobetaine and Octreotide Combination Delivery System to Overcome Multiple Barriers and Enhance Anti-Glioma Efficacy. *Colloids Surf. B Biointerfaces* **2017,** *156*, 330–339.

Chen, Y.; Liu, L. Modern Methods for Delivery of Drugs Across the Blood-Brain Barrier. Adv. *Drug Deliv. Rev.* **2012,** *64* (7), 640–665.

Chen, Z. L.; Huang, M.; Wang, X. R.; Fu, J.; Han, M.; Shen, Y. Q.; Xia, Z.; Gao, J. Q. Transferrin-Modified Liposome Promotes α-Mangostin to Penetrate the Bloodbrain Barrier. *Nanomed. Nanotechnol. Biol. Med.* **2016,** *12* (2), 421–430.

Comoglu, T.; Arisoy, S.; Akkus, Z. B. Nanocarriers for Effective Brain Drug Delivery. *Curr. Top. Med. Chem.* **2016,** 1490–1506

Dinda, S. C.; Pattnaik, G. Nanobiotechnology-Based Drug Delivery in Brain Targeting. *Curr. Pharm. Biotechnol.* **2013,** *14* (15), 1264–1274.

Feng, B.; Tomizawa, K.; Michiue, H.; Miyatake, S. ichi; Han, X. J.; Fujimura, A.; Seno, M.; Kirihata, M.; Matsui, H. Delivery of Sodium Borocaptate to Glioma Cells Using Immunoliposome Conjugated with anti-EGFR Antibodies by ZZ-His. *Biomaterials* **2009,** *30* (9), 1746–1755.

Gaillard, P. J.; Visser, C. C.; Appeldoorn, C. C. M.; Rip, J. Enhanced Brain Drug Delivery: Safely Crossing the Blood-Brain Barrier. *Drug Discov. Today Technol.* **2012,** *9* (2), e155–e160.

Gao, H. Progress and Perspectives on Targeting Nanoparticles for Brain Drug Delivery. *Acta Pharm. Sin. B* **2016,** *6* (4), 268–286.

Gao, J.; Chen, H.; Song, H.; Su, X.; Niu, F.; Li, W.; Li, B.; Dai, J.; Wang, H.; Guo, Y. Antibody-Targeted Immunoliposomes for Cancer Treatment. *Mini Rev. Med. Chem.* **2013,** *13* (14), 2026–2035.

Gaviani, P.; Corsini, E.; Salmaggi, A.; Lamperti, E.; Botturi, A.; Erbetta, A.; Milanesi, I.; Legnani, F.; Pollo, B.; Silvani, A. Liposomal Cytarabine in Neoplastic Meningitis from Primary Brain Tumors: A Single Institutional Experience. *Neurol. Sci.* **2013,** *34* (12), 2151–2157.

Glantz, M. J.; Jaeckle, K. A.; Chamberlain, M. C.; Phuphanich, S.; Recht, L.; Swinnen, L. J.; Maria, B.; LaFollette, S.; Schumann, G. B.; Cole, B. F.; et al., A Randomized Controlled trial Comparing Intrathecal Sustained-Release Cytarabine (DepoCyt) to Intrathecal Methotrexate in Patients with Neoplastic Meningitis from Solid Tumors. *Clin. Cancer Res.* **1999,** *5* (11), 3394–3402.

Glaser, T.; Han, I.; Wu, L.; Zeng, X. Targeted Nanotechnology in Glioblastoma Multiforme. *Front. Pharmacol.* **2017,** *8*, 1–14.

Grabrucker, A. M.; Ruozi, B.; Belletti, D.; Pederzoli, F.; Forni, F.; Vandelli, M. A.; Tosi, G. Nanoparticle Transport across the Blood Brain Barrier. *Tissue Barriers* **2016,** *4* (1), e1153568.

Green, M.; Ishino, M.; Loewenstein, P. M. Mutational analysis of HIV-1 Tat Minimal Domain Peptides: Identification of trans-Dominant Mutants that Suppress HIV-LTR-Driven Gene Expression. *Cell* **1989,** *58* (1), 215–223.

Harbi, I.; Aljaeid, B.; El-Say, K. M.; Zidan, A. S. Glycosylated Sertraline-Loaded Liposomes for Brain Targeting: QbD Study of Formulation Variabilities and Brain Transport. *AAPS PharmSciTech* **2016**, *17* (6), 1404–1420.

Hayward, S. L.; Wilson, C. L.; Kidambi, S. Hyaluronic Acid-Conjugated Liposome Nanoparticles for Targeted Delivery to CD44 Overexpressing Glioblastoma Cells. *Oncotarget* **2016**, *7* (23), 34158–34171.

Hu, Y.; Rip, J.; Gaillard, P. J.; de Lange, E. C. M.; Hammarlund-Udenaes, M. The Impact of Liposomal Formulations on the Release and Brain Delivery of Methotrexate: An In Vivo Microdialysis Study. *J. Pharm. Sci.* **2017**, *106* (9), 2606–2613.

Huang, Z.; Li, X.; Zhang, T.; Song, Y.; She, Z.; Li, J.; Deng, Y. Progress Involving New Techniques for Liposome Preparation. *Asian J. Pharm. Sci.* **2014**, *9* (4), 176–182.

Jaafar-Maalej, C.; Diab, R.; Andrieu, V.; Elaissari, A.; Fessi, H. Ethanol Injection Method for Hydrophilic and Lipophilic Drug-Loaded Liposome Preparation. *J. Liposome Res.* **2010**, *20* (3), 228–243.

Jiao, Z.; Li, Y.; Pang, H.; Zheng, Y.; Zhao, Y. Pep-1 Peptide-Functionalized Liposome to Enhance the Anticancer Efficacy of Cilengitide in Glioma Treatment. *Colloids Surfaces B Biointerfaces* **2017**, *158*, 68–75.

Kang, Y. S.; Jung, H. J.; Oh, J. S.; Song, D. Y. Use of PEGylated Immunoliposomes to Deliver Dopamine Across the Blood–Brain Barrier in a Rat Model of Parkinson's Disease. *CNS Neurosci. Ther.* **2016**, *22* (10), 817–823.

Koren, E.; Apte, A.; Sawant, R. R.; Grunwald, J.; Torchilin, V. P. Cell-penetrating TAT Peptide in Drug Delivery Systems: Proteolytic Stability Requirements. *Drug Deliv.* **2011**, *18* (5), 377–384.

Lapinski, M. M.; Castro-Forero, A.; Greiner, A. J.; Ofoli, R. Y.; Blanchard, G. J. Comparison of Liposomes formed by Sonication and Extrusion: Rotational and Translational Diffusion of an Embedded Chromophore. *Langmuir* **2007**, *23* (23), 11677–11683.

Li, L.; Di, X.; Zhang, S.; Kan, Q.; Liu, H.; Lu, T.; Wang, Y.; Fu, Q.; Sun, J.; He, Z. Large Amino Acid Transporter 1 Mediated Glutamate Modified Docetaxel-Loaded Liposomes for Glioma Targeting. *Colloids Surf. B Biointerfaces* **2016**, *141*, 260–267.

Li, X. Y.; Zhao, Y.; Sun, M. G.; Shi, J. F.; Ju, R. J.; Zhang, C. X.; Li, X. T.; Zhao, W. Y.; Mu, L. M.; Zeng, F.; et al., Multifunctional Liposomes Loaded with Paclitaxel and Artemether for Treatment Of Invasive Brain Glioma. *Biomaterials* **2014**, *35* (21), 5591–5604.

Lin, C. Y.; Hsieh, H. Y.; Chen, C. M.; Wu, S. R.; Tsai, C. H.; Huang, C. Y.; Hua, M. Y.; Wei, K. C.; Yeh, C. K.; Liu, H. L. Non-Invasive, Neuron-Specific Gene Therapy by Focused Ultrasound-Induced Blood-Brain Barrier Opening in Parkinson's Disease Mouse Model. *J. Control. Release* **2016**, *235*, 72–81.

Liu, C.; Liu, X. N.; Wang, G. L.; Hei, Y.; Meng, S.; Yang, L. F.; Yuan, L.; Xie, Y. A Dual-Mediated Liposomal Drug Delivery System Targeting the Brain: Rational Construction, Integrity Evaluation Across the Blood-Brain Barrier, and the Transporting Mechanism to Glioma Cells. *Int. J. Nanomedicine* **2017**, *12*, 2407–2425.

Mayer, L. D.; Hope, M. J.; Cullis, P. R. Vesicles of Variable Sizes Produced by a Rapid Extrusion Procedure. *BBA - Biomembr.* **1986**, *858* (1), 161–168.

Meng, J.; Agrahari, V.; Youm, I. Advances in Targeted Drug Delivery Approaches for the Central Nervous System Tumors: The Inspiration of Nanobiotechnology. *J. Neuroimmune Pharmacol.* **2017**, *12* (1), 84–98.

Mui, B.; Chow, L.; Hope, M. J. Extrusion Technique to Generate Liposomes of Defined Size. *Methods Enzymol.* **2003**, *367* (1980), 3–14.

Narayan, R.; Singh, M.; Ranjan, O.; Nayak, Y.; Garg, S. Development of Risperidone Liposomes for Brain Targeting through Intranasal Route. *Life Sci.* **2016,** *163,* 38–45.

National Brain Tumor Society. Tumor Types: Understanding Brain Tumors http://braintumor.org/brain-tumor-information/understanding-brain-tumors/tumor-types/ (accessed Aug 22, 2017).

Oller-Salvia, B.; Sanchez-Navarro, M.; Giralt, E.; Teixido, M. Blood-Brain Barrier Shuttle Peptides: an Emerging Paradigm for Brain Delivery. *Chem. Soc. Rev.* **2016,** *45* (17), 4690–4707.

Olson, F.; Hunt, C. A.; Szoka, F. C.; Vail, W. J.; Papahadjopoulos, D. Preparation of Liposomes of Defined Size Distribution by Extrusion through Polycarbonate Membranes. *BBA - Biomembr.* **1979,** *557* (1), 9–23.

Ordóñez-Gutiérrez, L.; Posado-Fernández, A.; Ahmadvand, D.; Lettiero, B.; Wu, L.; Antón, M.; Flores, O.; Moghimi, S. M.; Wandosell, F. ImmunoPEGliposome-Mediated Reduction of Blood and Brain Amyloid Levels in a Mouse Model of Alzheimer's Disease is Restricted to Aged Animals. *Biomaterials* **2017,** *112,* 141–152.

Paszko, E.; Senge, M. O. Immunoliposomes. *Curr. Med. Chem.* **2012,** *19* (31), 5239–5277.

Patel, M. M.; Patel, B. M. Crossing the Blood-Brain Barrier: Recent Advances in Drug Delivery to the Brain. *CNS Drugs* **2017,** *31* (2), 109–133.

Prince, M.; Comas-Herrera, A.; Knapp, M.; Guerchet, M.; Karagiannidou, M. World Alzheimer Report 2016 Improving Healthcare for People Living with Dementia. *Coverage, Quality and Costs Now and in the Future.* **2016,** 1–140.

Qin, J.; Yang, X.; Zhang, R. X.; Luo, Y. X.; Li, J. L.; Hou, J.; Zhang, C.; Li, Y. J.; Shi, J.; Lu, L.; et al., Monocyte Mediated Brain Targeting Delivery of Macromolecular drug for the Therapy of Depression. *Nanomed. Nanotechnol. Biol. Med.* **2015,** *11* (2), 391–400.

Qin, J.; Zhang, R. X.; Li, J. L.; Wang, J. X.; Hou, J.; Yang, X.; Zhu, W. L.; Shi, J.; Lu, L. CRGD Mediated Liposomes Enhanced Antidepressant-Like Effects of Edaravone in Rats. *Eur. J. Pharm. Sci.* **2014,** *58* (1), 63–71.

Qin, L.; Wang, C. Z.; Fan, H. J.; Zhang, C. J.; Zhang, H. W.; Lv, M. H.; Cui, S. De. A Dual-Targeting Liposome Conjugated with Transferrin and Arginine-Glycine-Aspartic Acid Peptide for Glioma-Targeting Therapy. *Oncol. Lett.* **2014,** *8* (5), 2000–2006.

Re, F.; Cambianica, I.; Sesana, S.; Salvati, E.; Cagnotto, A.; Salmona, M.; Couraud, P. O.; Moghimi, S. M.; Masserini, M.; Sancini, G. Functionalization with ApoE-Derived Peptides Enhances the Interaction with Brain Capillary Endothelial Cells of Nanoliposomes Binding Amyloid-Beta Peptide. *J. Biotechnol.* **2010,** *156* (4), 341–346.

Rip, J. Liposome Technologies and Drug Delivery to the CNS. *Drug Discov. Today Technol.* **2016,** *20,* 53–58.

Rip, J.; Chen, L.; Hartman, R.; van den Heuvel, A.; Reijerkerk, A.; van Kregten, J.; van der Boom, B.; Appeldoorn, C.; de Boer, M.; Maussang, D.; et al., Glutathione PEGylated Liposomes: Pharmacokinetics and Delivery of Cargo across the Blood-Brain Barrier in Rats. *J. Drug Target.* **2014,** *22* (5), 460–467.

Sarkar, A.; Fatima, I.; Jamal, Q. M. S.; Sayeed, U.; Khan, M. K. A.; Akhtar, S.; Kamal, M. A.; Farooqui, A. F.; Siddiqui, M. H. Nanoparticles as a Carrier System for Drug Delivery across Blood Brain Barrier. *Curr. Drug Metab.* **2017,** *18* (2), 129–137.

Sonar, S.; D'Souza, S. E.; Mishra, K. P. A simple one-step protocol for preparing small-sized doxorubicin-loaded liposomes. *J. Environ. Pathol. Toxicol. Oncol.* **2008,** *27* (3), 181–189.

Song, X. li; Liu, S.; Jiang, Y.; Gu, L. yan; Xiao, Y.; Wang, X.; Cheng, L.; Li, X. tao. Targeting Vincristine Plus Tetrandrine Liposomes Modified with DSPE-PEG2000-Transferrin in Treatment of Brain Glioma. *Eur. J. Pharm. Sci.* **2017,** *96*, 129–140.

Sun, X.; Pang, Z.; Ye, H.; Qiu, B.; Guo, L.; Li, J.; Ren, J.; Qian, Y.; Zhang, Q.; Chen, J.; et al., Co-delivery of pEGFP-hTRAIL and Paclitaxel to Brain Glioma Mediated by an Angiopep-Conjugated Liposome. *Biomaterials* **2011,** *33* (3), 916–924.

Tang, J.; Zhang, L.; Gao, H.; Liu, Y.; Zhang, Q.; Ran, R.; Zhang, Z.; He, Q. Codelivery of Doxorubicin and P-gp Inhibitor by a Reduction-Sensitive Liposome to Overcome Multi-drug Resistance, Enhance Anti-Tumor Efficiency and Reduce Toxicity. *Drug Deliv.* **2015,** *7544*, 1–14.

Tysnes, O.-B.; Storstein, A. Epidemiology of Parkinson's Disease. *J. Neural Transm.* **2017,** *124* (8), 901–905.

Vieira, D. B.; Gamarra, L. F. Getting into the brain: Liposome-Based Strategies for Effective Drug Delivery across the Blood–Brain Barrier. *Int. J. Nanomed.* **2016,** *11*, 5381–5414.

Wagner, A.; Vorauer-Uhl, K. Liposome Technology for Industrial Purposes. *J. Drug Deliv.* **2011,** *2011*, 1–9.

Wei, L.; Guo, X.-Y.; Yang, T.; Yu, M.-Z.; Chen, D.-W.; Wang, J.-C. Brain Tumor-Targeted Therapy by Systemic Delivery of siRNA with Transferrin Receptor-Mediated Core-Shell Nanoparticles. *Int. J. Pharm.* **2016,** *510* (1), 394–405.

Wong, H. L.; Wu, X. Y.; Bendayan, R. Nanotechnological Advances for the Delivery of CNS Therapeutics. *Adv. Drug Deliv. Rev.* **2012,** *64* (7), 686–700.

Xiang, Y.; Wu, Q.; Liang, L.; Wang, X.; Wang, J.; Zhang, X.; Pu, X.; Zhang, Q. Chlorotoxin-Modified Stealth Liposomes Encapsulating Levodopa for the Targeting Delivery Against the Parkinson's Disease in the MPTP-Induced Mice Model. *J. Drug Target.* **2012,** *20* (1), 67–75.

Yang, Z.; Xiang, B.; Dong, D.; Wang, Z.; Li, J.; Qi, X. Dual Receptor-Specific Peptides Modified Liposomes as VEGF siRNA Vector for Tumor-Targeting Therapy. *Curr. Gene Ther.* **2014,** *14* (4), 289–299.

Yin, T.; Wang, P.; Li, J.; Zheng, R.; Zheng, B.; Cheng, D.; Li, R.; Lai, J.; Shuai, X. Ultra-sound-Sensitive siRNA-Loaded Nanobubbles formed by Hetero-Assembly of Polymeric Micelles and Liposomes and Their Therapeutic Effect in Gliomas. *Biomaterials* **2013,** *34* (18), 4532–4543.

Yue, P.; He, L.; Qiu, S.; Li, Y.; Liao, Y.; Li, X.; Xie, D.; Peng, Y. OX26/CTX-Conjugated PEGylated Liposome as a Dual-Targeting Gene Delivery System for Brain Glioma. *Mol. Cancer* **2014,** *13* (1), https://doi.org/10.1186/1476-4598-13-191.

Zhang, F.; Xu, C.-L.; Lei, C.-M. Drug Delivery Strategies to Enhance the Permeability of the Blood-Brain Barrier for Treatment of Glioma. *Drug Des. Dev. Ther.* **2015,** *9*, 2089–2100.

CHAPTER 13

NIOSOMES FOR BRAIN TARGETING

DIDEM AG SELECI*, MUHARREM SELECI, REBECCA JONCZYK, FRANK STAHL, and THOMAS SCHEPER

Institute of Technical Chemistry, Leibniz University of Hannover, Callinstr. 5, 30167 Hannover, Germany

Corresponding author. E-mail: agseleci@iftc.uni-hannoverde

ABSTRACT

Delivering drug to the brain has still many obstacles. Especially crossing the brain barriers is a big challenge. The application of nanomaterials to medicine has provided the development of novel drug carriers, can facilitate the delivery of drugs to the brain. Niosomes are non-ionic surfactant based vesicles and has been used as a nanocarrier for different types of drugs. Moreover, niosome surfaces can be modified with targeting ligands to enable cell-specific targeting. Because of these promising features, niosomes have a great potential using as a carrier for the delivery of drugs to the brain. The present chapter provides the fundamental information about niosomal drug delivery systems and their recent applications brain targeting.

13.1 INTRODUCTION

A number of impediments are present for the effective treatment of central nervous system (CNS) diseases. The blood–brain barrier (BBB), which plays a key role in protecting and maintaining the homeostasis of the brain, prevents most drugs from entering the CNS from the bloodstream. Additionally, the blood–cerebrospinal fluid barrier (BCSF) and other specialized CNS barriers hinder the efficient delivery of therapeutic and diagnostic agents to diseased areas of the brain. Recent advances in nanotechnology caused a growing interest in using nanomaterials in medicine to solve a number of problems associated with BBB (Bhaskar et al., 2010). The development of a wide range of nanomaterials provides many advantages and new scientific approaches in CNS disease diagnosis, treatment, monitoring, and prevention. Multifunctional nanomaterials are able to penetrate the BBB and can be easily modified by active and passive targeting to enhance the concentration of the drug molecule inside the specific area of the brain (Srikanth and Kessler, 2012). Furthermore, the drug is protected by nanocarrier from degradation and/or interaction with the biological environment, before reaching the target tissue (Seleci et al., 2016).

Among various nanocarriers, vesicular systems have received growing attention in recent years for brain drug delivery (Lai et al., 2013). They can enhance bioavailability of encapsulated drug and provide therapeutic activity in a controlled manner for a prolonged period. Niosomes (non-ionic surfactant vesicles) are one of the promising vesicular drug carriers. They are assembled by self-association of non-ionic surfactants in an aqueous phase and have a bilayer structure. In recent years, a variety of non-ionic surfactants have been described to form niosomes and enable the encapsulation of numerous drugs (Tavano et al., 2016; Begum et at., 2015; Marianecci et al.; 2012). The non-ionic nature of non-ionic surfactants offers high biocompatibility and low toxicity that are important parameters for drug delivery applications. Moreover, niosomes can be produced with lower costs and have greater stability, longer shelf life and wider formulation versatility in comparison with traditional liposomes. These superiorities and advantages of niosomes, compared to other drug delivery devices, make them promising tools for brain targeting to produce commercially available therapeutics.

This chapter describes the using of niosomes as the potential drug delivery systems and provides up to date information regarding recent applications of niosomes for brain targeting.

13.2   NON-IONIC SURFACTANTS: ENHANCED PENETRATION THROUGH BBB

Non-ionic surfactants are amphiphilic molecules that are composed of two different regions: one of them is hydrophilic (water-soluble) and the other one is hydrophobic (organic-soluble). They are a class of surfactants that have no charged groups in their hydrophilic heads. Alkyl ethers, alkyl esters, alkyl amides, and fatty acids are the main non-ionic surfactant classes used for niosome production. They are generally less toxic, less hemolytic and less irritating to cellular surfaces compared to their anionic, amphoteric, or cationic counterparts. Non-ionic surfactants are used extensively in the chemical industry in such areas as detergents, health, and personal care, coatings, and polymers as solubilizers, wetting agents, emulsifiers and permeability enhancers(Kumar and Rajeshwarrao, 2011).

Apart from being a part of niosomes, surfactants play a key role in coating of the nanoparticle surface. Surface modifications lead to an increase of nanocarrier circulation time in the blood and facilitate penetration of nanoparticles through the BBB by recognizing cellular receptors (Kreuter, 2001). Especially polysorbate 80 (Tween 80) coated nanoparticles are able to transport the loaded drugs across BBB, which cannot cross the BBB without Tween 80 (Kreuter et al., 1995; Sun et al., 2004; Schroder et al., 1998). First in vivo experiments were performed by Kreuter et al. (2002) to enlighten the mechanism of this transport system. They investigated the possible involvement of apolipoproteins that bind to lipoprotein receptors on the brain capillary endothelial cells, in the transport of nanoparticle-bound drugs into the brain. Different apolipoproteins were adsorbed directly onto uncoated or Tween 80-precoated dalargin-loaded poly(butyl cyanoacrylate) (PBCA) nanoparticles. After intravenous injection of these samples to mice, the antinociceptive effect was measured. Results showed that especially apolipoproteins E and B yielded high antinociceptive effects that were similar to Tween 80-coating alone and even higher after Tween-80 coating plus apolipoprotein E and B overcoating. According to results, it was concluded that the Tween 80-coated PBCA nanoparticles adsorb apolipoproteins E and B from the blood, and these proteins stimulate receptor-mediated endocytosis of the particles by the endothelial cells forming the BBB. Moreover, coating PBCA with the non-ionic surfactant poloxamer 188 also enhanced the anti-tumor action of doxorubicin against intracranial glioblastoma (Petri et al., 2007; Gelperina et al. 2010). Consequently, these systems facilitate the delivery of the drug into the brain.

13.3   NIOSOMES: STRUCTURE, PREPARATION, AND CHARACTERIZATION

Niosomes are primarily composed of non-ionic surfactants, cholesterol and hydration medium. The self-assembly of non-ionic surfactants in aqueous media results in the formation of closed bilayered vesicles. Application of additional energy such as mechanical (physical shaking, ultrasound) or heat is needed for the formation of niosomes. Cholesterol interacts with non-ionic surfactants via hydrogen bonds in the bilayer and plays an important role in the physical properties and structure of niosomes. It prevents the vesicle aggregation and improves the rigidity of niosomes (Bouwstra et al., 1997). Besides, charged molecules such as dicetyl phosphate, phosphatidic acid, and stearylamine are added to the bilayer to increase the stability of vesicles (Uchegbu and Vyas, 1998). The addition amount of charged molecule to niosomal formulation needs to be 2.5–5.0 mol%. Adding the high amount of charged molecules may inhibit niosome formation. These vesicles can be categorized into three groups according to their size and number of bilayers: small unilamellar vesicles (SUV, 10–100 nm), large unilamellar vesicles (LUV, 100–3000 nm), and multilamellar vesicles (MLV, 1000–more nm) where more than one bilayer is present (Fig. 13.1).

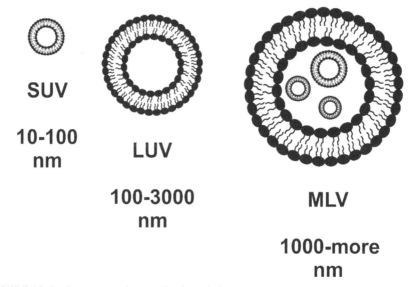

FIGURE 13.1   Structure and categorization of niosomes.

Preparation of niosomes requires simple methods. Preparation techniques comprise generally the hydration of a mixture of surfactant/lipid at elevated temperature using hydration medium including drug. Subsequently, size reduction methods such as sonication and extrusion are applied to obtain a colloidal dispersion. Finally, the unentrapped drug is removed from the niosomal dispersion by centrifugation, gel filtration, or dialysis.

The well-known protocols for niosome preparation are: thin-film hydration (Balakrishnan et al., 2009; Shirsand et al., 2012), ether injection (Marwa et al., 2013), reverse phase evaporation (Guinedi et al. 2005), trans membrane pH gradient (Moghassemi and Hadjizadeh, 2014) and proniosome (Yasam et al., 2014).

Physicochemical characterization parameters of niosomes are vesicle size, morphology, size distribution, zeta potential, number of lamellae. These parameters have a direct impact on the stability of niosomes. The characterization methods of niosomes are summarized in Table 13.1. Furthermore, entrapment efficiency (EE%), stability and drug release are critical factors for medical applications of niosomes. EE% is the percentage of the drug entrapped in niosomes referred to the initial amount of drug that is present in the non-purified sample. It is affected by niosome contents, physicochemical properties of the drug, and preparation methods (Rajera et al., 2011). The stability of niosomes can be tested by measuring mean vesicle size and size distribution or determining the entrapment efficiency over several month storage periods at different conditions. Sustained drug release from niosomes is another quite important issue to minimize the side effects of the drug in the human body and enhance the effects of the drug at the target location. The release rate of the drug from niosomes is generally determined via dialysis method.

TABLE 13.1 Characterization Methods of Niosomes.

Parameters	Methods used to determine the parameter	References
Size and morphology	Dynamic light scattering (DLS), Nanoparticle tracking analysis (NTA), Transmission electron microscopy (TEM), Scanning electron microscopy (SEM), Cryo-transmission electron microscopy (cryo-TEM), Atomic force microscopy (AFM)	Hua and Liu, 2007; Manosroi et al., 2008; Bayindir and Yuksel, 2010; Sharma et al., 2015
Size distribution	DLS	Tavano et al., 2016
Surface charge and zeta potential	Zetasizer, Microelectrophoresis, DLS, pH-sensitive fluorophores	Liu et al., 2007
Bilayer characteristics	Small angle X-ray scattering (SAXS), *In situ* energy-dispersive X-ray diffraction (EDXD)	Hua and Liu, 2007; Pozzi et al., 2009

13.4 NIOSOMAL DRUG DELIVERY SYSTEMS

Side effects, poor solubility, and chemical stability are the main problems of conventional drugs passing through different environments in the human body on their way to the target location. Thesecauseinefficient therapeutic effect. Niosomes have been used for the delivery of several pharmacological and diagnostic agents to overcome these problems. Because of their biocompatibility, low toxicity, and unique structure, they allow the development of effective novel drug delivery systems (Ag Seleci et al., 2016). They are able to load both hydrophilic and lipophilic drugs. Hydrophilic drugs are encapsulated into the aqueous core and lipophilic drugs are incorporated in the membrane bilayer of niosome. Moreover, they offer a great opportunity for loading both drugs together in one nanocarrier (Fig. 13.2). Another feature of the niosomes is the fluidity of their membrane, which allows the controlled release of a compound without destroying the vesicular structure. The drug release occurs by passive transport of the drug through the niosomal membrane bilayer. Physicochemical parameters of niosomes can be arranged to obtain the desired drug delivery system. Besides, the surface of niosomes can be easily modified to create targeted niosomes.

Hydrophilic drug   **Hydrophobic drug**   **Co-drug delivery**

FIGURE 13.2   Encapsulation drugs in niosome.

The application of niosomal technology is widely varied and can be used to treat a number of diseases. They have been used in pulmonary delivery (Moazeni et al., 2010), transdermal delivery (Muzzalupo et al., 2014), ophthalmic delivery (Khalil et al., 2017), vaccine delivery (Vyas et al., 2005), gene delivery (Puras et al., 2014), protein and peptide delivery (Pardakhty et al., 2007) and delivery of chemotherapeutics. The concept of loading anti-cancer drugs into niosomes for a better delivery of the drug to the specific target location is widely investigated by researchers. Anti-cancer drugs such as doxorubicin (Seleci et al., 2016; Tavano et al., 2013), paclitaxel

(Bayindir and Yuksel, 2010), methotrexate (Lakshmi et al., 2007), 5-fluoro-uracil (Paolino et al., 2008) were successfully entrapped in niosomes and characterized in detail to develop efficient drug carrier systems for cancer therapy. To test the efficiency and specificity of niosomal anti-cancer drugs, in vitro and in vivo investigations were performed (Kassem et al., 2017; Liu et al., 2017). Results from a number of studies suggest that niosomes have great potential in the application of several types of cancer therapy (Shaker et al., 2015; Dwivedi et al., 2015). Furthermore, antibiotics (Mahdiun et al., 2017), anti-inflammatory (Leelarungrayub et al., 2017), and antiviral drugs were entrapped in niosomes to improve their stability and reduce the dose of the drug.

13.5   TARGETING STRATEGIES

Transporting of drugs to the target site is a major drawback in the treatment of many diseases. Numbers of conventional drugs have limited effectiveness, poor biodistribution and a lack of selectivity. Especially, effective and specific delivery of drugs to the brain is a big challenge since most drugs cannot pass the BBB. Nanoparticles are promising tools to deliver drugs to the desired part of the body. There are two different strategies for targeting of the nanoparticles: passive and active targeting.

13.5.1   PASSIVE TARGETING

Tumor tissue has highly disorganized vascular architecture, irregular blood flow, and reduced lymphatic drainage. These properties provide the enhanced permeability and retention (EPR) effect for nanoparticles that can enhance the intracellular concentration of the drugs in cancer cells. The strength of the EPR effect is influenced by two factors. First, angiogenic tumors produce vascular endothelial growth factors. These growth factors increase the permeability of newly formed vessels associated with the tumor and cause infiltration of circulating particles. Second, because of the reduced lymphatic drainage of tumors, the permeating nanocarriers are not removed efficiently and thus are retained in the tumor tissue, which leads to the accumulation of nanoparticles.

Passive targeting of nanoparticles also enables the drugs to pass the BBB via different pathways (Bhaskar et al., 2010). A wide range of CNS drugs may enter into the brain with nanocarriers.

13.5.2   ACTIVE TARGETING

Active targeting is based on targeting ligands such as antibodies, peptides, aptamers and small molecules that bind specifically to an overexpressed target on the cell surface and trigger receptor-mediated endocytosis after binding. Nanoparticles are conjugated with targeting ligands thereby allowing accumulation of the drug within tissues or intracellular organelles specifically.

Active targeting of the BBB and brain tumors represent a promising non-invasive approach for enhanced drug delivery to the brain. The identified and commonly-targeted receptors for brain targeting are: transferrin, insulin, low-density lipoproteins (LDL), leptin, glutathione, folic acid, and neuro-pilin. Summary of brain targeting ligands and their receptors are listed in Table 13.2.

TABLE 13.2   Targeting Ligands and Their Receptors for Active Brain Targeting.

Targeting ligand	Targeted receptor	Reference
Folic acid	Folate receptor	Kuo and Chen, 2015; Kang et al., 2010
Transferrin	Transferrin receptor (TfR)	Li et al., 2012; Cui et al., 2013
Anti-TfR monoclonal anti-body (mAb), 7579	TfR	Xu et al., 2011
Angiopep-2	Low-density lipoprotein receptor (LDR)	Xin et al., 2011; Huang et al., 2011
Anti-insulin receptor mono-clonal antibody (29B4)	Insulin receptor	Ulbrich et al., 2011
Glutathione	Glutathione receptor	Birngruber et al., 2014
Peptides comprising amino acid residues 70–89 of leptin (Lep$_{70-89}$)	Leptin receptor (ObR)	Tamaru et al., 2010
tLyp-1 peptide	Neuropilin receptor	Hu et al., 2013; Miao et al., 2013{Ag Seleci, 2017 #75}

13.6   NIOSOMES FOR BRAIN TARGETING

In the literature, the advantages of niosomes have been already used to obtain efficient delivery of therapeutic agents to the brain. Morin hydrate (MH), which has a neuroprotective effect in Parkinson's disease, was encapsulated in niosomes composed of nonionic surfactants, cholesterol, and dicetyl phosphate. Niosomal formulations were optimized and injected to mouse via lateral tail vein to take real-time images. A non-invasive real-time imaging technique was applied to understand the in vivo biodistribution of MH niosomes. The ex vivo imaging of the excised organs demonstrated the capability of MH niosomes crosses the BBB(Waddad et al., 2013). Varshosaz et al. (2014) prepared niosomal formulation of α-tocopherol and ascorbic acid for enhanced brain delivery of these drugs in preventing neuronal cell damages during ischemia-reperfusion disorders. After characterization studies, neuroprotective effects of the niosomal formulations were investigated in an ischemiare-perfusion model in male rats. In vivo results showed that the effectiveness of the formulated new drug delivery system in protection of cerebral tissue against elevation in oxygen free radical concentration during cthe erebral ischemia-reperfusion course was more than the free ascorbic acid.

Moreover, delivery of the drugs to the brain via the nasal route provides some more advantages. Folic acid is a water-soluble vitamin having difficulty in crossing the BBB and the low blood level of folic acid is the main cause of depression in Alzheimer's disease. Ravouru et al. (2013) developed niosomal nasal drug delivery systems by using folic acid to target the brain. Ex-vivo perfusion studies were carried out using a rat model and results showed that about 48.15% of the drug was absorbed through the nasal cavity at the end of 6 hrs.

Recently, niosomes containing cationic lipids were used as a carrier for gene delivery in retina and brain. Niosome-DNA vectors (nioplexes) were prepared and characterized in detail. In vitro experiments were performed to evaluate transfection efficiency and cell viability in different cell lines. Subsequently, nioplexes were administrated to rat retina via intravitreal and subretinal injections and to rat brain within cerebral cortex. In vivo results demonstrated that after injections of nioplexes, the cells in rat retina and brain were transfected successfully (Ojeda et al., 2016).

These outcomes provide new insights into the development of niosome-based delivery systems for brain targeting. Furthermore, several approaches were used for active targeting of the niosomal drugs to the brain that is explained in the next Section.

13.6.1   MODIFICATION OF NIOSOMES WITH TARGETING LIGANDS

13.6.1.1   GLUCOSE DERIVATIVES

The large energetic demand of the brain is provided almost by β-D-glucose. Glucose in the blood must cross the BBB's luminal and abluminal membranes to reach neural tissue. It is transported to the brain via transporters, enzymes, and cell signaling processes. Glucose transporter (GLUT1) enables glucose transport across the BBB and it is overexpressed on BBB cells. Therefore, glucose derivatives are promising targeting ligands for drug transport through the BBB.

Dufes et al. (2004) synthesized N-Palmitoylglucosamine (NPG, the glucose-derivatized surfactant) niosomes entrapping vasoactive intestinal peptide (VIP) by shaking a mixture of NPG, non-ionic surfactants (Span 60 and Solulan C24) and cholesterol in PBS at 90°C for 30 min, followed by probe sonication for 5 min. VIP was entrapped into niosomes by probe sonicating them in ^{125}I-VIP and unlabelled VIP solution. VIP and ^{125}I-VIP-loaded glucose-bearing niosomes were intravenously injected to mice. After administration of VIP in solution or encapsulated in glucose-bearing niosomes or in control niosomes, brain uptake was determined by measuring the radioactivity of ^{125}I-labeled VIP. Results indicated that VIP encapsulation within glucose-bearing niosomes mainly allowed a significantly higher VIP brain uptake compared to control niosomes. In another study, Bragagni et al. (2012) investigated the development and characterization of a niosomal formulation functionalized with NPG to obtain a potential brain targeted delivery system for the anticancer drug doxorubicin. The developed doxorubicin NPG-niosomal formulation was injected in rats, in comparison with a commercial solution of drug in order to evaluate its effectiveness in enhancing doxorubicin brain delivery. After administration, significantly higher doxorubicin plasma levels were obtained with the NPG-niosomal dispersion with respect to the commercial solution at the same drug dosage. Their results showed that the developed niosomal formulation was able to keep the drug longer in the blood circulation system compared to the commercial drug solution. Moreover, NPG functionalized niosomal formulation was also used to targeted delivery of dynorphin-B, which is an endogenous neuropeptide with relevant pharmacological activities on the CNS (Bragagni et al., 2014). The optimized niosomal formulation with entrapped dynorphin-B was administered intravenously to mice. The antinociceptive effect of this niosomal formulation and a simple solution of the peptide were investigated. A significantly higher antinociceptive effect was obtained for targeted niosomal dynorphin-B, than for peptide solution. It can be concluded that encouraging and promising results were obtained in the previous studies by using niosomes bearing NPG as a drug carrier and this may trigger the usage of this system in further studies.

13.6.1.2 TRANSFERRIN

Transferrin receptor, which is an iron-binding transmembrane protein and facilitates iron uptake in cells, is highly expressed in brain endothelial cells. Several types of nanoparticles were functionalized with TfR binding ligands such as peptides (Dixit et al., 2015), antibodies (Ulbrich et al., 2009), or transferrin (Sonali et al., 2016) to deliver therapeutics to the brain.

TfR is also overexpressed in tumor cells. Niosomes were coupled with transferrin to improve tumor therapy. Hydroxycamptothecin (HCPT) was loaded into polyethylene glycolated niosomes (PEG-niosomes) and transferrin was conjugated to the surface of PEG-niosomes. Compared with HCPT injection, transferrin conjugated PEG-niosomes demonstrated stronger anti-tumor activity in mice (Hong et al., 2009).In another study, niosomes were prepared from Pluronic L64 surfactant and cholesterol by Tavano et al. (2013). After the preparation, transferrin was conjugated to niosomes and rhodamine was loaded the vesicles. The specific uptake of rhodamine-loaded transferrin conjugated-niosomes was evaluated on tumor cells via confocal microscopy. Results demonstrated that transferrin conjugated-niosomes were specifically uptaken by tumor cells. However, transferrin-bearing niosomes have not been applied for brain targeting yet, they have a great potential for future studies.

13.6.1.3 tLyp-1 PEPTIDE

Neuropilin-1 (NRP-1) is a transmembrane protein overexpressed on the surface of both glioma and endothelial cells of angiogenic blood vessels (Li and Rossman, 2001; Nasarre et al., 2010; Hu et al., 2013). tLyp-1 (tumor homing and penetrating peptide) peptide with 7 amino acid (CGNKRTR), is as a ligand-targeted to the NRP-1 receptor with high affinity and specificity. Hence, tLyp-1 has been used as a targeting ligand for the delivery of drugs to the brain tumor (Hu et al., 2013).

Recently, polyethylene glycolated niosomes (PEGNIO) were synthesized and doxorubicin and curcumin were encapsulated in niosomes via thin film hydration method. The surface of co-drug loaded PEGNIO was modified with tLyp-1. After characterization studies, in vitro investigations were carried out on human glioblastoma and human mesenchymal stem cells. The results clearly indicated that the strategy by co-administration of doxorubicin and curcumin with tLyp-1 functionalized niosomes could significantly improve anti-glioma treatment (Ag Seleci et al., 2017).

13.7 CONCLUSIONS AND FUTURE DIRECTIONS

The treatment of brain-related diseases presents a major challenge. Using nanoparticles may enable to overcome the difficulties of delivering therapeutic agents to specific regions of the brain. Niosomes are one of the promising drug carriers to design novel drug delivery systems for brain disease treatment. Their unique structure provides loading hydrophilic, or lipophilic drugs, or both drugs together in the same vesicle at the same time. Besides, their surface can easily be functionalized and modified with ligands. These features of the niosomes have been already applied to deliver different types of agents to the brain. However, presently there are no commercial niosomal drugs available for brain targeting. Therefore, further research studies needed to be performed. Especially, the design of new targeted and co-drug loaded niosomal delivery systems for brain targeting may contribute to producing commercially available products.

KEYWORDS

- brain targeting
- drug delivery
- niosomes
- nonionic surfactants
- passive and active targeting
- targeting ligands

REFERENCES

Ag Seleci, D.; Seleci, M.; Walter, J.-G.; Stahl, F.; Scheper, T. Niosomes as Nanoparticular Drug Carriers: Fundamentals and Recent Applications. *J. Nanomater.* **2016,** *2016,* 13.

Ag Seleci, D.; Seleci, M.; Stahl F.; Scheper, T. Tumor Homing and Penetrating Peptide-Conjugated Niosomes as Multi-Drug Carriers for Tumor-Targeted Drug Delivery. *RSC Adv.* **2017**, *7*, 33378–33384.

Balakrishnan, P.; Shanmugam, S.; Lee, W. S.; Lee, W. M.; Kim, J. O.; Oh, D. H.; Kim, D.-D.; Kim, J. S.; Yoo, B. K.; Choi, H.-G. Formulation and İn Vitro Assessment of Minoxidil Niosomes for Enhanced Skin Delivery. *Int. J. Pharm.* **2009**, *377*, 1–8.

Bayindir, Z. S.; Yuksel, N. Characterization of Niosomes Prepared with Various Nonionic Surfactants for Paclitaxel Oral Delivery. *J. Pharm. Sci.* **2010**, *99*, 2049–2060.

Begum, K.; Khan, A. F.; Hana, H. K.; Sheak, J.; Jalil, R. U. Rifampicin Niosome: Preparations, Characterizations and Antibacterial Activity Against *Staphylococcus aureus* and *Staphylococcus epidermidis* İsolated from Acne. *Dhaka Univ. J. Pharm. Sci.* **2015**, *14*, 117–123.

Bhaskar, S.; Tian, F.; Stoeger, T.; Kreyling, W.; de la Fuente, J. M.; Grazú, V.; Borm, P.; Estrada, G.; Ntziachristos, V.; Razansky, D. Multifunctional Nanocarriers for Diagnostics, Drug Delivery and Targeted Treatment Across Blood-Brain Barrier: Perspectives on Tracking and Neuroimaging. *Part. Fibre Toxicol.* **2010**, *7*, 3.

Birngruber, T.; Raml, R.; Gladdines, W.; Gatschelhofer, C.; Gander, E.; Ghosh, A.; Kroath, T.; Gaillard, P. J.; Pieber, T. R.; Sinner, F. Enhanced Doxorubicin Delivery to the Brain Administered Through Glutathione PEGylated Liposomal Doxorubicin (2B3-101) as Compared with Generic Caelyx,®/Doxil®—A Cerebral Open Flow Microperfusion Pilot Study. *J. Pharm. Sci.* **2014**, *103*, 1945–1948.

Bouwstra, J. A.; van Hal, D. A.; Hofland, H. E.; Junginger, H. E. Preparation and Characterization of Nonionic Surfactant Vesicles. *Colloids Surf. A.* **1997**, *123*, 71–80.

Bragagni, M.; Mennini, N.; Ghelardini, C.; Mura, P. Development and Characterization of Niosomal Formulations of Doxorubicin Aimed at Brain Targeting. *J. Pharm. Pharm. Sci.* **2012**, *15*, 184–196.

Bragagni, M.; Mennini, N.; Furlanetto, S.; Orlandini, S.; Ghelardini, C.; Mura, P. Development and Characterization of Functionalized Niosomes for Brain Targeting of Dynorphin-B. *Eur. J. Pharm. Biopharm.* **2014**, *87*, 73–79.

Cui, Y.; Xu, Q.; Chow, P. K.-H.; Wang, D.; Wang, C.-H. Transferrin-Conjugated Magnetic Silica PLGA Nanoparticles Loaded with Doxorubicin and Paclitaxel for Brain Glioma Treatment. *Biomaterials* **2013**, *34*, 8511–8520.

Dixit, S.; Novak, T.; Miller, K.; Zhu, Y.; Kenney, M. E.; Broome, A.-M. Transferrin Receptor-Targeted Theranostic Gold Nanoparticles for Photosensitizer Delivery in Brain Tumors. *Nanoscale* **2015**, *7*, 1782–1790.

Dufes, C.; Gaillard, F.; Uchegbu, I. F.; Schätzlein, A. G.; Olivier, J.-C.; Muller, J.-M. Glucose-Targeted Niosomes Deliver Vasoactive İntestinal Peptide (VIP) to the Brain. *Int. J. Pharm.* **2004**, *285*, 77–85.

Dwivedi, A.; Mazumder, A.; Du Plessis, L.; Du Preez, J. L.; Haynes, R. K.; Du Plessis, J. In Vitro Anti-Cancer Effects of Artemisone Nano-Vesicular Formulations on Melanoma Cells. *Nanomedicine* **2015**, *11*, 2041–2050.

Gelperina, S.; Maksimenko, O.; Khalansky, A.; Vanchugova, L.; Shipulo, E.; Abbasova, K.; Berdiev, R.; Wohlfart, S.; Chepurnova, N.; Kreuter, J. Drug Delivery to the Brain Using Surfactant-Coated Poly (Lactide-Co-Glycolide) Nanoparticles: İnfluence of the Formulation Parameters. *Eur. J. Pharm. Biopharm.* **2010**, *74*, 157–163.

Guinedi, A. S.; Mortada, N. D.; Mansour, S.; Hathout, R. M. Preparation and Evaluation of Reverse-Phase Evaporation and Multilamellar Niosomes as Ophthalmic Carriers of Acetazolamide. *Int. J. Pharm.* **2005**, *306*, 71–82.

Hong, M.; Zhu, S.; Jiang, Y.; Tang, G.; Pei, Y. Efficient Tumor Targeting of Hydroxycampto-thecin Loaded PEGylated Niosomes Modified with Transferrin. *J. Control. Release* **2009**, *133*, 96–102.

Hu, Q.; Gao, X.; Gu, G.; Kang, T.; Tu, Y.; Liu, Z.; Song, Q.; Yao, L.; Pang, Z.; Jiang, X. Glioma Therapy Using Tumor Homing and Penetrating Peptide-Functionalized PEG–PLA Nanoparticles Loaded With Paclitaxel. *Biomaterials* **2013**, *34*, 5640–5650.

Hua, W.; Liu, T. Preparation and Properties of Highly Stable İnnocuous Niosome in Span 80/ PEG 400/H$_2$O System. *Colloids Surf. A* **2007**, *302*, 377–382.

Huang, S.; Li, J.; Han, L.; Liu, S.; Ma, H.; Huang, R.; Jiang, C. Dual Targeting Effect of Angiopep-2-Modified, DNA-Loaded Nanoparticles for Glioma. *Biomaterials* **2011**, *32*, 6832–6838.

Kang, C.; Yuan, X.; Li, F.; Pu, P.; Yu, S.; Shen, C.; Zhang, Z.; Zhang, Y. Evaluation of Folate-PAMAM for the Delivery of Antisense Oligonucleotides to Rat C6 Glioma Cells İn Vitro and İn Vivo. *J. Biomed. Mater. Res. B* **2010**, *93*, 585–594.

Kassem, M. A.; El-Sawy, H. S.; Abd-Allah, F. I.; Abdelghany, T. M.; Khalid, M. Maximizing the Therapeutic Efficacy of Imatinib Mesylate–Loaded Niosomes on Human Colon Adeno-carcinoma Using Box-Behnken Design. *J. Pharm. Sci.* **2017**, *106*, 111–122.

Khalil, R. M.; Abdelbary, G. A.; Basha, M.; Awad, G. E.; El-Hashemy, H. A. Design and Evaluation of Proniosomes as a Carrier for Ocular Delivery of Lomefloxacin HCl. *J. Lipos. Res.* **2017**, *27*, 118–129.

Kreuter, J. Nanoparticulate Systems for Brain Delivery of Drugs. *Adv. Drug Deliver. Rev.* **2001**, *47*, 65–81.

Kreuter, J.; Alyautdin, R. N.; Kharkevich, D. A.; Ivanov, A. A. Passage of Peptides through the Blood-Brain Barrier with Colloidal Polymer Particles (Nanoparticles). *Brain Res.* **1995**, *674*, 171–174.

Kreuter, J.; Shamenkov, D.; Petrov, V.; Ramge, P.; Cychutek, K.; Koch-Brandt, C.; Alyautdin, R. Apolipoprotein-Mediated Transport of Nanoparticle-Bound Drugs across the Blood-Brain Barrier. *J. Drug Target.* **2002**, *10*, 317–325.

Kumar, G. P.; Rajeshwarrao, P. Nonionic Surfactant Vesicular Systems for Effective Drug Delivery—an Overview. *Acta Pharm. Sin B* **2011**, *1*, 208–219.

Kuo, Y.-C.; Chen, Y.-C. Targeting Delivery of Etoposide to İnhibit the Growth of Human Glioblastoma Multiforme Using Lactoferrin- and Folic Acid-Grafted Poly (Lactide-Co-Glycolide) Nanoparticles. *Int. J. Pharm.* **2015**, *479*, 138–149.

Lai, F.; Fadda, A. M.; Sinico, C. Liposomes for Brain Delivery. *Expert Opin. Drug. Del.* **2013**, *10*, 1003–1022.

Lakshmi, P.; Devi, G. S.; Bhaskaran, S.; Sacchidanand, S. Niosomal Methotrexate Gel in the Treatment of Localized Psoriasis: Phase I and Phase II Studies. *Indian J. Dermatol. Vene-reol. Leprol.* **2007**, *73*, 157.

Leelarungrayub, J.; Manorsoi, J.; Manorsoi, A. Anti-İnflammatory Activity of Niosomes Entrapped with Plai Oil (*Zingiber cassumunar* Roxb.) by Therapeutic Ultrasound in a Rat Model. *Int. J. Nanomed.* **2017**, *12*, 2469–2476.

Li, P.; Rossman, T. G. Genes Upregulated in Lead-Resistant Glioma Cells Reveal Possible Targets for Lead-İnduced Developmental Neurotoxicity. *Toxicol. Sci.* **2001**, *64*, 90-99.

Li, Y.; He, H.; Jia, X.; Lu, W.-L.; Lou, J.; Wei, Y. A Dual-Targeting Nanocarrier Based on Poly (Amidoamine) Dendrimers Conjugated with Transferrin and Tamoxifen for Treating Brain Gliomas. *Biomaterials* **2012**, *33*, 3899–3908.

Liu, T.; Guo, R.; Hua, W.; Qiu, J. Structure Behaviors of Hemoglobin in PEG 6000/Tween 80/ Span 80/H 2 O Niosome System. *Colloids Surf. A* **2007**, *293*, 255–261.

Liu, F.-R.; Jin, H.; Wang, Y.; Chen, C.; Li, M.; Mao, S.-j.; Wang, Q.; Li, H. Anti-CD123 Antibody-Modified Niosomes for Targeted Delivery of Daunorubicin Against Acute Myeloid Leukemia. *Drug Deliv.* **2017**, *24*, 882–890.

Mahdiun, F.; Mansouri, S.; Khazaeli, P.; Mirzaei, R. The Effect of Tobramycin İncorporated with Bismuth-Ethanedithiol Loaded on Niosomes on the Quorum Sensing and Biofilm Formation of *Pseudomonas aeruginosa*. *Microb. Pathog.* **2017**, *107*, 129–135.

Manosroi, A.; Chutoprapat, R.; Abe, M.; Manosroi, J. Characteristics of Niosomes Prepared by Supercritical Carbon Dioxide (scCO$_2$) Fluid. *Int. J. Pharm.* **2008**, *352*, 248–255.

Marianecci, C.; Rinaldi, F.; Mastriota, M.; Pieretti, S.; Trapasso, E.; Paolino, D.; Carafa, M. Anti-İnflammatory Activity of Novel Ammonium Glycyrrhizinate/Niosomes Delivery System: Human and Murine Models. *J. Control. Release* **2012**, *164*, 17–25.

Marwa, A.; Omaima, S.; Hanaa, E.-G.; Mohammed, A.-S. Preparation and İn-Vitro Evaluation of Diclofenac Sodium Niosomal Formulations. *Int. J. Pharm. Sci. Res.* **2013**, *4*, 1757–1765.

Miao, D.; Jiang, M.; Liu, Z.; Gu, G.; Hu, Q.; Kang, T.; Song, Q.; Yao, L.; Li, W.; Gao, X. Co-Administration of Dual-Targeting Nanoparticles with Penetration Enhancement Peptide for Antiglioblastoma Therapy. *Mol. Pharm.* **2013**, *11*, 90–101.

Moazeni, E.; Gilani, K.; Sotoudegan, F.; Pardakhty, A.; Najafabadi, A. R.; Ghalandari, R.; Fazeli, M. R.; Jamalifar, H. Formulation and İn Vitro Evaluation of Ciprofloxacin Containing Niosomes for Pulmonary Delivery. *J. Microencapsul.* **2010**, *27*, 618–627.

Moghassemi, S.; Hadjizadeh, A. Nano-Niosomes as Nanoscale Drug Delivery Systems: an İllustrated Review. *J. Control. Release* **2014**, *185*, 22–36.

Muzzalupo, R.; Tavano, L.; Lai, F.; Picci, N. Niosomes Containing Hydroxyl Additives as Percutaneous Penetration Enhancers: Effect on the Transdermal Delivery of Sulfadiazine Sodium Salt. *Colloids Surf. B* **2014**, *123*, 207–212.

Nasarre, C.; Roth, M.; Jacob, L.; Roth, L.; Koncina, E.; Thien, A.; Labourdette, G.; Poulet, P.; Hubert, P.; Cremel, G. Peptide-Based İnterference of the Transmembrane Domain of Neuropilin-1 İnhibits Glioma Growth İn Vivo. *Oncogene* **2010**, *29*, 2381–2392.

Ojeda, E.; Puras, G.; Agirre, M.; Zarate, J.; Grijalvo, S.; Eritja, R.; Martinez-Navarrete, G.; Soto-Sánchez, C.; Díaz-Tahoces, A.; Aviles-Trigueros, M. The influence of the Polar Head-Group of Synthetic Cationic Lipids on the Transfection Efficiency Mediated by Niosomes in rat Retina and Brain. *Biomaterials* **2016**, *77*, 267–279.

Paolino, D.; Cosco, D.; Muzzalupo, R.; Trapasso, E.; Picci, N.; Fresta, M. Innovative Bola-Surfactant Niosomes as Topical Delivery Systems of 5-Fluorouracil for the Treatment of Skin Cancer. *Int. J. Pharm.* **2008**, *353*, 233–242.

Pardakhty, A.; Varshosaz, J.; Rouholamini, A. In Vitro Study of Polyoxyethylene Alkyl Ether Niosomes for Delivery of İnsulin. *Int. J. Pharm.* **2007**, *328*, 130–141.

Petri, B.; Bootz, A.; Khalansky, A.; Hekmatara, T.; Müller, R.; Uhl, R.; Kreuter, J.; Gelperina, S. Chemotherapy of Brain Tumour Using Doxorubicin Bound to Surfactant-Coated Poly (Butyl Cyanoacrylate) Nanoparticles: Revisiting the Role of Surfactants. *J. Control. Release* **2007**, *117*, 51–58.

Pozzi, D.; Caminiti, R.; Marianecci, C.; Carafa, M.; Santucci, E.; De Sanctis, S. C.; Caracciolo, G. Effect of Cholesterol on the Formation and Hydration Behavior of Solid-Supported Niosomal Membranes. *Langmuir* **2009**, *26*, 2268–2273.

Puras, G.; Mashal, M.; Zárate, J.; Agirre, M.; Ojeda, E.; Grijalvo, S.; Eritja, R.; Diaz-Tahoces, A.; Navarrete, G. M.; Avilés-Trigueros, M. A Novel Cationic Niosome Formulation for Gene Delivery to the Retina. *J. Control. Release* **2014**, *174*, 27–36.

Rajera, R.; Nagpal, K.; Singh, S. K.; Mishra, D. N. Niosomes: a Controlled and Novel Drug Delivery System. *Biol. Pharm. Bull.* **2011**, *34*, 945–953.

Ravouru, N.; Kondreddy, P.; Korakanchi, D. Formulation and Evaluation of Niosomal Nasal Drug Delivery System of Folic Acid for Brain Targeting. *Curr. Drug Discov. Technol.* **2013**, *10*, 270–282.

Schroeder, U.; Sommerfeld, P.; Sabel, B. A. Efficacy of Oral Dalargin-Loaded Nanoparticle Delivery across the Blood–Brain Barrier. *Peptides* **1998**, *19*, 777–780.

Seleci, D. A.; Seleci, M.; Jochums, A.; Walter, J.-G.; Stahl, F.; Scheper, T. Aptamer Mediated Niosomal Drug Delivery. *RSC Adv.* **2016**, *6*, 87910–87918.

Seleci, M.; Ag Seleci, D.; Joncyzk, R.; Stahl, F.; Blume, C.; Scheper, T. Smart Multifunctional Nanoparticles in Nanomedicine. *BioNanoMat.* **2016**, *17*, 33-41.

Shaker, D. S.; Shaker, M. A.; Hanafy, M. S. Cellular Uptake, Cytotoxicity and İn-Vivo Evaluation of Tamoxifen Citrate Loaded Niosomes. *Int. J. Pharm.* **2015**, *493*, 285–294.

Sharma, V.; Anandhakumar, S.; Sasidharan, M. Self-degrading Niosomes for Encapsulation of Hydrophilic and Hydrophobic Drugs: an Efficient Carrier for Cancer Multi-Drug Delivery. *Mater. Sci. Eng. C* **2015**, *56*, 393–400.

Shirsand, S.; Para, M.; Nagendrakumar, D.; Kanani, K.; Keerthy, D. Formulation and Evaluation of Ketoconazole Niosomal Gel Drug Delivery System. *Int. J. Pharm. Investig.* **2012**, *2*, 201–207.

Sonali; Singh, R. P.; Singh, N.; Sharma, G.; Vijayakumar, M. R.; Koch, B.; Singh, S.; Singh, U.; Dash, D.; Pandey, B. L. Transferrin Liposomes of Docetaxel for Brain-Targeted Cancer Applications: Formulation and Brain Theranostics. *Drug Deliv.* **2016**, *23*, 1261–1271.

Srikanth, M.; Kessler, J. A. Nanotechnology—Novel Therapeutics for CNS Disorders. *Nat. Rev. Neurol.* **2012**, *8*, 307–318.

Sun, W.; Xie, C.; Wang, H.; Hu, Y. Specific Role of Polysorbate 80 Coating on the Targeting of Nanoparticles to the Brain. *Biomaterials* **2004**, *25*, 3065–3071.

Tamaru, M.; Akita, H.; Fujiwara, T.; Kajimoto, K.; Harashima, H. Leptin-Derived Peptide, a Targeting Ligand for Mouse Brain-Derived Endothelial Cells via Macropinocytosis. *Biochem. Biophys. Res. Commun.* **2010**, *394*, 587–592.

Tavano, L.; Mauro, L.; Naimo, G. D.; Bruno, L.; Picci, N.; Andò, S.; Muzzalupo, R. Further Evolution of Multifunctional Niosomes Based on Pluronic Surfactant: Dual Active Targeting and Drug Combination Properties. *Langmuir* **2016**, *32*, 8926–8933.

Tavano, L.; Muzzalupo, R.; Mauro, L.; Pellegrino, M.; Andò, S.; Picci, N. Transferrin-Conjugated Pluronic Niosomes as a New Drug Delivery System for Anticancer Therapy. *Langmuir* **2013**, *29*, 12638–12646.

Tavano, L.; Vivacqua, M.; Carito, V.; Muzzalupo, R.; Caroleo, M. C.; Nicoletta, F. Doxorubicin Loaded Magneto-Niosomes for Targeted Drug Delivery. *Colloids Surf. B* **2013**, *102*, 803–807.

Uchegbu, I. F.; Vyas, S. P. Non-ionic Surfactant Based Vesicles (Niosomes) in Drug Delivery. *Int. J. Pharm.* **1998**, *172*, 33–70.

Ulbrich, K.; Hekmatara, T.; Herbert, E.; Kreuter, J. Transferrin- and Transferrin-Receptor-Antibody-Modified Nanoparticles Enable Drug Delivery Across the Blood–Brain Barrier (BBB). *Eur. J. Pharm. Biopharm.* **2009**, *71*, 251–256.

Ulbrich, K.; Knobloch, T.; Kreuter, J. Targeting the İnsulin Receptor: Nanoparticles for Drug Delivery Across the Blood–Brain Barrier (BBB). *J. Drug Target.* **2011,** *19*, 125–132.

Varshosaz, J.; Taymouri, S.; Pardakhty, A.; Asadi-Shekaari, M.; Babaee, A. Niosomes of Ascorbic Acid and α-Tocopherol in the Cerebral İschemia-Reperfusion Model in Male Rats. *Biomed. Res. Int.* **2014,** *2014*, 9.

Vyas, S.; Singh, R.; Jain, S.; Mishra, V.; Mahor, S.; Singh, P.; Gupta, P.; Rawat, A.; Dubey, P. Non-İonic Surfactant Based Vesicles (Niosomes) for Non-İnvasive Topical Genetic İmmunization Against Hepatitis B. *Int. J. Pharm.* **2005,** *296*, 80–86.

Waddad, A. Y.; Abbad, S.; Yu, F.; Munyendo, W. L.; Wang, J.; Lv, H.; Zhou, J. Formulation, Characterization and Pharmacokinetics of Morin Hydrate Niosomes Prepared from Various Non-İonic Surfactants. *Int. J. Pharm.* **2013,** *456*, 446–458.

Xin, H.; Jiang, X.; Gu, J.; Sha, X.; Chen, L.; Law, K.; Chen, Y.; Wang, X.; Jiang, Y.; Fang, X. Angiopep-Conjugated Poly (Ethylene Glycol)-Co-Poly (ε-Caprolactone) Nanoparticles as Dual-Targeting Drug Delivery System for Brain Glioma. *Biomaterials* **2011,** *32*, 4293–4305.

Xu, G.; Wen, X.; Hong, Y.; Du, H.; Zhang, X.; Song, J.; Yin, Y.; Huang, H.; Shen, G. An Anti-Transferrin Receptor Antibody Enhanced the Growth İnhibitory Effects of Chemotherapeutic Drugs on Human Glioma Cells. *Int. Immunopharmacol.* **2011,** *11*, 1844–1849.

Yasam, V. R.; Jakki, S. L.; Natarajan, J.; Kuppusamy, G. A Review on Novel Vesicular Drug Delivery: Proniosomes. *Drug Deliv.* **2014,** *21*, 243–249.

SECTION IV
MISCELLANEOUS

CHAPTER 14

NANOEMULSION FOR BRAIN TARGETING

KHUSHWINDER KAUR* and SHIVANI UPPAL

Department of Chemistry and Centre of Advanced Studies in Chemistry, Punjab University, Chandigarh 160014, India

Corresponding author. E-mail: makkarkhushi@gmail.com

ABSTRACT

The typical anatomical, physiological, and histological characteristics of the brain make its targeting a more challenging and crucial task. The complex and compact architecture along with the powerful defense mechanism is the major Hiccup in brain targeting. The drug administration can be carried out via various routes of administration, such as intraventricular pathway, intra-cerebral, and intrathecal route. The drug transit via various routes needs to be safe as well as effective. Further, targeting at the cellular level is an arduous task. The field of nanotechnology offers a suitable and promising solution for the conveyance of medications into the cerebrum. The chapter centers upon the utility and advantage of nanoemulsions as a delivery vehicle for the administration of medicaments to the brain. Nanoemulsions are thus considered as an ideal delivery system as they offer inherent advantages of site-specific drug delivery, ability to dissolve hydrophobic drugs, protecting drug degradation thus enhancing long-term drug stability.

14.1 INTRODUCTION

Since the commencement of logical investigations on human body, the brain has been found to be the most vital and the most complicated organ of the human body. Every component of the brain must work together in order to keep the body function normally as it controls all senses and functions of the body. The most important physiological task of the brain is interpreting information collected from different parts and enabling the body for the production of a suitable response are some of the physiological tasks of the brain. The brain and the spinal cord make up the central nervous system, which alongside the peripheral nervous system is responsible for regulating all functions of the body. The brain is composed of three mail parts cerebrum, cerebellum, and brainstem. The cerebrum composed of right and left hemispheres is the largest part of the brain. It performs major functions of interpreting responses of touch, hearing, speech reasoning emotions and fine motor control. The part of the brain located under the cerebellum is known as cerebrum. It works to coordinate muscle movements, maintain posture, and balance. The brain stem (composed of midbrain, pons, and medulla) connects the cerebrum and cerebellum to the spinal cord thus acting as a relay center. It is involved in many routine functions such as breathing, heart rate, body temperature, wake and sleep cycles, digestion, sneezing, coughing, vomiting, and swallowing. Scheme 14.1 depicts the various parts of the brain.

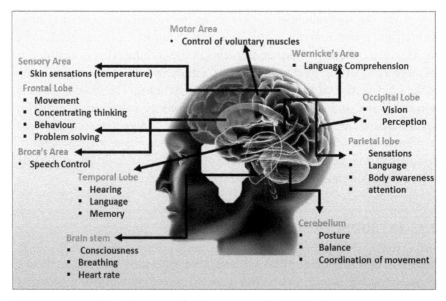

SCHEME 14.1 **(See color insert.)** Arius components of the brain.

14.2   DEFENSE MECHANISM OF BRAIN

A series of layers protect and enclose the brain to help keep it safe. Because of this protection probably why many of us survive the bumps and falls of childhood, and the crazy things we do as adolescents and beyond. The brain is protected from injury by the skull, meninges, cerebrospinal fluid (CSF) and the blood–brain barrier (BBB) (Hawkins and Davis, 2005).

14.2.1   SKULL (THE DOME)

It is a collection of 22 bones; protecting the brain along with providing supports the other soft tissues of the head.

14.2.2   MENINGES (THE DRAPE)

It is a series of three membranes that cover and enclose the brain (Scheme 14.2). They protect the brain by housing a fluid-filled space, and they function as a framework for blood vessels. The meninges have three layers: the dura mater, the arachnoid mater, and the pia mater.

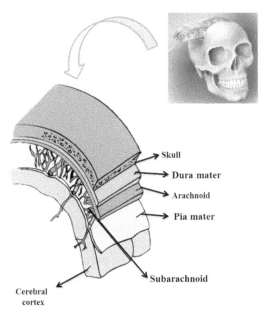

SCHEME 14.2   Protective layers of the brain.

14.2.3   DURA MATER (TOUGH MOTHER)

It is a strong external, fibrous membrane enriched with blood supply and wholly covers the brain. It is continuous with the dura that covers the spinal cord. The dura mater is the most superficial layer of the meninges, which contains folds, sinuses and is able to register pain. It contacts the endosteum that lines bones of the cranial cavity.

14.2.4   ARACHNOID

Arachnoid is a Greek word meaning "spider." This layer looks like a spider's web and is the middle layer of the meninges. These projections are the arachnoid called granulation/arachnoid villi projects into the sinuses formed by the dura mater at some places. They also transfer CSF from the ventricles back into the bloodstream. The arachnoid mater is further attached to the pia mater by arachnoid trabeculae, which is a weblike matrix of connective tissue. The space between the two layers, the subarachnoid space, is filled with CSF.

14.2.5   PIA MATER

Pia Mater means "tender mother." It is a skinny sheer membrane, covering the exterior of the brain and central nervous system (CNS). Compared with the other layers, this tissue holds closely to the brain, operating down into the sulci and fissures of the cortex. It forms choroid plexus by fusing with the ependyma, to produce CSF.

14.2.6   THE HELPFUL RIVER: CSF

This fluid is rightly defined as "helpful river" is still a bit of a mystery. Complete working of this fluid is still unknown. It is made in cavities within the brain and circulates around the brain, and the spinal cord. The brain, therefore, floats in the CSF.

Brain damage can occur when this system gets blocked, infected or a bit out of control. Damage can also happen when there is too much pressure within the head and not enough fluid can drain away. Drug delivery systems have been used to ferry drugs to the brain and CNS; however, CSF turnover can rapidly clear drugs from the CNS and reduce treatment efficacy.

14.2.7 THE INVISIBLE FENCE—OUR BLOOD–BRAIN BARRIER

BBB, interjected between the blood and the CSF by the endothelial cells of the capillaries and the choroid plexus protects the neurons of the brain and spinal cord from chemical damages and foreign biological substances. It has many elements, ranging from junctions between endothelial cells in the capillaries of the brain, restricting permeability of larger molecules to neuroglia. It is of great clinical importance as many drugs cannot penetrate the barrier. (Goasdoue et al., 2017)

This barrier works in several ways:

1. Filtering of toxic substances that might damage the brain
2. Maintaining "homeostasis" for the brain.
3. Protecting the brain from some unwanted substances made in other parts of the body such as hormones and neurotransmitters (Goasdoue et al., 2017).

14.3 STRUCTURE OF BLOOD–BRAIN BARRIER

BBB has a very compact and complex architecture. As the structure is related to the function, the structural integrity of BBB plays an important role. It is composed of various types of cells, primarily brain endothelial cells (BECs), which are appended to the pericytes by the peg and socket type junctions (Alam et al., 2010). Pericytes have processes into vascular basement membrane connecting astrocytic processes of glia limitans. The glia limitans form a protective sheath around the brain and spinal cord. Glia limitans superficial is located in the subarachnoid space. Glia limitansper-vascularis is a similar architecture located in the parenchymal region of the brain (Barcroft and Kennedy, 1939).

Neurovascular unit is the result of the interaction between neurons and microglia. Such interactions are vital for the maintenance of the blood flow in the cephalic region. Pericytes are believed to contribute to the structural integrity of the BBB (Bergers and Song, 2005). Gliovascular is another similar construct present in the BBB. It comprises blood vessels and astro-cytes. It is involved in the regulation of the blood flow by multiple interactions with various cellular physiologies like transport of the water-soluble substrates, tight junction (TJ) integrity between the BECs angiogenesis, metabolic homeostasis (Mack and Wolburg, 2013).

Abnormalities in the vascular function of the brain can contribute to the onset and/or progression of neurodegenerative conditions like AD. As the neuronal functions across the brain are quite specific as well as heterogeneous in nature, BEC and BBB are required for the maintenance of the homeostasis of the brain (Daneman and Prat, 2015).

14.4 BRAIN DISORDERS/DIAGNOSIS AND ON HAND TREATMENTS

The brain is a vital part of the nervous system. Given its complexity, it should not be surprising that abundant faults may occur in this system. Identifying pathophysiological pathways that underlie the development of neurological disorders has been an important research topic for many years due to the fact that the incidence of brain diseases is increasingly becoming popular (Ek et al., 2012; D'Andrea et al., 2017). When the brain is affected/damaged the functioning of the entire body is hampered in one or the other way (D'Andrea et al., 2017). Some of the brain disorders have been tabulated below Scheme 14.3–14.5:

SCHEME 14.3   Common brain disorders and their treatments.

Brain Tests
Computed tomography (CT scan): A scanner takes multiple X-rays, which a computer converts into detailed images of the brain and skull.
Magnetic resonance imaging (MRI scan): Using radio waves in a magnetic field, an MRI scanner creates highly detailed images of the brain and other parts of the head.
Angiography (brain angiogram): A special substance doctors call "a contrast agent" is injected into the veins, and travels into the brain. X-ray videos of the brain are taken, which can show problems in the brain's arteries.
Magnetic resonance angiography MRA): A special MRI scan of the brain's arteries. An MRA scan may show a blood clot or another cause for stroke.
Lumbar puncture (spinal tap): A needle is inserted into the space around the spinal nerves, and fluid is removed for analysis. Lumbar puncture is often done if meningitis is suspected.
Electroencephalogram (EEG): Brain activity is monitored through electrodes placed on the skin on the head. EEG can help diagnose seizures, or other brain problems.
Neurocognitive testing: Tests of problem-solving ability, short-term memory, and other complex brain functions. Usually, neurocognitive testing is done through questionnaires.

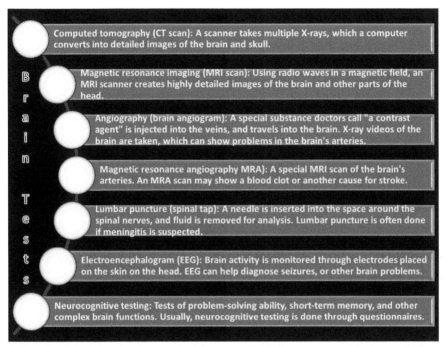

SCHEME 14.4 List of various methodologies available for treatment of the human brain.

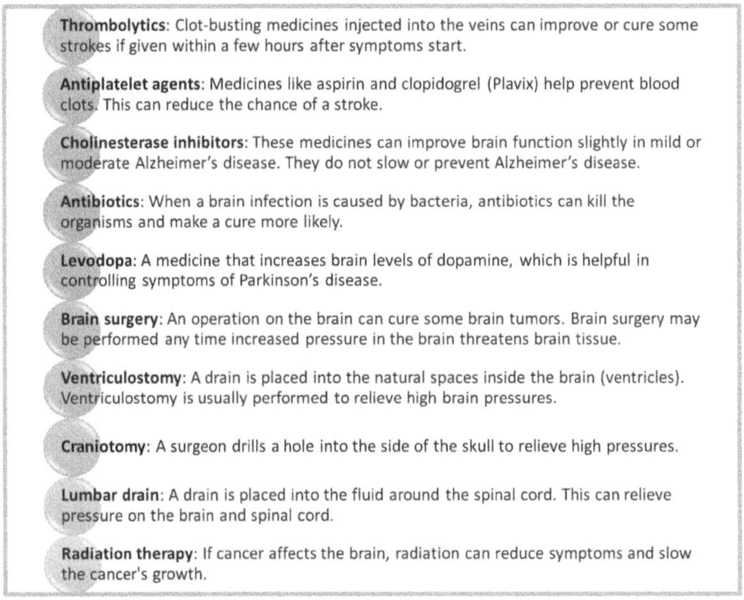

Thrombolytics: Clot-busting medicines injected into the veins can improve or cure some strokes if given within a few hours after symptoms start.
Antiplatelet agents: Medicines like aspirin and clopidogrel (Plavix) help prevent blood clots. This can reduce the chance of a stroke.
Cholinesterase inhibitors: These medicines can improve brain function slightly in mild or moderate Alzheimer's disease. They do not slow or prevent Alzheimer's disease.
Antibiotics: When a brain infection is caused by bacteria, antibiotics can kill the organisms and make a cure more likely.
Levodopa: A medicine that increases brain levels of dopamine, which is helpful in controlling symptoms of Parkinson's disease.
Brain surgery: An operation on the brain can cure some brain tumors. Brain surgery may be performed any time increased pressure in the brain threatens brain tissue.
Ventriculostomy: A drain is placed into the natural spaces inside the brain (ventricles). Ventriculostomy is usually performed to relieve high brain pressures.
Craniotomy: A surgeon drills a hole into the side of the skull to relieve high pressures.
Lumbar drain: A drain is placed into the fluid around the spinal cord. This can relieve pressure on the brain and spinal cord.
Radiation therapy: If cancer affects the brain, radiation can reduce symptoms and slow the cancer's growth.

SCHEME 14.5 List of various treatments available.

14.5   HICCUPS IN BRAIN TARGETING

For the management of neurological disorders, both invasive and noninvasive technologies each having their own advantages and disadvantages have been used. Drug transport from blood to CSF is regulated by the choroid plexus or blood–CSF barrier. Drug transport from blood to interstitial fluid (ISF) is regulated by the brain microvascular endothelium or BBB. The complex architecture of the BBB accompanied by a high degree of specificity and lipophilicity makes it difficult for the drugs to traverse across the BBB. Thus, BBB acts as a highly formidable barrier for the passage of xenobiotics. Literature reports (Alam et al., 2010; Krishnan et al., 2016; Wager et al., 2017) have ample examples of the use of surgical interventions for the management of neurological disorders. But, such approaches are associated with secondary complications like the scope of infections or seizures. Thus, the scientific explorations have been directed in finding noninvasive tools for the treatment.

14.6   ADMINISTRATION STRATEGIES

The drug administration can be carried out via various routes of administration, such as intraventricular pathway, intracerebral, and intrathecal route. The drug transit via various routes needs to be safe as well as effective. The main issue associated with such system was that there was poor permeability across BBB. The permeability acts as the rate limiting factor. Only a small fraction of the total drug payload is able to move across BBB (Alam et al., 2010).

14.6.1   NANOMEDICINES FOR BBB CROSSING: GENERAL CONSIDERATIONS

Examining the field of nanotechnology offers a promising future for focused medication conveyance by presenting devices, for example, nanoparticles (NPs), fit for the coordinated conveyance of medications into the cerebrum (Dordevic et al., 2015). Nonetheless, nanocarriers should be changed on their surface with reasonable ligands to guarantee focusing to a particular tissue or a particular organ, for example, the CNS.

To be sure, a few reviews demonstrate that these nanocarriers with an appropriate ligand can cross the BBB without obvious harm and can be utilized to convey drugs or hereditary material into the mind (Fornaguera et al., 2015). The method of transport of NPs over the BBB has been conjectured to be intervened by uninvolved dissemination as well as receptor-intervened endocytosis, liquid stage endocytosis or phagocytosis, transporter intervened transport or by absorptive-interceded transcytosis (Krishnan et al., 2016; Mizrahi et al., 2014).

14.6.2   ROUTES OF DELIVERY

The various routes of delivery have been summarized in Scheme 14.6.

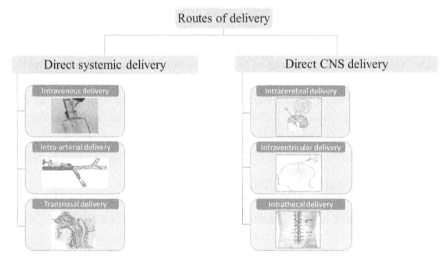

SCHEME 14.6   Routes of delivery.

14.6.2.1   INTRAVENOUS DELIVERY

Intravenous (IV) route or infusion is the most frequently exercised procedure because of the control it provides overdosage. It delivers a medication directly into bloodstream preventing its first-pass metabolism and has the potential to transport drugs to the brain. The brain is a complex structure of spreading capillaries hence this route has great potential drug targeting. However the medications given intravenously act on the body very quickly, so side effects, allergic reactions, and other effects can happen very fast. Also, there is an increase in side effects because of the accumulation of the drug in the brain and rapid clearance from the ECF (extracellular fluid). In addition, the brain availability of drug through IV route is largely affected by the half-life of the drug in the plasma, rapid metabolism, level of nonspecific binding to plasma proteins and the permeability of the compound across the BBB and into peripheral tissues (Patel et al., 2003).

14.6.2.2 INTRA-ARTERIAL DELIVERY

It is a form of the systemic parenteral route for the administration of drug directly into the bloodstream. It provides desired blood concentrations and allows the drug to access the brain vasculature before entering the peripheral tissue by avoiding first pass metabolism. It is a form of regional delivery to brain tumors, designed to enhance the intra-tumoral concentrations of a given drug. The basic mechanism behind the bioavailability of drug in brain may be due to movement of drug in capillaries, then to choroid plexus epithelium and finally reaching CSF or by falling into arterial blood and then going to CSF through white matter and perivascular pathway (Patel et al., 2003; Rautio and Chikhale, 2004). BBB disrupting agents further enhance the efficacy of the intraarterial route of drug delivery.

14.6.2.3 TRANSNASAL DELIVERY

Nasal delivery is a route of delivery in which *therapeutics* are insufflated through the nasal mucosa. The intranasal route has been deployed for administrating drugs for a long time, as the drug directly reaches to blood bypassing first pass metabolism (Muntimadugu et al., 2016). Effective brain targeting can be attributed to the direct movement of the drug from the submucosa space of the nose into the CSF compartment of the brain. It is a noninvasive method of bypassing the BBB to deliver the drug substances and peptides to the CNS. The highly permeable nasal epithelium allows rapid drug absorption to the brain because of high total blood flow, porous endothelial membrane, large surface area and avoidance of the first-pass metabolism. A wide variety of therapeutic agents (small molecules and macromolecules) can be delivered to the CNS by Transnasal method (Krag et al., 2007). Many agents active in the CNS are more effective when given nasally and provide the advantage of a small dose, self-administration and avoidance of sterile techniques. It does not need a variation of the therapeutic agent nor has the drug to be coupled with any carrier. Transnasal delivery has some limitations including damage of nasal mucosa on frequent use of this route, rapid clearance from nasal cavity by mucociliary clearance system, interference due to nasal congestion, elimination of some quantity of drug absorbed systemically via normal clearance mechanism and possibility of partial degradation or irritation to the nasal mucosa (Ali et al., 2010; Illum 2000)

14.6.2.4   DIRECT CNS DELIVERY

Assorted contemporary approaches have been conferred for the direct delivery of drug molecules to the CNS. Only a few have been discussed in detail.

14.6.2.4.1   Intracerebral (Intraparenchymal) Delivery

In the intracerebral mode of delivery is the mode in which the drug is administered directly into parenchymal space of the brain. This can be done by using controlled release matrices intrathecal catheters, microencapsulated chemicals or recombinant cells. However, the method reports a major drawback involving the slow movement of the drug due to the diffusion coefficient. The impediment is because of intently packed adjustment of cells in both gray and white matter and due to the concentration-dependent diffusion phenomena in the brain. (Nicholson and Syková, 1998) Therefore, a high dose of the drug is needed to maintain optimum drug level in the parenchyma. (Kawakami et al., 2004)

Polymer/drug formulations have been explored as delivery vesicles for drugs into the cerebral environment in the tumor cavity of the brain. It presents a persistent release of drugs by the biodegradation of polymer. The release profile of the drug depends upon the thickness of the matrix, type of polymer used in matrix and loading concentration of the drug (Brem et al., 1995). The shortcoming of low release was overcome by the placement of monolithic depots in an effectual way. The most commonly used monolithic polymeric depots are made up of copolymer like poly(lactide-co-glycolide) which is of recent interest in the delivery of drugs and peptides into CNS across BBB due to size, biodegradability and biocompatibility nature (Huynh et al., 2006). Monolithic depots help increase the active transport to endothelial cells with decreased chances of drug elimination from parenchyma by interstitial fluid flow (Gabikian and Brem, 2001).

14.6.2.4.2   Intraventricular Delivery (Transcranial Drug Delivery)

The particular and rare organization of blood vessels and sinuses in the brain and the communication between the extracranial veins and brain via emissary's veins makes intraventricular administration viable and most suited route for meningioma treatment and metastatic cells of CSF. It apportions drug primarily into ventricles and subarachnoidal area of the brain after bypassing the BBB. The edge of this route over other routes is its ability to achieve a higher concentration in the brain which credits to its lack of interconnection with interstitial fluid of brain. At the same time, this higher concentration of drug can increase the probability of causing a subependymalastrogliatic reaction due to high drug exposure.

14.6.2.4.3   Intrathecal Delivery (Intra-CSF Drug Delivery)

In this method of delivery, the neurotherapeutic agents are administered into the spinal theca which directly sends the agents into the cisterna magna of the brain. The drug is infused this way to prevent the BBB. This is relatively a less invasive administration route but it fails to accumulate adequate drug concentration in parenchymal structures. This leads to disrupted release. The radical drawback of this route is the possibility of drug scattering along the distal space of the spinal canal. Therefore, this route is most relevant for drug delivery for remediation of spinal diseases but not for large parenchymal diseases like parenchymal tumors such as glioblastoma (Kerr et al., 2001).

14.7   NANOTECHNOLOGY AS THE PROMISING TOOL IN NEUROMEDICINE

Targeting at the cellular level is an arduous task. It requires a high degree of precision which as quite difficult to be attained used the conventional drug delivery systems (Chung et al., 2014). The cargo needs to combat with the physiological barriers. At such checkpoints, it has to interact with the various intrinsic, extrinsic and integrated proteins, and phospholipids of the animal cell membrane.

In this quest, nanotechnology has emerged as a promising and highly capable tool for overcoming the hiccups of BBB and managing the neurological disorders (Obermeier et al., 2013). A diverse arena has been unveiled with the advent of nanotechnology, known has nanobiomedicine. The potential of such systems is quite impressive in comparison to the conventional drug delivery systems. Nanotechnology is being used as nanowagon to carry a diverse array of drugs and bioactives (Sun et al., 2017). It grants us the much-desired flexibility and freedom in the generation of the nanoparticles. The surface area to volume ratio is very high for the nanoscaled systems. Moreover, there is a lot of scope for the surface engineering possible, ranging from biologics to various chemical species. This makes nanosystems as versatile carrier systems for the delivery of various types of cargo. Due to the very small size of the NPs, they are not uptaken by the RES. (Warnken et al., 2016) Therefore, they have a very high circulation time in the biological milieu. Surface engineering provides the NPs with some added attributes. These include the exacerbation of the targetability of the system, reduction in the cytotoxicity and genotoxicity of the nanomaterials, and the other related adverse effects. The particles having the small size in the nano range provides it the freedom to traverse through the various biological barriers and target the most distant organs. Nanoscaled systems exhibit the stellar property of EPR (enhanced permeation and retention) which provides a higher endocytic property to the nanodevice (Fang et al., 2016). The complex vasculature and unique biochemistry of the tumor tissue are responsible for the inability of the conventional systems to act on the target site. The different size range has been a crucial factor for the mechanics for the action. For the particles in the size range of 120–200 nm, the uptake is via spleen; particles within the size range of <120 nm, the particles can bypass RES. The uptake mechanics involved also is dependent on the size of the cargo involved. For the particle having the size less than 200 nm, the preferred pathway is clathrin-mediated endocytosis, whereas for the larger particles having the size of order 500 nm the mechanism of caveolae-mediated endocytosis is more preferred.

Different nanoassemblies have been used, for example, , solid lipid nanoparticles (SLNs), nanostructured lipid carriers (NLCs), polymeric nanoparticles (PNPs), lipid-polymer hybrid nanoparticles (LPHNPs), dendrimers, carbon nanotubes (CNTs), emulsions and nanoemulsions, etc. The small size of these nanosystems adds to enhanced attributes for brain targeting (Tığlı Aydın et al., 2016; Xu et al., 2016).

Various mechanisms have been hypothesized and proved which are involved in the smuggling of the cargo across BBB (Saunders et al., 2016). Various types of nanosystems alone or in combination are used for delivery of therapeutic agents. Nanoscaled engineered probes have been found to be useful as they act at the cellular level using the various nanofluidic channel as pathways for transit, thereby improved efficacy of drug delivery to the brain. Recently, such systems have been explored for the management of brain tumors, neurodegenerative disorders, psychological disorders, etc.

The chapter centers upon the utility and advantage of nanoemulsions (NEms) as a delivery vehicle for the administration of medicaments to the brain. NEms belong to a special stratum of transparent/translucent systems with average droplet size ranging from 50 to 1000 nm. (Sekhon 2014; Capek, 2004) The formation of this special category of nano wagons requires external energy since they cannot be formed spontaneously by simple mixing. Since an emulsion is a mixture of two immiscible liquids, they can be classified as oil in water (o/w) and water in oil (w/o) NEms.(Capek 2004) These systems are thermodynamically unstable on account of the free positive energy colligated with the formation of the oil–water interface (Dickinson 2009). However, the random Brownian motion is sufficient in overcoming gravitational separation, flocculation, coalescence, and Ostwald ripening thence, owing to their remarkable kinetic stability and making them eligible for commercial applications. (Friberg et al., 2003) A further enhancement in the kinetic stability can be achieved by manipulating their composition (e.g., oil, surfactant, and water phase) and microstructure (e.g., particle size distribution), or by encompassing additives known as stabilizers such as emulsifiers, texture modifiers, weighing agents or ripening retarders. (Lidich et al., 2016)

Various physicochemical properties of the droplets such as droplet size, size distribution, morphology, and viscosity greatly impact the pharmacological characteristics of NEms and thus, the formulation of extremely fine emulsion droplets continues to be a challenge. Simple mixing, grinding, colloidal mills, static mixers, etc., are some of the conventional methodologies that were practiced to generate emulsions. The fabrication of a formulation with uniform droplet size along with the selection of its preparation method plays a pivotal role in the present pharmaceutical industry. (McClements, 2015)

NEms can be fabricated using low energy or high energy emulsification approaches. Low energy emulsification banks upon the spontaneous formation of tiny oil droplets with alteration in the chemical or environmental conditions of the system. These include spontaneous emulsification, phase inversion temperature (PIT), phase inversion composition (PIC), and emulsion inversion point (EIP) methodologies. High energy emulsification requires the input of energy to produce disruptive forces that mechanically break up large oil droplets into fine droplets that are suspended within the aqueous phase. Commonly used high energy techniques for the preparation of NEms include high-pressure homogenization, microfluidization, and probe ultrasonication. (Weiss et al., 2006; McClements, 2015)

For the design and formulation of food-based or pharmaceutical NEms, the common emphasis is laid on a low energy emulsification method, that is, phase inversion temperature (PIT) technique where chemical potential is generated through the use of surfactants. However, certain limitations accompany this method such as, the use of a large quantity of surfactant, mindful choice of both the surfactant and co-surfactant, high polydispersity index (PDI), which bring about instability after long-term storage and make it cumbersome to identify the system inversion temperature.

The other method reported in literature is high energy homogenization method which basically involves two steps. In the first step, a high-speed blender prepares a coarse emulsion by homogenizing the aqueous and organic phases for two minutes. Secondly, the prepared coarse emulsion is instantly passed through a high-pressure homogenizer. Recently, ultra-sonication has gained popularity since the homogenization method is quite expensive and sensitive. The ultrasonication cavitation method allows us to obtain a more stable nanoemulsion with a reduced rate of Ostwald ripening than those prepared using PIT method. The high energy input generated from the micro-turbulent implosions of cavitation bubbles provides enormous forces to deform and break-off the droplets into the nanometer scale (provided the Laplace pressure is overcome). It has been found that the ultrasonically-generated NEms have much smaller droplets with improved physical stability than the ones prepared by a homogenizer. Though micro-fluidizer can be employed to obtain NEms with much smaller droplet sizes, a comparative study unfolded that microfluidization is not as convenient as ultrasound probe sonication. The latter uses a probe and is much easier to clean and is free of line blockage, while the contamination of materials still poses a limitation in case of microfluidizer. All the above facts stand in favor of ultrasonic cavitation as the most effectual and propitious methodology for the production of pharmaceutical nanoemulsions.

14.8 NANOEMULSIONS IN DRUG DELIVERY TO BRAIN

Cyclosporine-A (CsA), is often used as an immunosuppressive agent in high dosage and chronic administration (Borel et al., 1994). However, such a high dose produces negative side effects, such as immune suppression, hepatotoxicity, and nephrotoxicity. So an alternative noninvasive strategy using nanoemulsion was employed to overcome the hurdles in exploring the potential of CsA for various brain disorders (Caicco et al., 2013). Based on the various reports in literature where omega-3 polyunsaturated fatty acid (PUFA) has been reported to have a considerable effect on BBB penetration, Yadav et al. (2015) prepared an oil-in-water nanoemulsion using flaxseed oil. The formulation (CsA-NE) consisted of flax-seed oil/Lipoid E80/Tween 80/ water/stearylamine. Blank nanoemulsions were prepared similarly without the addition of CsA. The morphology of the oil droplets in the nanoemulsion formulations was spherical.

The main objective of the study was to evaluate comparative bio-distribution and pharmacokinetics of cyclosporine-A (CsA) following intranasal (IN) administration versus IV administration in Sprague–Dawley rats using an oil-in-water nanoemulsion. For this, the prepared CsA-NE and solution formulations (CsA-S) were characterized for drug content, encapsulation efficiency, globule size, and zeta potential. The comparison of IN and IV methods were done by mapping brain regions and peripheral organs following IN and IV administration using LC-MS/MS method. The maximum concentrations obtained in the olfactory bulb, midbrain, and hindbrain were similar, in the range of 200–300 ng/g of tissue. This has been depicted in Figure 14.1. However, the observed kinetics was different in the three brain parts, namely, olfactory bulb, midbrain, and hindbrain. Whole brain concentrations were calculated based on the sum of the amount of CsA in the three regions of the brain divided by the sum of the weights of the three brain subparts. It was observed that IN CsA-NE resulted in the highest accumulation compared to that with any other treatment and route of administration. Further, it was also found to be consistent for all three regions of the brain. The study further supports the fact that CsA-NE is capable of direct nose-to-brain transport, bypassing the BBB as indicated by the brain/ blood exposure ratios, that is, 4.49, 0.01, 0.33, and 0.03 for CsA-NE (IN), CsA-NE (IV), CsA-S (IN), and CsA-S (IV). CsA-NE administration further reduces nontarget organ exposure (Yadav et al., 2015).

In another report by Musa et. al. (Musa et al., 2013) Chloramphenicol has been used as a drug for the treatment of bacterial meningitis (Franco-Paredes et al., 2008). Meningitis can lead to acute inflammation and subsequent brain damage. It carries a high risk of death and severe neurologic sequelae, especially when there is a delay in diagnosis and antibiotic administration (Fitch and van de Beek, 2007). Nowadays, one of the major challenges is bacterial resistance to antibiotics, so there is a problem of increased dosage for a definite action. Further, most of the drugs are hydrophobic in nature and therefore poorly dissolve poorly in water resulting in a higher drug dosage.

To overcome the issue a NEm-based chloramphenicol carrier composed of palm kernel oil esters, lecithin amount and glycerol was prepared by high-pressure homogenization method. This was found to improve the solubility of the drug in the dispersed phase along with a reported increase in drug penetration into target cells due to extremely small size. As a result, an effective formulation with a lower dosage and efficient penetration was achieved.

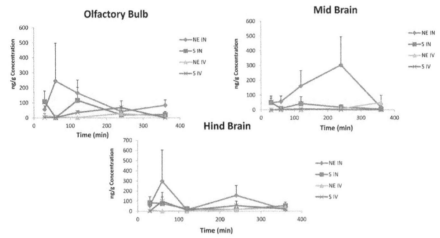

FIGURE 14.1   **(See color insert.)** Mean (±SD) concentration–time curves of cyclosporine-A (CsA) in different regions of rat brain after intranasal (IN) or intravenous (IV) administration of CsA-nanoemulsion (CsA-NE) or CsA-solution (CsA-S) at a dose of 5 mg/kg.
Source: Reprinted with permission from Yadav et al., 2015. © American Chemical Society.

Neurocysticercosis (NCC) NCC, a contagious zoonotic disease, caused by the cystic larval stage of the parasite Taeniasolium (tapeworm) (Nash et al., 2006). However, if the disease is left untreated, it could also be fatal (Mahanty et al., 2011). A combination drug of Albendazole sulfoxide (ABZ-SO) and Curcumin (CUR) has been used in the treatment of NCC using microemulsions (MEs) whereas standard therapy involves the combination of albendazole (ABZ) and corticosteroids (Mwape et al., 2013; Nash et al., 2006). ABZ; however, exhibits poor oral bioavailability with a limited concentration in the brain due to the insurmountable BBB. Targeted nose to brain delivery using MEs, therefore, provides a promising strategy to enhance drug concentration in the brain and thereby improved cure rates of NCC (Lawrence and Rees, 2000). MEs composed of 60% tween 80:ethanol (3:1) and 30% water by weight 10% DHA rich oil:Capmul MCM (1:1) or Capmul MCM and has been used in the treatment (Shinde et al., 2015). The emulsion was prepared as a simple solution exhibited a globule size <20 nm, negative zeta potential, and good stability. The drug-loaded ME formulations revealed high and rapid ex vivo permeation of drugs through sheep nasal mucosa. Intranasal ME formulation resulted in high brain concentrations and 10.76 (ABZ-SO) and 3.24(CUR) fold enhancement in brain area-under-the-curve as compared to IV DHA MEs at the same dose. The observed nose to brain Direct transport of both the drugs was >95%. High drug targeting efficiency to the brain compared to Capmul ME and drug solution (p < 0.05) suggested the role of DHA in aiding nose to brain delivery. However, Histopathology data showed no significant changes. High efficacy of ABZ-SO: CUR (100:10 ng/mL) DHA ME in vitro on Taeniasolium cysts was confirmed by complete ALP inhibition and disintegration of cysts at 96 h (Fig. 14.2). Considering that the brain concentration at 24 h was 1400 ± 160.1 ng/g (ABZ-SO) and 120 ± 35.2 ng/g (CUR), the in vitro efficacy seen at a 10 fold lower concentration of the drugs strongly supported the assumption of clinical efficacy.

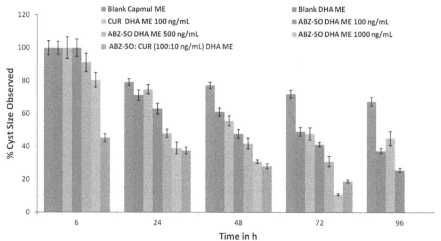

FIGURE 14.2   Changes in the *T. solium* cysts size during treatment with blank Capmul ME, blank DHA ME, CUR DHA ME (100 ng/mL), ABZ-SO DHA ME (100, 500, 1000 ng/mL), ABZSO and CUR (100:1).
Source: Reprinted with permission from Shinde et al., 2015. © Elsevier.

An in vitro efficacy study of the MEs on Taeniasolium (*T. solium*) cysts (Cysticercuscellulosae) was also undertaken as an indicator of possible efficacy in vivo. The decrease in cyst size and alkaline phosphatase (ALP) levels was retained as parameters of in vitro efficacy study. It is evident that while the MEs provide advantage for targeted delivery to the brain, the DHA ME provided an even greater benefit, confirming that DHA in the MEs played a crucial role not only in enhancing nose to brain delivery but also in maintaining sustained level up to 24 h and also for its possible role in enhancing efficacy. The DHA ME proved to be a superior intranasal delivery system for the targeted nose to brain delivery. (Shinde et al., 2015)

Infection with the human immunodeficiency virus (HIV) often results in progression to the acquired immune deficiency syndrome (AIDS). The HIV retrovirus is found in the brain soon after infection and leads to a variety of central nervous system adverse effects (Chearskul et al., 2006). Highly active anti-retroviral therapy (HAART) strategy involves the use of combination anti-retroviral agents for synergistic therapeutic outcomes (Frezzini et al., 2005). With the adoption of HAART, the average survival of HIV/AIDS patients has increased from less than 1 year to over 10 years. Despite the success of HAART in the clinics, HIV/AIDS therapy is far from optimal (Gallo, 2006). One of the major problems in the chronic treatment is the sub-therapeutic concentration of the present drugs in the brain which leads to failure of treatment and results in the development of drug-resistant viral strains in the brain despite the presence of adequate plasma concentrations. The reason for the sub-therapeutic concentration in the brain is due to efflux by P-glycoprotein (P-gp) expressed at the BBB. Increasing the brain concentration of drugs by improving its permeation of the BBB is, therefore, a key to reducing the viral burden in the brain during antiretroviral therapy.

To improve levels of drugs in brain different plans can be adopted, this includes co-administration of P-gp inhibitors and utilization of colloidal carriers. Prabhakar et al.(Prabhakar et al., 2013) have reported the use of, lipid nanoemulsions (LNEs) as carriers for drugs with poorly aqueous solubility. This involves drugs such as Indinavir. The use of colloidal carriers not only improves the oral bioavailability but also helps in sustained release and targeting. LNEs prepared with different compositions were characterized for globule size, polydispersity index, zeta-potential and in vitro drug release. Five formulations, as given in Table 14.1, were formulated and then evaluated for drug content, entrapment efficiency and stability.

TABLE 14.1 Composition of Lipid Nanoemulsions of Indinavir.

Formulation ingredient	No cholesterol	Cholesterol		Tween 80			
	F_1	F_2	F_3	F	F	F	F
Indinavir	0.1	0.1	0.1	0.1	0.1	0.1	0.1
Soyabean oil	10	10	10	10	10	10	10
Egg lecithin	1.2	1.2	1.2	1.2	1.2	1.2	1.2
Cholesterol	0	0.3	0.3	0.3	0.3	0.3	0.3
α-tocopherol acetate	0.25	0.25	0.25	0.25	0.25	0.25	0.25
Oleic acid	0.3	0.3	0.3	0.3	0.3	0.3	0.3
Tween 80	0	0	0	0.2	0.6	1.0	1.0
Glycerol	2.25	2.25	2.25	2.25	2.25	2.25	2.25
Did oil	–	–	0.1	–	–	–	0.1
Doubly distilled water	10	10	10	10	10	10	10

The brain uptake studies were carried out using fluorescent labeled LNEs and Pharmacokinetic (PK) and tissue distribution studies were conducted after IV administration in mice. The results of fluorescence microscopy and pharmacokinetic studies in mice clearly demonstrated improved uptake of LNEs by brain tissue in comparison to control formulations (Fig. 14.3).

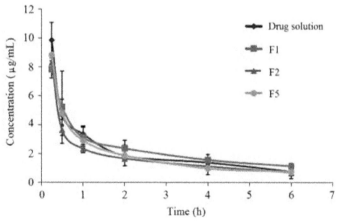

FIGURE 14.3 Plasma profiles of indinavir from drug solution and formulations F1, F2, and F5 after IV administration in mice ($n = 3$).

Brain uptake of indinavir, as reported by the authors (Fig. 14.4) shows improved uptake for a 1% Tween 80 containing formulation (F5) as compared to a formulation containing 0.3% cholesterol (F2). In PK studies, the brain level of indinavir subsequent to administration of F5 was significantly (Po 0.05) higher than produced by administration of a drug solution (2.44-fold) or a control nanoemulsion (F1) (1.48-fold) or formulation F2 (1.6-fold). The increased brain-specific accumulation of indinavir from F5 was probably due to enhanced low-density lipoprotein-mediated endocytosis and P-gp inhibition by Tween80 at the BBB. These results suggested Tween 80-containing LNEs could provide a simple but effective means of delivering Indinavir to the brain.

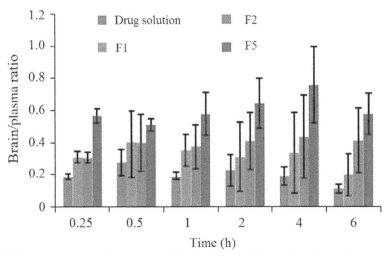

FIGURE 14.4 Brain-to-plasma ratios of indinavir after IV administration of drug solution and formulations F1, F2, and F5 in mice ($n = 3$).

Another interesting report on anti-HIV protease inhibitor using NEm as a delivery vehicle was given by Vyas et al (Vyas et al., 2008). The oral bioavailability and distribution to the brain, were significantly enhanced with SQV delivered in NEms formulations. The results of this studies showed that o/w NEms made with PUFA-rich oils are very promising for HIV/AIDS therapy, in particular, for reducing the viral load in important anatomical reservoir sites (King et al., 2004). The authors targeted the development of novel oil-in-water (o/w) NEMm containing Saquinavir (SQV), an anti-HIV protease inhibitor, for enhanced oral bioavailability and brain disposition. SQV was dissolved in different types of edible oils rich in essential polyunsaturated fatty acids (PUFA) to constitute the internal oil phase of the NEms. The external phase consisted of Lipoid®-80 and deoxycholic acid dissolved in water (Groves and Herman, 1993). The NEms with an average oil droplet size of 100–200 nm, containing tritiated [3H]-SQV, were administered orally and intravenously to male Balb/c mice. This was followed by investigations of SQV bioavailability as well as distribution in different organ systems. It was observed that SQV concentrations in the systemic circulation administered in flax-seed oil NEms were three fold higher as compared to the control aqueous suspension. In comparing SQV in flax-seed oil NEm with an aqueous suspension, the maximum concentration (C_{max}) and the AUC values were found to be five and threefold higher in the brain, respectively, suggesting enhanced rate and extent of SQV absorption following oral administration of NEms.

Oxidative stress is one of the primary factors that exacerbate damage by cerebral Ischemia. Brain tissues are particularly susceptible to oxidative damage, therefore, it is believed that pharmacologic modification of oxidative damage is one of the most promising avenues for stroke therapy (Ahmad et al., 2013; Katsura et al., 1994). Thymoquinone (TQ), the main constituents of the volatile oil was incorporated in Nanoemulsion.(Ganea et al., 2010) The formulated nanoemulsion loaded with thymoquinone through intranasal administration in middle cerebral artery occlusion-induced focal cerebral ischemia in Wistar rats. Nanoemulsions with a size range (10–200 nm) exhibited more stability in suspension due to small particle size. Thymoquinone nanoemulsion (TNE) was prepared by titration method using oleic acid as oil, carbitol as co-surfactant and tween20/labrasol/cremophore as a surfactant and purified water as the continuous phase (Ahmad et al., 2016).

In order to elucidate the disposition of nanoemulsions in the nasal mucosa, the authors examined cross-sections of the nasal mucosa by CLSM (Fig. 14.5). The confocal images of different cross-sections of the goat Nasal mucosa post washing with buffer solution exposed to the nanoemulsion. Qualitative assessment of confocal images revealed intense red colored fluorescent areas located in between and inside the mucosal cells. Due to the mucoadhesive nature, it was observed that mucoadhesive formulation TMNE showed more red colored intense areas as compared to nonmucoadhesive nanoemulsion TNE as shown in Figure 14.5. UPLC-PDA-based bioanalytical method was developed, validated, and successfully applied for the pharmacokinetic and biodistribution studies. Pharmacodynamic performance of the developed NPs was assessed on the MCAO-induced cerebral ischemia model in rats based on locomotor and grip strength studies.

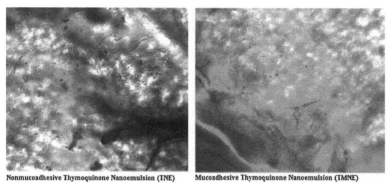

Nonmucoadhesive Thymoquinone Nanoemulsion (TNE)   Mucoadhesive Thymoquinone Nanoemulsion (TMNE)

FIGURE 14.5   (See color insert.) CLSM images of thymoquinone nonmucoadhesive nanoemulsion (TNE) and thymoquinone mucoadhesive nanoemulsion (TMNE).
Source: Reprinted with permission from Ahmad et al., 2016. © Elsevier.

Intranasal nanoemulsion-based brain targeting drug delivery system of Risperidone (RSP), (D'Souza et al., 2013) an approved antipsychotic drug belonging to the chemical class of benzisoxazole derivative, was formulated by Kumar and his coworkers (Kumar et al., 2008). The objective of the investigation was to prepare nanoemulsion containing RSP to accomplish the delivery of drug to the brain via the nose. Risperidone nanoemulsion (RNE) consisting of capmul MCM (8%, w/w) and Tween 80 29.33% w/w as surfactant. A mixture of transcutol and propylene glycol (1:1, w/w) was used as co-surfactant (CoS, 14.66%, w/w) and distilled water (48%,w/w) as the aqueous phase (Đorđević et al., 2015). Mucoadhesive nanoemulsion (RMNE) was prepared by addition of chitosan (0.50%,w/w) to RNE and the dispersion stirred for 1 h. The formed nanoemulsions were characterized for drug content, pH, percentage transmittance, globule size, and zeta potential. Biodistribution of RNE, RMNE, and risperidone solution (RS) in the brain and blood of Swiss albino rats following intranasal (i.n.) and intravenous (i.v.) administrationwas examined using optimized technetium labeled (99mTc-labeled) RSP formulations.

Gamma scintigraphy imaging of rat brain following i.v. and i.n. administrations were performed to ascertain the localization of drug in the brain (Fig. 14.6). The brain/blood uptake ratio of 0.617, 0.754, 0.948, and 0.054 for RS (i.n.), RNE (i.n.), RMNE (i.n.), and RNE (i.v.), respectively, at 0.5 h are indicative of direct nose to brain transport bypassing the BBB. Higher drug transport efficiency (DTE%) and direct nose to brain drug transport (direct transport percentage, DTP%) for mucoadhesive nanoemulsions indicated more effective and best brain targeting of RSP amongst the prepared nanoemulsions. Studies conclusively demonstrated rapid and larger extent of transport of RSP by RMNE (i.n.) when compared to RS (i.n.), RNE (i.n.) and RNE (i.v.) into the rat brain. (Kumar et al., 2008)

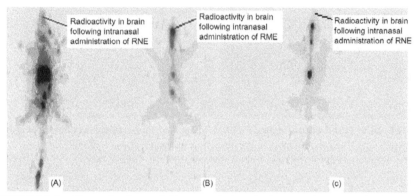

FIGURE 14.6   (See color insert.) Gamma scintigraphy images of rat (A/P view) showing the presence of radioactivity (A) RNE (i.v.), (B) RMNE (i.n.), and (C) RNE (i.n.). *Source*: Reprinted with permission from Kumar et al., 2008. © Elsevier.

Another Parenteral Nanoemulsion was formed for the same drug, that is, Risperidone (RSP) by Dordevic and coworkers (Đorđević et al., 2015). The work elucidated the design and evaluation of parenteral lecithin-based nanoemulsions intended for brain delivery of risperidone, a poorly water-soluble psychopharmacological drug. The nanoemulsions were prepared through cold/hot high-pressure homogenization and characterized regarding droplet size, polydispersity, surface charge, morphology, drug–vehicle interactions, and physical stability. To estimate the simultaneous influence of nanoemulsion formulation and preparation parameters—co-emulsifier type, aqueous phase type, homogenization temperature—on the critical quality attributes of developed nanoemulsions, a general factorial experimental design was applied. From the established design space and stability data, promising risperidone loaded nanoemulsions (mean size about 160 nm, size distribution <0.15, zeta potential around −50 mV), containing sodium oleate in the aqueous phase and polysorbate 80, poloxamer 188 or Solutol1 HS15 as co-emulsifier, were produced by hot homogenization and their ability to improve risperidone delivery to the brain was assessed in rats (Fig. 14.7). A pharmacokinetic study demonstrated erratic brain profiles of risperidone following intraperitoneal administration in selected nanoemulsions, most probably due to their different droplet surface properties (different composition of the stabilizing layer). Namely, polysorbate 80-co-stabilized nanoemulsion showed increased (1.4–7.4-fold higher) risperidone brain availability compared to other nanoemulsions and drug solution, suggesting this nanoemulsion as a promising carrier worth exploring further for brain targeting. (Đorđević et al., 2015)

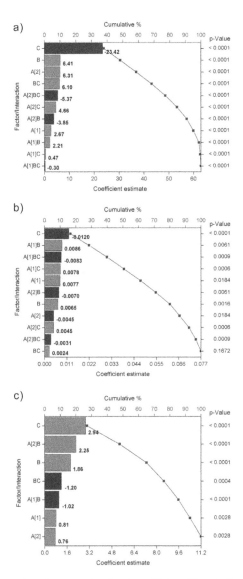

FIGURE 14.7   Pareto plots for mean droplet size—Z-Ave (a), polydispersity index (b), and zeta potential (c) of designed nanoemulsions, along with the coefficient values and p values for the studied variables—A: co-emulsifier type; B: aqueous phase type; C: HPH method.
Source: Reprinted with permission from Dordevic et al., 2015. © Elsevier.

Another work was published by the same group on the same drug. The work aimed to deepen the lately acquired knowledge about parenteral nanoemulsions as carriers for brain delivery of risperidone, a poorly water-soluble antipsychotic drug, through establishing the prospective relationship between their physicochemical, pharmacokinetic, biodistribution, and behavioral performances (Silva et al., 2012). For this purpose, two optimized risperidone-loaded nanoemulsions, stabilized by lecithin or lecithin/polysorbate 80 mixture, and co-stabilized by sodium oleate, were produced by high-pressure homogenization (Dordevic et al., 2017). The characterization revealed the favorable droplet size, narrow size distribution, high surface charge, with proven stability to autoclaving and long-term stability for at least one year at 25°C. Pharmacokinetic and tissue distribution results demonstrated improved plasma, liver, and brain pharmacokinetic parameters, resulting in 1.2–1.5-fold increased relative bioavailability, 1.1–1.8-fold decreased liver distribution, and about 1.3-fold improved brain uptake of risperidone active moiety following intraperitoneal administration of nanoemulsions relative to the solution in rats. In a behavioral study, investigated nanoemulsions showed a pronounced reduction in basal and, more pertinently, amphetamine-induced locomotor activity in rats, with an early onset of antipsychotic action, and this effect lasted at least 90 min after drug injection.

In vivo pharmacokinetic study complemented the behavioral results, (Fig. 14.8) demonstrating subtle differences in RSP pharmacokinetics after drug incorporation in parenteral nanoemulsions. These changes were characterized by higher AUC0–24 h, longer MRT and reduced clearance in plasma, lower AUC0–24 h and reduced MRT in liver, and increased AUC0–24 h in brain, accompanied by increased relative bioavailability, decreased distribution to the liver, and improved brain uptake of RSP total active moiety following i.p. administration of RSP nanoemulsions in comparison to RSP solution in rats. From the targeting point of view, it could be deduced that tested nanoemulsions improved the RSP ability to penetrate the BBB, thus, making them potentially advantageous over conventional vehicles.

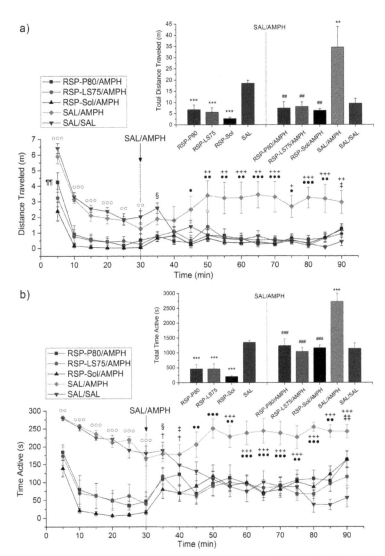

FIGURE 14.8 The effects of two RSP-loaded nanoemulsions (RSP-P80, RSP-LS75) and RSP solution (RSP-Sol) on basal and AMPH-induced locomotor activity, expressed as distance traveled (a) and time active (b), following i.p. injection to rats (mean SEM, $n = 7$–8); [**]p <0.01 and [***]p <0.001, compared to control saline (SAL or SAL/SAL) group; [##]p <0.01 and [###]p <0.001, compared to AMPH group (SAL/AMPH); p <0.01 and p <0.001, RSP-P80, RSP-LS75, and RSP-Sol versus SAL group; {{p <0.01, RSP-P80 vs. RSP-Sol; xp<0.05, RSP-Sol/AMPH vs. SAL/SAL; yp<0.05, RSP-Sol/AMPH vs. SAL/AMPH; zp<0.05 and zzp<0.01, RSP-LS75/AMPH vs. SAL/AMPH; [*]p <0.05, [**]p <0.01, and [***]p <0.001, RSP-P80/AMPH, RSP-LS75/AMPH, and RSP-Sol/AMPH vs. SAL/AMPH group; ++p <0.01 and +++p <0.001, SAL/SAL vs. SAL/AMPH group.

Source: Reprinted with permission from Dordevic et al., 2017. © Elsevier.

Another interesting finding was reported by Fornaguera et al. (Fornaguera et al., 2015) where PLGA nanoparticles were prepared by nanoemulsion templating using low energy methods. The nano-emulsification approach was found to be simple and very versatile technology, which allowed a precise size control and it was performed at mild process conditions. Drug-loaded PLGA Nanoparticles were obtained using safe components by solvent evaporation of template nano-emulsions. Characterization of PLGA nanoparticles was performed, together with the study of the BBB crossing. The nanoparticle cytotoxicity produced on a HeLa cell culture was assessed by the MTT assay (Fig. 14.9). Loperamide-loaded nanoparticles (NP (0.1)), 8D3-functionalized nanoparticles (NP (0, 8D3)), and loperamide-loaded8D3-functionalized nanoparticles, at a N/P ratio of 12.5/1 (NP (0.1,8D3)),were studied at the as-prepared concentration (3 mg/mL),without a concentration step and also at the required therapeutic concentration(30 mg/mL), to ensure their safety in vivo. As Figure 14.8 shows, cell viability was higher than 80% for all sets of nanoparticles studied, independently on the concentration, thus confirming that the formulated nanoparticles were noncytotoxic up to a concentration of 30 mg/mL.

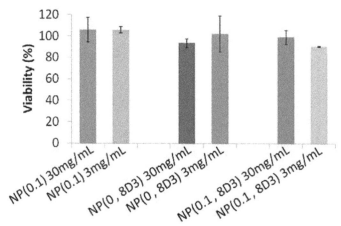

FIGURE 14.9   Viability (in %) of HeLa cells, after 24 h of incubation with different sets of nanoparticles, at the as-prepared and use concentrations.
Source: Reprinted with permission from Fornaguera et al., 2015. © Elsevier.

The in vivo results of measuring the analgesic effect using the hot-plate test evidenced that the designed PLGA loperamide-loaded nanoparticles were able to efficiently cross the BBB, with high crossing efficiencies when their surface was functionalized with an active targeting moiety (a monoclonal antibody against the transferrin receptor). These results, together with the nanoparticle characterization performed provided sufficient evidence to end up to clinical trials in the near future.

Neuroinflammation is a hallmark of acute and chronic neurodegenerative disorders. The main aim of the study conducted by Yadav et al. was to evaluate the therapeutic efficacy of intranasal cationic nanoemulsion encapsulating an anti-TNFα siRNA, for potential anti-inflammatory therapy. TNFα siRNA nanoemulsions were prepared and characterized for particle size, surface charge, morphology, and stability and encapsulation efficiency. Qualitative and quantitative intracellular uptake studies by confocal imaging and flow cytometry, respectively, showed higher uptake compared to Lipofectamine® transfected siRNA. Nanoemulsion significantly lowered TNFα levels in LPS-stimulated cells. Upon intranasal delivery of cationic nanoemulsions, almost fivefold higher uptake was observed in the rat brain compared to nonencapsulated siRNA. More importantly, intranasal delivery of TNFα siRNA nanoemulsions in vivo markedly reduced the unregulated levels of TNFα in an LPS-induced model of neuro-inflammation. These results indicated that intranasal delivery of cationic nanoemulsions encapsulating TNFα siRNA offered an efficient means of gene knockdown and this approach has significant potential in the prevention of neuroinflammation (Yadav et al., 2016).

Neurodegenerative diseases generate the accumulation of specific misfolded proteins, such as PrPSc prions or A-beta in Alzheimer's diseases, and share common pathological features, like neuronal death and oxidative damage (Barchet and Amiji, 2009). To test whether reduced oxidation alters disease manifestation, Mizrahi and his co-workers (Mizrahi et al., 2014) treated TgMHu2ME199K mice, modeling for genetic prion disease, with Nano-PSO, a nano-droplet formulation of pomegranate seed oil (PSO). PSO comprises large concentrations of a unique polyunsaturated fatty acid, Punicic acid, among the strongest natural antioxidants. Figure 14.10 shows the effect of PSO administration on the clinical presentation and advance of the disease in TgMHu2ME199K mice. Results were similar for treatment of heterozygous and homozygous TgMHu2ME199K mice (panels AI and BI), consistent with our recent results indicating that wtPrP does not participate in disease presentation in the genetic mice and show a significant delay in disease presentation in the treated mice (P b 0.02 for both panels A and B, see methods for description of statistical analysis). Nano-PSO significantly delayed disease presentation when administered to asymptomatic TgMHu2ME199K mice and postponed disease aggravation in already sick mice. Analysis of brain samples revealed that Nano-PSO treatment did not decrease PrPSc accumulation, but rather reduced lipid oxidation and neuronal loss, indicating a strong neuroprotective effect. (Mizrahi et al., 2014).

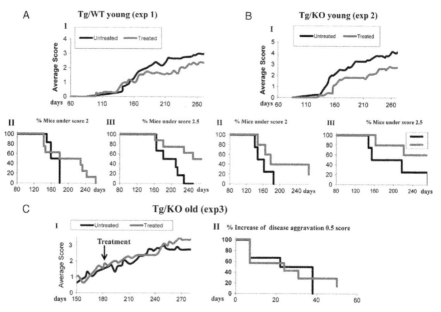

FIGURE 14.10   Natural PSO delays disease onset in TgMHu2ME199K mice. Young TgMHu2ME199K/wt and TgMHu2ME199K/ko mice or 190 days old TgMHu2ME199K/KO mice were treated with regular or with PSO-enriched food for the designated time course. Mice were scored for disease signs as described in Table 14.1. Figures (I) in panels (A), (B), and (C): Average group score as related to age of mice. Figures (II) in panels (A) and (B): percentage of mice under score 2 as related to the age of mice. Figures (III) in panels (A) and (B): percentage of mice under score 2.5 as related to the age of mice. Figure (II) in panel (C): percentage of mice aggravated by 0.5 scores.
Source: Reprinted with permission from Mizrahi et al., 2014. © Elsevier.

14.9   CONCLUSION

Neurological research has always been a challenging domain in the arena of biomedical research. There are various hurdles that stand in the successful detection and the management of the neurological disorders. But, the biologics and physiology of BBB act as the roadblock in the progress of this domain. Nanotechnology has emerged as a potent tool to overcome all the hiccups in the way of neuromedicine. Scientists have achieved meticulous results after employing nanotechnology in the biomedical therapeutics, diagnostics and imaging sector. But, there are certain issues that need to be addressed and much more needs to be done in this domain so that the untamed potential of nanotechnology can be harnessed for the cure and management of the neurological disorders.

KEYWORDS

- **brain targeting**
- **delivery to the brain**
- **hyrdophobic drugs**
- **nanoemulsions**
- **neurodegenerative diseases**
- **structure of brain**

REFERENCES

Ahmad, N., et al. Quantification and Evaluation of Thymoquinone Loaded Mucoadhesive Nanoemulsion for Treatment of Cerebral Ischemia. *Int. J. Biol. Macromol.* **2016,** *88,* 320–332.

Ahmad, N., et al. A Comparative Study of PNIPAM Nanoparticles of Curcumin, Demethoxycurcumin, and Bisdemethoxycurcumin and Their Effects on Oxidative Stress Markers in Experimental Stroke. *Protoplasma* **2013,** *250,* 1327–1338.

Alam, M. I., et al. Strategy for Effective Brain Drug Delivery. *Eur. J. Pharm. Sci.* **2010,** *40,* 385–403.

Ali, J., et al. Potential of Nanoparticulate Drug Delivery Systems by Intranasal Administration. *Curr. Pharm. Des.* **2010,** *16,* 1644–1653.

Barchet, T. M.; Amiji, M. M. Challenges and Opportunities in CNS Delivery of Therapeutics for Neurodegenerative Diseases. *Expert. Opin. Drug. Deliv.* **2009,** *6,* 211–225.

Barcroft, J.; Kennedy, J. The Distribution of Blood between the Foetus and the Placenta in Sheep. *J. Physiol.* **1939,** *95,* 173–186.

Bergers, G.; Song, S. The Role of Pericytes in Blood-Vessel Formation and Maintenance. *Neuro-Oncology* **2005,** *7,* 452–464.

Borel, J. A., et al. Biological Effects of Cyclosporin A: a New Antilymphocytic Agent. *Agents Actions* **1994,** *43,* 179–186.

Caicco, M. J., et al. A Hydrogel Composite System for Sustained Epi-Cortical Delivery of Cyclosporin A to the Brain for Treatment of Stroke. *J. Control. Release* **2013,** *166,* 197–202.

Capek, I. Degradation of Kinetically-Stable o/w Emulsions. *Adv. Colloid Interface Sci.* **2004,** *107,* 125–155.

Chearskul, P., et al. New Antiretroviral Drugs in Clinical Use. *Indian J. Pediatr.* **2006,** *73,* 335–341.

Chung, E. J., et al. Fibrin-Binding, Peptide Amphiphile Micelles for Targeting Glioblastoma. *Biomaterials* **2014,** *35,* 1249–1256.

D'Andrea, G., et al. Brain Metastases: Surgical Treatment and Overall Survival. *World Neurosurg.* **2017,** *97,* 169–177.

D'Souza, S., et al. Microsphere Delivery of Risperidone as an Alternative to Combination Therapy. *Eur. J. Pharm. Biopharm.* **2013,** *85,* 631–639.

Daneman, R.; Prat, A. The Blood–Brain Barrier. *Cold Spring Harb. Perspect. Biol.* **2015,** *7,* https://doi.org/10.1101/cshperspect.a020412..

Dickinson, E. Hydrocolloids as Emulsifiers and Emulsion Stabilizers. *Food Hydrocoll.* **2009,** *23,* 1473–1482.

Đorđević, S. M., et al. Parenteral Nanoemulsions as Promising Carriers for Brain Delivery of Risperidone: Design, Characterization and In Vivo Pharmacokinetic Evaluation. *Int. J. Pharm.* **2015,** *493,* 40–54.

Dordevic, S. M., et al. Parenteral Nanoemulsions of Risperidone for Enhanced Brain Delivery in Acute Psychosis: Physicochemical and In Vivo Performances. *Int. J. Pharm.* **2017,** *533,* 421–430.

Ek, C. J., et al. Barriers in the Developing Brain and Neurotoxicology. *NeuroToxicology* **2012,** *33,* 586–604.

Fang, Z., et al. Pluronic P85-Coated Poly (Butylcyanoacrylate) Nanoparticles Overcome Phenytoin Resistance in P-Glycoprotein Overexpressing Rats with Lithium-Pilocarpine-Induced Chronic Temporal Lobe Epilepsy. *Biomaterials* **2016,** *97,*110–121.

Fitch, M. T. and van de Beek, D. Emergency Diagnosis and Treatment of Adult Meningitis. *Lancet Infect. Dis.* **2007,** *7,* 191–200.

Fornaguera, C., et al. PLGA Nanoparticles Prepared by Nano-Emulsion Templating Using Low-Energy Methods as Efficient Nanocarriers for Drug Delivery Across the Blood-Brain Barrier. *J. Control. Release* **2015,** *211,* 134–143.

Franco-Paredes, C., et al. Epidemiology and Outcomes of Bacterial Meningitis in Mexican Children: 10-Year Experience (1993–2003). *Int. J. Infect. Dis.***2008,** 12, 380–386.

Frezzini, C., et al. Current Trends of HIV Disease of the Mouth. *J. Oral Pathol. Med.* **2005,** *34,* 513–531.

Friberg, S., et al. Food Emulsions, 4th ed.; CRC Press, 2003.

Gallo, R. C. A Reflection on HIV/AIDS Research after 25 Years. *Retrovirology* **2006,** *3,* 72.

Ganea, G. M., et al. Delivery of Phytochemical Thymoquinone Using Molecular Micelle Modified Poly (d,l-Lactide-Co-Glycolide)(PLGA) Nanoparticles. *Nanotechnology* **2010,** *21,* 285104.

Goasdoue, K., et al. Review: The Blood-Brain Barrier; Protecting the Developing Fetal Brain. *Placenta* **2017,** *54,* 111–116.

Groves, M.; Herman, C. The Redistribution of Bulk Aqueous Phase Phospholipids During Thermal Stressing of Phospholipid stabilized Emulsions. *J. Pharm. Pharmacol.* **1993,** *45,* 592–596.

Hawkins, B. T.; Davis, T. P. The Blood-Brain Barrier/Neurovascular Unit in Health and Disease. *Pharmacol. Rev.* **2005,** *57,* 173–185.

Illum, L. Transport of Drugs from the Nasal Cavity to the Central Nervous System. *European J. Pharm. Sci.* **2000,** *11,* 1–18.

Katsura, K. -I., et al. Acidosis Induced by Hypercapnia Exaggerates Ischemic Brain Damage. *J. Cereb. Blood Flow. Metab.* **1994**, *14*, 243–250.

Kawakami, K., et al. Distribution Kinetics of Targeted Cytotoxin in Glioma by Bolus or Convection-Enhanced Delivery in a Murine Model. *J. Neurosurg.* **2004**, *101*, 1004–1011.

King, J. R., et al. Pharmacokinetic Enhancement of Protease Inhibitor Therapy. *Clin. Pharmacokinet.* **2004**, *43*, 291–310.

Krag, D. N., et al. Technical Outcomes of Sentinel-Lymph-Node Resection and Conventional Axillary-Lymph-Node Dissection in Patients with Clinically Node-Negative Breast Cancer: Results from the NSABP B-32 Randomised Phase III Trial. *Lancet Oncol.* **2007**, *8*, 881–888.

Krishnan, J. K. S., et al. Intranasal Delivery of Obidoxime to the Brain Prevents Mortality and CNS Damage from Organophosphate Poisoning. *NeuroToxicology* **2016**, *53*, 64–73.

Kumar, M., et al. Intranasal Nanoemulsion Based Brain Targeting Drug Delivery System of Risperidone. *Int. J. Pharm.* **2008**, *358*, 285–291.

Lawrence, M. J.; Rees, G. D. Microemulsion-Based Media as Novel Drug Delivery Systems. *Adv. Drug Deliv. Rev.* **2000**, *45*, 89–121.

Lidich, N., et al. Structural Characteristics of Oil-Poor Dilutable Fish Oil Omega-3 Microemulsions for Ophthalmic Applications. *J. Colloid Interface Sci.* **2016**, *463*, 83–92.

Mack, A. F.; Wolburg, H. A Novel Look at Astrocytes: Aquaporins, Ionic Homeostasis, and the Role of the Microenvironment for Regeneration in the CNS. *The Neuroscientist* **2013**, 19, 195–207.

Mahanty, S., et al. Sensitive In Vitro System to Assess Morphological and Biochemical Effects of Praziquantel and Albendazole on Taenia Solium Cysts. *Antimicrob. Agents Chemothe.* **2011**, *55*, 211–217.

McClements, D. J. Food Emulsions: Principles, Practices, and Techniques. *2nd ed.*; CRC press, 2004.

Mizrahi, M., et al. Pomegranate Seed Oil Nanoemulsions for the Prevention and Treatment of Neurodegenerative Diseases: the Case of Genetic CJD. *Nanomedicine* **2014**, *10*, 1353–1363.

Muntimadugu, E., et al. Intranasal Delivery of Nanoparticle Encapsulated Tarenflurbil: A Potential Brain Targeting Strategy for Alzheimer's Disease. *Eur. J. Pharm. Sci.* **2016**, *92*, 224–234.

Musa, S. H., et al. Formulation Optimization of Palm Kernel Oil Esters Nanoemulsion-Loaded with Chloramphenicol Suitable for Meningitis Treatment. *Colloids Surf. B Biointerfaces* **2013**, *112*, 113–119.

Mwape, K. E., et al. The Incidence of Human Cysticercosis in a Rural Community of Eastern Zambia. *PLOS Negl. Trop. Dis.* **2013**, *7*, 2142.

Nash, T., et al. Treatment of Neurocysticercosis Current Status and Future Research Needs. *Neurology* **2006**, *67*, 1120–1127.

Nicholson, C.; Syková, E. Extracellular Space Structure Revealed by Diffusion Analysis. *Trends Neurosci.* **1998**, *21*, 207–215.

Obermeier, B., et al. Development, Maintenance and Disruption of the Blood-Brain Barrier. *Nat. Med.* **2013**, *19*, 1584–1596.

Prabhakar, K., et al. Tween 80 Containing Lipid Nanoemulsions for Delivery of Indinavir to Brain. *Acta Pharm. Sin. B.* **2013**, *3*, 345–353.

Saunders, N. R., et al. The Biological Significance of Brain Barrier Mechanisms: Help or Hindrance in Drug Delivery to the Central Nervous System? *F1000Research*, **2016**, *5*, 1–15.

Sekhon, B. S. Nanotechnology in Agri-Food Production: an Overview. *Nanotechnol. Sci. Appl.* **2014**, *7*, 31.

Shinde, R. L., et al. Intranasal Microemulsion for Targeted Nose to Brain Delivery in Neurocysticercosis: Role of Docosahexaenoic Acid. *Eur. J. Pharm. Biopharm.* **2015**, *96*, 363–379.

Silva, A., et al. Long-Term Stability, Biocompatibility and Oral Delivery Potential of Risperidone-Loaded Solid Lipid Nanoparticles. *Int. J. Pharm.* **2012**, *436*, 798–805.

Sun, C., et al. Noninvasive Nanoparticle Strategies for Brain Tumor Targeting. *Nanomed. Nanotech. Biol. Med.* **2017**, *13*, 2605–2621.

Tığlı Aydın, R. S., et al. Salinomycin Encapsulated Nanoparticles as a Targeting Vehicle for Glioblastoma Cells. *J. Biomed. Mater. Res. A* **2016**, *104*, 455–464.

Vyas, T. K., et al. Improved Oral Bioavailability and Brain Transport of Saquinavir Upon Administration in Novel Nanoemulsion Formulations. *Int. J. Pharm.* **2008**, *347*, 93–101.

Wager, M., et al. Operating Environment for Awake Brain Surgery–Choice of Tests. *Neurochirurgie* **2017**, *63*, 150–157.

Warnken, Z. N., et al. Formulation and Device Design to Increase Nose to Brain Drug Delivery. *J. Drug. Deliv. Sci. Technol.* **2016**, *35*, 213–222.

Weiss, J., et al. Functional Materials in Food Nanotechnology. *J. Food Sci.* **2006**, *71*, R107–R116.

Xu, H.-L., et al. Glioma-Targeted Superparamagnetic Iron Oxide Nanoparticles as Drug-Carrying Vehicles for Theranostic Effects. *Nanoscale* **2016**, *8*, 14222–14236.

Yadav, S., et al. Intranasal Brain Delivery of Cationic Nanoemulsion-Encapsulated TNF Alpha siRNA in Prevention of Experimental Neuroinflammation. *Nanomedicine* **2016**, *12*, 987–1002.

Yadav, S., et al. Comparative Biodistribution and Pharmacokinetic Analysis of Cyclosporine-A in the Brain upon Intranasal or Intravenous Administration in an Oil-in-Water Nanoemulsion Formulation. *Mol. Pharm.* **2015**, *12*, 1523–1533.

CHAPTER 15

NANOGELS FOR BRAIN TARGETING: INTRODUCTION, FORMULATION ASPECTS, AND APPLICATIONS

NAGARJUN RANGARAJ and SUNITHA SAMPATHI*

Department of Pharmaceutics, National Institute of Pharmaceutical Education & Research (NIPER-H), Balanagar, Hyderabad 500037, India

Corresponding author. E-mail: sunithaniper10@gmail.com

ABSTRACT

In spite of the technological advancements, crossing blood brain barrier and achieving sufficient drug concentrations in the brain has been one of the biggest challenges for drug delivery for central nervous systems disorders. Several formulation strategies have been employed to address the issue one such formulation is nanogel, which had displayed promising results in delivering therapeutics to brain. Nanogels are three-dimensional hydrogel particles bearing the sub-micron particle size with the combined advantages of both nanoparticulate system and hydrogels. Properties like biocompatibility, high hydophilicity, high drug loading, stimuli sensitivity, targeting and controlled release make these agents appropriate for drug delivery to brain. This chapter attempts to discuss the various aspects like physiology of brain, nanogels classification, formulation (synthesis), characterization, and finally the application for brain delivery of drugs.

15.1   NANOGELS FOR BRAIN TARGETING: INTRODUCTION, FORMULATION ASPECTS, AND APPLICATIONS

15.1.1   INTRODUCTION

15.1.1.1   BRAIN MORPHOLOGY AND FUNCTION

The most sensitive organ of the body is Brain which is protected by the three barriers such as blood–brain barrier (BBB), blood–cerebrospinal fluid barrier (BCSFB) and ependyma, of which BBB is considered as the strictest one formed by highly specialized blood vessels, endothelial cells lining the cerebral micro-vessels and neighboring perivascular components (pericytes, astrocytes, microglia). BBB screens the blood before it enters the brain thereby restricting the entry of molecules and ions. For any thera-peutic agent to show its activity on the brain it needs to surpass the BBB and blood-cerebrospinal fluid barrier (BCSFB) (Azadi et al., 2012, Blanchette et al., 2015, Chen et al., 2012, de Boer et al., 2007, Li et al., 2017, Lockman et al., 2002, Patel et al., 2009, Soni et al., 2006, Yang, 2010). These endothe-lial cell proteins such as occludin, junctional adhesion molecule-1, claudins, zonula occludens-1 and -2, cingulin AF-6/afadin, and 7H6 complex make up the Tight junction (TJ). This TJ blocks the paracellular route of endothe-lial cells resulting in high trans-endothelial electrical resistance (TEER) of 8,000 Ω cm^2 thereby obstructing the entry of small ions and charged particle (Chen et al., 2012, Yang 2010). BBB restricts approximately 98% and 100% of small and large molecules, respectively (Pardridge, 2001, Pardridge, 2002). Efflux transporters like ATP-binding cassette (ABC) transporters, for example, P-glycoprotein (Pgp), multidrug resistance-associated proteins (MRP) and ABC-sub-family G member 2 present on the membrane play a significant role in pumping the substances back into the blood circulation thereby preventing the amassing of molecules (Chen et al., 2012, Koziara et al., 2006, Linnet et al., 2008).

15.1.1.2   TRANSPORT ACROSS BBB

Transport of a molecule across the BBB occurs by passive diffusion, adsorp-tion mediated transcytosis, carrier-mediated transport, receptor-mediated transcytosis and cell-mediated transport as discussed below.

15.1.1.2.1 Passive Diffusion

Passive diffusion is driven by the concentration gradient across the membrane and does not require energy. For a drug molecule to get passively diffused it should meet the criteria as given in Table 15.1 (Chen et al., 2012, Fischer et al., 1998).

TABLE 15.1 Criteria for Passive Diffusion of Drug Molecule.

Property	Limit
Molecular weight	< 500 Da
Log p	Close to 2
Cumulative number of H bonds	< 10
Cross-sectional area A_D	< 80 Å
Ionization constant	$pK_a > 4$ for acids $pK_a < 10$ for bases

15.1.1.2.2 Adsorption-Mediated Transcytosis

Adsorption-mediated transcytosis (AMT) or pinocytosis is steered by electrostatic interaction amidst the cationic ligand and the anionic luminal surface of cerebral endothelial cells. The presence of proteoglycans, sulfated mucopolysaccharides, glycoproteins and glycolipids containing sulfate group and sialic acid renders negative charge to the luminal surface of cerebral endothelial cells. AMT is nonspecific, has an inferior affinity and greater capacity compared to receptor-mediated transcytosis (RMT) (Chen et al., 2012, Poduslo et al., 1996, Vorbrodt, 1987, Vorbrodt et al., 1990, Yang, 2010).

15.1.1.2.3 Carrier-Mediated Transport

It occurs by membrane restricted carriers that transport small molecules (nucleoside, glucose, amino acids, and purines) across the cell membrane. Carriers are membrane-bound proteins that transport the substance across the membrane by a conformational change upon binding with the solute. The transport of a solute aided by concentration gradient is called facilitated diffusion (no energy is required) and the transport against the concentration gradient is termed as active (require energy ATP). These transporters are saturable and substrate-specific hence substances which resemble endogenous molecules are transported across the membrane. CMT is not only responsible for transport to the brain, but also responsible for efflux of drugs from the brain by means of efflux transporters (ATP binding cassette (ABC) transporter P-GP and multidrug resistant protein (MRP) (de Boer et al., 2007, Koziara et al., 2006, Lee et al., 2001, Oldendorf, 1971, Wong et al., 2012).

15.1.1.2.4   Receptor-Mediated Transcytosis

Endothelial cells possess receptors through which large endogenous molecules (growth factors, enzymes, neuropeptides, and plasma proteins) cross the BBB in three steps: (1) binding of the ligand to the receptor at specialized areas called coated pits, (2) subsequent formation of endosomal vesicles, and (3) exocytosis at the other side (once the endosome gets acidified, the ligand dissociates from the receptor and crosses the other side of the membrane). This transport is extensively studied for targeting, owing to their high specificity, ability to carry large molecules and rapid transport. The receptors such as transferrin (Fishman et al., 1987), insulin-like growth factors (IGF-1 and IGF-2) (Reinhardt et al., 1994), leptin (Golden et al., 1997), low-density lipoprotein receptor-related protein (LRP) (Ke et al., 2009), diphtheria toxin receptor and glutathione transporter make use of this transport (Yang 2010).

15.1.1.2.5   Cell-Mediated Transport

cMT relies on the phagocytic cells of the innate immune system (monocytes and macrophages). Afergan. (Afergan et al., 2008) explored the phagocytic brain delivery of serotonin aided by anionic nano-sized liposomes. Hydrophobic/ hydrophilic, low molecular weight drugs, gene constructs, and peptides or proteins can be transported by this pathway. Since this transport is based on mononuclear phagocytic system (MPS) it may cause drugs to accumulate in cell/organs of MPS, hence its clinical relevance is limited (Afergan et al., 2008, Chen et al., 2012, Park 2008).

15.1.1.3   STRATEGIES TO DELIVER DRUG ACROSS BBB

Following are the various strategies employed to deliver a drug to the brain.

15.1.1.3.1 Drug Modification

Chemical modifications such as Lipidation (increase in lipophilicity, masking of hydrogen bonds) (Greig 1989, Sawynok 1986, Temsamani et al., 2000), Pro-drug approach (Bodor et al., 1987), lock in systems (Bodor 1994) can be employed to modify the drug such that it can cross the BBB.

15.1.1.3.2 BBB Modifications

Modification of BBB may alter the barrier permeability, thus allowing the transport of molecules. These modifications can be carried out by approaches such as Physical (ultrasound (Hynynen, 2008), Microwave (Moriyama et al., 1991), Electromagnetic (Kuo et al., 2008, Qiu et al., 2010)), Chemical (hyperosmotic agent (Doolittle et al., 2000, Greenwood et al., 1988, Kroll et al., 1998), sodium dodecyl sulphate (SDS) (Saija et al., 1997), Cyclodextrin (Tilloy et al., 2006), Polaxomers (Batrakova et al., 2001)) and Biological (vasoactive compounds, bradykinin and VEGF (Abbott, 2000, Kumar et al., 2009), Cereport (synthetic peptide analogue of bradykinin) (Emerich et al., 2000, Mackic et al., 1999)) approaches.

15.1.1.3.3 By-Pass BBB

Bypassing BBB is an alternative strategy for the drugs that cannot cross BBB. Some approaches to bypass BBB are Intranasal delivery, Intra-cerebroventricular (ICV) injections and Infusions (Gozzi et al., 2005) and Interstitial delivery (Alam et al., 2010, Koziara et al., 2006) among which intranasal is the most commonly employed approach owing to its non-invasiveness.

15.1.1.3.4 Active Targeting

Significant enhancement of CNS delivery can be achieved by surface modifications of nanocarriers with the "targeting molecules" such as cell penetrating peptides (trans-activator of transcription (TAT), angiopep, SynB, etc.) (Stalmans et al., 2015), monoclonal antibodies (OX26, 8D3, etc.) (Neves et al., 2016), endogenous molecules (apolipoprotein A (ApoA), apolipoprotein E (ApoE), transferrin, etc.) (Georgieva et al., 2014).

15.1.1.3.5   Cationic Nanocarrier

As discussed in adsorption-mediated transcytosis cell membranes are negatively charged and show a better interaction with the positively charged (cationic) nanocarrier. In addition, lectins present on the surface of endothelial cells bind to the positively charged nanocarriers promoting the endocytosis (Gonatas et al., 1984).

15.1.2   NANOPARTICULATE SYSTEM

Despite the advancements in the delivery of therapeutic agents problems still persist, which render focus on developing novel technologies such as "nanoparticulate drug delivery system". Nanoparticulate systems are drug carriers with nanoscale size range (1–1000 nm) (Wong et al., 2012) that comprise numerous approaches like (solid lipid nanoparticles (Singh et al., 2016), nanospheres (Cai et al., 2016), micelles (Kanazawa et al., 2011), liposomes (Qin et al., 2011) etc). Though these systems have proven to be beneficial several problems do persist such as high cost, complex preparation, low drug loading (Yallapu et al., 2007), off-target toxicity, instability, rapid elimination (Raemdonck et al., 2009) and exposure to organic solvents. This has led to the development of an alternative approach which can deliver drugs to the brain (Vinogradov et al., 2004).

15.1.3   NANOGELS

Nanogel drug delivery systems are three-dimensional hydrogel particles bearing the sub-micron particle size (Li et al., 2015) fabricated by physical/chemical crosslinking of the polymer (Soni et al., 2016b). Crosslinking of polymers allows nanogels to absorb fluid, maintain structural stability without undergoing dissolution (Hamidi et al., 2008, Raemdonck et al., 2009). Nanogels are entitled as next-generation drug delivery systems pertaining to their benefits like tunable size, ease of preparation, swelling, biocompatibility, hydrophilicity and stimuli responsiveness (temperature, pH, light, biological agent etc.) (Mura et al., 2013, Neamtu et al., 2017). The ability of nanogels to absorb large quantities of water and biological fluids can be attributed to the presence of –OH, –COOH, –NH$_2$, –CONH$_2$, –CONH, –SO$_2$H, etc. (Yallapu et al., 2007). Hydrophilic polymers can show hydration up to 90% whereas hydrophobic polymers show hydration of about 5–10% (Gonçalves et al., 2010). Nanogels are the integration of nano with hydrogel delivery system thereby presenting the merits of both hydrogel system and very small size of the nanoparticulate system (Azadi et al., 2012).

Though the concept of hydrogel nano-particulate system for the delivery of insulin was used by Kazunari Akiyoshi (Akiyoshi et al., 1998) in the year 1998, the term "nanogel" (NanoGel™) was introduced by Vinogradov (Vinogradov et al., 1999) and Lemieux (Lemieux et al., 2000) to define the networks of a polyion and a nonionic polymer for delivery of polynucleotides crosslinked polyethyleneimine (PEI) and polyethyleneglycol (PEG).

15.2 PROPERTIES OF NANOGELS

a) Solubility: In addition to the presence of hydrophobic pockets formed by crosslinks, nanosize of the nanogels render them with high surface area to volume ratio enhancing the solubility of hydrophobic drugs (Adamoa et al., 2014).

b) Size and shape: Bottom-up approach generally yield spherical nanogels, while in order to obtain nanogels of other shapes Top-down approach can be used. Authors have clearly defined the size ranges of nanogels as < 100 nm (Sasaki et al., 2010), < 200 nm (Yallapu et al., 2011) and 1–1000 nm (Akiyama et al., 2007).

c) Incorporation of guest molecules: Nanogels can incorporate a plethora of guest molecules with diverse chemical and functional properties like SiRNA (Blackburn et al., 2009), nasal vaccines (Fukuyama et al., 2015), nucleic acid (Li et al., 2015), anti-cancer drugs (Soni et al., 2016a) and even drugs for autoimmune diseases (Neamtu et al., 2017).

d) Swelling: Swelling behavior of a non-ionic gel is described by two divergent free-energy forces, firstly swelling is caused by the interaction of the polymer with solvent and secondly the elasticity of the network which opposes the swelling. Whereas, in the ionic gel along with the above two forces, ionization of groups also contributes in swelling (Peppas et al., 1993). Swelling of the polymer can be controlled by structural features, such as the chemistry of the polymer matrix, the degree of crosslinking, charge concentration in polyelectrolyte gels and environmental variables (Kabanov et al., 2009). As a thumb rule, an increase in the number of crosslinks decreases the swelling ratio of hydrophilic polymer (Bontha et al., 2006, McAllister et al., 2002). Swelling of polyelectrolytes decreases with an increase in ionic strength due to the influence of the crosslinker (Kabanov et al., 2009).

e) Colloidal stability: The hamker constant of nanogel and the surrounding solvents is more or less same, therefore there is no influence of driving forces for the gel nanoparticles cohesion (Vincent 2006, Vinogradov et al., 2002).

f) Stimuli sensitivity: Stimuli-sensitive nanogels possess the ability to respond to external stimuli and show a change in the physicochemical properties (swelling, permeability, viscosity, hydrophobicity/hydrophilicity) facilitating the drug release (Zha et al., 2011).

g) Targeting: Both active and passive targeting can be achieved by nanogels. Active targeting of nanogels is accomplished by surface functionalization with ligands to enhance the site-specific delivery. Increase in circulation time of nanogel by PEGylation or manipulation of surface properties increases the Enhanced Permeability and Retention (EPR) which in turn increases the passive targeting (Chacko et al., 2012, Eckmann et al., 2014).

h) Biocompatibility: Nanogels are made up of natural or synthetic polymers which degrade to nontoxic biodegradable products. Moreover, as they absorb a large quantity of water/biological fluids and do not provoke any immune response, they serve to be biocompatible (An et al., 2011).

i) Extended circulation time: Nanogels are soft materials showing excessive swelling that are too large to be eliminated by renal clearance. Hendrickson (Hendrickson et al., 2010) demonstrated the ability of nanogels to get squeezed through the pores of 10 times smaller size. Zhang (Zhang et al., 2012) prepared nanogels with varying flexibility whose *in vivo* study revealed that softer nanogels pass through the splenic filtration with lesser splenic accumulation in comparison to the stiffer nanogels.

15.3 CLASSIFICATION OF NANOGELS

Various authors have classified nanogels into different classes as shown in Figure 15.1

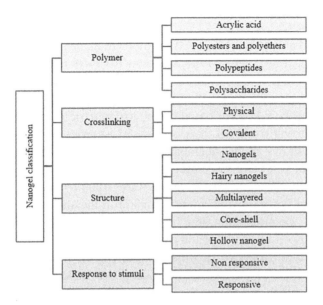

FIGURE 15.1 Classification of polymers.

15.3.1 *BASED ON POLYMER*

15.3.1.1 *POLYACRYLATES*

Polyacrylates as shown in Figure 15.2, are the derivatives of acrylic acid (CH_2=CHCOOH), formed by the modifications of the vinyl and/or carboxyl hydrogen. These polymers lose the proton and acquire positive charge rendering in polyelectrolytic PAA sensitive to temperature, pH, and charge. The polymers show swelling due to the exchange of counterions with positive ions from the external environment (Argentiere et al., 2009, Kim et al., 2003, Xiong et al., 2011).

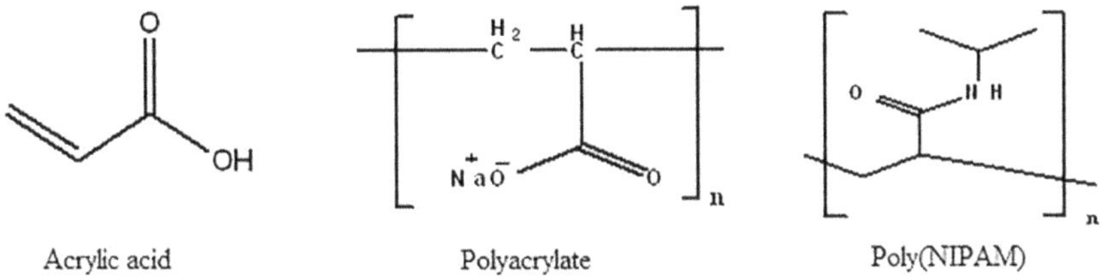

Acrylic acid   Polyacrylate   Poly(NIPAM)

FIGURE 15.2 Acrylic acid polymers.

15.3.1.2   POLYETHERS AND POLYESTERS

Polyethers and polyesters (Fig. 15.3) are biodegradable polymers formed by monomers linked by ether (R-O-R) and ester (RCOOR) bonds respectively. Polyglycolic acid (PGA), Polylactic acid (PLA) and their copolymers- poly (lactic acid-co-glycolic acid) (PLGA), are the commonly used polyesters. PLA may be in L or D form based on the asymmetric α-carbon. Since the naturally occurring form is L it is considered as more biodegradable than D form. PLA shows greater degradation rate in comparison to PGA. Poly-ethylene glycol is a commonly used polyether polymer, owing to its high biocompatibility (98% excreted by man), water solubility, high stability, and no toxicity and is regarded as a gold standard in the area of drug delivery. PEG increases the circulation time by extending the endocytosis, phagocy-tosis, liver uptake and clearance and other adsorptive processes. Poloxamers are blocked co-polymers of two hydrophilic chains of polyethyleneoxide and one hydrophobic polypropylene oxide. The trade name is Pluronic which is available in various grades based on the varying length of polymer blocks (Devi et al., 2013, Gupta et al., 2012, Maurus et al., 2004, Uhrich et al., 1999).

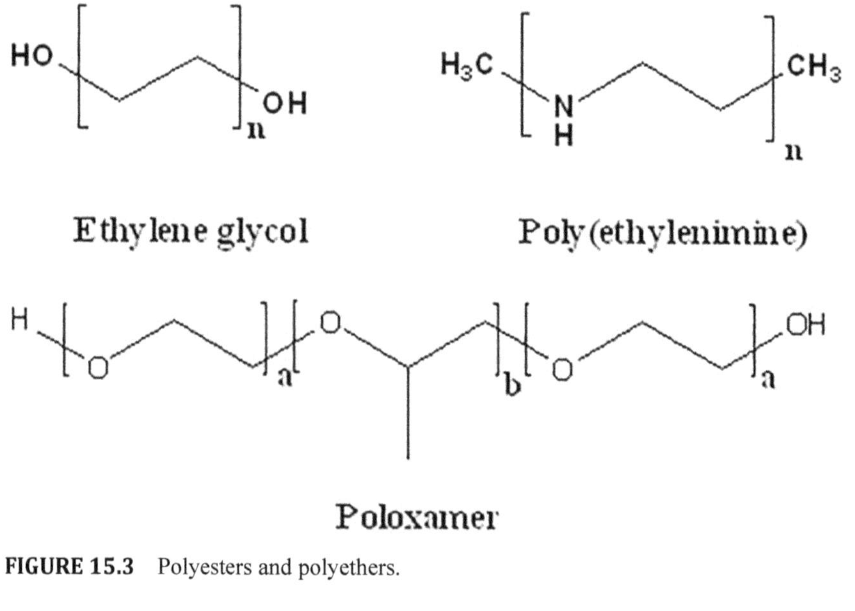

Ethylene glycol   Poly (ethylenimine)

Poloxamer

FIGURE 15.3   Polyesters and polyethers.

15.3.1.3 POLYPEPTIDES

Polypeptides, as given in Figure 15.4, are biodegradable polymers linked by amide bonds hence are also known as polyamides. Their ability to undergo manipulation of amino acid sequence during synthesis makes them ideal polymers for delivering low molecular weight compounds. Most of these polymers are homopolymers such as poly (glutamic acid), poly (aspartic acid) and poly (L-lysine). Achieving controlled release by these polymers is challenging since they are stable to hydrolysis and the degradation is enzyme dependent which in turn is dependent on the hydrophilic nature of amino acids. Biodegradation can be altered by incorporation of benzyl, hydroxyl, or methyl groups during polymerization. Difficulty in peptide sequence synthesis, cost, and mild antigenic nature makes their use uncertain. The most commonly used polymer is poly (glutamic acid) which is water-soluble and exists in D/L-optically active forms (Anderson et al., 1974, Kamaly et al., 2016, Marck et al., 1977, Martin et al., 1971, Shih et al., 2004, Uhrich et al., 1999).

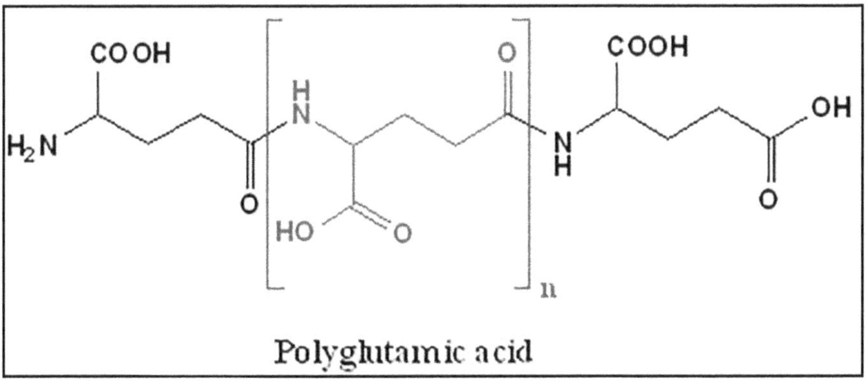

FIGURE 15.4 Polypeptides.

15.3.1.4 POLYSACCHARIDES

Polysaccharides are polymers with glycosidic bond or linkage of long-chain monosaccharide units. If the monosaccharide units are same then they are termed as Homo-polymers (e.g., glycogen, starch, cellulose, pullulan, pectin, etc.) and if they are different they are called Hetero-polymers (e.g., heparin, chitosan, hyaluronic acid, chondroitin sulfate, keratin sulfate, and dermatan sulfate) (Debele et al., 2016, Drogoz et al., 2007, Miller et al.,

2014, Posocco et al., 2015). Polysaccharides may be obtained from various plants, animal, microbial and algae sources (Leung et al., 2006). Polysaccharides possess multiple reactive groups (hydroxyl, amino, and carboxylic acid) which can be derived chemically as well as biochemically to produce derivatives exhibiting structural and functional diversity. (Liu et al., 2008, Zhang et al., 2013). These polymers are abundantly available in nature and are found to be biocompatible, biodegradable, non-toxic, water soluble and bioactive (Abed et al., 2011, Arnfast et al., 2014, Jian et al., 2012, Rodrigues et al., 2015). Figure 15.5 shows some examples of homo and heteropolysaccharides used in the synthesis of nanogels.

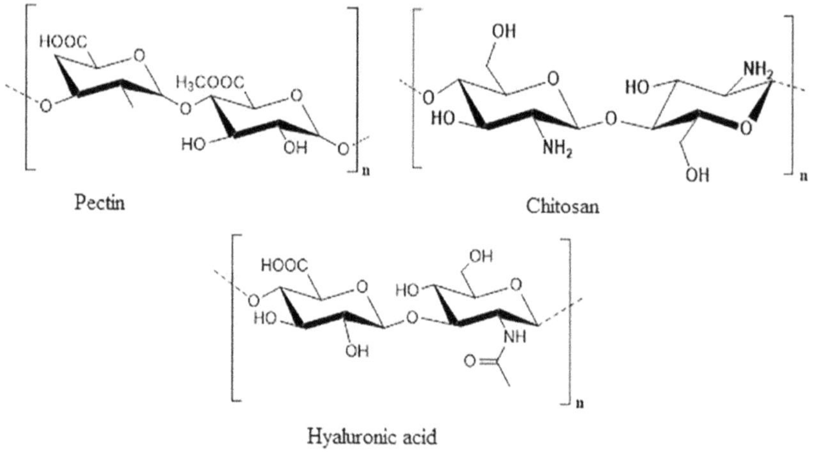

Pectin

Chitosan

Hyaluronic acid

FIGURE 15.5 Polysaccharides.

15.3.2 BASED ON STRUCTURE

Based on the structure, nanogels can be classified into six major types as shown in Table 15.2.

TABLE 15.2 Classification of Nanogels Based on Structure.

Type of nanogel	Structure	Polymer	References
Simple nanogel	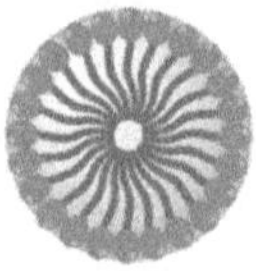	Oligoethyleneglycol (OEG) and Pyridyldisulfide (PDS)	Asadi et al., 2016

TABLE 15.2 *(Continued)*

Hairy nanogel		Poly(3-dimethyl-(methacryloyloxyethyl) ammonium propane-sulfonate) crosslinked with N,N′-methylenebis (acrylamide)	Fu et al., 2017
Functionalized nanogels		Oligoethylene glycol (OEG) units (29%) and Pyridyldisulfide (PDS) moieties (71%)	Ryu et al., 2012
Multi-layered nanogels		Poly(N-isopropylacryl-amide) (PNIPAm) and Poly(N-isopropylmethar-cylamide) (PNIPMAm)	Schmid et al., 2016
Core crosslinked nanogels		Poly(ethylene oxide)- b -poly(methacrylic acid)	Kim et al., 2009
Hollow nanogel		Acrylic acid (AAc) and 2 methacryl oyl ethyl acry-late (MEA) units as the backbone and poly(N-isopropyl acrylamide) (PNIPAAm) alone or both PNIPAAm and monomethoxy poly(ethylene glycol) (mPEG) as grafts	Chiang et al., 2012

15.3.3 BASED ON CROSSLINKING

15.3.3.1 PHYSICAL CROSSLINKING

These crosslinking are self-assembled by physical interactions that does not require any crosslinking agents (Chiang et al., 2012). Physical crosslinks have advantages of reversibility, absence of chemical reactions as well as potentially harmful bioactive agents or cells. On the other hand, their stability is hindered owing to weak linkages (Teixeira et al., 2012).

15.3.3.2   CHEMICAL CROSSLINKING

Chemical crosslinking involves covalent bonds which can be done in a highly precise manner to produce stable and rigid linkages. It enables to introduce components with the ability to change in response to the environment, thus release can be controlled (Argentiere et al., 2011, Kim et al., 2011).

15.3.4   BASED ON RESPONSE TO STIMULI

Based on the response to stimuli nanogels are classified as responsive and non-responsive. Responsive nanogels show a change in behavior and trigger the drug release in response to the external stimuli, whereas the non-responsive nanogels do not.

15.3.4.1   STIMULI NON-RESPONSIVE

Nanogels which show drug release irrespective of the external stimuli are known as stimuli non-responsive nanogels (Li et al., 2008).

15.3.4.2   STIMULI-RESPONSIVE GELS

These nanogels have the capability of responding to external stimuli and show a change in the physicochemical properties which can be brought back by removing or reversing the stimuli (Wu et al., 2016).

15.3.4.2.1   Physical Stimuli

15.3.4.2.1.1   Temperature Sensitive

Temperature is a common factor that triggers the drug release, thus temperature sensitive polymers show characteristic temperature called volume phase transition temperature (VPTT) which causes changes in the polymer and thus drug release (Zha et al., 2011). The VPTT of polymers is of two types.

15.3.4.2.1.2   Volume Phase Transition of Stimuli-Sensitive Polymers

Stimuli-sensitive polymers exhibit volume phase transition in response to the external stimuli. These transitions fall into one of the four bio-relevant forces as suggested by Ilmain (Ilmain et al., 1991) like Vander Waal interaction, Hydrophobic interaction, Hydrogen bonding, and Attractive interaction.

Lower critical solution temperature (LCST) is the temperature below which the polymer is soluble and above which the polymer shows phase transition from soluble state (random coil form) to an insoluble state (collapsed or globule form) (Chaterji et al., 2007). There exists an inverse relationship between the temperature and the solubility hence these gels are called negative temperature responsive nanogels (Zha et al., 2011). Conversely, the temperature above which the polymers solubilize is called as Upper critical solution temperature (UCST). These polymers are called positive temperature sensitive owing to their direct relation between the temperature and solubility (Bajpai et al., 2008). Polymers exhibiting LCST and UCST behavior are quoted in Table 15.3.

TABLE 15.3   Various Polymers and Their Transition Temperatures.

Polymer	Temperature range	Reference
LCST		
Poly(N isopropyl acrylamide) (NIPAAm)	32°C	Yoo et al., 2000
Poly(N,N-diethyl acrylamide)	32–34°C	Idziak et al., 1999
Poly(dimethylamino ethylmethacrylate)	50°C	Cho et al., 1997
Poly(N-(L)- 1-hydroxymethyl propylmethacrylamide)	30°C	Aoki et al., 2001
Poly(methyl vinyl ether)	37°C	Arndt et al., 2001
Poly(N-vinylcaprolactam)	30–50°C	Van Durme et al., 2004
PEO-b-PPO (Pluronics, poloxamers, tetronics)	20–85°C	Gandhi et al., 2015
UCST		
Poly(acrylic acid) PAA and polyacrylamide (PAAm)	25°C	(Aoki et al., 1994)

15.3.4.2.1.3   Light Sensitive

Light-responsive nanogels can be prepared by two approaches:

a. Incorporation of light-sensitive polymers possessing photoactive groups (azobenzene, spirobenzopyran, triphenylmethane, or cinnamonyl). Irradiation causes the change in size, shape, ionization, and polarity of these functional groups. Azobenzene shows the transition from *cis* to *trans* and *trans* to *cis* upon irradiation at 300–380 nm and visible region respectively. Limited penetration depth (up to 10 mm) and need of ultra violet or shorter wavelength light as an irradiation source are the major drawbacks of this system and so these polymers are rarely used for drug delivery (Alvarez-Lorenzo et al., 2009, Li et al., 2009a, Mura et al., 2013).

b. Hybrid systems which contain plasmonic materials (Au or Ag metal nanoparticles) and thermosensitive polymer. These metal particles have the capability to absorb the Near-infrared (NIR) light (700–1000 nm) and transform the energy into heat causing a phase transition in the polymer (Kang et al., 2011, Lajunen et al., 2015, Zha et al., 2011).

Though conceptually light sensitive nanogels sound good, their biodegradability and safety are questionable (Mura et al., 2013, You et al., 2012).

15.3.4.2.2   Chemical Stimuli

15.3.4.2.2.1   pH-Sensitive

pH is another property which can be employed as a stimulus for the release of the drug since the pH of various body organs and compartments possess difference in pH values (Table 15.4). pH-sensitive nanogels contain cross-linked electrolytes of weakly acidic (carboxylic) or weakly basic (Amino) groups which dissociate into ions upon contact with water or solvent. Ionization causes the gel to act as a semi-permeable barrier for counter ions thereby increasing or decreasing the osmotic pressure which further causes the polymer to swell or de-swell. The ionization is dependent upon the solution pH called as critical pH value (pH_c). pH_c should be below the pK_b value for cationic nanogels and above the pK_a for anionic nanogels (i.e., Anionic gels show swelling with an increase in pH and the opposite with cationic gels). As shown in Table 15.5 polyelectrolytic polymers may accept or donate protons depending on the pH (Peppas et al., 1993, Schmaljohann 2006).

TABLE 15.4 pH of Various Tissue/Cellular Compartment.

Region	pH	Reference
Stomach	1.0–3.0	
Colon	7.0–7.5	Schmaljohann, 2006
Tumor	6.5–7.2	
Lysosomes	4.5–5	
Endosomes	5.5–6.0	Karimi et al., 2016
Cytosol	7.4	
Chronic wounds	5.4–7.4	Zha et al., 2011

TABLE 15.5 Effect of pH on Polymeric Electrolyte.

Type of polymeric electrolyte	Acidic pH	Basic pH	pH_c relation
Weakly acidic	Proton acceptors	Proton donors	$pH > pK_a$
Weakly basic	Proton donors	Proton acceptors	$pH < pK_b$

15.3.4.2.2.2 Biomolecule Sensitive

Nanogels which mimic the body's closed feedback loop mechanism of recognizing specific molecules and then releasing the biological agent is called biomolecule sensitive nanogels (Chaterji et al., 2007, Zha et al., 2011). These are further classified into three types.

Redox Type: Drug release is based on the redox reaction occurring in presence of biomolecule. Glutathione (GSH) sensitive nanogels are an example of this system; disulfide bond linked nanogels are sensitive to GSH and hence show drug release in its presence. Giorgia Adamo (Adamoa et al., 2014) developed carboxyl-functionalized nanogel (poly N-vinyl pyrrolidone (PVP) variant in which the drug (doxorubicin) was linked through a linker containing a cleavable disulfide bond that gets reduced in presence of GSH.

Self-Regulated: Self-regulated nanogels release the drug in response to the levels of biomolecules, whereas no drug release is observed under normal physiological condition. Glucose-sensitive nanogels are classical examples of self-regulated nanogels. Most of the glucose sensitive nanogels possess phenylboronic acid which shows a transition from hydrophobic to the hydrophilic state upon complexing with glucose which requires glucose levels of about 50 mg mL^{-1} way beyond the physiological conditions (1–3 mg mL^{-1}). This phenomenon was employed for insulin release from poly (ethylene glycol)-b-poly (acrylic acid-co-acrylamidophenylboronic acid) (Hoare et al., 2007, Mura et al., 2013, Zhang et al., 2006). Sometimes Glucose oxidase (GOx) based systems are also employed for glucose sensed insulin release (Rehor et al., 2005). GOx attached to the polymer senses the glucose and causes drug release by oxidation of disulfide links and/or pH sensitive release by decreasing the local pH by the production of hydrogen peroxide and gluconic acid (Motornov et al., 2008, Zhao et al., 2011).

Enzyme-Sensitive: Enzymes like proteases, phospholipases, and glycosidases are mostly used for drug delivery in cancer and inflammation. Xiong et al. (2012a) prepared a triple layer nanogels which shows drug release in presence of lipase secreted by microbes.

Apart from the above-mentioned drug release can also be stimulated by electric, magnetic, ultrasound stimulus.

Multi responsive nanogels are the newer class of stimuli-responsive nanogels that possess the ability to release the drug response to more than one stimuli (Xiong et al., 2011).

15.4   SYNTHESIS

Approaches for the synthesis of nanogel can be divided as

15.4.1   *TOP DOWN*

This approach generates nanogels from large particles by means of physical, chemical and mechanical forces. The imperfection of particle surface and more suitability toward micro ranged particle production are the undesirable problems of the top-down approach. (Amamoto et al., 2011, Kim et al., 2007, Nie et al., 2005).

15.4.1.1   *PHOTOLITHOGRAPHIC TECHNIQUES*

This technique is used to fabricate 3D nanogel/microgel rings for drug delivery and tissue engineering (Sultana et al., 2013). The development of techniques for surface treatment of stamps or new materials for the replica molds is required in order to permit the release of molded gel (Oh et al., 2008). Five steps involved in this method are:

i) Pre-baked photo-resist coated water is exposed to UV cross-linkable polymer possessing low surface energy, as a substrate.

ii) The polymer is molded into patterns on the silicon wafer by pressing the quartz template on the polymer and then exposing it to intense UV light.

iii) The quartz template is removed so as to uncover the thin residual interconnecting film layer.

iv) The residual thin layer is removed due to its oxidation by plasma that contains oxygen.

v) Fabricated particles are collected by dissolving the substrate in water or buffer (Akiyoshi et al., 1999, Glangchai et al., 2008).

In order to mold, release and stack gels into 3D structures, Poly (dimethylsiloxane) (PDMS) stamps are generally used (Oh et al., 2008). To fabricate submicron-sized microgels with a control over particle size, shape and composition, a Top-down method called (Particle Replication Nonwetting Templates) "PRINT" was developed (Rolland et al., 2005) and incorporation of DNA, proteins and small molecules into PEG-based microgels by a simple encapsulation technique was established by PRINT, which showed its compatibility with biomolecules. Also, various monodisperse microgels of polyethylene glycol diacrylate (PEGDA), triacrylate resin, PLA and polypyrrole were prepared using PRINT, whose sizes ranged from 200 nm to micron-scale in diameter possessing various shapes like bar, conical, arrow and trapezoidal (Oh et al., 2008).

15.4.1.2 MICROMOLDING TECHNIQUE

This method is similar to the photolithographic technique, without the requirement of expensive lithographic equipment and clean room facilities. In this procedure, cells are suspended in a hydrogel precursor solution containing either methacrylated hyaluronic acid (MeHA) or PEGDA or a photoinitiator in water. Followed by deposition of the mixture onto the plasma-cleaned hydrophilic PDMS patterns. The UV light is exposed to it and crosslinking is performed. (i.e., photocrosslinking). The cell-embedded microgels are then removed, hydrated, harvested and also molded into a variety of shapes like square prisms, disks, and strings (Sultana et al., 2013).

15.4.2   BOTTOM UP

This methodology is applied for the preparation of nanogels from atomic or molecular species (precursors) which are chemically or physically cross-linked allowing the precursor particles to grow in size (Amamoto et al., 2011, Neamtu et al., 2017). This method is further classified as;

15.4.2.1   SYNTHESIS BASED ON THE STARTING MATERIAL

15.4.2.1.1   Synthesis Using Polymer Precursors

The polymer precursors may be mono, di or triblock copolymers. Polymer precursors have the capability to self-assemble to form nanogels or can be modified with groups which can be subsequently used to form physical or chemical crosslinks. To formulate nanogels with high stability, synthesis using preformed polymers is carried out in dilute aqueous solutions, preferably in combination with ultrasonication in organic solvents or with the aid of surfactants in microemulsion based heterogeneous systems (An et al., 2011, Chacko et al., 2012, Zhang et al., 2015).

15.4.2.1.2   Synthesis Using Monomer

This method has the ability to produce nanogel during the polymerization process (*in situ* formation) referred as one-pot protocol and in some cases, it requires a template for size control of nanogels (An et al., 2011). Polymer production by monomer is efficient than the production from precursors (Zhang et al., 2016).

15.4.2.2   BASED ON TYPE OF CROSSLINKING

15.4.2.2.1   Physical/Supramolecular Crosslinking

Various nanogels were produced by physical self- assembly of polymers via noncovalent bonds, (hydrogen, ionic) and co-ordinate covalent bond (Khoee et al., 2016). Nanogels formed by this approach have a wide application in the drug delivery systems as they possess specific characteristics of sol-gel reversibility (upon external stimulus) (Amamoto et al., 2011). This method does not make the use of any crosslinking agents thereby preventing the unwanted interactions with the bioactive substances during the encapsulation of biomacromolecules. It is carried out under mild conditions, mostly water, which also prevents unwanted effects. However, poor stability (i.e., harsh conditions) and low mechanical strength make these crosslinks less applicable compared to covalently crosslinked gels (Teixeira et al., 2012).

15.4.2.2.1.1 Hydrogen Bonding

It is a standard method for the formation of physical nanogels formed between the functional groups which include -OH, -NH$_2$, etc. (Zhang et al., 2015). Hydrogen bonding depends on the types of solvents and pH of the solution resulting in low stability. For example, nanogels of PMMA with dendritic benzamide groups were prepared using RAFT polymerization, which showed that it has been formed by hydrogen bonding in toluene (Seo et al., 2008). Supramolecular hydrogen bonding of hyperbranched polycarbonate has also been utilized in the synthesis of pH-sensitive nanohydrogels (Jia et al., 2012). Resulting in drug release in an acidic environment (pH 6.6) and the nanogels expanded to a high degree due to charge repulsion (Zhang et al., 2015). Figure 15.6 shows hydrogen bonding between two monomers used to synthesize nanogel.

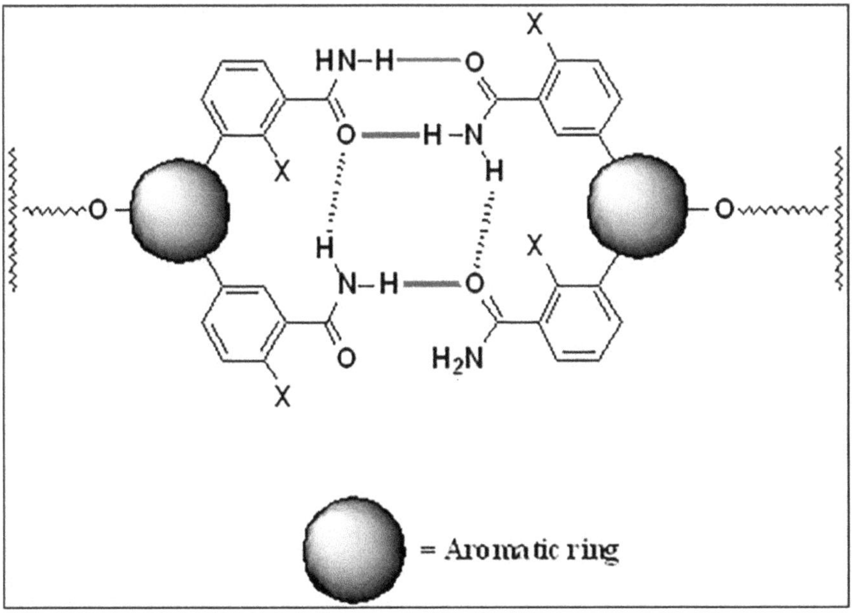

FIGURE 15.6 Hydrogen bonding.

15.4.2.2.1.2 Ionic Bonding and Co-Ordination Bonding

Electrostatic interaction between positively charged surfactant micelles and negative polypeptides result in the formation of self-assembled supramolecular nanogels as depicted in Figure 15.7 (Kim et al., 2013). Above critical micellar concentration (CMC), small positive surfactants self-assemble into micelles which interact with negatively charged gelatin B to form core–shell-like aggregates. Diffusion of hydrophobic guests is avoided by stable hydrophobic pockets of core-shell-like nanogel aggregates (Zhang et al., 2015).

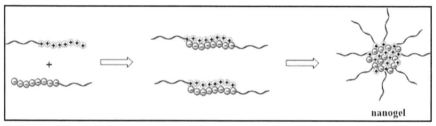

FIGURE 15.7   Ionic bonding and co-ordination bonding.

15.4.2.2.1.3 Hydrophobic Interactions

These are the intermolecular interactions in which the hydrophobic molecules form assemblies in an aqueous solution by preventing the entropy loss of water (Amamoto et al., 2011). The most commonly used hydrophobic unit to induce the self-assembly of amphiphilic copolymer to form a nanogel is cholesterol (Fujioka-Kobayashi et al., 2012, Hasegawa et al., 2009, Hasegawa et al., 2005, Hirakura et al., 2010, Noh et al., 2012, Sasaki et al., 2011, Sekine et al., 2012, Shimoda et al., 2011). Cholesterol-bearing pullulan (CHP), consisting of hydrophilic pullulan in the main chain and hydrophobic cholesterol in the side chain, can form nanogels in aqueous media (Akiyoshi et al., 1993, Akiyoshi et al., 1997). The quasi-crosslinking points were formed due to the hydrophobic interaction among the cholesterol units and the formed nanogels were so stable that they could be measured by GPC (Akiyoshi et al., 1993). Moreover, CHP can also form complexes with proteins, which can be destroyed at their crosslinking points by β-cyclodextrin (Inomoto et al., 2009, Nishikawa et al., 1996). These materials have found great applicability in both protein and drug delivery systems (Kobayashi et al., 2009).

15.4.2.2.2 Covalent Crosslinking

15.4.2.2.2.1 Polymerization

Radical Polymerization: This is a widely used method owing to its ease of manipulation, efficiency, tolerance of functionality, and adaptability to water-based heterogeneous systems (An et al., 2011). Radical polymerization is further classified into free radical and controlled radical polymerization.

Free Radical Polymerization: These are two-phased systems in which the monomer or the final polymer exist as a fine dispersion in liquid. In this method monomer, initiator and additives are added to the medium for emulsification or stabilization of the monomer droplets or final polymer. Based on the kinetics of polymerization, initial polymerization state, mechanism of particle formation and size of the polymer particles formed by FRP process can be further classified as dispersion, precipitation, mini-emulsion (oil in water), inverse (mini) emulsion (water in oil), and polymerization (Arshady 1992, Kim et al., 2005, Ulijn et al., 2007). The emulsion polymerization requires surfactant, removal of which is costly and not possible all the time. Hence surfactant free polymerization methods such as suspension polymerization and precipitation are developed (Amamoto et al., 2011).

Controlled/Living Radical Polymerization: CRP produces copolymers with narrow weight distribution (M_w/M_n or PDI <1.5). Stable free radical polymerization (SFRP), atom transfer radical polymerization (ATRP), and reversible addition fragmentation chain transfer (RAFT) were successful among the various methods developed (An et al., 2011, Khoee et al., 2016, Oh et al., 2008).

15.4.2.2.2.2 Click Chemistry

CuAAC (Copper-Catalyzed Azide-Alkyne Huisgen Cyclo Addition): It is a fast, versatile, region-specific and a highly efficient method/approach for nanogel synthesis which is carried out under mild reaction conditions affording to quantitative yields and functional group tolerance. This modular approach requires an alkyne and an azide in presence of a Copper catalyst and sodium ascorbate (reducing agent) as given in Figure 15.8 (Khoee et al., 2016, Kolb et al., 2001).

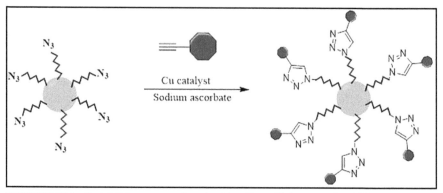

FIGURE 15.8   CuAAC (copper-catalyzed azide–alkyne Huisgen cyclo addition).

Copper-Free Click Chemistry: Owing to the disadvantages of CuAAC such as removal of copper catalyst and in vivo cytotoxicity, Copper-free click chemistry was introduced (i.e., SPAAC- Strain promoted azide-alkyne cycloaddition and Thiol-ene click chemistry)

a)   *SPAAC (Strain promoted azide-alkyne cycloaddition)*

As depicted in Figure 15.9, the cyclooctyne molecules were found to react quickly with azides without a copper catalyst as a result of ring strain and electron-withdrawing fluorine substituents offering to the formation of biomedically applicable nanogels (Laughlin et al., 2008, Prescher et al., 2004). SPAAC is thus a copper-free and non-toxic approach for obtaining nanogels.

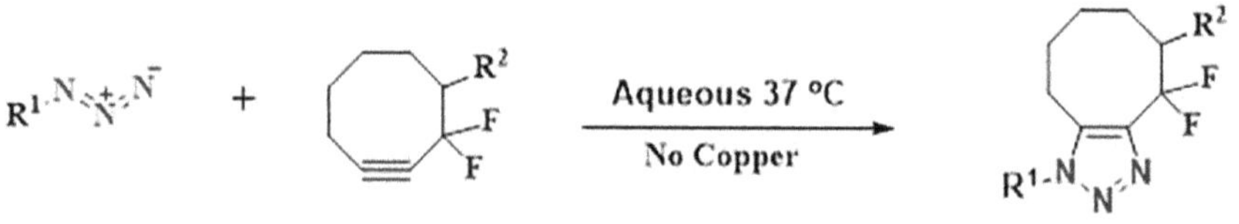

$$R^1-N^+{=}N^- \quad + \quad \bigcirc \begin{matrix} R^2 \\ F \\ F \end{matrix} \quad \xrightarrow[\text{No Copper}]{\text{Aqueous 37 °C}} \quad \bigcirc \begin{matrix} R^2 \\ F \\ F \\ R^1-N_{N{\nwarrow}N} \end{matrix}$$

FIGURE 15.9   Strain promoted azide-alkyne cycloaddition.

b)   *Thiol-ene click chemistry*

It is yet another high yielding, efficient and functional group tolerant, Copper-free, photoinduced radical mediated addition in which a thiol is added to a double bond (alkene) (Campos et al., 2008, Hoyle et al., 2004). This copper-free click chemistry-based approache to aid in the cell encapsulation, drug delivery, dye separation, etc.

15.4.2.2.2.3   Schiff Base Reaction

It is a catalyst-free approach used to generate biocompatible gels (due to its mild reaction conditions) that employ aldehydes and amines or hydrazide-containing compounds by the formation of an imine covalent bond as shown in Figure 15.10. Stability of these imine bonds under physiological conditions makes the nanogels promising candidates for intracellular protein delivery.

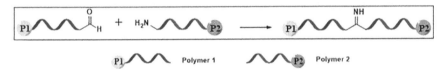

FIGURE 15.10   Schiff base reaction.

15.4.2.2.2.4   Thiol-Disulfide Exchange Reaction Based Crosslinking or Disulfide Based Crosslinking

Formation of nanogels in aqueous solution was achieved by crosslinking amphiphilic copolymers. Figure 15.11 depicts a hydrophilic unit of polyethylene glycol and a hydrophobic unit of Pyridyl disulfide (PDS) along with a cross-linkable unit which was used under mild reaction conditions, wherein the thiol-disulfide exchange occurred (i.e., the deprotonated free thiol displaces one sulfur of the disulfide bond in the oxidized species) (Ryu et al., 2010a, Ryu et al., 2010b, Sevier et al., 2002). Owing to the stability of the disulfide bonds, disulfide-thiol chemistry has benefitted in conventional polymer synthesis and also in the synthesis of recyclable crosslinking of micelles (Khoee et al., 2016, Klibanov et al., 1990, Ryu et al., 2010a). It was found that a catalytic amount of dithiothreitol (DTT) was used in order to initiate the thiol-disulfide exchange reaction thereby affording cross-linked nanogels (Li et al., 2009b).

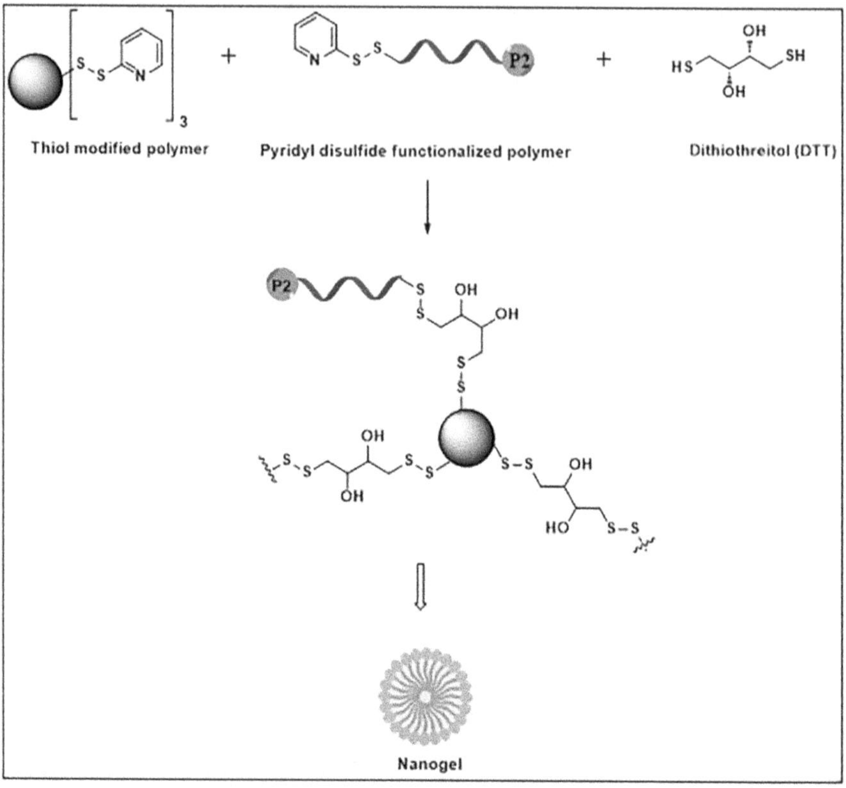

FIGURE 15.11   Thiol-disulfide exchange reaction based crosslinking or disulfide-based crosslinking.

15.4.2.2.2.5   Photo-Induced Crosslinking

It is a highly efficient and clean technique, as it does not involve the utilization of crosslinking agents and also no by-products are formed at the end of the reaction. This method is utilized to stabilize polymer assemblies functionalized with polymerizable and dimerizable units (Christensen et al., 2006, Kakizawa et al., 1999, Piogé et al., 2011). A photo-initiator needs to be incorporated in order to initiate the reaction. This might cause cyto-toxicity, hence the choice of photo-initiator becomes necessary for the synthesis of bio-compatible nanogels (Williams et al., 2005). Figure 15.12 shows an example of nanogel formation by photo-induced crosslinking.

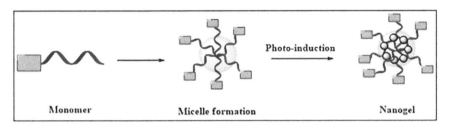

FIGURE 15.12   Photo-induced crosslinking.

15.4.2.2.2.6   Amide-Based Crosslinking

Preparation of amide-based nanogels is driven by the amine group's high reactivity toward carboxylic acids, activated esters, isocyanates, iodides, etc, (Chacko et al., 2012). No additives are required as such in this method. This approach has applicability in drug delivery and protein delivery. This approach was utilized by incorporating pentafluorophenyl (PFP)-activated hydrophobic esters along with diamine crosslinker to synthesize hydrophilic nanogels. Here in PFP was used as a crosslinker and also for functionalizing the nanogels and thereby providing lipophilic domains within the nanogel for encapsulation of hydrophobic drugs (Chacko et al., 2012, Khoee et al., 2016). Figure 15.13 gives an example of amide-based crosslinking in order to synthesize nanogel.

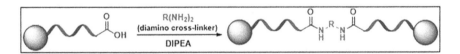

FIGURE 15.13   Amide-based crosslinking.

15.4.2.2.2.7   Boronic Acid-Diol Complexation

In this approach, the boronic acid and 1,2 or 1,3-diols undergo complexation in order to fabricate nanogels and core-crosslinked micelles (Fig. 15.14). This approach was utilized in cell encapsulation wherein the encapsulated cells remained in the viable state. Nowadays, boronate esters have also been utilized for nanogel synthesis (Chen et al., 2013, He et al., 2011, Jay et al., 2009, Mahalingam et al., 2011, Piest et al., 2011, Roberts et al., 2007).

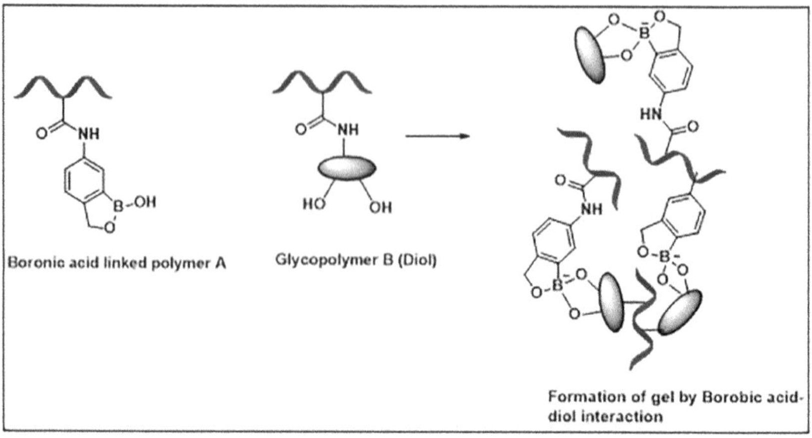

FIGURE 15.14   Boronic acid-diol complexation.

15.4.2.2.2.8   *Enzyme Catalysed Crosslinking*

This is an innovative approach with greater efficiency in comparison to other cross-linking methods owing to the shorter reaction time, higher biocompatibility and mild reaction conditions. Horseradish peroxidase (HRP) has been utilized as an enzyme-catalyzed approach to synthesize redox-sensitive disulfide crosslinked nanogels (without using hydrogen peroxide), which thus keeps the viability of the cells in an intact state. This method of cross-linking provides good control over the ratio of reactants, the composition of macromere and enzyme concentration (DeVolder et al., 2013, Menzies et al., 2013, Park et al., 2012).

Hence by utilizing the above mentioned covalent crosslinking approaches, the stability of nanogels can be maintained even in a harsh and complex condition, leakage of the encapsulated payload of drugs can be prevented and the therapeutic efficacy can be enhanced. By using the traditional covalent crosslinking techniques, sometimes unwanted toxic effects can emerge due to the usage of crosslinking agents, which can cause damage to the entrapped substances including proteins and cells. This leads to the necessity of exploration of more and more biocompatible and biorthogonal reactions.

15.4.3   *CLEAVABLE CROSSLINKERS AND THEIR MECHANISM*

Nanogels after reaching the target site needs to be cleaved in order to release the drug and impart the therapeutic action. Thus, labile linkers are introduced into the nanogels scaffolds. Under the influence of certain stimuli, the covalently crosslinked nanogels are broken down. The covalently cleavable links are of following type: Acid-cleavable, pH-cleavable, photo-cleavable, redox-cleavable and enzyme-cleavable. Table 15.6 provides the structure and conditions for cleavage of various above-mentioned linkers.

TABLE 15.6   Various Cleavable Crosslinkers and Their Condition for Cleavage.

Crosslinkers	Structure	Conditions for cleavage
Acid cleavable		
Acetal linker		pH 5
Ketal linker		pH 5.5–5.8
Hydrazone linker		Hydrolytically cleavable, particularly at low pH values, i.e., 10 times faster at pH 5.5 than at pH 7.4. Degradation rate depends on H$^+$ concentration of acid
Crosslinkers	Structure	Conditions for cleavage
Imine linker		Stable under physiological conditions and dissociated at acidic pH
Vinyl ether		Hydrolysis at pH < 5
Carboxylate ester		Hydrolysed under physiological conditions
Boronate ester		Nanogel swells at pH 5 after 24 h incubation and is completely degraded at pH 4
Carbonate		Hydrolyzed at pH 7.4
Photo cleavable		

TABLE 15.6   *(Continued)*

Crosslinkers	Structure	Conditions for cleavage
O-Nitrobenzyl ester		UV 315–390 nm
Biscoumarin		UV < 260 nm which led to selling of nanogels
Enzyme cleavable		
Phosphoester		Phosphatase Phospholipase (bacterial enzyme)
Redox cleavable		
Disulfide		GSH DTT

15.4.3.1   ACID CLEAVABLE

The nanogels containing acid-cleavable crosslinks are cleaved under acidic pH such as; tumor or inflammatory cells (0.5–1.0 pH), endosomes (5.5–6.5 pH) or lysosomes (4.5–5.5 pH) possess acidic pH, which triggers the cleavage of the acid-sensitive crosslinks.

15.4.3.2   PHOTO CLEAVABLE

Photochemical reactions are potentially cytocompatible, however, toxicity may be elicited in the mid UV (280–315 nm) and far UV (200–280 nm) ranges. Thus when photochemistry is applied in biomedical areas, it becomes important to use the functional wavelength (DeForest et al., 2011).

15.4.3.3   ENZYME CLEAVABLE

The enzyme concentration is cell and tissue type dependent. Thus, enzymatically degradable nanogels that have been synthesized in biomedical applications (site-specific drug delivery) are cleaved under its influence enabling the local drug release (Kharkar et al., 2013, Wang et al., 2010, Xiong et al., 2012b).

15.4.3.4   REDOX CLEAVABLE

Redox cleavable crosslinks are governed by the redox gradient between the oxidative extracellular medium and reductive intracellular environments (Saito et al., 2003). The cancer tumors are found to have high levels of Glutathione (GSH) as compared to the normal cells (Gupte et al., 2009, Russo et al., 1986) The disulfide bonds are efficiently cleaved under intracellular reductive conditions enabling the rapid release of drug and hence site-specific delivery of the antitumor drug (Qiao et al., 2011, Whittell et al., 2011). Dithiothreitol (DTT) also showed the cleavage of the disulfide bonds leading to the formation of water-soluble polymers which were assumed to promote kidney elimination (Zhang et al., 2015).

15.5   DRUG LOADING

Drug loading is an important aspect of any drug delivery system, higher the drug load ability lower is the nanocarrier administered for medication. Drug loading of 5–25% is considered reasonable. As compared to other conventional dosage forms nanogels hold a key advantage of high "payloads" (up to 50 wt %), which is mainly due to its swelling property that permits the formulation to imbibe an enormous quantity of water (Vinogradov et al., 2004, Yadav et al., 2017).

A wide range of approaches can be employed for loading drugs into the nanogels. These include:

1) Covalent conjugation of biological agents which is achieved either during or following the nanogels synthesis. For instance: Copolymerization of modified enzymes with acrylamide in both inverse microemulsion and dilute aqueous solutions were employed to obtain nanosized hydrogels (Khmelnitsky et al., 1992, Yan et al., 2006).

2) Physical entrapment is referred to as the linkage between hydrophilic chains and hydrophobic regions of the polymer or dissolution of hydrophobic molecules in hydrophilic vehicles (Yadav et al., 2017).

Example: Cholesterol modified pullulan nanogels were incorporated with proteins employing this strategy (Akiyoshi et al., 1999).

3) Passive or diffusion based loading: Dextran lysozyme nanogels have been loaded with silver nanoparticles (Ferrer et al., 2014a, Ferrer et al., 2013) and dexamethasone (Ferrer et al., 2014b) separately by using this strategy.

Composition, molecular weight, the possible interactions between the drug and employed polymer and the different functional groups in each of the polymeric unit are other factors which contribute to high drug loading capacity (Gonçalves et al., 2010, Sultana et al., 2013).

15.6   DRUG RELEASE FROM NANOGELS

The release of biomolecules or small molecules from nanogels depends on the properties of the polymer used in the preparation of the nanogels. Mesh size of the matrix plays a prominent role in controlling the release. The release can either be simple diffusion from the matrix system or through stimuli mediated (also called as 'triggered release'). Basic release mechanisms include: (1) simple diffusion, (2) degradation of the nanogels, (3) changes in surrounding pH, (4) release by displacement through counterions present in the external environment, or (5) triggered release from external energy sources like a magnetic field, light etc. Type of polymer and respective functionalization determine the drug release kinetics. These systems were mainly designed to attain controlled release of the drug. Thus, their main objective is to achieve zero order kinetics. Since nanogels are both kinetically and thermodynamically stable, these systems are expected to exhibit controlled release properties. Unlike micelles which release the drug faster, nanogels release drugs slowly, owing to their superior stability in biological fluids. Apart from the polymer properties the degree of crosslinking also affects the drug release from the nanogels. Since these systems are hydrophilic, lower degree of crosslinking aids rapid release of the drug. Majority of the formulations comprising nanogels employs multiple crosslinking cycles to fine tune the drug release. Basic kinetic models to evaluate the drug release from the nanogels include zero-order kinetic model, Higuchi model, Korsmeyer-Peppas model, Hixon-Crowell model, Weibull model etc. Majority of the systems follow Higuchi and Peppas kinetics as the release is primarily diffusion mediated. Recently a phenomenon termed as 'shape memory effect' (i.e., polymers which exhibit exceptional elastic properties regain their original shape after the triggering stimulus is removed) can be employed in nanogels to attain stimuli mediated pulsatile release of the drug. E. S. Lee (Lee et al., 2008) made use of the shape memory effect to achieve the pulsatile release of doxorubicin. They prepared nanogels comprising of the hydrophobic core of poly (L-histidine-co-phenylalanine) held by the hydrophilic double shell. The drug release is triggered by a change in pH. The polymer swells rapidly at endosomal pH of 6.8 but is stable at neutral pH. Both the buffering effect and swelling of poly (L-histidine-co-phenylalanine) led to endosomal disruption thus releasing the drug into the cytoplasm. Since the pH of cytoplasm is neutral the nanogels shrink to their original size with the remaining amount of doxorubicin and is then released to affect other cells thereby attaining a pulsatile release (Azadi et al., 2013, Jafari et al., 2016, Kabanov et al., 2009, Zhang et al., 2016). Figure 15.15 gives the idea about various mechanisms by which the drug is released from the nanogels after reaching the target site.

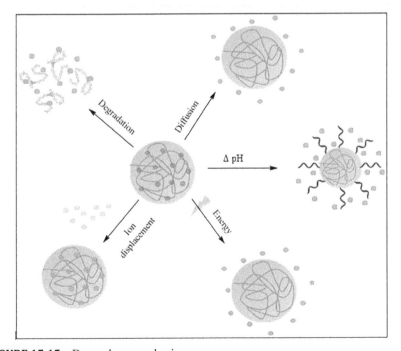

FIGURE 15.15 Drug release mechanism.

15.7   CHARACTERIZATION

Numerous analytical techniques are available for the characterization of nanogels as shown in Table 15.7.

TABLE 15.7   Characterization Techniques for Nanogels.

Physicochemical properties	Analytical technique	Inference	References
Nanogel formation	Darkfield microscopy	Gives direct image of nanogel	Gref et al., 2006
	Nuclear magnetic resonance (NMR)	Gives physical and chemical properties as well as structure, reaction state, chemical environments, electronic structure of molecules.	Huang et al., 2009
	RAMAN spectroscopy	To determine chemical bonds, the symmetry of molecules and crystallographic orientations.	Kettel et al., 2011
	FTIR	Confirms the structure of the major functional group in nanogels formulation, also use to determine absorption, emission, and photoconductivity.	Li et al., 2011
Structure/ morphology	Scanning electron microscopy (SEM)	Detects surface topography of nanogels, composition, 3D structure, chemical composition, crystalline structure and orientation of formulation	Malzahn et al., 2014
	Transmission electron microscopy (TEM)	Provides information about surface features, shape, size and structure of nanogels. It also determines the flaws, fractures, damages in nanogels formulation	Xiong et al., 2011
	Atomic force microscopy (AFM)	Determines the size and structure of nanogel	Li et al., 2008

TABLE 15.7   *(Continued)*

Mean diameter and size distribution	Dynamic light scattering (DLS)	Determines the size and distribution of nano-molecules	Soni et al., 2006
		The particle size is calculated by	
		Stoke-Einstein equation:	
		D = Translational diffusion coefficient	
		d = particle diameter	
		η = viscosity of the liquid in which particles were suspended	
		k = Boltzmann's constant	
		T = Absolute temperature.	
Rheology	Rheometer	Rheological properties of the gel	Yao et al., 2016
Cellular uptake	Flow cytometry	Fluorescent intensity of each well can be determined by flow cytometry.	Xiong et al., 2011
	Confocal Microscopy.	Fluorescein isothiocyanate-conjugated nanogels are visualized under a microscope.	
Hydrophobic/ hydrophilic nature	Fluorescence spectroscopy using hydrophobic dyes such as Pyrene and Nile red	Solubility and fluorescence of these dyes are weak in water but high in hydrophobic environments	Ferreira et al., 2011
Net surface charge	Zeta potential (zetasizer)	Directly relates to the net charges on the surface of the macromolecules and particles	Li et al., 2008
Swelling ratio	Soaking studies	Determines the swelling property of the nanogels W_2 is the weight of swollen gel W_1 is the weight of initial polymer content	Li et al., 2011
Drug loading (L) and Drug entrapment (EE)	UV spectrophotometry	Determined by W_0 = Feed amount of drug W_f = Free drug W_n = Weight of nanogel	Huang et al., 2009

TABLE 15.7 *(Continued)*

Drug release	Dialysis bag technique Franz-diffusion cells	Determines the drug release pattern from nanogels	Yao et al., 2016 Azadi et al., 2012
Physicochemical properties		Analytical technique	Inference
In vitro brain permeation study Murine brain capillary endothelial cells (bEnd3)		Bovine brain microvessel endothelial cells (BBMEC) model	$V\,dC/dt$ = Rate of appearance of the drug at the basolateral side of the monolayers C_0 = Initial drug concentration at the apical side A = Surface area of the membrane.
		dC/dt = Rate of appearance of the drug at the basolateral side of the monolayers V_a = fluid volume in the basolateral side. C_0 = Initial drug concentration at the apical side A = Surface area of the membrane	Li et al., 2010
Biodistribution/ organ targeting		Radioscintigraphy, Magnetic resonance imaging (MRI), Positron emission tomography (PET) Single-photon emission computed tomography (SPECT), Computed X-ray tomography (CT), Optical imaging techniques (e.g., multiphoton microscopy)	Radiolabeling can be calculated by

15.8 APPLICATIONS OF NANOGELS IN BRAIN TARGETING

15.8.1 CATIONIC GELS

S.V. Vinogradov (Vinogradov et al., 2002) studied the delivery of oligo-nucleotides across the BBB using polarized monolayers of Bovine brain microvessel endothelial cells (BBMEC). Transport of oligonucleotides across cell membrane was significantly increased after incorporation into crosslinked poly(ethylene oxide) (PEO) and polyethyleneimine (PEI) nanogels.

15.8.2 SURFACE MODIFICATION OF NANOGELS

15.8.2.1 TRANSFERRIN AND INSULIN

S.V. Vinogradov (Vinogradov et al., 2004) delivered oligonucleotides (ODN) to the brain using crosslinked poly (ethylene glycol) and polyethyleneimine (cationic nanogel). Surface modification of nanogels with transferrin or insulin was found to show a 15-fold increase in brain deposition and 2-fold decreases in liver and spleen compared to the free ODN. Nanogels not only increase the brain deposition but also decrease the degradation of ODN.

Gil and lowe (Gil et al., 2008) synthesized the poly (β-aminoester) and β-Cyclodextrin and proved 20% increase in permeability of insulin across the Bovine brain microvessel endothelial cell.

15.8.2.2 PEPTIDE

Gerson (Gerson et al., 2014) achieved a 10-fold suppression of retroviral activity and reduced virus-associated inflammation in a humanized mouse model of HIV-1 infection in the brain which was achieved by administering brain-specific ApoE peptide conjugated polyethyleneimine (PEI) nanogel with nucleoside reverse transcriptase inhibitors (NRTIs). Similarly, Warren (Warren et al., 2015) delivered the NRTIs using cholesterol-epsilon-poly-lysine (CEPL) nanogels attached with brain-specific peptide (BP-2) and anticipated 5-fold higher therapeutic levels of nanodrugs in the brain and a 10-fold suppression of retroviral activity.

Targeting of glioblastoma was achieved by Chen (Chen et al., 2017) where disulfide (SS) linked poly (vinyl alcohol) nanogel decorated with cyclo arginine-glycine-aspartame (c-RGD) peptide was used to deliver doxorubicin (DOX). The outcome of the study was promising with increased targeting toward the tumor and effective tumor growth inhibition.

15.8.2.3   MONOCLONAL ANTIBODIES

V. P. Baklaushev (Baklaushev et al., 2015) studied the effect of cisplatin-loaded PEG-b-PMAA nanogels conjugated with monoclonal antibodies to Cx43 and BSAT1 in treating the intra-cranial glioma. Protein connexin 43 (Cx43) and brain-specific anion transporter (BSAT1) in the tumor and peri-tumoral area are key components for targeted drug delivery. Conjugation of Cx43 and BSAT1 increased the medial survival rates of rats by 27 and 26.6 days than the control group.

15.8.2.4   POLYSORBATE

Kreuter (Kreuter, 2004) studied the effect of surface functionalization with polysorbates (20, 40, 60, and 80), poloxamers (188) on the transport to the brain and concluded an induction in the uptake occurring by LDL receptor. Azadi (Azadi et al., 2012, Azadi et al., 2013) prepared methotrexate (MTX) loaded chitosan and sodium tripolyphosphate (TPP) nanogels by ionic gelation method, the surface of nanogels were functionalized by polysorbate 80. *In vitro* drug release studies show that the drug release was governed by both diffusion of the drug and swelling and/or dissolution of the polymeric network. After the *in vivo* study significant differences in the plasma concentration levels of free MTX, unmodified nanogel and surface coated nanogel were observed. On the other hand, no significant difference was observed in the brain concentration of MTX from unmodified nanogel and modified nanogel. The brain drug concentration of nanogels was 10- to 15-fold higher compared to free MTX. Sheetal Soni (Soni et al., 2006) prepared the N-hexylcarbamoyl-5-fluorouracil (HCFU) loaded nanogels prepared by free radical polymerization of N-isopropylacrylamide (NIPAAM) and N-vinylpyrrolidone (VP) cross-linked by N, N-methylenebisacrylamide (MBA). *In vivo* results show that the brain accumulation of surface coated nanogels was 0.52% compared to un-coated nanogels 0.18%.

Both the above studies proved that surface modification with polysorbate 80 causes retention in the reticuloendothelial system resulting in prolonged residence time and no significant brain delivery. But Kalaiarasi (Kalaiarasi et al., 2016) concluded that surface functionalization of poly N-isopropyl acryl-amide (PNIPAM) nanogel with polysorbate 80 resulted in a sustainable and controlled release of donepezil to the brain. However, these results were based on the locomotive and swimming behavior of adult zebrafishes.

15.8.3 PH GRADIENT

W. Cui (Cui et al., 2016) has made use of pH gradient around ischemic brain tissue for the pH-triggered the release of Urokinase from polyethylene glycol conjugated nanogels. *In vivo* drug release at low pH was studied in permanent middle cerebral artery occlusion (pMCAO) rat model. They concluded that the decrease in the severity of ischemic stroke was associated with PEG-Urokinase.

15.9 LIMITATIONS

Though nanogels are superior to other nanoparticulate systems there are some limitation with this carrier systems such as

1. Nanogels have limited drug-loading and suboptimal regulation of drug release (Sharma et al., 2016).
2. In most of the cases, the accumulation of drug is higher in reticuloendothelial system (kidney, liver, spleen) rather than the targeted region (Soni et al., 2016b).
3. Interaction of nanogel with serum proteins causes opsonization, to avoid this PEGylation is done but PEG specific IgM antibodies are known to be generated after administration (Ishihara et al., 2009, Shiraishi et al., 2013).
4. Nanogel prepared from the natural polymer may trigger the immune response owing to their high positive charge (Yallapu et al., 2011).
5. Reduction in the hydrophilicity of nanogels occurs due to a strong interaction between drug and polymer which causes the structure to collapse leading to irreversible entrapping of the drug (Vinogradov et al., 2002).
6. Adverse effects may occur due to the surfactant or monomers present in the nanogel formulation (Sharma et al., 2016).

15.10   CONCLUSION

In conclusion, nanogels possess the ability to incorporate a wide range of guest molecules making them highly versatile carriers for brain targeting. The properties of nanogels such as higher drug loading, swelling, biocompatibility, and colloidal stability make them superior over other carrier systems. The "intelligent nanogels" retain the drug molecule in an intact form until the internalization by the target cell. This stimulus sensitive drug release rendered by various types of crosslinkers play a key role in minimizing the off-target effects and provides controlled release at the intended site of action. Advancements in the synthesis of nanogels offer precise control over the size and shape which impart control over the bio-distribution. Furthermore, advanced characterization techniques enable the prediction of *in vivo* behavior of the nanogels.

Various studies have shown promising outcomes in nanogel aided drug delivery to the brain which has laid a platform for delivery of various drug molecules for the treatment of ailments such as CNS disorders, Cancers, HIV, Ischemia etc. Like any other nano particle surface functionalization of nanogels by various ligands has been proven efficient in brain targeting. Thus, brain targeting by nanogel is a potential area of research for further exploration.

Though nanogels have gained importance over the past decade there are only a few clinical studies reported owing to their complexity and obligatory careful engineering. In spite of the extensive research pharmacodynamics and pharmacokinetics of these carriers remain a gray area which needs to be explored prior to clinical translation. Since no drug delivery system is perfect with all the advantages; careful engineering, a thorough study of pharmacokinetics, long-term accumulation and degradation profile would make nanogels the next generation drug delivery systems and possibly be a revolution in this area.

KEYWORDS

- brain targeting
- chemical stimuli
- drug release
- nanogel
- polymer
- physical stimuli
- stimuli sensitive
- synthesis

REFERENCES

Abbott, N. J. Inflammatory Mediators and Modulation of Blood–Brain Barrier Permeability. *Cell. Mol. Neurobiol.* **2000**, *20*, 131–147.

Abed, A.; Assoul, N.; Ba, M., et al. Influence of Polysaccharide Composition on the Biocompatibility of Pullulan/Dextran-Based Hydrogels. *J. Biomed. Mater Res. A* **2011**, *96*, 535–542.

Adamoa, G.; Grimaldib, N.; Camporaa, S.; Antonietta, M.; Sabatinob, C. D.; Ghersia, G. Glutathione-Sensitive Nanogels for Drug Release. *Chem. Eng.* **2014**, *38*, 457–462.

Afergan, E.; Epstein, H.; Dahan, R., et al. Delivery of Serotonin to the Brain by Monocytes Following Phagocytosis of Liposomes. *J. Control. Release* **2008**, *132*, 84–90.

Akiyama, E.; Morimoto, N.; Kujawa, P.; Ozawa, Y.; Winnik, F. M.; Akiyoshi, K. Self-Assembled Nanogels of Cholesteryl-Modified Polysaccharides: Effect of the Polysaccharide Structure on Their Association Characteristics in the Dilute and Semidilute Regimes. *Biomacromolecules* **2007**, *8*, 2366–2373.

Akiyoshi, K.; Deguchi, S.; Moriguchi, N.; Yamaguchi, S.; Sunamoto, J. Self-Aggregates of Hydrophobized Polysaccharides in Water. Formation and Characteristics of Nanoparticles. *Macromolecules* **1993**, *26*, 3062–3068.

Akiyoshi, K.; Deguchi, S.; Tajima, H.; Nishikawa, T.; Sunamoto, J. Microscopic Structure and Thermoresponsiveness of a Hydrogel Nanoparticle by Self-Assembly of a Hydrophobized Polysaccharide. *Macromolecules* **1997**, *30*, 857–861.

Akiyoshi, K.; Kobayashi, S.; Shichibe, S., et al. Self-Assembled Hydrogel Nanoparticle of Cholesterol-Bearing Pullulan as a Carrier of Protein Drugs: Complexation and Stabilization of Insulin. *J. Control. Release* **1998**, *54*, 313–320.

Akiyoshi, K.; Sasaki, Y.; Sunamoto, J. Molecular Chaperone-Like Activity of Hydrogel Nanoparticles of Hydrophobized Pullulan: Thermal Stabilization with Refolding of Carbonic anhydrase B. *Bioconjug. Chem.* **1999**, *10*, 321–324.

Alam, M. I.; Beg, S.; Samad, A., et al. Strategy for Effective Brain Drug Delivery. *Eur. J. Pharm. Sci.* **2010**, *40*, 385–403.

Alvarez-Lorenzo, C.; Bromberg, L.; Concheiro, A. Light-Sensitive Intelligent Drug Delivery Systems. *Photochem. Photobiol.* **2009**, *85*, 848–860.

Amamoto, Y.; Otsuka, H.; Takahara, A. Synthesis and Characterization of Polymeric Nanogels. Wiley, 2011.

An, Z.; Qiu, Q.; Liu, G. Synthesis of Architecturally Well-Defined Nanogels Via RAFT Polymerization for Potential Bioapplications. *Chem. Commun.* **2011**, *47*, 12424–12440.

Anderson, J.; Gibbons, D.; Martin, R.; Hiltner, A.; Woods, R. The Potential for Poly-α-Amino Acids as Biomaterials. *J. Biomed. Mater Res. A* **1974**, *8*, 197–207.

Aoki, T.; Kawashima, M.; Katono, H., et al. Temperature-Responsive Interpenetrating Polymer Networks Constructed with Poly (Acrylic Acid) and Poly (N, N-Dimethylacrylamide). *Macromolecules* **1994**, *27*, 947–952.

Aoki, T.; Muramatsu, M.; Torii, T.; Sanui, K.; Ogata, N. Thermosensitive Phase Transition of an Optically Active Polymer in Aqueous Milieu. *Macromolecules* **2001**, *34*, 3118–3119.

Argentiere, S.; Blasi, L.; Ciccarella, G.; Barbarella, G.; Cingolani, R.; Gigli, G. Synthesis of Poly (Acrylic Acid) Nanogels and Application in Loading and Release of an Oligothiophene Fluorophore and its Bovine Serum Albumin conjugate. *Macromol. Symp.* **2009**, *281*, 69–76.

Argentiere, S.; Blasi, L.; Morello, G.; Gigli, G. A Novel pH-Responsive Nanogel for the Controlled Uptake and Release of Hydrophobic and Cationic Solutes. *J. Phys. Chem. C* **2011**, *115*, 16347–16353.

Arndt, K.-F.; Schmidt, T.; Reichelt, R. Thermo-Sensitive Poly (Methyl Vinyl Ether) Micro-Gel Formed by High Energy Radiation. *Polymer* **2001**, *42*, 6785–6791.

Arnfast, L.; Madsen, C.G.; Jorgensen, L.; Baldursdottir, S. Design and Processing of Nanogels as Delivery Systems for Peptides and Proteins. *Ther. Deliv.* **2014**, *5*, 691–708.

Arshady, R. Suspension, Emulsion, and Dispersion Polymerization: A Methodological Survey. *Colloid Polym. Sci.* **1992**, *270*, 717–732.

Asadi, H.; Khoee, S. Dual Responsive Nanogels for Intracellular Doxorubicin Delivery. *Int. J. Pharm.* **2016**, *511*, 424–435.

Azadi, A.; Hamidi, M.; Khoshayand, M. R.; Amini, M.; Rouini, M. R. Preparation and Optimization of Surface-Treated Methotrexate-Loaded Nanogels Intended for Brain Delivery. *Carbohydr Polym.* **2012**, *90*, 462–471.

Azadi, A.; Hamidi, M.; Rouini, M.-R. Methotrexate-Loaded Chitosan Nanogels as 'Trojan Horses' for Drug Delivery to Brain: Preparation and In Vitro/In Vivo Characterization. *Int. J. Biol. Macromol.* **2013**, *62*, 523–530.

Bajpai, A.; Shukla, S. K.; Bhanu, S.; Kankane, S. Responsive Polymers in Controlled Drug Delivery. *Prog. Polym. Sci.* **2008**, *33*, 1088–1118.

Baklaushev, V. P.; Nukolova, N. N.; Khalansky, A. S., et al. Treatment of Glioma by Cisplatin-Loaded Nanogels Conjugated with Monoclonal Antibodies Against Cx43 and BSAT1. *Drug Deliv.* **2015**, *22*, 276–285.

Batrakova, E. V.; Miller, D. W.; Li, S.; Alakhov, V. Y.; Kabanov, A. V.; Elmquist, W. F. Pluronic P85 Enhances the Delivery of Digoxin to the Brain: In Vitro and In Vivo Studies. *J. Pharmacol. Exp. Ther.* **2001**, *296*, 551–557.

Blackburn, W. H.; Dickerson, E. B.; Smith, M. H.; McDonald, J. F.; Lyon, L. A. Peptide-Functionalized Nanogels for Targeted siRNA Delivery. *Bioconjug. Chem.* **2009**, *20*, 960–968.

Blanchette, M.; Daneman, R. Formation and Maintenance of the BBB. *Mech. Dev.* **2015**, *138*, 8–16.

Bodor, N. Drug Targeting and Retrometabolic Drug Design Approaches Introduction. *Adv. Drug Deliv. Rev.* **1994**, *14*, 157–166.

Bodor, N.; Kaminski, J. J. Prodrugs and Site-Specific Chemical Delivery Systems. *Annu. Rep. Med. Chem.* **1987**, *22*, 303–313.

Bontha, S.; Kabanov, A. V.; Bronich, T. K. Polymer Micelles with Cross-Linked Ionic Cores for Delivery of Anticancer Drugs. *J. Control. Release* **2006**, *114*, 163–174.

Cai, Q.; Ruan, C.; Jiang, L.; Ma, Y.; Pan, H. Preparation of Lactoferrin Modified Poly (Vinyl Alcohol) Nanospheres for Brain Drug Delivery. *Nanomed. Nanotechnol. Biol. Med.* **2016**, *12*, 542–543.

Campos, L. M.; Killops, K. L.; Sakai, R., et al. Development of Thermal and Photochemical Strategies for Thiol− Ene Click Polymer Functionalization. *Macromolecules* **2008**, *41*, 7063–7070.

Chacko, R. T.; Ventura, J.; Zhuang, J.; Thayumanavan, S. Polymer Nanogels: a Versatile Nanoscopic Drug Delivery Platform. *Adv. Drug Deliv. Rev.* **2012**, *64*, 836–851.

Chaterji, S.; Kwon, I. K.; Park, K. Smart Polymeric Gels: Redefining the Limits of Biomedical Devices. *Prog. Polym. Sci.* **2007**, *32*, 1083–1122.

Chen, W.; Zheng, M.; Meng, F., et al. In Situ Forming Reduction-Sensitive Degradable Nanogels for Facile Loading and Triggered Intracellular Release of Proteins. *Biomacromolecules* **2013**, *14*, 1214–1222.

Chen, W.; Zou, Y.; Zhong, Z.; Haag, R. Cyclo (RGD)-Decorated Reduction-Responsive Nanogels Mediate Targeted Chemotherapy of Integrin Overexpressing Human Glioblastoma In Vivo. *Small* **2017**, *13*, 1–9.

Chen, Y.; Liu, L. Modern Methods for Delivery of Drugs Across the Blood–Brain Barrier. *Adv. Drug Deliv. Rev.* **2012**, *64*, 640–665.

Chiang, W.-H.; Ho, V. T.; Huang, W.-C.; Huang, Y.-F.; Chern, C.-S.; Chiu, H.-C. Dual Stimuli-Responsive Polymeric Hollow Nanogels Designed as Carriers for Intracellular Triggered Drug Release. *Langmuir* **2012**, *28*, 15056–15064.

Cho, S.H.; Jhon, M.S.; Yuk, S.H.; Lee, H.B. Temperature-Induced Phase Transition of Poly (N, N-Dimethylaminoethyl Methacrylate-Co-Acrylamide). *J. Polym. Sci. Part B Polym. Phys.* **1997**, *35*, 595–598.

Christensen, L.V.; Chang, C.-W.; Kim, W.J., et al. Reducible Poly (Amido Ethylenimine) s Designed for Triggered Intracellular Gene Delivery. *Bioconjug. Chem.* **2006**, *17*, 1233–1240.

Cui, W.; Liu, R.; Jin, H., et al. pH Gradient Difference Around Ischemic Brain Tissue can Serve as a Trigger for Delivering Polyethylene Glycol-Conjugated Urokinase Nanogels. *J. Control. Release* **2016**, *225*, 53–63.

de Boer, A. G.; Gaillard, P. J. Strategies to Improve Drug Delivery Across the Blood-Brain Barrier. *Clin. Pharmacokinet.* **2007**, *46*, 553–576.

Debele, T. A.; Mekuria, S. L.; Tsai, H. -C. Polysaccharide Based Nanogels in the Drug Delivery System: Application as the Carrier of Pharmaceutical Agents. *Mater. Sci. Eng. C.* **2016**, *68*, 964–981.

DeForest, C. A.; Anseth, K. S. Cytocompatible Click-Based Hydrogels with Dynamically Tunable Properties Through Orthogonal Photoconjugation and Photocleavage Reactions. *Nat. Chem.* **2011**, *3*, 925–931.

Devi, D. R.; Sandhya, P.; Hari, B. V. Poloxamer: a Novel Functional Molecule for Drug Delivery and Gene Therapy. *J. Pharm. Sci. Res.* **2013**, *5*, 159–165.

DeVolder, R.; Antoniadou, E.; Kong, H. Enzymatically Cross-Linked Injectable Alginate-G-Pyrrole Hydrogels for Neovascularization. *J. Control. Release* **2013**, *172*, 30–37.

Doolittle, N. D.; Miner, M. E.; Hall, W. A., et al. Safety and Efficacy of a Multicenter Study Using Intraarterial Chemotherapy in Conjunction with Osmotic Opening of the Blood-Brain Barrier for the Treatment of Patients with Malignant Brain Tumors. *Cancer* **2000**, *88*, 637–647.

Drogoz, A.; David, L.; Rochas, C.; Domard, A.; Delair, T. Polyelectrolyte Complexes from Polysaccharides: Formation and Stoichiometry Monitoring. *Langmuir* **2007**, *23*, 10950–10958.

Eckmann, D.; Composto, R.; Tsourkas, A.; Muzykantov, V. Nanogel Carrier Design for Targeted Drug Delivery. *J. Mater. Chem. B* **2014**, *2*, 8085–8097.

Emerich, D. F.; Dean, R. L.; Marsh, J., et al. Intravenous Cereport (RMP-7) Enhances Delivery of Hydrophilic Chemotherapeutics and Increases Survival in rats with Metastatic Tumors in the Brain. *Pharm. Res.* **2000**, *17*, 1212–1219.

Ferreira, S. A.; Coutinho, P. J.; Gama, F. M. Synthesis and Characterization of Self-Assembled Nanogels Made of Pullulan. *Materials* **2011**, *4*, 601–620.

Ferrer, M. C. C.; Ferrier, R. C.; Eckmann, D. M.; Composto, R. J. A Facile Route to Synthesize Nanogels Doped with Silver Nanoparticles. *J. Nanopart. Res.* **2013**, *15*, 1–7.

Ferrer, M. C. C.; Dastgheyb, S.; Hickok, N. J.; Eckmann, D. M.; Composto, R. J. Designing
Nanogel Carriers for Antibacterial Applications. *Acta Biomater.* **2014a**, *10*, 2105–2111.

Ferrer, M. C. C.; Shuvaev, V. V.; Zern, B. J.; Composto, R. J.; Muzykantov, V. R.; Eckmann,
D. M. Icam-1 Targeted Nanogels Loaded with Dexamethasone Alleviate Pulmonary
Inflammation. *PloS one* **2014b**, e102329.

Fischer, H.; Gottschlich, R.; Seelig, A. Blood-Brain Barrier Permeation: Molecular Parameters Governing Passive Diffusion. *J. Membr. Biol.* **1998**, *165*, 201–211.

Fishman, J.; Rubin, J.; Handrahan, J.; Connor, J.; Fine, R. Receptor-Mediated Transcytosis of
Transferrin across the Blood-Brain Barrier. *J. Neurosci. Res.* **1987,** *18*, 299–304.

Fu, W.; Luo, C.; Morin, E. A.; He, W.; Li, Z.; Zhao, B. UCST-Type Thermosensitive Hairy
Nanogels Synthesized by RAFT Polymerization-Induced Self-Assembly. *ACS Macro Lett.*
2017, *6*, 127–133.

Fujioka-Kobayashi, M.; Ota, M. S.; Shimoda, A., et al. Cholesteryl Group- and Acryloyl
Group-Bearing Pullulan Nanogel to Deliver BMP2 and FGF18 for Bone Tissue Engineering. *Biomaterials* **2012**, *33*, 7613–7620.

Fukuyama, Y.; Yuki, Y.; Katakai, Y., et al. Nanogel-Based Pneumococcal Surface Protein
A Nasal Vaccine Induces MicroRNA-Associated Th17 Cell Responses with Neutralizing
Antibodies Against Streptococcus Pneumoniae in Macaques. *Mucosal Immunol.* **2015,** *8*,
1144–1153.

Gandhi, A.; Paul, A.; Sen, S. O.; Sen, K. K. Studies on Thermoresponsive Polymers: Phase
Behaviour, Drug Delivery and Biomedical Applications. *Asian J. Pharm. Sci.* **2015,** *10*,
99–107.

Georgieva, J. V.; Hoekstra, D.; Zuhorn, I. S. Smuggling Drugs into the Brain: an Overview of
Ligands Targeting Transcytosis for Drug Delivery across the Blood–Brain Barrier. *Pharmaceutics* **2014,** *6*, 557–583.

Gerson, T.; Makarov, E.; Senanayake, T. H.; Gorantla, S.; Poluektova, L. Y.; Vinogradov, S.
V. Nano-NRTIs Demonstrate Low Neurotoxicity and High Antiviral Activity Against HIV
Infection in the Brain. *Nanomed. Nanotechnol. Biol. and Med.* **2014,** *10*, 177–185.

Gil, E. S.; Lowe, T. L. Invention of Polysaccharide-Based Nanoparticles for Enhancing Drug
Permeability across the Blood Brain Barrier. *NSTI-Nanotechnol.* **2008,** *2*, 379.

Glangchai, L. C.; Caldorera-Moore, M.; Shi, L.; Roy, K. Nanoimprint Lithography Based
fabrication of Shape-Specific, Enzymatically-Triggered Smart Nanoparticles. *J. Control.
Release* **2008,** *125*, 263–272.

Golden, P. L.; Maccagnan, T. J.; Pardridge, W. M. Human Blood-Brain Barrier Leptin
Receptor. Binding and Endocytosis in isolated Human Brain Microvessels. *J. Clin. Invest.*
1997, *99*, 14–18.

Gonatas, N. K.; Stieber, A.; Hickey, W. F.; Herbert, S. H.; Gonatas, J. Endosomes and Golgi
Vesicles in Adsorptive and Fluid Phase Endocytosis. *J. Cell Biol.* **1984,** *99*, 1379–1390.

Gonçalves, C.; Pereira, P.; Gama, M. Self-Assembled Hydrogel Nanoparticles for Drug
Delivery Applications. *Materials* **2010,** *3*, 1420–1460.

Gozzi, A.; Schwarz, A. J.; Reese, T., et al. Functional Magnetic Resonance Mapping of Intracerebroventricular Infusion of a Neuroactive Peptide in the Anaesthetised Rat. *J. Neurosci.
Methods* **2005,** *142*, 115–124.

Greenwood, J.; Luthert, P.; Pratt, O.; Lantos, P. Hyperosmolar Opening of the Blood-Brain Barrier in the Energy-Depleted Rat Brain. Part 1. Permeability Studies. *J. Cereb. Blood Flow Metab.* **1988**, *8*, 9–15.

Gref, R.; Amiel, C.; Molinard, K., et al. New Self-Assembled Nanogels Based on Host–Guest Interactions: Characterization and Drug Loading. *J. Control. Release* **2006**, *111*, 316–324.

Greig, N. H. Drug Delivery to the Brain by Blood-Brain Barrier Circumvention and Drug Modification. In: Neuwelt E.A. (Ed.), Implications of the Blood-Brain Barrier and Its Manipulation. Springer: Boston, MA 1989; pp 311–367.

Gupta, S.; Tyagi, R.; Parmar, V.S.; Sharma, S.K.; Haag, R. Polyether Based Amphiphiles for Delivery of Active Components. *Polymer* **2012**, *53*, 3053–3078.

Gupte, A.; Mumper, R. J. Elevated Copper and Oxidative Stress in Cancer Cells as a Target for Cancer Treatment. *Cancer Treat. Rev.* **2009**, *35*, 32–46.

Hamidi, M.; Azadi, A.; Rafiei, P. Hydrogel Nanoparticles in Drug Delivery. *Adv. Drug Deliv. Rev.* **2008**, *60*, 1638–1649.

Hasegawa, U.; Sawada, S.-I.; Shimizu, T., et al. Raspberry-Like Assembly of Cross-Linked Nanogels for Protein Delivery. *J. Control. Release* **2009**, *140*, 312–317.

Hasegawa, U.; Shin-Ichiro, M. N.; Kaul, S. C.; Hirano, T.; Akiyoshi, K. Nanogel-Quantum Dot Hybrid Nanoparticles for Live Cell Imaging. *Biochem. Biophys. Res. Commun.* **2005**, *331*, 917–921.

He, L.; Fullenkamp, D. E.; Rivera, J. G.; Messersmith, P. B. pH Responsive Self-Healing Hydrogels Formed by Boronate–Catechol Complexation. *Chem. Commun.* **2011**, *47*, 7497–7499.

Hendrickson, G. R.; Lyon, L. A. Microgel Translocation through Pores Under Confinement. *Angew. Chem. Int. Ed.* **2010**, *49*, 2193–2197.

Hirakura, T.; Yasugi, K.; Nemoto, T., et al. Hybrid Hyaluronan Hydrogel Encapsulating Nanogel as a Protein Nanocarrier: New System for Sustained Delivery of Protein with a Chaperone-Like Function. *J. Control. Release* **2010**, *142*, 483–489.

Hoare, T.; Pelton, R. Engineering Glucose Swelling Responses in Poly (N-Isopropylacrylamide)-Based Microgels. *Macromolecules* **2007**, *40*, 670–678.

Hoyle, C. E.; Lee, T. Y.; Roper, T. Thiol–Enes: Chemistry of the Past with Promise for the Future. *J. Polym. Sci. A Polym. Chem.* **2004**, *42*, 5301–5338.

Huang, S.-J.; Sun, S.-L.; Feng, T.-H.; Sung, K.-H.; Lui, W.-L.; Wang, L.-F. Folate-Mediated Chondroitin Sulfate-Pluronic® 127 Nanogels as a Drug Carrier. *Eur. J. Pharm. Sci.* **2009**, *38*, 64–73.

Hynynen, K. Ultrasound for Drug and Gene Delivery to the Brain. *Adv. Drug Deliv. Rev.* **2008**, *60*, 1209–1217.

Idziak, I.; Avoce, D.; Lessard, D.; Gravel, D.; Zhu, X. Thermosensitivity of Aqueous Solutions of Poly (N, N-Diethylacrylamide). *Macromolecules* **1999**, *32*, 1260–1263.

Ilmain, F.; Tanaka, T.; Kokufuta, E. Volume Transition in a Gel Driven by Hydrogen Bonding. *Nature* **1991**, *349*, 400–401.

Inomoto, N.; Osaka, N.; Suzuki, T., et al. Interaction of Nanogel with Cyclodextrin or Protein: Study by Dynamic Light Scattering and Small-Angle Neutron Scattering. *Polymer* **2009**, *50*, 541–546.

Ishihara, T.; Takeda, M.; Sakamoto, H., et al. Accelerated Blood Clearance Phenomenon upon Repeated Injection of PEG-Modified PLA-Nanoparticles. *Pharm. Res.* **2009**, *26*, 2270–2279.

Jafari, M.; Kaffashi, B. Mathematical Kinetic Modeling on Isoniazid Release from Dex-HEMA-PNIPAAm Nanogels. *Nanomed. Res. J.* **2016,** *1,* 90–96.

Jay, J. I.; Shukair, S.; Langheinrich, K., et al. Modulation of Viscoelasticity and HIV Transport as a Function of pH in a Reversibly Crosslinked Hydrogel. *Adv. Funct. Mater.* **2009,** *19,* 2969–2977.

Jia, H.-Z.; Wang, H.-F.; Liu, C.-W., et al. A pH-Sensitive Macro- and Nanohydrogel Constructed from Cationic Hydroxyl-Containing Hyperbranched Polycarbonate. *Soft Matter* **2012,** *8,* 6906–6912.

Jian, F.; Zhang, Y.; Wang, J., et al. Toxicity of Biodegradable Nanoscale Preparations. *Curr. Drug Metab.* **2012,** *13,* 440–446.

Kabanov, A. V.; Vinogradov, S. V. Nanogels as Pharmaceutical Carriers: Finite Networks of Infinite Capabilities. *Angew. Chem. Int. Ed.* **2009,** *48,* 5418–5429.

Kakizawa, Y.; Harada, A.; Kataoka, K. Environment-Sensitive Stabilization of Core– Shell Structured Polyion Complex Micelle by Reversible Cross-Linking of the Core through Disulfide Bond. *J. Am. Chem. Soc.* **1999,** *121,* 11247–11248.

Kalaiarasi, S.; Arjun, P.; Nandhagopal, S., et al. Development of Biocompatible Nanogel for Sustained Drug Release by Overcoming the Blood Brain Barrier in Zebrafish Model. *J. Appl. Biomed.* **2016,** *14,* 157–169.

Kamaly, N.; Yameen, B.; Wu, J.; Farokhzad, O.C. Degradable Controlled-Release Polymers and Polymeric Nanoparticles: Mechanisms of Controlling Drug Release. *Chem. Rev.* **2016,** *116,* 2602–2663.

Kanazawa, T.; Taki, H.; Tanaka, K.; Takashima, Y.; Okada, H. Cell-Penetrating Peptide-Modified Block Copolymer Micelles Promote Direct Brain Delivery Via Intranasal Administration. *Pharm. Res.* **2011,** *28,* 2130–2139.

Kang, H.; Trondoli, A. C.; Zhu, G., et al. Near-Infrared Light-Responsive Core–Shell Nanogels for Targeted Drug Delivery. *ACS Nano* **2011,** *5,* 5094–5099.

Karimi, M.; Eslami, M.; Sahandi-Zangabad, P., et al. pH-Sensitive Stimulus-Responsive Nanocarriers for targeted delivery of therapeutic agents. *Wiley Interdiscip. Rev Nanomed. Nanobiotechnol.* **2016,** *8,* 696–716.

Ke, W.; Shao, K.; Huang, R., et al. Gene Delivery Targeted to the Brain Using an Angiopep-Conjugated Polyethyleneglycol-Modified Polyamidoamine Dendrimer. *Biomaterials* **2009,** *30,* 6976–6985.

Kettel, M. J.; Dierkes, F.; Schaefer, K.; Moeller, M.; Pich, A. Aqueous Nanogels Modified with Cyclodextrin. *Polymer* **2011,** *52,* 1917–1924.

Kharkar, P. M.; Kiick, K. L.; Kloxin, A. M. Designing Degradable Hydrogels for Orthogonal Control of Cell Microenvironments. *Chem. Soc. Rev.* **2013,** *42,* 7335–7372.

Khmelnitsky, Y. L.; Neverova, I. N.; Gedrovich, A. V.; Polyakov, V. A.; Levashov, A. V.; Martinek, K. Catalysis by α-Chymotrypsin Entrapped into Surface-Modified Polymeric Nanogranules in Organic Solvent. *FEBS J.* **1992,** *210,* 751–757.

Khoee, S.; Asadi, H. Nanogels: Chemical Approaches to Preparation. In: Encyclopedia of Biomedical Polymers and Polymeric Biomaterials. *Taylor Francis* **2016,** 5266–5293, https://doi.org/ 10.1201/b19038-60.

Kim, J.; Nayak, S.; Lyon, L.A. Bioresponsive Hydrogel Microlenses. *J. Am. Chem. Soc.* **2005,** *127,* 9588–9592.

Kim, J. O.; Kabanov, A. V.; Bronich, T. K. Polymer Micelles with Cross-Linked Polyanion Core for Delivery of a Cationic Drug Doxorubicin. *J. Control. Release* **2009,** *138,* 197–204.

Kim, J. W.; Utada, A. S.; Fernández-Nieves, A.; Hu, Z.; Weitz, D. A. Fabrication of Monodisperse Gel Shells and Functional Microgels in Microfluidic Devices. *Angew. Chem. Int. Ed.* **2007**, *119*, 1851–1854.

Kim, K.; Bae, B.; Kang, Y. J.; Nam, J.-M.; Kang, S.; Ryu, J.-H. Natural Polypeptide-Based Supramolecular Nanogels for Stable Noncovalent Encapsulation. *Biomacromolecules* **2013**, *14*, 3515–3522.

Kim, S. J.; Lee, K. J.; Kim, S. I.; Lee, Y. M.; Chung, T. D.; Lee, S. H. Electrochemical Behavior of an Interpenetrating Polymer Network Hydrogel Composed of Poly (Propylene Glycol) and Poly (Acrylic Acid). *J. Appl. Polym. Sci.* **2003**, *89*, 2301–2305.

Kim, Y.; Thapa, M.; Hua, D. H.; Chang, K.-O. Biodegradable Nanogels for Oral Delivery of Interferon for Norovirus Infection. *Antiviral Res.* **2011**, *89*, 165–173.

Klibanov, A. L.; Maruyama, K.; Torchilin, V. P.; Huang, L. Amphipathic Polyethyleneglycols Effectively Prolong the Circulation Time of Liposomes. *FEBS Lett.* **1990**, *268*, 235–237.

Kobayashi, H.; Katakura, O.; Morimoto, N.; Akiyoshi, K.; Kasugai, S. Effects of Cholesterol-Bearing Pullulan (CHP)-nanogels in Combination with Prostaglandin E1 on Wound Healing. *J. Biomed. Mater. Res. Part B Appl. Biomater* **2009**, *91*, 55–60.

Kolb, H. C.; Finn, M.; Sharpless, K. B. Click Chemistry: Diverse Chemical Function from a Few Good Reactions. *Angew. Chem. Int. Ed.* **2001**, *40*, 2004–2021.

Koziara, J.; Lockman, P.; Allen, D.; Mumper, R. The Blood-Brain Barrier and Brain Drug Delivery. *J. Nanosci. Nanotechnol.* **2006**, *6*, 2712–2735.

Kreuter, J. Influence of the Surface Properties on Nanoparticle-Mediated Transport of Drugs to the Brain. *J. Nanosci. Nanotechnol.* **2004**, *4*, 484–488.

Kroll, R. A.; Neuwelt, E. A.; Neuwelt, E. A. Outwitting the Blood-Brain Barrier for Therapeutic Purposes: Osmotic Opening and Other Means. *Neurosurgery* **1998**, *42*, 1083–1099.

Kumar, P.; Shen, Q.; Pivetti, C. D.; Lee, E. S.; Wu, M. H.; Yuan, S. Y. Molecular Mechanisms of Endothelial Hyperpermeability: Implications in Inflammation. *Expert Rev. Mol. Med.* **2009**, *11* e19.

Kuo, Y.-C.; Kuo, C.-Y. Electromagnetic Interference in the Permeability of Saquinavir Across the Blood–Brain Barrier Using Nanoparticulate Carriers. *Int. J. Pharm.* **2008**, *351*, 271–281.

Lajunen, T.; Viitala, L.; Kontturi, L.-S., et al. Light Induced Cytosolic Drug Delivery from Liposomes with Gold Nanoparticles. *J. Control. Release* **2015**, *203*, 85–98.

Laughlin, S. T.; Baskin, J. M.; Amacher, S. L.; Bertozzi, C. R. In Vivo Imaging of Membrane-Associated Glycans in Developing Zebrafish. *Science* **2008**, *320*, 664–667.

Lee, E. S.; Kim, D.; Youn, Y. S.; Oh, K. T.; Bae, Y. H. A Virus-Mimetic Nanogel Vehicle. *Angew. Chem. Int. Ed.* **2008**, *120*, 2452–2455.

Lee, G.; Dallas, S.; Hong, M.; Bendayan, R. Drug Transporters in the Central Nervous System: Brain Barriers and Brain Parenchyma Considerations. *Pharmacol. Rev.* **2001**, *53*, 569–596.

Lemieux, P.; Vinogradov, S.; Gebhart, C., et al. Block and Graft Copolymers and Nanogel™ Copolymer Networks for DNA Delivery Into Cell. *J. Drug Target.* **2000**, *8*, 91–105.

Leung, M.; Liu, C.; Koon, J.; Fung, K. Polysaccharide Biological Response Modifiers. *Immunol. Lett.* **2006**, *105*, 101–114.

Li, G.; Simon, M. J.; Cancel, L. M., et al. Permeability of Endothelial and Astrocyte Cocultures: In Vitro Blood–Brain Barrier Models for Drug Delivery Studies. *Ann. Biomed. Eng.* **2010**, *38*, 2499–2511.

Li, J.; Yu, S.; Yao, P.; Jiang, M. Lysozyme– Dextran Core– Shell Nanogels Prepared Via a Green Process. *Langmuir* **2008**, *24*, 3486–3492.

Li, M.-H.; Keller, P. Stimuli-Responsive Polymer Vesicles. *Soft Matter* **2009**, *5*, 927–937.

Li, N.; Wang, J.; Yang, X.; Li, L. Novel Nanogels as Drug Delivery Systems for Poorly Soluble Anticancer Drugs. *Colloids Surf. B* **2011**, *83*, 237–244.

Li, X.; Tsibouklis, J.; Weng, T., et al. Nano Carriers for Drug Transport across the Blood–Brain Barrier. *J. Drug Target.* **2017**, *25*, 17–28.

Li, Y.; Maciel, D.; Rodrigues, J.o.; Shi, X.; Tomás, H. Biodegradable Polymer Nanogels for Drug/Nucleic Acid Delivery. *Chem. Rev.* **2015**, *115*, 8564–8608.

Li, Y. L.; Zhu, L.; Liu, Z., et al. Reversibly Stabilized Multifunctional Dextran Nanoparticles Efficiently Deliver Doxorubicin into the Nuclei of Cancer Cells. *Angew. Chem. Int. Ed.* **2009**, *48*, 9914–9918.

Linnet, K.; Ejsing, T. B. A Review on the Impact of P-Glycoprotein on the Penetration of Drugs into the Brain. Focus on Psychotropic Drugs. *Eur. Neuropsychopharmacol.* **2008**, *18*, 157–169.

Liu, Z.; Jiao, Y.; Wang, Y.; Zhou, C.; Zhang, Z. Polysaccharides-Based Nanoparticles as Drug Delivery Systems. *Adv. Drug Deliv. Rev.* **2008**, *60*, 1650–1662.

Lockman, P.; Mumper, R.; Khan, M.; Allen, D. Nanoparticle Technology for Drug Delivery across the Blood-Brain Barrier. *Drug Dev. Ind. Pharm.* **2002**, *28*, 1–13.

Mackic, J. B.; Stins, M.; Jovanovic, S.; Kim, K.; Bartus, R. T.; Zlokovic, B. V. Cereport™(RMP-7) Increases the Permeability of Human Brain Microvascular Endothelial Cell Monolayers1. *Pharm. Res.* **1999**, *16*, 1360–1365.

Mahalingam, A.; Jay, J. I.; Langheinrich, K., et al. Inhibition of the Transport of HIV In Vitro Using a pH-Responsive Synthetic Mucin-Like Polymer System. *Biomaterials* **2011**, *32*, 8343–8355.

Malzahn, K.; Jamieson, W. D.; Dröge, M., et al. Advanced Dextran Based Nanogels for Fighting *Staphylococcus aureus* Infections by Sustained Zinc Release. *J. Mater. Chem. B* **2014**, *2*, 2175–2183.

Marck, K.; Wildevuur, C. R.; Sederel, W.; Bantjes, A.; Feijen, J. Biodegradability and Tissue Reaction of Random Copolymers of L-leucine, L-Aspartic Acid, and L-Aspartic Acid Esters. *J. Biomed. Mater Res. A* **1977**, *11*, 405–422.

Martin, E.; May, P.; McMahon, W. Amino Acid Polymers for Biomedical Applications. I. Permeability Properties of L-Leucine DL-Methionine Copolymers. *J. Biomed. Mater Res. A* **1971**, *5*, 53–62.

Maurus, P. B.; Kaeding, C. C. Bioabsorbable Implant Material Review. *Oper. Tech. Sports Med.* **2004**, *12*, 158–160.

McAllister, K.; Sazani, P.; Adam, M., et al. Polymeric Nanogels Produced Via Inverse Microemulsion Polymerization as Potential Gene and Antisense Delivery Agents. *J. Am. Chem. Soc* **2002**, *124*, 15198–15207.

Menzies, D. J.; Cameron, A.; Munro, T.; Wolvetang, E.; Grøndahl, L.; Cooper-White, J. J. Tailorable Cell Culture Platforms from Enzymatically Cross-Linked Multifunctional Poly (Ethylene Glycol)-Based Hydrogels. *Biomacromolecules* **2013**, *14*, 413–423.

Miller, T.; Goude, M. C.; McDevitt, T. C.; Temenoff, J. S. Molecular Engineering of Glycosaminoglycan Chemistry for Biomolecule Delivery. *Acta Biomater.* **2014**, *10*, 1705–1719.

Moriyama, E.; Salcman, M.; Broadwell, R. D. Blood-Brain Barrier Alteration After Microwave-Induced Hyperthermia is Purely a Thermal Effect: I. Temperature and Power Measurements. *Surg. Neurol.* **1991**, *35*, 177–182.

Motornov, M.; Zhou, J.; Pita, M., et al., "Chemical transformers" from Nanoparticle Ensembles Operated with Logic. *Nano Lett.* **2008**, *8*, 2993–2997.

Mura, S.; Nicolas, J.; Couvreur, P. Stimuli-Responsive Nanocarriers for Drug Delivery. *Nat. Mater.* **2013**, *12*, 991–1003.

Neamtu, I.; Rusu, A. G.; Diaconu, A.; Nita, L. E.; Chiriac, A. P. Basic Concepts and Recent Advances in Nanogels as Carriers for Medical Applications. *Drug Deliv.* **2017**, *24*, 539–557.

Neves, V.; Aires-da-Silva, F.; Corte-Real, S.; Castanho, M. A. Antibody Approaches to Treat Brain Diseases. *Trends Biotechnol.* **2016**, *34*, 36–48.

Nie, Z.; Xu, S.; Seo, M.; Lewis, P. C.; Kumacheva, E. Polymer Particles with Various Shapes and Morphologies Produced in Continuous Microfluidic Reactors. *J. Am. Chem. Soc* **2005**, *127*, 8058–8063.

Nishikawa, T.; Akiyoshi, K.; Sunamoto, J. Macromolecular Complexation between Bovine Serum Albumin and the Self-Assembled Hydrogel Nanoparticle of Hydrophobized Polysaccharides. *J. Am. Chem. Soc.* **1996**, *118*, 6110–6115.

Noh, Y.-W.; Kong, S.-H.; Choi, D.-Y., et al. Near-Infrared Emitting Polymer Nanogels for Efficient Sentinel Lymph Node Mapping. *ACS Nano* **2012**, *6*, 7820–7831.

Oh, J. K.; Drumright, R.; Siegwart, D. J.; Matyjaszewski, K. The Development of Microgels/Nanogels for Drug Delivery Applications. *Prog. Polym. Sci.* **2008**, *33*, 448–477.

Oldendorf, W. H. Brain Uptake of Radiolabeled Amino Acids, Amines, and Hexoses after Arterial Injection. *Am. J. Physiol.—Leg. Content* **1971**, *221*, 1629–1639.

Pardridge, W. M. Brain Drug Targeting: The Future of Brain Drug Development. The press Syndicate of the University of Cambridge, United Kingdom, **2001**, 353.

Pardridge, W. M. Drug and Gene Delivery to the Brain: the Vascular Route. *Neuron* **2002**, *36*, 555–558.

Park, K. Trojan Monocytes for Improved Drug Delivery to the Brain. *Nanomed. Nanotechnol. Biol. Med.* **2008**.

Park, K. M.; Lee, Y.; Son, J. Y.; Bae, J. W.; Park, K. D. In Situ SVVYGLR Peptide Conjugation into Injectable Gelatin-Poly (Ethylene Glycol)-Tyramine Hydrogel Via Enzyme-Mediated Reaction for Enhancement of Endothelial Cell Activity and Neo-Vascularization. *Bioconjug. Chem.* **2012**, *23*, 2042–2050.

Patel, M. M.; Goyal, B. R.; Bhadada, S. V.; Bhatt, J. S.; Amin, A.F. Getting into the Brain. *CNS Drugs* **2009**, *23*, 35–58.

Peppas, N. A.; Khare, A. R. Preparation, Structure and Diffusional Behavior of Hydrogels in Controlled Release. *Adv. Drug Deliv. Rev.* **1993**, 11, 1–35.

Piest, M.; Zhang, X.; Trinidad, J.; Engbersen, J. F. pH-Responsive, Dynamically Restructuring Hydrogels Formed by Reversible Crosslinking of PVA with Phenylboronic Acid Functionalised PPO–PEO–PPO Spacers (Jeffamines®). *Soft Matter* **2011**, *7*, 11111–11118.

Piogé, S.; Nesterenko, A.; Brotons, G., et al. Core Cross-Linking of Dynamic Diblock Copolymer Micelles: Quantitative Study of Photopolymerization Efficiency and Micelle Structure. *Macromolecules* **2011**, *44*, 594–603.

Poduslo, J.F.; Curran, G.L. Polyamine Modification Increases the Permeability of Proteins at the Blood-Nerve and Blood-Brain Barriers. *J. Neurochem.* **1996**, *66*, 1599–1609.

Posocco, B.; Dreussi, E.; De Santa, J., et al. Polysaccharides for the Delivery of Antitumor Drugs. *Materials* **2015**, *8*, 2569–2615.

Prescher, J. A.; Dube, D. H.; Bertozzi, C. R. Chemical Remodelling of Cell Surfaces in Living Animals. *Nature* **2004**, *430*, 873–877.

Qiao, Z.-Y.; Zhang, R.; Du, F.-S.; Liang, D.-H.; Li, Z.-C. Multi-Responsive Nanogels Containing Motifs of Ortho Ester, Oligo (Ethylene Glycol) and Disulfide Linkage as Carriers of Hydrophobic Anti-Cancer Drugs. *J. Control. Release* **2011**, *152*, 57–66.

Qin, Y.; Chen, H.; Zhang, Q., et al. Liposome Formulated with TAT-Modified Cholesterol for Improving Brain Delivery and Therapeutic Efficacy on Brain Glioma in Animals. *Int. J. Pharm.* **2011**, *420*, 304–312.

Qiu, L.-B.; Ding, G.-R.; Li, K.-C., et al. The Role of Protein Kinase C in the Opening of Blood–Brain Barrier Induced by Electromagnetic Pulse. *Toxicology* **2010**, *273*, 29–34.

Raemdonck, K.; Demeester, J.; De Smedt, S. Advanced Nanogel Engineering for Drug Delivery. *Soft Matter* **2009**, *5*, 707–715.

Rehor, A.; Botterhuis, N.; Hubbell, J.; Sommerdijk, N. A.; Tirelli, N. Glucose Sensitivity through Oxidation Responsiveness. An Example of Cascade-Responsive Nano-Sensors. *J. Mater. Chem.* **2005**, *15*, 4006–4009.

Reinhardt, R.R.; Bondy, C. Insulin-Like Growth Factors Cross the Blood-Brain Barrier. *Endocrinology* **1994**, *135*, 1753–1761.

Roberts, M. C.; Hanson, M. C.; Massey, A. P.; Karren, E. A.; Kiser, P. F. Dynamically Restructuring Hydrogel Networks Formed with Reversible Covalent Crosslinks. *Adv. Mater.* **2007**, *19*, 2503–2507.

Rodrigues, S.; Cardoso, L.; da Costa, A. M. R.; Grenha, A. Biocompatibility and Stability of Polysaccharide Polyelectrolyte Complexes Aimed at Respiratory Delivery. *Materials* **2015**, *8*, 5647–5670.

Rolland, J. P.; Maynor, B. W.; Euliss, L. E.; Exner, A. E.; Denison, G. M.; DeSimone, J. M. Direct Fabrication and Harvesting of Monodisperse, Shape-Specific Nanobiomaterials. *[J. Am. Chem. Soc* **2005**, *127*, 10096–10100.

Russo, A.; DeGraff, W.; Friedman, N.; Mitchell, J.B. Selective Modulation of Glutathione Levels in Human Normal Versus Tumor Cells and Subsequent Differential Response to Chemotherapy Drugs. *Cancer Res.* **1986**, *46*, 2845–2848.

Ryu, J.-H.; Bickerton, S.; Zhuang, J.; Thayumanavan, S. Ligand-Decorated Nanogels: fast One-Pot Synthesis and Cellular Targeting. *Biomacromolecules* **2012**, *13*, 1515–1522.

Ryu, J.-H.; Chacko, R.T.; Jiwpanich, S.; Bickerton, S.; Babu, R. P.; Thayumanavan, S. Self-Cross-Linked Polymer Nanogels: a Versatile Nanoscopic Drug Delivery Platform. *[J. Am. Chem. Soc* **2010a**, *132*, 17227–17235.

Ryu, J.-H.; Jiwpanich, S.; Chacko, R.; Bickerton, S.; Thayumanavan, S. Surface-Functionalizable Polymer Nanogels with Facile Hydrophobic Guest Encapsulation Capabilities. *[J. Am. Chem. Soc* **2010b**, *132*, 8246–8247.

Saija, A.; Princi, P.; Trombetta, D.; Lanza, M.; Pasquale, A. D. Changes in the Permeability of the Blood-Brain Barrier Following Sodium Dodecyl Sulphate Administration in the Rat. *Exp. Brain Res.* **1997**, *115*, 546–551.

Saito, G.; Swanson, J.A.; Lee, K.-D. Drug Delivery Strategy Utilizing Conjugation via Reversible Disulfide Linkages: Role and Site of Cellular Reducing Activities. *Adv. Drug Deliv. Rev.* **2003**, *55*, 199–215.

Sasaki, Y.; Akiyoshi, K. Nanogel Engineering for New Nanobiomaterials: from Chaperoning Engineering to Biomedical Applications. *Chem. Rec.* **2010**, *10*, 366–376.

Sasaki, Y.; Asayama, W.; Niwa, T., et al. Amphiphilic Polysaccharide Nanogels as Artificial Chaperones in Cell-Free Protein Synthesis. *Macromol. Biosci.* **2011**, *11*, 814–820.

Sawynok, J. The Therapeutic Use of Heroin: a Review of the Pharmacological Literature. *Can. J. Physiol. Pharmacol.* **1986**, *64*, 1–6.

Schmaljohann, D. Thermo- and pH-Responsive Polymers in Drug Delivery. *Adv. Drug Deliv. Rev.* **2006**, *58*, 1655–1670.

Schmid, A. J.; Dubbert, J.; Rudov, A. A., et al. Multi-Shell Hollow Nanogels with Responsive Shell Permeability. *Sci. Rep.* **2016**, *6*, 22736.

Sekine, Y.; Moritani, Y.; Ikeda-Fukazawa, T.; Sasaki, Y.; Akiyoshi, K. A Hybrid Hydrogel Biomaterial by Nanogel Engineering: Bottom-Up Design with Nanogel and Liposome Building Blocks to Develop a Multidrug Delivery System. *Adv. Healthcare Mater.* **2012**, *1*, 722–728.

Seo, M.; Beck, B. J.; Paulusse, J. M.; Hawker, C. J.; Kim, S. Y. Polymeric Nanoparticles Via Noncovalent Cross-Linking of Linear Chains. *Macromolecules* **2008**, *41*, 6413–6418.

Sevier, C. S.; Kaiser, C. A. Formation and Transfer of Disulphide Bonds in Living Cells. *Nat. Rev. Mol. Cell Biol.* **2002**, *3*, 836–847.

Sharma, A.; Garg, T.; Aman, A., et al. Nanogel—an Advanced Drug Delivery Tool: Current and Future. *Artif Cells Nanomed. Biotechnol.* **2016**, *44*, 165–177.

Shih, I.-L.; Van, Y.-T.; Shen, M.-H. Biomedical Applications of Chemically and Microbiologically Synthesized Poly (Glutamic Acid) and Poly (Lysine). *Mini Rev. Med. Chem.* **2004**, *4*, 179–188.

Shimoda, A.; Sawada, S. I.; Akiyoshi, K. Cell Specific Peptide-Conjugated Polysaccharide Nanogels for Protein Delivery. *Macromol. Biosci.* **2011**, *11*, 882–888.

Shiraishi, K.; Hamano, M.; Ma, H., et al. Hydrophobic Blocks of PEG-Conjugates Play a Significant Role in the Accelerated Blood Clearance (ABC) Phenomenon. *J. Control. Release* **2013**, *165*, 183–190.

Singh, I.; Swami, R.; Pooja, D.; Jeengar, M. K.; Khan, W.; Sistla, R. Lactoferrin Bioconjugated Solid Lipid Nanoparticles: a New Drug Delivery System for Potential Brain Targeting. *J. Drug Target.* **2016**, *24*, 212–223.

Soni, G.; Yadav, K. S. Nanogels as Potential Nanomedicine Carrier for Treatment of Cancer: A Mini Review of the State-of-the-Art. *Saudi Pharm J.* **2016a**, *24*, 133–139.

Soni, K. S.; Desale, S. S.; Bronich, T. K. Nanogels: an Overview of Properties, Biomedical Applications and Obstacles to Clinical Translation. *J. Control. Release* **2016b**, *240*, 109–126.

Soni, S.; Babbar, A. k.; Sharma, R. K.; Maitra, A. Delivery of Hydrophobised 5-Fluorouracil Derivative to Brain Tissue Through Intravenous Route Using Surface Modified Nanogels. *J. Drug Target.* **2006**, *14*, 87–95.

Stalmans, S.; Bracke, N.; Wynendaele, E., et al. Cell-Penetrating Peptides Selectively Cross the Blood-Brain Barrier In Vivo. *PLoS One* **2015**, *10*, e0139652.

Sultana, F.; Imran-Ul-Haque, M.; Arafat, M.; Sharmin, S. An Overview of Nanogel Drug Delivery System. *J. App. Pharm. Sci.* **2013**, *3*, s95–s105.

Teixeira, L. S. M.; Feijen, J.; van Blitterswijk, C. A.; Dijkstra, P. J.; Karperien, M. Enzyme-Catalyzed Crosslinkable Hydrogels: Emerging Strategies for Tissue Engineering. *Biomaterials* **2012**, *33*, 1281–1290.

Temsamani, J.; Scherrmann, J.-M.; Rees, A. R.; Kaczorek, M. Brain Drug Delivery Technologies: Novel Approaches for Transporting Therapeutics. *Pharm. Sci. Technolo. Today* **2000**, *3*, 155–162.

Tilloy, S.; Monnaert, V.; Fenart, L.; Bricout, H.; Cecchelli, R.; Monflier, E. Methylated β-Cyclodextrin as P-gp Modulators for Deliverance of Doxorubicin across an In Vitro Model of Blood–Brain Barrier. *Bioorg. Med. Chem. Lett.* **2006**, *16*, 2154–2157.

Uhrich, K. E.; Cannizzaro, S. M.; Langer, R. S.; Shakesheff, K. M. Polymeric Systems for Controlled Drug Release. *Chem. Rev.* **1999**, *99*, 3181–3198.

Ulijn, R. V.; Bibi, N.; Jayawarna, V., et al. Bioresponsive Hydrogels. *Mater. Today* **2007**, *10*, 40–48.

Van Durme, K.; Verbrugghe, S.; Du Prez, F. E.; Van Mele, B. Influence of Poly (Ethylene Oxide) Grafts on Kinetics of LCST Behavior in Aqueous Poly (N-Vinylcaprolactam) Solutions and Networks Studied by Modulated Temperature DSC. *Macromolecules* **2004**, *37*, 1054–1061.

Vincent, B. Microgels and Core-Shell Particles. *Surface Chem. Biomed. Environ. Sci.* **2006**, *228*, 11.

Vinogradov, S.; Batrakova, E.; Kabanov, A. Poly (Ethylene Glycol)–Polyethyleneimine NanoGel™ Particles: Novel Drug Delivery Systems for Antisense Oligonucleotides. *Colloids Surf. B* **1999**, *16*, 291–304.

Vinogradov, S. V.; Batrakova, E. V.; Kabanov, A. V. Nanogels for Oligonucleotide Delivery to the Brain. *Bioconjug. Chem.* **2004**, *15*, 50–60.

Vinogradov, S. V.; Bronich, T. K.; Kabanov, A. V. Nanosized Cationic Hydrogels for Drug Delivery: Preparation, Properties and Interactions with Cells. *Adv. Drug Deliv. Rev.* **2002**, *54*, 135–147.

Vorbrodt, A. W. Demonstration of Anionic Sites on the Luminal and Abluminal Fronts of Endothelial Cells with Poly-l-Lysine-Gold Complex. *J. Histochem. Cytochem.* **1987**, *35*, 1261–1266.

Vorbrodt, A. W.; Lossinsky, A. S.; Dobrogowska, D. H.; Wisniewski, H. M. Sequential Appearance of Anionic Domains in the Developing Blood-Brain Barrier. *Brain Res. Dev. Brain Res.* **1990**, *52*, 31–37.

Wang, Y.-C.; Wu, J.; Li, Y.; Du, J.-Z.; Yuan, Y.-Y.; Wang, J. Engineering Nanoscopic Hydrogels Via Photo-Crosslinking Salt-Induced Polymer Assembly for Targeted Drug Delivery. *Chem. Commun.* **2010**, *46*, 3520–3522.

Warren, G.; Makarov, E.; Lu, Y., et al. Amphiphilic Cationic Nanogels as Brain-Targeted Carriers for Activated Nucleoside Reverse Transcriptase Inhibitors. *J. Neuroimmune Pharmacol.* **2015**, *10*, 88.

Whittell, G. R.; Hager, M. D.; Schubert, U. S.; Manners, I. Functional Soft Materials from Metallopolymers and Metallosupramolecular Polymers. *Nat. Mater.* **2011**, *10*, 176–188.

Williams, C. G.; Malik, A. N.; Kim, T. K.; Manson, P. N.; Elisseeff, J. H. Variable Cytocompatibility of Six Cell Lines with Photoinitiators used for Polymerizing Hydrogels and Cell Encapsulation. *Biomaterials* **2005**, *26*, 1211–1218.

Wong, H. L.; Wu, X. Y.; Bendayan, R. Nanotechnological Advances for the Delivery of CNS Therapeutics. *Adv. Drug Deliv. Rev.* **2012**, *64*, 686–700.

Wu, H.-Q.; Wang, C.-C. Biodegradable Smart Nanogels: a New Platform for Targeting Drug Delivery and Biomedical Diagnostics. *Langmuir* **2016**, *32*, 6211–6225.

Xiong, M.-H.; Bao, Y.; Yang, X.-Z.; Wang, Y.-C.; Sun, B.; Wang, J. Lipase-Sensitive Polymeric Triple-Layered Nanogel for "On-Demand" Drug Delivery. *J. Am. Chem. Soc.* **2012a**, *134*, 4355–4362.

Xiong, M. H.; Li, Y. J.; Bao, Y.; Yang, X. Z.; Hu, B.; Wang, J. Bacteria-Responsive Multifunctional Nanogel for Targeted Antibiotic Delivery. *Adv. Mater.* **2012b**, *24*, 6175–6180.

Xiong, W.; Wang, W.; Wang, Y., et al. Dual Temperature/pH-Sensitive Drug Delivery of Poly (N-Isopropylacrylamide-Co-Acrylic Acid) Nanogels Conjugated with Doxorubicin for Potential Application in Tumor Hyperthermia Therapy. *Colloids Surf. B* **2011**, *84*, 447–453.

Yadav, H.; Al Halabi, N.; Alsalloum, G. Nanogels as Novel Drug Delivery Systems-A Review. *Insight Pharm. Res.* **2017**, *1*, 1.

Yallapu, M. M.; Jaggi, M.; Chauhan, S. C. Design and Engineering of Nanogels for Cancer Treatment. *Drug Discov. Today* **2011**, *16*, 457–463.

Yallapu, M. M.; Reddy, M. K.; Labhasetwar, V. Nanogels: Chemistry to Drug Delivery. *J. Biomed. Nanotechnol.* **2007**, 131–171, https://doi.org/ 10.1002/9780470152928.ch6.

Yan, M.; Ge, J.; Liu, Z.; Ouyang, P. Encapsulation of Single Enzyme in Nanogel with Enhanced Biocatalytic Activity and Stability. *J. Am. Chem. Soc* **2006**, *128*, 11008–11009.

Yang, H. Nanoparticle-Mediated Brain-Specific Drug Delivery, Imaging, and Diagnosis. *Pharm. Res.* **2010**, *27*, 1759–1771.

Yao, Y.; Xia, M.; Wang, H., et al. Preparation and Evaluation of Chitosan-Based Nanogels/ Gels for Oral Delivery of Myricetin. *Eur. J. Pharm. Sci.* **2016**, *91*, 144–153.

Yoo, M.; Sung, Y.; Lee, Y.; Cho, C. Effect of Polyelectrolyte on the Lower Critical Solution Temperature of Poly (N-Isopropyl Acrylamide) in the Poly (NIPAAm-Co-Acrylic Acid) Hydrogel. *Polymer* **2000**, *41*, 5713–5719.

You, J.; Zhang, R.; Xiong, C., et al. Effective Photothermal Chemotherapy Using Doxorubicin-Loaded Gold Nanospheres that Target EphB4 Receptors in Tumors. *Cancer Res.* **2012**, *72*, 4777–4786.

Zha, L.; Banik, B.; Alexis, F. Stimulus Responsive Nanogels for Drug Delivery. *Soft Matter* **2011**, *7*, 5908–5916.

Zhang, H.; Zhai, Y.; Wang, J.; Zhai, G. New Progress and Prospects: The Application of Nanogel in Drug Delivery. *Mater. Sci. Eng. C.* **2016**, *60*, 560–568.

Zhang, L.; Cao, Z.; Li, Y.; Ella-Menye, J.-R.; Bai, T.; Jiang, S. Softer Zwitterionic Nanogels for Longer Circulation and Lower Splenic Accumulation. *ACS Nano* **2012**, *6*, 6681–6686.

Zhang, N.; Wardwell, P. R.; Bader, R. A. Polysaccharide-Based Micelles for Drug Delivery. *Pharmaceutics* **2013**, *5*, 329–352.

Zhang, X.; Malhotra, S.; Molina, M.; Haag, R. Micro- and Nanogels with Labile Crosslinks– from Synthesis to Biomedical Applications. *Chem. Soc. Rev.* **2015**, *44*, 1948–1973.

Zhang, Y.; Guan, Y.; Zhou, S. Synthesis and Volume Phase Transitions of Glucose-Sensitive Microgels. *Biomacromolecules* **2006**, *7*, 3196–3201.

Zhao, W.; Zhang, H.; He, Q., et al. A Glucose-Responsive Controlled Release of Insulin System Based on Enzyme Multilayers-Coated Mesoporous Silica Particles. *Chem. Commun.* **2011**, *47*, 9459–9461.

APPLICATIONS OF NANOCARRIERS IN EMERGING AND RE-EMERGING CENTRAL NERVOUS SYSTEM TROPICAL INFECTIONS

S. YASRI[1,*] and V. WIWANITKIT[2]

[1]*KMT Primary Care Center, Bangkok 10140, Thailand*

[2]*Hainan Medical University, Haikou 570100, China*

Corresponding author. E-mail: sorayasri@outlook.co.th

ABSTACT

The brain is a vital organ. To manage the problem of the brain is usually difficult. The diagnosis and treatment of brain become the two important complex processes in the management of brain diseases. In clinical neurology, there are several tools for therapy. The neurological drug is the basic chemicals use aiming at therapeutic effect. To achieve new drug with increased efficacy, the new biotechnology technique application is interesting. Advanced biotechnology technique can be used in this case and the novel carriers for drug delivery for targeting the brain is the new technique that should be focused. The application of novel carrier for drug delivery for management of brain disorder is a new approach in clinical neurology. In this specific chapter, the authors summarize and discuss on the new carrier for drug delivery for management of brain diseases. In brief, there are many reports on using new carriers for drug delivery to target brain in several brain diseases. In this specific chapter, the authors specifically focus interest on the tropical emerging and re-emerging infections. At present, there are many new ongoing researches on new drug findings against tropical neurological emerging and reemerging infectious disease using novel carriers for drug delivery. The results from published reports are usually successful; however, the exact use of the novel carriers for drug delivery is still in the very early phase in clinical neurology at present. It is no doubt that the new drug carrier technique can be useful in tropical neurology for managing of emerging and reemerging neurological infections. Nevertheless, there are many concerns about the cost-effectiveness, safety and adverse effects of the new techniques.

16.1   INTRODUCTION

The disease is a common problem for everyone. The disease is basically due to disorder at cellular or organ levels. The disease can result in unwanted morbidity and mortality and it is the role of medicine to control and manage the unwanted outcomes. The diagnosis and treatment become the two important processes in medicine regarding diseases. For treatment, several tools are available for therapy such as drug, vaccine, radio management, and surgical manipulation. However, the most widely used in medicine is using "drug" for management of the disease. The drug is the generally administered aiming at the therapeutic effect of the medical problem.

To get a successful treatment outcome, there must be these basic requirements; (1) effective drug, (2) safe drug, (3) sufficient drug, (4) good drug administration, (5) good drug transportation in the body, and (6) good drug distribution to the desired target. The delivery of the drug is an important concern in the therapy of many diseases. The delivery of drug becomes the important issues in modern pharmacotherapy. How to have a good drug delivery system to the focused target site is the basic question in pharmacotherapy in the management of any diseases in the present day. With the advent pharmaco-biotechnology, the new modes of drug delivery are developed and can be the hope for pharmacotherapy. Of several novel technologies, the use of nanotechnology is an interesting example.

Basically, nanotechnology is the technology on the application of extremely "small" objects that cannot be visible by naked eye (at the nanoscale). The nano-object in nanoscale is extremely small and has many interesting biochemical and physical properties. An object is in a nanoscale have different new electrostatic and biochemical properties comparing to its twin in supra-nanoscale. The new characteristics of nano-objects are the main properties that the technologists use them in the application for many purposes. In medicine, the specific new branch of medicine, nanomedicine, is the present new emerging medical science that has been launched for only a few years. The nanotechnology in drug delivery is very interesting medical biotechnology. The early example of the applications of new nanotechnology for drug delivery is in the treatment of previously unable-to-treat or difficult-to-manage diseases. In the early phase, the application in use in medical diabetology for management of diabetes mellitus, which is the most common metabolic disorder worldwide. In clinical medicine, insulin treatment for the patient with diabetes mellitus is usually problematic and this is a common problem leads to poor control of disease and unwanted disease complications. For a long time, medical researchers have performed searching appropriate technique for insulin delivery. The applied nanocarrier technology for insulin delivery is a good example of successful application of novel pharmacy technology. In the present clinical medicine, nanocarrier is the new technique for the effective management of patients with diabetes mellitus (Khafagy et al., 2007).

Focusing on the use of nanotechnology for drug delivery, the newly developed nano-object is usually used for attachment to a drug that is called nanoconjugation. The conjugatio of nano-object is the basic applicatio in designing of new drug delivery system (Jain et al, 2013). Jain et al. (2013) noted that this technique was an important tool in pharmaceutical and biotechnological research and could be applied for drug delivery systems for further use in various therapeutic purposes. At present, there are many conjugation applications (Table 16.1). At present, many new available nano-objects are available and applicable for drug delivery in clinical medicine. The use of nanocarriers can be the big advent for present management of medical disorders. The present common application of nanocarriers is for management of the difficult-to-manage and complex disorders such as malignant tumors (Charron et al, 2015; Jin et al., 2014). There are many reports on successful use of medial nanocarriers in medical oncology. Nevertheless, there are also applications of novel nanocarriers in another group of disease. In this short chapter, the authors will specifically summarize and discuss the application of nanocarriers in tropical medicine focusing on tropical emerging and re-emerging infections. The special focus is on the brain targeting in those diseases.

TABLE 16.1 Examples of Some Important Nano-Objects for Therapeutic Purposes.

Examples	Details
Aquasome (Jain et al., 2013)	Aquasome is an interesting nano-object. The aquasome is soluble nano-object and can be effectively used in present in drug delivery
Dendrimer (Rodríguez Villanueva et al., 2016)	A dendrimer is a well specifically designed nano-object. The dendrimer production is more difficult than that of simple nano-object. Hence, the use of dendrimer in nanomedicine is limited but there are many researches on the use of dendrimer in cancer diagnosis and cancer therapy
Graphene (Li et al., 2016)	Graphene is a complex nano-object that is limited used in medicine. Similar to the case of the dendrimer, there are some ongoing studies on using graphene in cancer diagnosis and cancer therapy
Liposome (Weissig, 2017)	A liposome is an interesting nano-object that is widely used in pharmacology. As a lipid-based nano-object, it can be effectively used in carrying purpose. Due to its lipidic nature, the permeability permission into a living cellular compartment can be observed. Hence, the liposome is widely applied as a new modern technique for drug delivery
Nanocapsule (Kreuter, 1978)	Nanocapsule is a basic nano-object. The nanocapsule is the new nanocarrier technique in pharmacology that can be used in medical therapy. Basically, nanocapsule acts as the specific container for protective delivery of the intracapsular object. Hence, the nanocapsule can be used for drug delivery. In general, apart from a tablet, the capsule is a common drug form widely used in medicine
Nanofabric (Feinberg, and Parker, 2010)	Nanofabric is another limited used nanotechnique in clinical pharmacology
Nanogel Nanobiomaterial (Soni and Yadav, 2016)	Nanogel nanobiomaterial is another interesting nano-object. Basically, a gel form of the drug is clinically used for local application and mainly used in medical dermatology. The nanogel technique can be designed aiming at the same purpose as basic standard extra-nanogel for basic local application. After the application, nanogel nanobiomaterial will effectively penetrate and reach the desired target site

TABLE 16.1   *(Continued)*

Examples	Details
Nanorod (Kumar et al., 2016)	Nanorod is a new nano-object that is still limited used in clinical pharmacology. The rod form drug is generally used for a specific therapeutic purpose such as contraceptive rod and chemo-radiotherapy rod. The nanorod is more difficult to prepare comparing to nanocapsule, hence, it is not widely used. Nevertheless, the rod should be stronger than the capsule and a more protection of the drug content inside can be expected. For the medical purpose, nanorod can be applied in the same way as general drug rod is prepared, mainly for cancer therapy
The nanotube (Kumar et al., 2016)	Nanotubes (Kumar et al., 2016) is a specific kind of nanotube with its main composition as carbon. It is usually designed aiming at organic and inorganic integration purpose. Similar to nanofabric, for medical treatment, nanotube application can be seen in some very specific cases such as medical oncology work
Quantum Dot (Kagan et al., 2016)	Quantum dot is another complexly designed nano-object. It is widely studied for its possible application in diagnosis and therapy in medicine. However, at present, it is mostly used for diagnostic purpose, not for treatment

16.2   APPLICATION OF NANOCARRIERS IN NEUROLOGY

The medical application of nanocarriers is a really interesting present biotechnology (Jain et al., 2014; Jin et al., 2014). The application can be either in diagnosis and treatment. The nano-object might act as a part of diagnostic reagent. The good example is the quantum rod application in diagnostic imaging. For treatment, an application as a technique for drug delivery is a good example. Some specific kinds of nanocarriers such as liposome and nanogel can also be used in drug formulation. Also, the nanocarriers can be used for vaccine formulation. Many new vaccines are produced based on the nano-object technology, especially for liposome-based technology.

In the drug delivery system, nanocarrier can answer the basic requirements on drug stability, permeability, and distribution. Hence, it is no doubt that the nanocarrier technology is widely used in present clinical medicine. Applications in several branches of medicine can be seen in the present day. In this specific short chapter, the authors will further briefly discuss the application in tropical neurology. The cases of nanocarriers for brain targeting as an application in tropical emerging and re-emerging infections will be focused. Indeed, most of the people in our world live in tropical regions; hence, tropical diseases become an important group of disease in global public health. Infection is a common problem for the tropical world. The brain infection in the tropical world is usually deadly and becomes the present global public health threat. Application of nanotechnology in tropical neurology can very be useful for the tropical world.

Here, the first thing to be discussed is on the basic application of nano-object as a drug carrier in neurology. When we talk about neurology, the main focuses are on the brain, spinal cord, and nerves. It is no doubt that the brain is the vital organ of everyone. Any problems in the brain are usually serious and become big problems in medicine. The diagnosis and treatment of brain disorder are usually hard and a big challenge for a clinical neurologist. For treatment, the drug is also a min mode of therapy for brain disease. Similar to any organ, the effective treatment for brain disease is based on the good drug. This implies the requirement on (1) effective drug, (2) safe drug, (3) sufficient drug, (4) good drug administration, (5) good drug transportation in the body and (6) good drug distribution to the desired target, the brain. Nevertheless, the brain is a vital organ in everyone and highly protective.

Anatomically brain is a soft organ filled in the human skull. The brain consists of forebrain, midbrain, and hindbrain. The hindbrain further connects to the spinal cord that is downward into the spinal canal. Both brain and spinal cord form the specific part of a neurological system called central nervous system (CNS). All CNS parts are wrapped by meninges. In human, the brain plays an important physiological role in regulating and controlling of all other parts in the human body. Sensory and motor are the main physiological functions of the brain. In addition, the brain also controls memory and homeostasis. Brain death is considered the death of the whole human body. As a vital organ, the human body has a specific mechanism for brain protection. The anatomical and physiological defense namely blood–brain barrier can be seen. In neurology, blood–brain barrier is an important natural defense for any human beings. In general, blood–brain barrier acts as a border control for prevention of unwanted objects to enter the brain. In general, the unwanted objects might enter the brain in several ways including direct penetrating via direct injury and via bloodstream. The blood circulation from all over the body can pass to the neurological system but the blood–brain barrier plays important role in the prevention of the unwanted objects to enter into the vital organ, brain. For sure, this is also applied to the case of drugs which is also considered an alien chemical (Fig. 16.1). As noted by Theodorakis et al. (2017), *"While candidate drugs for many of these diseases are available, most of these pharmaceutical agents cannot reach the brain rendering most of the drug therapies that target the CNS inefficient."* In prevention of entry, the extremely small size of the pore at blood–brain barrier plays an important role and acts as a sieving system to prevent the unwanted object to enter into the brain. Also, there is a complex and dynamic interface which directly the influx and efflux of substances through various different translocation mechanisms (Theodorakis et al., 2017). In many cases with defects or incomplete or immature blood–brain barrier (such as in infant), the entry of unwanted objects such as germ can be easy and this can result in brain disorder. Hence, it is no doubt that there are several infectious brain diseases in infants. Also, the accumulation of excessive chemicals and biometals usually cause the neurological problem in infants (Wiwanitkit, 2011).

The difficulty in drug treatment of brain disease is due to several reasons. The blood–brain barrier plays an important role and becomes an important determinant. The concentration of drug in the brain is usually low due to the difficulty of many drugs to cross the blood–brain barrier. Hence, in pharmacology, the great concern on the drug for the management of brain disorders is the ability to cross the blood–brain barrier.

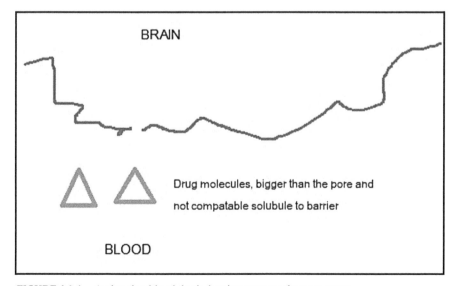

FIGURE 16.1   Action that blood–brain barrier prevents drugs to enter.

The use of novel carriers to help drug crosses the blood–brain barrier is the exact application of advanced pharmaco-biotechnology in clinical neurology (Hersh et al., 2016). Patel et al. (2009) noted for several new possible strategies that might increase drug delivery into the brain include chemical delivery systems (lipid-mediated transport, the prodrug approach, and the lock-in system) and new biological delivery systems, in which are re-engineered to cross the blood–brain barrier via new specific endogenous transporters localized within the barrier. The use of nanocarrier is a good example of an attempt to use applied nanotechnology to serve as a new chemical delivery system. The nanoparticle- and liposome-carried drugs can be effective because there are an increased cellular uptake and reduced efflux through ATP-binding cassette transporters, which plays role in reducing drug entry (Pinzón-Daza et al., 2013). Gabathuler (2010) noted that the novel carrier might be effective for increasing the transport of therapeutics from the blood into the brain parenchyma. Wohlfart et al. (2012) reported that the mechanism that nanocarrier carries the drug into the brain *"appears to be receptor-mediated endocytosis in brain capillary endothelial cells. Modification of the nanoparticle surface with covalently attached targeting ligands or by coating with certain surfactants enabling the adsorption of specific plasma proteins are necessary for this receptor-mediated uptake."* After successfully enter, the nanoparticle will be further diffused to targeted brain interior or maybe transcytosed (Kreuter, 2001). Using novel nanocarriers for drug delivery is a new technology and emerging concept to overcome the blood–brain barrier and it is the present new alternative drug delivery strategies targeting brain (Hersh et al., 2016). For example, Grabrucker et al. (2016) noted that using nanocarriers can help manage several brain diseases including to Alzheimer, Parkinson and Huntington's disease. Focusing on tropical disease, especially for infections, there are also some interesting reports which will be further summarized and discussed in this chapter.

16.3   USEFULNESS OF NANOCARRIERS FOR BRAIN TARGETING IN SOME IMPORTANT EMERGING AND RE-EMERGING INFECTIONS

As already mentioned, tropical medicine is important medical subjects that directly relate to diagnosis, treatment, and prevention of diseases for more than half of world population. Due to the poverty, hot climate and limited infrastructure, there are many infectious infections in tropical countries and it is called tropical infection. The tropical infections cause a considerable lost each year and it is a big global public health concern. The application of new drug delivery technology for the management of tropical infection in tropical medicine is possible as previously mentioned. The application of novel nanocarriers technology can be for diagnostic or therapeutic purposes. In clinical neurology, the use of the new technology to correspond to the emerging and re-emerging tropical brain infections is a very interesting clinical issue. Here, the authors will further discuss in each specific important tropical and re-emerging brain infections

16.3.1   TROPICAL PARASITIC BRAIN DISEASE

There are many parasitic diseases in medicine. Some are common or endemic in the tropical region and it is called tropical parasitosis. Many diseases become more important due to its rapid spreading worldwide. Some parasitic diseases can also induce brain disorders and becomes an important consideration in clinical neurology. As already noted, the role of nanomedicine in the management of infections are confirmed, hence, it is no doubt that there are many interesting reports on using novel nanocarriers for drug delivery in clinical neurology. For the specific tropical and re-emerging parasitic brain infections, there are also some interesting reports as will be further described.

16.3.1.1   CEREBRAL MALARIA

Malaria is one of the most important mosquito-borne infections that is still the public health threat. Malaria presently affects more than 3.2 billion people around the world annually (data from WHO at http://www.who.int/gho/malaria/en/). The disease is caused by protozoa in the group of *Plasmodium* spp. Malaria is an important blood infection that causes the public health problem in several tropical countries and causes many deaths each year. Emergence and re-emergence of malaria are often reported from both tropical and non-tropical countries around the world. In malariology, the control and management of malaria are widely studied and researched. The effective drug to treat malaria is a common basic research question in clinical malariology. For a long time, the antimalarial drugs have been available but there was no success in getting rid of malaria. The emerging problem of drug-resistant malaria can be observed worldwide and becomes an important problem to manage tropical medicine. There are many ongoing researches on the new antimalarial drug. The use of nanocarriers for drug delivery in the treatment of malaria is very interesting (Aditya et al., 2013).

Focusing on neurological complication due to malaria, a well-known condition is a cerebral malaria, which is a deadly condition that requires good clinical management. To treat cerebral malaria is usually difficult. However, the role of new novel nanocarrier for targeting brain in cerebral malaria is widely mentioned in the present literature. Some important reports on the use of nanocarriers for drug delivery for cerebral malarial therapy are shown in Table 16.2.

TABLE 16.2 Some Important Reports on Using Nanocarriers for Drug Delivery in Treatment of Cerebral Malaria.

Authors	Details
Portnoy, et al. (2016)	Portnoy, et al. (2016) reported on using Indocyanine green (ICG) liposomes for diagnosis and therapeutic monitoring of cerebral malaria. Portnoy, et al. (2016) concluded that " liposomal ICG offers a valuable diagnostic tool and a biomarker for the effectiveness of CM treatment"
Waknine-Grinberg et al. (2013)	Waknine-Grinberg et al. (2013) discussed using novel glucocorticosteroids in nano-sterically stabilized liposomes that were proved efficacious for the elimination of the acute clinical symptoms in experimental cerebral malaria

16.3.1.2 CYSTICERCOSIS

Cysticercosis is an important parasitic infestation. It is common problem in many tropical countries and can be seen around the world. The cysticercosis usually presents as a cystic lesion and can be seen at any organs in the body. In some cases, the central nervous system infection can be detected and it is specifically called neurocysticercosis. Neurocysticercosis can be silent and accidental detected. Nevertheless, it is usually related to neurological problems such as seizure. In a severe case, mortality due to neurocysticercosis in the brain can be expected (Rajshekhar, 2016). The management of cysticercosis is usually based on surgical removal of the cyst. For the neurocysticercosis in the deep brain, the surgical procedure is usually impossible and this is usually dependent on medical therapy. The use of the antiparasitic drug is the common practice (Ahmad et al., 2017; Rajshekhar, 2016). Using the novel nanocarrier for management of neurocysticercosis is a new approach for management of the brain infection. Recently, there is an interesting report on *"Dual Drug Delivery Using Lactic Acid Conjugated Solid*

Lipid Nanoparticles (SLN) for Effective Management of Neurocysticercosis
(Devi et al., 2015)." Devi et al. (2015) developed Lactic acid-conjugated
SLNs bearing albendazole and prednisolone for effective management
of neurocysticercosis and found that the approach was effective. The drug
could be effectively carried into the brain and there was no significant
cellular toxicity (Devi et al., 2015).

16.3.1.3   TRYPANOSOMIASIS

Trypanosomiasis is an important fly-borne infection. It is a blood infection
caused by a parasite in Trypanosoma spp. Emergence and re-emergence of
this disease can be seen in several countries around the world. The disease
is still prevalent in many tropical countries (about 0.3 million new cases
annually; data from WHO at www.who.int/trypanosomiasis_african/en/).
There are many new studies on searing new drug against trypanosomiasis.
Romero and Morilla (2010) noted that new drug with "more efficient pharmacotherapy that (1) eradicates the scarce amastigotes present at the indeter-
minate/chronic form and (2) employs less toxic drugs than benznidazole or
nifurtimox" was necessary for successful management of trypanosomiasis.
To support that mentioned aim, the use of nanocarriers for drug delivery in
the treatment of trypanosomiasis is very interesting. For the specific brain
involvement in trypanosomiasis, the use of novel nanocarriers is a really
new issue in clinical neurology. There are some recent reports on the use
of new nanocarriers for drug delivery targeting brain for trypanosomiasis
therapy. The important reports are shown in Table 16.3.

TABLE 16.3   Some Important Reports on Using New Nanocarriers for Drug Delivery
Targeting Brain for Trypanosomiasis Therapy.

Authors	Details
Baldissera et al. (2017)	Baldissera et al. (2017) reported that Nerolidol-loaded nanospheres (N-NS) could help prevent behavioral impairment via ameliorating Na+, K+-ATPase and AChE activities as well as reducing oxidative stress in the brain of Trypanosoma evansi-infected mice. Baldissera et al. (2017) concluded that "N-NS treatment may be a useful strategy to treat memory dysfunction and oxidative stress caused by Trypanosoma evansi infection"
Kroubi et al. (2010)	Kroubi et al. (2010) reported on the development of a nanoparticulate formulation of diminazene to treat African trypanosomiasis that can overcome the fundamental problem of poor affinity for brain tissue

16.3.1.4   LEISHMANIASIS

Leishmaniasis is another important fly-borne infection. It is a blood infection caused by a parasite in *Leishmania* spp. Similar to trypanosomiasis, emergence and re-emergence of this disease can be seen in several countries around the world. The disease is still prevalent in many tropical countries There are many new studies on searing new drug against leishmaniasis. There are many ongoing studies on the searing new antileishmanial drug. The use of nanocarriers for drug delivery in the treatment of leishmaniasis is very interesting. To support that mentioned aim, the use of nanocarriers for drug delivery in the treatment of leishmaniasis is very interesting. For the specific brain involvement in leishmaniasis, the use of novel nanocarriers is a really new issue in clinical neurology. This is similar to the current situation of a similar infection, trypanosomiasis. Of interest, the liposomal amphotericin B (AmB) is the first and only one available novel antiparasitic drug against a tropical disease that is produced based on nanocarriers for drug delivery (Wiwanitkit, 2012). There are some recent reports on the use of new nanocarriers for drug delivery targeting brain for leishmaniasis therapy. The important reports are shown in Table 16.4.

TABLE 16.4 Some Important Reports on Using New Nanocarriers for Drug Delivery Targeting Brain for Leishmaniasis Therapy.

Authors	Details
Veerareddy et al. (2009)	Veerareddy et al. (2009) studied on antileishmanial activity, pharmacokinetics and tissue distribution studies of mannose-grafted amphotericin B lipid nanospheres and found that the new drug could entry brain tissue but less than liver and spleen tissue

16.3.2   TROPICAL VIRAL BRAIN DISEASE

There are many viral diseases in medicine. Some diseases are common and endemic in the tropical region and those diseases are called tropical viral infections. Many diseases become more important due to its rapid spreading and worldwide emergence. An important group of tropical viral infection is tropical viral encephalitis. These diseases are usually transmitted by tropical insects or vectors (mosquito is the best example). This group of the viral disease usually induces severe brain disorders and becomes an important consideration in clinical neurology. As already noted, the role of nanomedicine in the management of infections are confirmed; hence, it is no doubt that there are many interesting reports on using novel nanocarriers for drug delivery targeting brain in tropical viral brain infection. For the specific tropical and re-emerging viral brain infections, there are also some interesting reports as will be further described.

16.3.2.1 JAPANESE ENCEPHALITIS

Japanese encephalitis is an important arbovirus infection. This disease is common in several tropical countries. The disease might be asymptomatic but can result in serious complications, encephalitis, and death. The prevention of this disease is the best management and there is already an available vaccine for Japanese encephalitis (Wiwanitkit, 2009). It is the recommendation for the local people in endemic areas to get the vaccination. Also, a traveler who wants to visit the endemic area in the tropical regions should get a vaccination for the prevention of Japanese encephalitis. For management of the case with encephalitis due to this infection, symptomatic management is routinely used. There is still no specific effective antiviral drug. The use of the nanocarriers for drug delivery is limited. Nevertheless, the role of the technology for Japanese encephalitis vaccine development can be seen. The summary of important reports is given in Table 16.5.

TABLE 16.5 Some Interesting Reports on Using Nanoparticle as Carrier for Development of New Japanese Encephalitis Vaccine.

Authors	Details
Huang et al. (2009)	Huang et al. (2009) reported on a new transdermal immunization with low-pressure-gene-gun mediated chitosan-based DNA vaccines against Japanese encephalitis virus
Zhai et al. (2014)	Zhai et al. (2014) reported on the effects of cell-mediated immunity induced by intramuscular chitosan-pJME/ GM-CSF nano-DNA vaccine in BAlb/c mice
Zhai et al. (2015)	Zhai et al. (2015) reported on the immune-enhancing effect of nano-DNA vaccine encoding a gene of the prME protein of Japanese encephalitis virus and BALB/c mouse granulocyte-macrophage colony-stimulating factor

16.3.2.2   RABIES

Rabies is an important arbovirus infection. This disease is common in several tropical countries. New emergence of rabies is usually a big concern for local public health. Rabid animal, especially for the dog, is the main cause of the infection. Being bitten is the mode of viral receiving. The disease results in serious neurological problems, encephalitis, and death. The prevention of this disease is the best management and there is already an available vaccine for rabies that should be given after exposure to the rabid animal. For management of the case with encephalitis due to this infection, symptomatic management is routinely used but it is hopeless to save the life of the patient when there is already an overt neurological symptom. There is still no specific effective antiviral drug. The role of the technology for rabies vaccine development can be seen (Shah et al., 2014). The summary of important reports is given in Table 16.6.

TABLE 16.6   Some Interesting Reports on Using Nanoparticle as a Carrier for Development of the New Rabies Vaccine.

Authors	Details
Shah et al. (2014)	Shah et al. (2014) noted that nanoparticle is "a valuable alternative for the production of cheaper rabies vaccines"
Ullas et al. (2014)	Ullas et al. (2014) reported on the enhancement of immunogenicity and efficacy of a plasmid DNA rabies vaccine by nanoformulation with a fourth-generation amine-terminated polyether imine) dendrimer

16.4   FUTURE PROSPECTS

The nanotechnology will become more widely used around the world in the near future. The nanotechnology will play important role in therapy of medical disorders. There will be many new pharmaceutical manufacturers that will apply nanotechnology for production of the new nanocarriers. Based on the continuous improvement of nanotechnology, many new nano-carriers are expected to be available in the near future. The new nanocarriers can be useful for medical therapy of several disorders including to CNS tropical infections. The improved drug activity due to better drug pene-trating property can be expected. The newly available drugs based on the new nanocarriers will be the new hopes for management of the difficult-to-manage CNS tropical infections.

With the possible new influx of the new nanocarriers technology for management of tropical CNS infections, the better outcome of treatment can be expected. Nevertheless, there will be also new concerns on the possible unknown adverse effects of the newly used nanocarriers (such as short-term and long-term toxicity). With the rapid growth of the new technology, the closed monitoring of the negative side, the unwanted effect of the nanocarrier technology is needed. There should be new standards and legal control of the production of the new nanocarriers for medical use.

16.5   CONCLUSION

The use of nanocarrier for drug delivery is the new interesting nanomedicine technique that can also be applied in clinical neurology. There are many nanocarriers that can be useful as a new drug delivery system. The main advantages offered by novel nanocarrier are increasing efficacy in targeting the desired pathogens or pathological tissue and increasing basic drug properties (such as stability, solubility, and absorption). In clinical neurology, the effective targeting at the brain is usually problematic and the use of novel nanocarrier becomes the present new hope. As already shown and discussed in this chapter, the application of nanocarriers for brain targeting for management of tropical neurological infectious diseases is possible. The advantages of using new carriers for brain targeting are (1) effective drug, (2) safe drug, (3) sufficient drug, (4) good drug administration, (5) good drug transportation in the body, and (6) good drug distribution to the desired target, the brain. At present, there are many reports on using new carriers for brain targeting in emerging and re-emerging neurological tropical infections. The results in those reports are usually favorable. However, although there are many reports in the present day, there are still only a few approved drugs for use (such as liposomal AmB). Hence, further researches are still needed. It is no doubt that the new drug carrier technique can be useful in tropical neurology for managing of emerging and reemerging neurological infections. There are some important concerns for future development of nanocarriers for drug delivery targeting brain for management of tropical infectious diseases. First, the production of the new system requires high technology, which is still not available in the poor tropical developing tropical countries where the diseases are endemic. Nevertheless, there are many concerns about the cost-effectiveness, safety and adverse effects of the new techniques. There is still no proof for the long-term safety of using nanomaterials. The toxicity of the nanomaterial is the big issue that needs further studies for clarification. Finally, the cost-effectiveness of the new technology has to be clarified. How to increase the cost-effective is the next thing to be thought.

KEYWORDS

- **CNS**
- **emerging**
- **infections**
- **nanocarriers**
- **re-emerging**
- **tropical**

REFERENCES

Aditya, N. P.; Vathsala, P. G.; Vieira, V.; Murthy, R. S.; Souto, E. B. Advances in Nanomedicines for Malaria Treatment. *Adv. Colloid Interface Sci.* **2013**, *201-202*, 1–17.

Ahmad, R.; Khan, T.; Ahmad, B.; Misra, A.; Balapure, A. K. Neurocysticercosis: a Review on Status in India, Management, and Current Therapeutic Interventions. *Parasitol Res.* **2017**, *116* (1), 21–33.

Baldissera, M. D.; Souza, C. F.; Grando, T. H.; Moreira, K. L.; Schafer, A. S.; Cossetin, L. F.; da Silva, A. P.; da Veiga, M. L.; da Rocha, M. I.; Stefani, L. M.; da Silva, A. S.; Monteiro, S. G. Nerolidol-loaded Nanospheres Prevent Behavioral Impairment via Ameliorating Na+, K+-ATPase and AChE Activities as well as Reducing Oxidative Stress in the Brain of Trypanosoma Evansi-Infected Mice. *NaunynSchmiedebergs Arch. Pharmacol.* **2017**, *390* (2), 139–148.

Devi, R.; Jain, A.; Hurkat, P.; Jain, S. K. Dual Drug Delivery Using Lactic Acid Conjugated SLN for Effective Management of Neurocysticercosis. *Pharm. Res.* **2015**, *32* (10), 3137–3148. Feinberg, A. W.; Parker, K. K. Surface-Initiated Assembly of Protein Nanofabrics. *Nano Lett.* **2010**, *10* (6), 2184–2191.

Gabathuler R. Approaches to Transport Therapeutic Drugs across the Blood-Brain Barrier to Treat Brain Diseases. *Neurobiol. Dis.* **2010**, *37* (1), 48–57.

Grabrucker, A. M.; Ruozi, B.; Belletti, D.; Pederzoli, F.; Forni, F.; Vandelli, M. A.; Tosi, G. Nanoparticle Transport across the Blood Brain Barrier. *Tissue Barr.* **2016**, *4* (1), e1153568.

Hersh, D. S.; Wadajkar, A. S.; Roberts, N. B.; Perez, J. G.; Connolly, N. P.; Frenkel, V.; Winkles, J. A.; Woodworth, G. F.; Kim, A. J. Evolving Drug Delivery Strategies to Overcome the Blood Brain Barrier. *Curr Pharm Des.* **2016**, *22* (9), 1177–1193.

Huang, H. N.; Li, T. L.; Chan, Y. L.; Chen, C. L.; Wu, C. J. Transdermal Immunization with Low-Pressure-Gene-Gun Mediated Chitosan-Based DNA Vaccines Against Japanese Encephalitis Virus. *Biomaterials* 2009, *30* (30), 6017–6025.

Jain, A.; Jain, A.; Gulbake, A.; Shilpi, S.; Hurkat, P.; Jain, S. K. Peptide and Protein Delivery Using New Drug Delivery Systems. *Crit. Rev. Ther. Drug Carrier Syst.* **2013**, *30* (4), 293–329.

Jain, K.; Mehra, N. K.; Jain, N. K. Potentials and Emerging Trends in Nanopharmacology. *Curr. Opin. Pharmacol.* **2014**, *15*, 97–106.

Jin, S. E.; Jin, H. E.; Hong, S. S. Targeted Delivery System of Nanobiomaterials in Anticancer Therapy: from Cells to Clinics. *Biomed. Res. Int.* **2014**, *2014*, https://doi.org/10.1155/2014/814208.

Kagan, C. R.; Lifshitz, E.; Sargent, E. H.; Talapin, D. V. Building Devices from Colloidal Quantum Dots. *Science* **2016**, *353* (6302). pii: aac5523, https://doi.org/10.1126/science.aac5523.

Khafagy, E. S.; Morishita, M.; Onuki, Y.; Takayama, K. Current Challenges in Noninvasive Insulin Delivery Systems: a Comparative Review. *Adv. Drug Deliv. Rev.* **2007**, *59* (15), 1521–1546.

Kreuter, J. Nanoparticles and Nanocapsules--New Dosage Forms in the Nanometer Size Range. *Pharm. Acta Helv.* **1978**, *53* (2), 33–39.

Kreuter J. Nanoparticulate Systems for Brain Delivery of Drugs. *Adv. Drug Deliv. Rev.* **2001**, *47* (1), 65–81.

Kroubi, M.; Daulouede, S.; Karembe, H.; Jallouli, Y.; Howsam, M.; Mossalayi, D.; Vincendeau, P.; Betbeder, D. Development of a Nanoparticulate Formulation of Diminazene to Treat African Trypanosomiasis. *Nanotechnology* **2010**, *21* (50), https://doi.org/10.1088/0957-4484/21/50/505102.

Kumar, S.; Rani, R.; Dilbaghi, N.; Tankeshwar, K.; Kim, K. H. Carbon Nanotubes: a Novel Material for Multifaceted Applications in Human Healthcare. *Chem. Soc. Rev.* **2017**, *46* (1), 158–196.

Li, D., Zhang, W., Yu, X., Wang, Z., Su, Z.; Wei, G. When Biomolecules Meet Graphene: from Molecular Level Interactions to Material Design and Applications. *Nanoscale* **2016**, *8*, 19491–19509.

Patel, M. M.; Goyal, B. R.; Bhadada, S. V.; Bhatt, J. S.; Amin A. F. Getting into the Brain: Approaches to Enhance Brain Drug Delivery. *CNS Drugs* **2009**, *23* (1), 35–58.

Pinzón-Daza, M. L.; Campia, I.; Kopecka, J.; Garzón, R.; Ghigo, D.; Riganti, C. Nanoparticle- and Liposome-Carried Drugs: New Strategies for Active Targeting and Drug Delivery across Blood-Brain Barrier. *Curr. Drug Metab.* **2013**, *14* (6), 625–640.

Portnoy, E.;Vakruk, N.; Bishara, A.; Shmuel, M.; Magdassi, S.; Golenser, J.; Eyal, S. Indocyanine Green Liposomes for Diagnosis and Therapeutic Monitoring of Cerebral Malaria. *Theranostics.* **2016**, *6* (2), 167–176.

Rajshekhar, V. Neurocysticercosis: Diagnostic Problems & Current Therapeutic Strategies. *Indian J. Med. Res.* **2016**, *144* (3), 319–326.

Rodríguez Villanueva, J.; Navarro, M. G.; Rodríguez Villanueva, L. Dendrimers as a Promising Tool in Ocular Therapeutics: Latest Advances and Perspectives. *Int. J. Pharm.* **2016**, *511* (1), 359–366.

Romero, E. L.; Morilla, M. J. Nanotechnological Approaches Against Chagas Disease. *Adv. Drug Deliv. Rev.* **2010**, *62* (4–5), 576–588.

Shah, M. A.; Khan, S. U.; Ali, Z.; Yang, H.; Liu, K.; Mao, L. Applications of Nanoparticles for DNA Based Rabies Vaccine. *J. Nanosci. Nanotechnol.* **2014**, *14* (1), 881–891.

Soni, G.; Yadav, K. S. Nanogels as Potential Nanomedicine Carrier for Treatment of Cancer: a Mini Review of the State of the Art. *Saudi Pharm. J.* **2016,** *24*, 133–139.

Theodorakis, P. E.; Müller, E. A.; Craster, R. V.; Matar, O. K. Physical Insights into the Blood-Brain Barrier Translocation Mechanisms. *Phys. Biol.* **2017,** *14* (4), 041001.

Ullas, P. T.; Madhusudana, S. N.; Desai, A.; Sagar, B. K.; Jayamurugan, G.; Rajesh, Y. B.; Jayaraman, N. Enhancement of Immunogenicity and Efficacy of a Plasmid DNA Rabies Vaccine by Nanoformulation with a Fourth-Generation Amine-Terminated Poly(Ether Imine) Dendrimer. *Int. J. Nanomed.* **2014,** *9*, 627–634.

Veerareddy, P. R.; Vobalaboina, V.; Ali, N. Antileishmanial Activity, Pharmacokinetics and Tissue Distribution Studies of Mannose-Grafted Amphotericin B Lipid Nanospheres. *J. Drug Target.* **2009,** *17* (2), 140–147.

Waknine-Grinberg, J. H., Even-Chen, S., Avichzer, J.; Turjeman, K.; Bentura-Marciano, A.; Haynes, R. K.; Weiss, L.; Allon, N.; Ovadia, H.; Golenser, J.; Barenholz, Y. Glucocorticosteroids in Nano-Sterically Stabilized Liposomes are Efficacious for Elimination of the Acute Symptoms of Experimental Cerebral Malaria. *PLoS One* **2013,** *8* (8), e72722.

Weissig, V. Liposomes Came First: The Early History of Liposomology. *Methods Mol. Biol.* **2017,** 1522, 1–15.

Wiwanitkit, V. Vaccination for Tropical Mosquito Borne Encephalitis. *Acta Neurol. Taiwan* **2009,** *18* (1),60–63.

Wiwanitkit, V. Iron and Neurodevelopment in Infant. *Brain Dev.* **2011,** *33* (5), 448.

Wiwanitkit, V. Interest in Paromomycin for the Treatment of Visceral Leishmaniasis (Kala-Azar).*Ther. Clin. Risk Manag.* **2012,** 8, 323–328.

Wohlfart, S.; Gelperina, S.; Kreuter, J. Transport of Drugs across the Blood-Brain Barrier by Nanoparticles. *J. Control Release.* **2012,** *161* (2), 264–273.

Zhai, Y.; Zhou, Y.; Li, X.; Feng, G. Immune-Enhancing Effect of Nano-DNA Vaccine Encoding a Gene of the prME Protein of Japanese Encephalitis Virus and BALB/c Mouse Granulocyte-Macrophage Colony-Stimulating Factor. *Mol. Med. Rep.* **2015,** *12* (1), 199–209.

Zhai, Y. Z.; Zhou, Y.; Ma, L.; Feng, G. H. Effects of Cell-Mediated Immunity Induced by Intramuscular Chitosan-pJME/GM-CSF Nano-DNA Vaccine in BAlb/c Mice. *Z Bing Du Xue Bao.* **2014,** *30* (4), 423–428.

INDEX